S0-BLB-832

EXTRACHROMOSOMAL DNA

Academic Press Rapid Manuscript Reproduction

Proceedings of the 1979 ICN–UCLA Symposia on
Molecular and Cellular Biology held in Keystone,
Colorado, March 11–16, 1979

ICN–UCLA Symposia on Molecular and Cellular Biology
Volume XV, 1979

EXTRACHROMOSOMAL DNA

edited by

DONALD J. CUMMINGS
Department of Microbiology and Immunology
University of Colorado Medical Center
Denver, Colorado

PIET BORST
Jan Swammerdam Institute
University of Amsterdam
Amsterdam, Netherlands

IGOR B. DAWID
Laboratory of Biochemistry
National Cancer Institute
National Institutes of Health
Bethesda, Maryland

SHERMAN M. WEISSMAN
School of Medicine
Yale University
New Haven, Connecticut

C. FRED FOX
Department of Bacteriology and Molecular Biology Institute
University of California, Los Angeles
Los Angeles, California

ACADEMIC PRESS 1979
A Subsidiary of Harcourt Brace Jovanovich, Publishers
New York London Toronto Sydney San Francisco

ACADEMIC PRESS, INC.
111 Fifth Avenue, New York, New York 10003

United Kingdom Edition published by
ACADEMIC PRESS, INC. (LONDON) LTD.
24/28 Oval Road, London NW1 7DX

ISBN 0–12–198780–9

PRINTED IN THE UNITED STATES OF AMERICA

79 80 81 82 9 8 7 6 5 4 3 2 1

CONTENTS

IX. BIOLOGICAL AND STRUCTURAL FEATURES OF KINETOPLAST AND *PODOSPORA* MITOCHONDRIAL DNA

CONTRIBUTORS

Numbers in parentheses refer to chapter numbers

ALZNER-DEWEERD B. (24), Department of Biology, Massachusetts Institute of Technology, Cambridge, Massachusetts 02139

ARNBERG, A. C. (20), Biochemistry Laboratories, University of Groningen, Zernikelaan, Groningen, Netherlands

ASTIN, ANNE (17), Department of Biochemistry, Monash University, Clayton, Victoria 3168 Australia

ATTARDI, GIUSEPPE (28), Division of Biology, California Institute of Technology, Pasadena, California 91125

BATTEY, JIM (27), Department of Pathology, Stanford Medical Center, Stanford California 94305

BEALE, G. H. (1), Institute of Animal Genetics, West Mains Road, Edinburgh, EH9 3JN, Scotland

BELCOUR, L. (34), Centre de Genetique Moleculaire, CNRS, Gif sur Yvette, France

BOER, P. H. (20), University of Amsterdam, Box 60.000, 1005 GA, Amsterdam, Netherlands

BOGORAD, LAWRENCE (8), Biological Laboratories, Harvard University, Cambridge, Massachusetts 02138

BORST, P. (20, 32), University of Amsterdam, Box 60.000, 1005 GA, Amsterdam, Netherlands

BOS, J. L. (20), University of Amsterdam, Box 60.000, 1005 GA, Amsterdam, Netherlands

BOYNTON, J. E. (6), Department of Botany, Duke University, Durham, North Carolina 27706

BROWN, WESLEY M. (30), Department of Biochemistry, University of California, Berkeley, California 94720

BUKHARI, AHMAD, I. (10), Cold Spring Harbor Laboratory, Cold Spring Harbor, New York 11724

BUTOW, RONALD A. (18), University of Texas, Health Science Center at Dallas, Dallas, Texas 75235

CANTATORE, P. (28), Division of Biology, California Institute of Technology, Pasadena, California 91125

CHACONAS, G. (10), Cold Spring Harbor Laboratory, Cold Spring Harbor, New York 11724

CHALLBERG, MARK D. (12), Department of Microbiology, 725 North Wolfe Street, Baltimore, Maryland 21205

CHING, E. (28), Division of Biology, California Institute of Technology, Pasadena, California 91125

CHING, HO (31), Laboratory of Cell Biology, National Cancer Institute, NIH, Bethesda, Maryland 20205

CLAYTON, DAVID A. (27), Department of Pathology, Stanford Medical Center, Stanford, California 94305

COON, HAYDEN G. (31), Laboratory of Cell Biology, National Cancer Institute, NIH, Bethesda, Maryland 20205

CREWS, S. (28), Division of Biology, California Institute of Technology, Pasadena, California 91125

CUMMINGS, DONALD J. (3, 34), Department of Microbiology and Immunology, University of Colorado Medical Center, Denver, Colorado 80262

DAWID, I. B. (25), Laboratory of Biochemistry, National Cancer Institute, NIH, Bethesda, Maryland 20205

DE WAARD, ADRIAN (15), Sylvius Laboratories, Physiological Chemistry, 2333, AL Leiden, Netherlands

DIN, N. (4), University of Copenhagen, DK-2200, Copenhagen, Denmark

DONELSON, JOHN E. (23), Department of Biochemistry, University of Iowa, Iowa City, Iowa 52240

DROUIN, JACQUES (29), Department of Biochemistry and Biophysics, University of California, San Francisco, California 94143

DuBOW, M. (10), Cold Spring Harbor Laboratory, Cold Spring Harbor, New York 11724

EDWARDS. J. (19), Department of Medicine, University of Chicago, Chicago, Illinois 60637

ENGBERG, J. (4), Biochemical Institute, University of Copenhagen, DK-2200, Copenhagen, Denmark

FAURON CHRISTIANE, M.-R. (26), Department of Biology, University of Utah, Salt Lake City, Utah 84112

FAUST, EMANUEL A. (13), Department of Human Genetics, Yale University, New Haven, Connecticut 06510

GELFAND, R. (28), Division of Biology, California Institute of Technology, Pasadena, California 91125

GHOSH, PRABAT K. (14), Department of Human Genetics, Yale University School of Medicine, New Haven, Connecticut 06510

GILLHAM, NICHOLAS W. (6), Department of Zoology, Duke University, Durham, North Carolina 27706

GODDARD, JUDY M. (26), Department of Biology, University of Utah, Salt Lake City, Utah 84112

GOODMAN, HOWARD M. (30), Department of Biochemistry and Biophysics, University of California, San Francisco, California 94143

GRANDCHAMP, C. (34), Centre de Genetique Moleculaire, CNRS, Gif sur Yvette, France

GRANT, D. M. (6), Department of Zoology, Duke University, Durham, North Carolina 27706

GRAY, PATRICK W. (9), Department of Biochemistry and Biophysics, University of California, San Francisco, California 94143

GRINDLEY, NIGEL D. F. (11), Department of Biological Science, University of Pittsburgh, Pittsburgh, Pennsylvania 15260

GRIVELL, LESLIE A. (20), Laboratory of Biochemistry, University of Amsterdam, 1005 GA, Amsterdam, Netherlands

GROOT, G. S. P. (20), Laboratory of Biochemistry, University of Amsterdam, 1005 GA, Amsterdam, Netherlands

HALLICK, RICHARD B. (9), Department of Chemistry, University of Colorado, Boulder, Colorado 80309

HARTLEY, JAMES L. (23), Department of Biochemistry, University of Iowa, Iowa City, Iowa

HECHT, N. B. (20), University of Amsterdam, Box 60.000, 1005 GA, Amsterdam, Netherlands

HECKMAN, J. E. (24), Department of Biology, Massachusetts Institute of Technology, Cambridge, Massachusetts 02139

HENSGENS, L. A. M. (20), University of Amsterdam, Box 60.000, 1005 GA, Amsterdam, Netherlands

HOEIJMAKERS, J. H. J. (32), Jan Swammerdam Institute, 1005 GA, Amsterdam, Netherlands

HOLLENBERG, CORNELIS P. (21), Max Planck Institut, Spemannstr. 34, 7400 Tubingen, FRG

HU, W. W. L. (5), Crop Science, North Carolina State University, Raleigh, North Carolina 27650

JOCHEMSEN, HENK (15), Sylvius Laboratories, Physiological Chemistry, 2333 AL Leiden, Netherlands

JOLLY, SETSUKO O. (8), Harvard University, Cambridge, Massachusetts 02138

KELLY, JR., THOMAS J. (12), Department of Microbiology, 725 North Wolfe Street, Baltimore, Maryland 21205

KHATOON, H. (10), Cold Spring Harbor Laboratory, Cold Spring Harbor, New York 11724

KIDANE, G. (33), Biology Department, University of California, Los Angeles, California 90024

LEBOWITZ, PAUL (14), Department of Human Genetics, Yale University School of Medicine, New Haven, Connecticut 06510

LEVENS, D. (19), Department of Medicine, University of Chicago, Chicago, Illinois 60637

LEVINGS III, C. S. (5), Department of Genetics, North Carolina State University, Raleigh, North Carolina 27650

LI, MAY (22), Department of Biological Sciences, Columbia University, New York, New York 10027

LINK, GERHARD (8), Harvard University, Cambridge, Massachusetts 02138

LINNANE, ANTHONY W. (17), Department of Biochemistry, Monash University, Clayton, Victoria, 3168, Australia

LOCKER, J. (19), University of Chicago, Department of Medicine, Chicago, Illinois 60637

LUPKER, JAN H. (15), Sylvius Laboratories, Physiological Chemistry, 2333 AL Leiden, Netherlands

LUSTIG, A. (19), Department of Medicine, University of Chicago, Chicago, Illinois 60637

MAAT, JAN (15), Sylvius Laboratories, Physiological Chemistry, 2333 AL Leiden, Netherlands

MACINO, GIUSEPPE (22), Instituto di Fisiologia Generale, Universita degli Studi di Roma, Roma, Italy

McINTOSH, LEE (8), Harvard University, Cambridge, Massachussetts 02138

MAHLER, HENRY R. (2), Department of Chemistry, Indiana University, Bloomington, Indiana 47405

MAKI, R. A. (3), University of Colorado Medical Center, Denver Colorado 80262

MARTIN, NANCY C. (23), Department of Biochemistry, University of Minnesota, Minneapolis, Minnesota 55455

MARZUKI, SANGKOT (17), Department of Biochemistry, Monash University, Clayton, Victoria 3168. Australia

MASUDA, H. (33), University of Rio de Janeiro, Cidade Universitaria 20.000, Rio de Janeiro, Brazil

MERKEL, C. (28), Division of Biology, California Institute of Technology, Pasadena, Calfornia 91125

MERTEN, S. (19), Department of Medicine, University of Chicago, Chicago, Illinois 60637

MICHAEL, N. (33), Biology Department, University of California, Los Angeles, California 90024

MILLER, DENNIS L. (23), Department of Biology, University of Iowa, Iowa City, Iowa

MORIMOTO, R. (19), Department of Medicine, University of Chicago, Chicago, Illinois 60637

MOYNIHAN, PATRICK S. (23), Department of Biochemistry, University of Iowa, Iowa City, Iowa

MURPHY, MARK (17), Department of Biochemistry, Monash University, Clayton, Victoria, 3168. Australia

NOBREGA, MARINA P. (22), Department of Biological Sciences, Columbia University, New York, New York 10027

OJALA, D. (28), Division of Biology, California Institute of Technology, Pasadena, California 91125

OROZCO, JR., EMIL M. (9), University of Colorado, Box 215, Boulder, Colorado 80309

PAGANO, JOSEPH S. (16), Cancer Research Center, University of North Carolina, Chapel Hill, North Carolina 27514

PERLMAN, PHILIP S. (2), Department of Genetics, Ohio State University, Columbus, Ohio 43210

PHAM, HUNG DINH (23), Department of Biochemistry, University of Minnesota, Minneapolis, Minnesota

PIATAK, MICHAEL (14), Department of Human Genetics, Yale University School of Medicine, New Haven, Connecticut 06510

PRING, D. R. (5), Department of Plant Pathology, University of Florida, Gainesville, Florida 32611

PRITCHARD, A. E. (3), University of Colorado Medical Center, Denver, Colorado 80262

RABINOWITZ, MURRAY (19), Department of Medicine, University of Chicago, Chicago, Illinois 60637

RAJBHANDARY, U. L. (24), Department of Biology, Massachusetts Institute of Technology, Cambridge, Massachusetts 02139

RASTL, E. (25), National Cancer Institute, NIH, Bethesda, Maryland 20205

REDDY, V. BHASKARA (14), Department of Human Genetics, Yale University School of Medicine, New Haven, Connecticut 06510

ROSENBLATT, H. (33), Biology Department, University of California, Los Angeles, California 90024

RUBENSTEIN, JOHN L. R. (27), Department of Pathology, Stanford Medical Center, Stanford, California 94305

RUSHLOW, KEITH E. (9), University of Colorado, Box 215, Boulder, Colorado 80309

SAGER, RUTH (7), Sidney Farber Cancer Institute, Boston, Massachusetts 02115

SCHRIER, PETER I. (15), Sylvius Laboratories, Physiological Chemistry, 2333 AL Leiden, Netherlands

SHAH, D. M. (5), Department of Microbiology, University of Chicago, Chicago, Illinois 60637

SHEPHERD, H. S. (6), Department of Biology, University of California, La Jolla, California 92093

SIGURDSON, CHRIS (23), Department of Biochemistry, University of Minnesota, Minneapolis, Minnesota

SIMPSON, A. M. (33), Biology Department, University of California, Los Angeles, California 90024

SIMPSON, LARRY (33), Biology Department, University of California, Los Angeles, California 90024

SMITH, STUART C. (17), Department of Biochemistry, Monash University, Clayton, Victoria, 3168. Australia

STIEGLER, GARY L. (9), University of Colorado, Box 215, Boulder, Colorado 80309

STRAUSBERG, R. L. (18), University of Texas, Dallas, Texas 75235

SYMONS, ROBERT H. (29), Department of Biochemistry, University of Adelaide, Adelaide, South Australia 5001

SYNENKI, R. (19), Department of Medicine, University of Chicago, Chicago, Illinois 60637

TABAK, H. F. (20), University of Amsterdam, Box 60.000, 1005 GA, Amsterdam, Netherlands

TERPSTRA, P. (18), University of Texas, Dallas, Texas 75235

TIMOTHY, D. H. (5), Crop Science, North Carolina State University, Raleigh, North Carolina 27650

TZAGOLOFF, ALEXANDER (22), Department of Biological Sciences, Columbia University,

VAN BEVEREN, CHARLES P. (15), Sylvius Laboratories, Physiological Chemistry, 2333 AL Leiden, Netherlands

VAN BRUGGEN, E. F. J. (20), Biochemistry Laboratories, University of Gronigen, Zernikelaan, Groningen, Netherlands

VAN DEN ELSEN, PETER J. (15), Sylvius Laboratories, Physiological Chemistry, 2333 AL Leiden, Netherlands

VAN DER EB, ALEX J. (15), Sylvius Laboratories, Physiological Chemistry, 2333 AL Leiden, Netherlands

VAN OMMEN, G. J. B. (20), University of Amsterdam, Box 60.000, 1005 GA Amsterdam, Netherlands

VAN ORMONDT, HANS (15), Sylvius Laboratories, Physiological Chemistry, 2333 AL Leiden, Netherlands

WARD, DAVID C. (13), Department of Human Genetics, Yale University, New Haven, Connecticut 06510

WEISSMAN, SHERMAN M. (14), Department of Human Gentics, Yale University School of Medicine, New Haven, Connecticut 06510

WOLSTENHOLME, DAVID R. (26), Department of Biology, University of Utah, Salt Lake City, Utah 84112

WURTZ, E. A. (6), Department of Physiology and Biophysics, University of Illinois, Urbana, Illinois 61801

YIN, S. (24), Department of Biology, Massachusetts Institute of Technology, Cambridge, Massachusetts 02139

PREFACE

In some respects, the organization and fundamental concept of this symposium on extrachromosomal DNA resembles the circular nature of some of the extrachromosomal elements themselves. It is difficult to separate the beginning from the end. The symposium was organized on the premise that the diversity and complexity of primitive mitochondrial and perhaps chloroplast DNA structure and replication had more in common with many viral systems than with either prokaryotic or eukaryotic systems. This is especially striking in the case of so-called split genes. Intervening sequences in DNA were first discovered in the small DNA viruses and later in nuclear genes. But it is in yeast mitochondrial DNA that the extent of their involvement in RNA processing is most noteworthy. As reported at this symposium, it is possible to isolate mutants in some intervening sequences and analyze their effect in loci of known genetic function. Not only will such analyses in yeast mitochondrial split genes lead to a basic understanding of intervening sequences in general, but their very presence will have to be dealt with in evaluating theories on mitochondrial evolution.

It should not be surprising that the most active area of research represented at this meeting is the biogenesis of yeast mitochondria. At the close of the symposium, Pyotr Slonimski presented a brief overview that put the meeting in historical perspective. Much to the surprise of the organizers, Slonimski pointed out that the timing coincided almost to the month with the 30th anniversary of Boris Ephrussi's announcement on the petite mutation (Ephrussi, B., Hottinguer, H., and Tavlitzki, *J. Ann. Inst. Pasteur* 76, 351, April 1949). In a series of seven articles, Ephrussi and his collaborators reported on many aspects of the petite mutation with only passing reference to DNA. Since this was some 14 years before the first demonstration of mitochondrial DNA, this was certainly not a startling omission. The ensuing years have shown that yeast mitochondria have occupied a signal position in elucidating the function of extrachromosomal DNA. As we shall see, the analysis of intervening genes is especially important in yeast mitochondria, as well as the sequencing of a variety of genes of known function.

This symposium witnessed the gathering of seemingly disparate groups of researchers involved in mitochondrial, chloroplast, plasmid, and viral DNA function and replication. As will be apparent, however, great similarities exist in these systems at both the molecular and phenomenological levels. In future years, these similarities may well lead to a more basic understanding of organelle evolution and biogenesis.

I want to thank my coorganizers Piet Borst, Igor Dawid, and Sherman Weissman for their enthusiasm and assistance in organizing this symposium. We also wish to thank the Life Sciences Division of ICN Pharmaceuticals, Inc., for their continued support of the ICN–UCLA Symposia series and the National Institutes of Health for contract #263-MD-912641 (jointly sponsored by the Fogarty International Center, National Cancer Institute and National Institute for Allergy and Infectious Diseases).

EXTRANUCLEAR GENETICS

G.H. Beale[1]

ABSTRACT A brief survey of the development of our knowledge of extranuclear genetics is presented. The material is grouped under three headings: (1) DNA-containing cell organelles; (2) endosymbionts, and (3) virus-like particles. The extremely uneven development of research on the different examples is pointed out. Comparisons between examples in the different groups are made regarding their DNA, the presence of histone-like proteins, and the relative control of extranuclear structures by nuclear and extranuclear genes. An attempt is made to establish homologies between members of different groups, and some evolutionary hypotheses are considered. It is suggested that use of the terms "prokaryote" and "eukaryote" may be inappropriate for extranuclear DNA and should be applied to whole cells or organisms. It is also pointed out that even the meaning of the word "extranuclear" has some obscurity. The need for research on a wider range of materials than have been studied hitherto is stressed.

INTRODUCTION

Extranuclear genetics is now such a large and diversified subject that it is well-nigh impossible to write a coherent account in the short space of a single contribution to a symposium. Nevertheless there are good reasons for making the attempt: it offers an opportunity to allow one's mind to roam over the whole field before concentrating on the minutia of particular examples. It shows that though a great deal has been learnt about certain parts of the subject, others have been seriously neglected. We know far more about mitochondria than about other extranuclear structures with genetic properties, except possibly some viruses, and amongst mitochondria, most of our information comes from one organism - Saccharomyces cerevisiae.

[1]Present address: Institute of Animal Genetics, West Mains Road, Edinburgh EH9 3JN, Scotland

ISBN 0-12-198780-9

To illustrate the variability of mitochondria amongst different organisms, the following facts should be noted. The DNA of yeast mitochondria is in the form of a circle of about 25 μm in circumference, while in other organisms the size may vary from 5 μm to 35 μm or more. In protozoa the mitochondrial DNA may be circular, linear or (in kinetoplasts) catenated (1). The mitochondrial ribosomes of different organisms also vary a great deal in size (from 55S in animals to 80S in plants or ciliates), and there are important differences in the genes (nuclear and mitochondrial) coding for the mitochondrial ribosomal proteins in yeast, *Neurospora* and *Paramecium* (2). Thus even within the one category of mitochondria a misleading impression is created if all conclusions are based on yeast, and still greater disparity is evident when one takes into consideration cytoplasmic structures other than mitochondria.

By contrast with this diversity of extranuclear genetic systems, the classical chromosomal mechanism controlling Mendelian heredity is remarkably uniform over the whole range of eukaryotic organisms, from protista to mammals, and there we were not led far astray by basing the whole theory on *Drosophila*, more or less. My aim therefore in presenting this paper is to draw attention to the diversity of extranuclear phenomena, and to the necessity of studying a wider range of materials.

To illustrate the development of our knowledge of extranuclear genetics, it is interesting to compare our present programme with the proceedings of a symposium held in Paris in 1948 (3). The earlier meeting was organised by André Lwoff and entitled "Unités biologiques douées de continuité génétique". At that meeting Ephrussi presented his first results on the "petite colonie" mutants of yeast, and tentatively ascribed their determination to a cytoplasmic,non-genic, factor. L'Héritier described his CO_2-sensitive strains of *Drosophila*, and showed them to be controlled by cytoplasmic elements which were called at that time "génoïdes". Rhoades discussed the plastids, which had long been known to show non-Mendelian properties - ever since the observations of Correns and Baur in 1909-, though in 1948 there was still a lot of argument about whether the non-Mendelian determinants were actually inside the chloroplasts, or somewhere else in the cytoplasm, or possibly comprised some ill-defined entity known as the "plasmon".

In the discussion of these and other non-Mendelian phenomena the word "plasmagene" was used by several speakers at the Paris meeting, as a kind of alternative to real chromosomal genes. Some participants speculated that there might be a whole range of sub-cellular particles, including genes, plasmagenes, viruses and so on; and some people had doubts about the existence of extranuclear genes altogether. (I remember a postcard was written to Tracy Sonneborn, who did not attend the meeting, stating that "Les plasmagènes n'existent pas"). Delbrück put forward a hypothetical scheme based on a system of steady states involving enzymes and inhibiting metabolites, thus avoiding the need for any particulate determinant at all. This was before current ideas of protein synthesis based on translation of m-RNA had been put forward, and hence Delbrück's scheme could only be expressed in very general terms.

DNA was not referred to very much at the Paris meeting, except in connection with bacterial transforming principles, and this omission is not surprising since the theory of the double helix had not yet been conceived.

In the years between the Paris meeting and the present one, great progress has been made in certain areas, and no one any longer doubts the existence of extranuclear genetic determinants. Yeast and Chlamydomonas stand out as exceptionally favourable materials for this work and have been brilliantly exploited. However, some of the matters discussed in 1948 have not progressed very much, and some are not now thought to involve gene-line determinants. For example Lwoff discussed the notion of genetic continuity of kinetosomes - organelles situated just beneath the surface of ciliate cells. It is now thought that kinetosomes lack their own DNA and whatever hereditary properties they have (- and they certainly have something, as shown by the work of Beisson and Sonneborn (4)-) must be based on some other mechanism. Centrioles also, which resemble kinetosomes to some extent, are likewise probably devoid of gene-line components.

A crucial step in the development of our understanding of extranuclear genetics was the discovery, some 15 years after the Paris meeting, of DNA in mitochondria and plastids (5,6). This was the signal for many workers to claim immediately that organelle DNA had a genetic role, and conversely that presence of this DNA proved that structures containing it had genetic properties. It took some time however to substantiate these conjectures. I must confess that I found the discovery of organelle DNA rather depressing, since - seeking variety - some of us had previously hoped that some other system, not involving DNA, might emerge as a basis for extranuclear genetics. When this thought had to be relinquished, we faced the boring possibility that cytoplasmic genes would turn out to be merely nuclear genes in a new situation. After all DNA is DNA, always basically the same, chemically speaking.

This leads us to a question of current interest: is there a fundamental difference (organizational if not chemical) between nuclear and extranuclear DNA? I expect this question will be discussed later at this meeting. Whether the answer is "yes" or "no", however, it is now quite clear that extranuclear genetic particles show a number of idiosyncrasies which are not typical of nuclear genes. For example, with extranuclear genes there are usually numerous copies of a single structure in a single cell, whereas there are usually only one or two nuclear genes of a given kind in a cell; nuclear genes are arranged on a number of chromosomes, each individually distinct, whereas cytoplasmic genes are usually on a single structure, a circular piece of DNA; nuclear genes are subject to constraints imposed by the chromosomal mechanisms of mitosis and meiosis, which do not affect cytoplasmic genes; mutation of cytoplasmic genes shows differences from that of nuclear genes, and cytoplasmic genes may be transferred from cell to cell by infection and other mechanisms, unlike chromosomal genes. Many other differences could be listed.

Following this introduction, I plan to make an attempt to define what is meant by the word "extranuclear", and then classify the different types of extranuclear genetic phenomena, describing a few characteristics of each group. This will lead on to a discussion of problems of homology and evolution of extranuclear factors, and to consideration of the terms "prokaryotic" and "eukaryotic".

THE MAIN GROUPS OF EXTRANUCLEAR FACTORS

The problem of defining "extranuclear" factors turns out to be unexpectedly difficult. Obviously "extranuclear" means "outside the nucleus" and in eukaryotic cells the meaning is unambiguous (except possibly for some viruses and symbionts which are located within the nucleus, though separate from the chromosomes). In prokaryotes however there is no sharp boundary between the DNA-containing region (or nucleoid) and the rest of the cell: in *E. coli*, the prototype of all prokaryotes, there is a main DNA structure, often called "the chromosome", and in addition one or more smaller circles (plasmids, F factors, etc.) which may be considered as "extranuclear". However, some extranuclear components of eukaryotes (e.g. endosymbionts and organelles) may be considered homologous with the "nuclei" of prokaryotes, and if one accepts this, then a given type of structure might be "extranuclear" in one situation and "nuclear" in another, which is absurd. The problem is seen to be still more complicated when we consider the kappa particles of

Paramecium, which are located in the cytoplasm, and therefore "extranuclear" so far as *Paramecium* is concerned, but themselves contain virus-like particles. What are they? Extra-extranuclear? Here we have a kind of 3-tiered hierarchical system. If one wishes the definition to embrace both eukaryotes and prokaryotes, the word "extranuclear" can only refer to the position of one structure in relation to another structure denoted a "nucleus". I do not want to propose a more elaborate nomenclature inventing new terms, but merely wish to point out that we should face up to the reality of a complicated range of biological structures.

Extranuclear genetic factors may be put into three groups, as follows:

1. DNA-containing cell organelles,
2. endosymbionts, and
3. virus-like particles, including plasmids.

No doubt objections can be raised to lumping three lots of disparate materials in this crude way (especially in group (3)), but such a grouping is convenient for purposes of discussion. In group (1) there are only two examples: mitochondria and plastids, both of which are of course essential for life as it now exists on the earth. Group (2), the endosymbionts, comprises particles of a more restricted occurrence by comparison with organelles, though there are vast numbers and many different kinds of endosymbionts, and some are essential for the normal growth of their host organisms (e.g. the chlorophyll containing zoochlorellae of *Paramecium bursaria*, and the omikron particles of some freshwater *Euplotes* species (7)). As for group 3, the virus-like particles, they are so varied that one cannot say whether they are essential or not: many are of course pathological and therefore not essential at all to their host cells, but others e.g. some phages and the F factors of *E. coli*, play a role very closely integrated into that of the host bacterium. I would like to compare and contrast members of the three groups in regard to the following features: (1) their DNA; (2) the association of this DNA with basic proteins, and (3) the relative control by nuclear and extranuclear genes of extranuclear characters.

As for the DNA of members of the three groups, I mentioned above the variation in size and structural organisation of mitochondrial DNAs. Not much is known about the DNA of endosymbionts. Kappa particles have been reported to contain a main component consisting of long linear strands of DNA, and in addition small DNA circles of 13.75 μm contour length in the virus-like elements which are present, as already mentioned,within the kappa particles (8). Little work has been done on the DNA of other endosymbionts. Considering viruses in general, there is of course a vast amount of information about virus and plasmid DNA, and it would be pointless to go into that here. It is however relevant to point out that some viruses and plasmids contain DNA circles roughly equivalent in size to those of mitochondria.

The determinants of the CO_2-sensitive strains of Drosophila, now denoted sigma viruses and known to be very similar to a known vertebrate virus (VSV), contain as a genetic element not DNA but RNA. Likewise the virus-like particles controlling another example of non-Mendelian genetics - the "killer" phenomenon in Saccharomyces and Ustilago, also contain RNA, this time double stranded (2). I suppose these two examples should be excluded from discussion at a symposium entitled "Extranuclear DNA", but we have to face the facts of life, and these particles are undoubtedly extranuclear genetic factors.

That is all I want to say about DNA, (and RNA as a genetic element), and I now turn to the question of the association of DNA with histones or histone-like proteins, such as occurs as a characteristic feature of eukaryotic chromosomes. Until recently it was generally accepted that the DNA of prokaryotes was "naked", though recently there have been reports of the occurrence of one or two histone-like proteins even in bacteria (9,10). However it has been suggested that the histone-like proteins of E. coli are present in too small an amount to complex E. coli DNA, as they do in eukaryotes forming nucleosomes (11). What is the situation in regard to mitochondrial DNA?

Formerly it was stated that mitochondrial DNA, like that of bacteria, was "naked", but this appearance was probably due to the drastic methods used in preparing the DNA. More delicate procedures have revealed the presence in mitochondria of Xenopus (12), Physarum (13) and Paramecium (14), of a beaded, chromatin-like structure. There is at present no certain proof that histones are present in these beads, but provisional study of the Paramecium material indicates that there may be four or five such basic proteins, which seem to be distinct from the histones in the nuclear chromatin of the same organism. We have no information about the genetic determination of these proteins: they could be nuclearly or mitochondrially coded, and it is hoped that this matter can be pursued in Paramecium which has certain technical advantages for studying the problem. Although this work is still at an early stage, present indications are that, in regard to possession of histone-like substances, and possibly nucleosomes, mitochondrial chromatin appears to have some eukaryotic features.

Nothing is known about histones in endosymbionts, so far

as I am aware. With regard to viruses, it should be noted that some animal viruses (adenovirus and SV-40) have the ability of associating with histones and forming nucleosome-like structures (15). This is discussed below.

The next matter which I would like to discuss is concerned with the relative importance of nuclear and extranuclear genes in the development and functioning of extranuclear structures. Members of each of the three groups referred to above are controlled by both nuclear and extranuclear genes, but the relative contributions of the two types of determinant vary a great deal in the different examples. We have extensive data about this matter for mitochondria, plastids and some viruses and plasmids, but very little for endosymbionts. Unfortunately, refined genetic analysis and gene mapping has not yet been achieved with any endosymbiont. However, there is crude evidence indicating that kappa particles and other ciliate endosymbionts are largely governed by their own genes, and relatively little by genes in the ciliate nuclei (2). An indication that this is so is given by the finding that some of these endosymbionts (though so far not kappa) have been cultured in a synthetic medium.

By contrast, cell organelles are largely governed by nuclear genes, though organelle genes are certainly essential too. Perhaps 90% of gene products in mitochondria are coded in the nucleus, and only 10% in the mitochondrial DNA. No one has cultured mitochondria _in vitro_, and I don't imagine this will be done for a long time, if ever.

What is remarkable about the genetic control of cell organelles is the extraordinarily tight integration of the products of nuclear and organelle gene activity. The replication and transcription of mitochondrial DNA is largely controlled by proteins coded in nuclear DNA; the translational mechanims comprising r-RNA, ribosomal proteins and t-RNAs, etc., involves an extremely intimate association of nuclear and extranuclear gene products, and the same applies to some proteins in the inner membranes, such as ATP-ase, cytochrome oxidase, etc. A similar situation exists in chloroplasts, where the major protein (fraction I or ribulose bisphosphate carboxylase)is a mixture of nuclear and plastid gene products. Thus there is an intricate mixing of nuclear and extranuclear gene products in many parts of the organelles.

As for the third group - the virus-like particles - there is too great a variety amongst them for us to be able to make a general statement concerning the relative importance of host and virus genes. However in some cases (e.g. bacterial plasmids) there is overriding control by the bacterial genome (15) and the plasmid genome does very little except determine the structure of replicas of itself. In the most extreme situation where there is integration of the plasmid in the bacterial chromosome, the plasmid practically loses its identity altogether.

Thus, in a very general way one can arrange the three groups of determinants in the following order: endosymbionts showing most control by their own genes, and relatively little by the nuclear genes of the host cell; organelles showing a preponderance of control by nuclear genes, but an intimate interaction between nuclear and organelle gene products; virus-like particles showing - in some cases - the most extreme situation of almost total control by the host genome, though of course there are many exceptions to these statements.

HOMOLOGIES AND EVOLUTION OF EXTRANUCLEAR STRUCTURES

The three groups - organelles, endosymbionts and viruses - are so dissimilar that it may seem pointless to try and establish homologies between them, even in an evolutionary sense. While it is true that all - or nearly all - contain DNA which can be called extranuclear, this might be their only common feature. However, there are a number of considerations - mainly of a speculative nature - indicating relationships between members of the different groups. I will briefly mention a few, at random.

1. If the symbiotic theory (16) for the evolution of organelles is true (and of course it may not be), mitochondrial and bacterial endosymbionts have a common origin, as have chloroplasts and blue-green algae.
2. As Raff and Mahler (17) have hypothesized, mitochondrial DNA may have originated from a plasmid.
3. Mitochondria have a number of virus-like properties, e.g. the size of their DNA circles, and their mechanism of recombination, (where it occurs).
4. Viruses occur in some endosymbionts.

Other inter-connections could certainly be traced. Nevertheless to attempt to establish an evolutionary order amongst this welter of extranuclear structures is a daunting task. It is perhaps simplest to start with the endosymbionts, many of which show resemblances to known free-living microorganisms. For example, kappa shows a number of bacterial characteristics, and has been awarded a bacterial, binomial designation (*Caedobacter taeniospiralis*)(18). Other endosymbionts resemble rickettsiae, spiroplasma, green algae and other microbial groups (2). Some are naturally or artificially infectious and there seems no conceptual difficulty in accepting them as descendants of various microorganisms.

No such easy solution is offered to us for the evolution of cell organelles, which in my view are very distinct from any known free-living organism. Two distinct hypotheses have been proposed (17,18), the symbiotic and non-symbiotic, as well as a number of variants of each. In connection with these hypotheses there has been much discussion of the supposed prokaryotic or eukaryotic nature of cell organelles. Formerly the prokaryotic features were stressed, e.g. the 'naked' DNA,which I have discussed above, and which was thought to be characteristic of organelles and bacteria; and the sensitivity of organelle ribosomes to antibiotics inhibiting ribosomes of bacteria but not of higher organisms. Recently opinion is veering in the opposite, i.e. eukaryotic, direction (19), as facts have accumulated showing presence of repetitive DNA sequences and of "split" genes in organelle DNA, poly-A on organelle m-RNA, and finally the possibility of nucleosome-like structures on organelle DNA as mentioned above.

It seems to me however that discussion as to whether organelle DNA is prokaryotic or eukaryotic is rather futile, since any structure in a eukaryotic cell is liable to show some eukaryotic features. Even some viruses (adenovirus and SV 40), when in animal cells, are capable of forming associations with four histones and form nucleosome-like beads; and the same viruses show leader or spliced segments of m-RNA (15). So far as I know, no one has described these viruses as "eukaryotic". It seems likely that cell organelles, like viruses inhabiting eukaryotic cells, are able to exploit the production by the host cell (ultimately from its nuclear genes) of certain enzymes which are required for such processes as the excision and ligation of the m-RNA produced by split genes, or the attachment of poly-A to messenger, or other "eukaryotic" functions. If we are right in believing that mitochondria contain histones, the latter could be coded by the nuclear genome and transported into the organelles. Thus all eukaryotic features of organelles or viruses could be due, ultimately, to the genetic capabilities of the host cell.

In my view, the terms "prokaryotic" and "eukaryotic" should not be applied to DNA, or even to individual constituents of cells, but to larger structures, whole cells, with their heterogeneous contents of nuclei, cytoplasm, organelles etc. This line of argument does not help us establish the homologies and evolutionary history of organelles, but perhaps might have the advantage of compelling us to seek alternative hypotheses.

CONCLUSION

In this article I have attempted to review briefly the development of our knowledge of extranuclear genetics, stress the uneven development of different parts of the subject, and point out what seem to me to be a number of inconsistencies. I conclude that, contrary to the optimistic views sometimes put forward by some workers in the field, prophesying an imminent completion of our studies, there are many unsolved problems in this fascinating branch of biology.

ACKNOWLEDGMENT

I am grateful to Andrew Tait for his comments.

REFERENCES

1. Borst, P. (1979). In "Biochemistry and Physiology of Protozoa". 2nd ed. Academic Press, New York.
2. Beale, G.,and Knowles, J. (1978). "Extranuclear Genetics" Arnold, London.
3. Lwoff, A. (ed)(1949). "Unités biologiques douées de continuité génétique". C.N.R.S., Paris.
4. Beisson, J., and Sonneborn, T.M. (1965). Proc.Natl. Acad.Sci. 53, 275.
5. Ris, H., and Plaut, W. (1962). J.Cell Biol. 13, 383.
6. Nass, M.M.K., and Nass, S. (1963). J.Cell Biol. 19, 593.
7. Heckmann, K. (1975). J.Protozool. 22, 97.
8. Dilts, J.A. (1976). Genet.Res. 27, 161.
9. Varshavsky, A.J., Bakayev, U.V., Nedospasov, S.A. and Georgiev, G.P. (1977). Cold Spring Harbor Symp. 42, 457.
10. Rouvière-Yaniv, J. and Gros, F. (1975). Proc.Natl.Acad. Sci. 72, 3428.
11. Griffith, J.D. (1976). Proc.Natl.Acad.Sci. 73, 563.
12. Pinon, H., Barat, M., Tourte, M., Dufresne, C. and Mounolou, J.C. (1978). Chromosoma, 65, 383.
13. Kuroiwa, T., Kawano, S. and Hizume, M. (1976). Exptl Cell Res. 97, 435.
14. Olszewska, E. (Personal Communication).
15. Chambon, P. (1977). Cold Spring Harbor Symp. 42, 1209.
16. Novick, R.P., Wyman, L., Bouanchaud, D. and Murphy, E. (1975). In "Microbiology 1974" (D. Schlessinger, ed.) Amer.Soc.Microbiol., Washington, D.C.
17. Margulis, L. (1970). "Origin of eukaryotic cells". Yale University Press, New Haven and London.
18. Raff, R.A. and Mahler, H.R. (1972). Science 177, 575.
19. Preer, J.R., Preer, L.B. and Jurand, A. (1974). Bact. Rev. 38, 113.
20. Bernardi, G. (1978). Nature 276, 558.

MITOCHONDRIAL BIOGENESIS: EVOLUTION AND REGULATION

Henry R. Mahler

Department of Chemistry[1], Indiana University, Bloomington, IN

Philip S. Perlman

Department of Genetics, Ohio State University, Columbus, OH

ABSTRACT The polypeptide gene products expressed in and by the mitochondria of respiration-competent eucaryotic cells appear to be few in number and conserved in kind: They consist of three subunits of cytochrome oxidase, of cytochrome *b* of the respiratory chain, and a more variable number (≥2) of subunits of the membrane-integrated (F_0) portion of the ATP synthase (oligomycin-sensitive ATPase). In spite of this apparent invariance, the resident genophore and its means of expression appear to have been - and perhaps still are - subject to profound and relatively rapid evolutionary changes. A particularly instructive example is provided by various yeast strains of the species *Saccharomyces cerevisiae* and *carlsbergensis* in which the mtDNA, although topologically and functionally homologous, can vary in size by as much as 12 kbp (15% of the total). These insertions/deletions can be as large as 900-3000 bp and are located in specific gene regions. The organization and expression of the mitochondrial gene for apocytochrome *b*, investigated in several laboratories, is highly complex: four coding regions (exons) appear to be interspersed among at least three non-coding regions (introns). This sequence organization resembles that found in nuclear and viral eucaryotic genes, but in the mitochondrial case mutations in introns can readily be isolated. They have been shown to result in i) the appearance of novel polypeptides not present in the wild type and ii) regulation of the expression of a second, unlinked mt gene, i.e. the one responsible for the specification of the largest subunit of cytochrome oxidase. This gene probably also exhibits a complex sequence organization. However, other mt genes may not share this property. Thus the mt genophore in *Saccharomyces* is unusual in several respects, among them i) an unusually large proportion and length of AT-rich sequences, ii) non-contiguity of genes for rRNA precursors, and iii) the simultaneous presence of genes exhibiting both a pro- and a eucaryotic pattern of sequence organization.

[1]Publication No. 3267

ISBN 0-12-198780-9

INTRODUCTION

The appearance of mitochondria in the emerging protoeucaryotic cell was an essential and determinative event in its evolution. Mitochondria have ever since constituted an integral and characteristic element of eucaryotic cellular structure and function, and - so far as can be ascertained from a study of their contemporary exemplars - so has a separate intramitochondrial system for the maintenance, transmission and expression of a discrete set of genes vital for mitochondrial biogenesis and hence for cellular survival.

In this presentation we shall endeavor to deal with some aspects of the continuity of this system, within the confines of present-day eucaryotic organisms. From them we draw the inference that in spite of strong conservative constraints, possibly of teleological significance, tending toward uniformity, the system also exhibits unmistakable signs of great diversity both between species and within a single species. As a result of these evolutionary developments assembly of mitochondrial genes and their mode of expression exhibit features that are unique and distinct from both their procaryotic and nuclear counterparts.

In this paper we will have occasion to introduce certain themes that will receive exposition and elaboration in depth by others in the course of this conference.

MITOCHONDRIAL GENE PRODUCTS

All mitochondria of respiration-competent cells[2] appear to contain an apparatus designed for the controlled expression of a set of genes specifying a restricted set of polypeptides of their inner membrane (1-5). This includes the information transcribed into the precursors of their mRNAs, as well as the tRNAs and rRNAs required for their translation, but not that for the proteins participating in these events. There is currently no strong evidence in favor of the import of any RNA (particularly mRNA) into mitochondria, and certainly none for export from the organelle.

[2]We thus explicitly exclude in this context mutant strains (such as *rho*⁻, *syn*⁻ and certain *mit*⁻ mutants in *Saccharomyces cerevisiae*) which have been of the greatest utility in establishing the very properties of the system which we are considering. For their detailed description see Refs. 1-5 and the papers by Linnane, Levens and Rabinowitz, Tzagoloff *et al.* and Slonimski in this volume.

TABLE 1

MITOCHONDRIAL POLYPEPTIDES ENCODED IN MT DNA

RESPIRATORY COMPLEXES

DESIGNATION	REACTION	M_R ($x10^{-3}$)	POLYPEPTIDES			
			NUMBER	MITOCHONDR.	M_R ($x10^{-3}$)[a]	% OF PROTEIN IN COMPLEX
I	NADH:CoQ	850	≥20	0		0
II	Succ:CoQ	2 x 100	2 x 2	0		0
III	$CoQH_2$:c	275	7 - 9	1[a]	30	20
IV	c:0_2	2 x 250	2 x 14	3[b]	40,31,22	65
V	ATP SYNTHASE (OS-ATPASE)	450	≥14	4*	29,22 12,7,5	16 (100% OF F_0)
		2300	75	18		22

OTHER

VAR1*	RIBOSOMAL(?)			1	45	

*IN SACCHAROMYCES †≥9 SPECIES [a] 2 COPIES/300,000 [b] 2 COPIES/250,000

Nature of Polypeptides Specified by mtDNA. This question has been investigated in a large number of species varying from fungi to man with concordant results. The most conclusive evidence has been derived from studies on the fungi *Neurospora* and *Saccharomyces*, which have the advantage of permitting a genetic approach to its resolution. There is general agreement concerning the outcome summarized in Table 1, which states that the contribution of mitochondrial gene products to the biogenesis of the organelle is restricted to a participation in the construction of its inner membrane, to which they supply a maximum of about 20% of its total polypeptides or protein mass. Furthermore these contributions are confined to only three of the five complexes of the respiratory chain, all of which, in addition also contain proteins encoded in nuclear genes and synthesized on cytoplasmic ribosomes.

Are There Additional, As Yet Unidentified, Mitochondrial Gene Products? As concerns major polypeptides, synthesized on mt ribosomes, the answer, based on the data shown in Table 2 is clearly in the negative. This inference holds both for *S. cerevisiae*, an organism with one of the largest mtDNAs, as well as for vertebrate animals, with the smallest functional mtDNA known. The types of mitochodnrial genes studied in the former instance are the *mit* genes, responsible for the specification of the polypeptides described in Table 1: *oxi3, 1* and *2* for the cytochrome oxidase subunits of $M_R \simeq$ 40-42.5, 31-32.5 and 22-23.5 x 10^3 respectively; *cob-box* for cytochrome *b* with

TABLE 2

PROTEINS AND THEIR mRNAs ENCODED IN MITOCHONDRIAL DNA

S. cerevisiae			*Animal Cells*
mit	genes	6	
oli	genes[a]	2	
var	genes	1	
	Total	9	
Major transcripts		10[b]	8[c]
Major polypeptides		6-8[d]	8-10
		[M_R=≳55,<u>45</u>,<u>42.5</u>,<u>32.5</u>, <u>30</u>,<u>23.5</u>,~<u>10</u>,<10]	[M_R=45,39,35,33, 27.5,25,23, 19.5,~15,~12]

[a]two are probably identical with *mit* genes

[b]2-3 may have originated from a single transcription unit (4,Grivell *et al.*, this symposium)

[c]Battey and Clayton (8)

[d]most recent studies show six such products (underlined)

M_R ≃ 30,000; *pho2* (*oli1* or O_I - Ref. 9) for one of the F_0 ATPase subunits of M_R ≃ 7500, and *pho1* for a second such subunit; the *var* gene; and perhaps two additional (*oli*) loci responsible for oligomycin resistance. Additional genes may exist; in fact studies in our laboratory have provided evidence for two additional, regulatory loci, one designated 2-34, mapping between *oli1* (O_I) and *var1*, and the other, called E65, between *C* and *oxi1* on the linearized map shown in Fig. 1 (11). The latter map position is close to that of another, thermosensitive, mitochondrial mutation, *tsm8* (12), exhibiting a somewhat similar phenotype at its non-permissive temperature (13). However, the nature of the primary defect in these cases has not yet been ascertained: some or all of them could affect a tRNA species essential for the synthesis of a restricted class of proteins rather than being in a structural gene. An explicit search for novel genes involved in the synthesis of *essential* polypeptides has failed to disclose

COMBINED PHYSICAL (RESTRICTION ENDONUCLEASE) AND GENETIC MAP OF W.T. STRAIN ID41-6/161

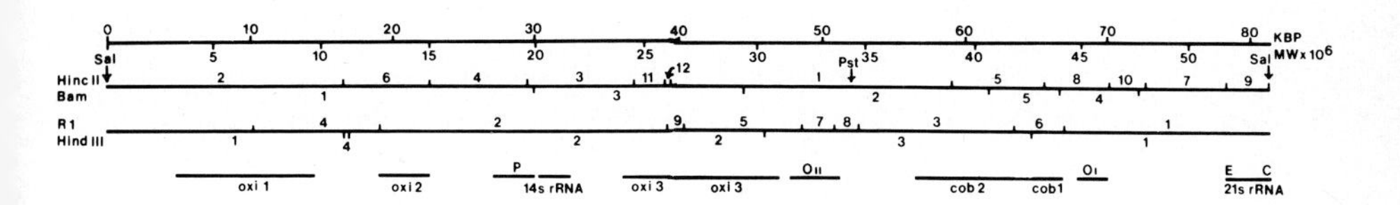

such entities (14). It thus appears likely that the total number of major, mitochondrially encoded polypeptides will not exceed twelve.

EVIDENCE FOR SEQUENCE DIVERGENCE IN MT DNA

Variability of MtDNA Size and Base Composition. Existing compilations of these properties of mtDNA isolated from different sources (1,7,15) provide clear evidence for extensive evolutionary variation of mtDNA. The selected data summarized in Table 3 show such changes in its characteristic size, within rather broad limits, depending on the kingdom (16) and family of origin; thus the mtDNAs from metaphyta are the longest, and those of the metazoan animals among the smallest. The smallest mtDNA known so far is found in *Chlamydomonas* (17) and consists of fewer than 15,000 base pairs (bp). The mtDNA of protists and fungi fall generally within the limits of 30-55 kbp. The base composition of the molecules spans a wide range between 20% G + C (density = 1.684 g/ml) and ≥60% G + C (density 1.708 g/ml) with the lower limit probably set by the requirements of the genetic code (18). This variability is readily apparent, even if we restrict ourselves to the animal kingdom (cf. *Drosophila* vs. vertebrates).

However, more persuasive evidence for mitochondrial diversity exists. As also shown in Table 3, it extends beyond

TABLE 3

PROPERTIES OF MITOCHONDRIAL DNA

ORGANISM	DENSITY IN CSCL (g/cm³)	COMPLEXITY (M daltons)	LENGTH IN K BASE PAIRS (complex.)	(restr. analy.)
Euglena gracilis	1.690	100,40	151,61	
Acanthamoeba castellani	1.694,1.690	34,26	51,39	
Paramecium aurelia† *4-51*	1.693	35	53	
T	1.685	30	45	
Tetrahymena pyriformis† (diff. strains)	1.685	25.8,28.4 31.0,32.6	36.3,38.0 44 ,49.5	37.8,40.6 46 ,50.6
Chlamydomonas reinhardtii	1.706	9.78	14.8	
Pisum sativum	1.706	74	110	
Dicteostelium discoideum	1.688	40,35	60,53	
S. cerevisiae† (diff. strains)	1.684	50		70.8,75.8 78.3
S. carlsbergensis†	1.683	50		68.0
Kluyveromyces lactis	1.691	20	30.3	
Neurospora crassa	1.701	40 ± 0.5		60.0
Aspergillus nidulans		21.6	32.6	31.5
Urechis caupo	1.699	11.5	17.4	
Drosophila melanogaster†	1.680	12	18	18.4
Drosophila viridis†	1.711			
Xenopus laevis†	1.700	11.7	17.7	19.5
Xenopus borealis†	1.700	11.5	17.4	17.0
Rat (liver)	1.701	10.3	15.1	15.3
Sheep*, Goat*	1.700			15.81,15.76
Chicken†	1.708	10.5	15.5	16.4

†base sequence divergence in different strains

*base sequence divergence in different individuals

closely related genes and species (e.g. the yeasts *Saccharomyces* and *Kluyveromyces*, different species of *Drosophila* and *Xenopus* and subspecies of *Paramecium*) to differences within the same species. The former case is particularly instructive since here the difference in number of base pairs between closely related organisms is greater than the total number retained in the smaller molecule (19). A further refinement, providing evidence for more profound evolutionary change, has been introduced by comparisons of different strains of both unicellular and multicellular species and of different individuals in the latter case. In several instances evidence for divergence (e.g. *Tetrahymena pyriformis* and *S. cerevisiae*) becomes evident already at the level of molecular size, determined either from hydrodynamic or sequence complexity measurements (19,20).

Sequence Divergence in mtDNA. These conclusions have been substantiated, and additional, compelling evidence for sequence divergence has been provided by construction of physical maps using restriction endonucleases. This has previously been accomplished for different strains of *Tetrahymena* (19), *Saccharomyces* (4,11,12,21,22), two closely related species of *Xenopus* (23), chickens, laboratory rats (24), different human cell lines (25), and for individual sheep and goats (26). These last studies have suggested base sequence divergences between individual animals at about 1%, and 6-11% between the two species. Similarly, Brown *et al.* (27) have found sequence divergences of 0.02 substitutions/base pair/ 10^6 years by comparing restriction endonuclease maps of mtDNA from four higher primates. In some of the instances cited the differences can be explained in terms of insertions and deletions of blocks of DNA sequences. An extreme example of this kind of variation is provided by *Saccharomyces* (4,21,22) where in addition to small inserts (25-50 bp) at a variety of sites, six large (900-3000 bp) insertions/deletions have been identified and precisely localized in defined regions of the mtDNA (4). Among them, insert VI (1000 bp) - closely linked to the polarity locus ω - is located within the gene for 21S RNA, insert III (3000 bp) as well as the small insert IX is within the *cob-box* region (see below), and inserts I (900 bp), II (2000 bp) and IV (1500 bp) are all located within the *oxi3* gene. This region can thus vary from 6400 bp in *S. carlsbergensis* to 11,400 bp in different strains of *S. cerevisiae.*

Divergence in Mt rRNAs and tRNAs. Evidence for divergence in mitochondrial gene products is provided by an examination of these stable RNA species (4,7) (Table 4). The large rRNA varies between the limits of 1.30 and 0.48 x 10^6 daltons - the latter being the value for *Drosophila melanogaster* (28) - and corresponds to chains 4060 and 1500 nucleotides long, respectively. Similarly, the small rRNA varies from 0.70 to 0.28 Mdaltons, or 2150 and 860 nucleotides, respectively. The base composition of the two sets of RNAs extends from a minimum of 17.5% (G + C) to a maximum of 54% (G + C) with substantial variation even within the animal kingdom. In spite of these differences, particularly of size, in the rRNAs, the number (but not the kind!) of associated polypeptides required to form a functional ribosomal subunit, appears relatively unaffected. Since construction of a functional ribosome demands that changes in rRNA be accompanied by obligatory co-ordinate alterations in ribosomal proteins, these observations suggest strong evolutionary convergence of entities specified by mitochondrial and nuclear DNA, respectively. Other significant observations concerning these rRNA species are their unusually low degree of methylation and - except for the molecules from plants - the apparent absence of a

TABLE 4

EVOLUTIONARY DIVERGENCE OF rRNAs

SPECIES	LARGE SUBUNIT rRNA G + C (%)	LARGE SUBUNIT rRNA Mass (Md)	LARGE SUBUNIT associated proteins	SMALL SUBUNIT rRNA G + C (%)	SMALL SUBUNIT rRNA Mass (Md)	SMALL SUBUNIT associated proteins
Euglena	(28)*			(28)*		
Paramacium						
Tetrahymena	28	0.90		31	0.47	
higher plants	54	1.10-1.25		54	0.60-0.75	
Saccharomyces	25,33	1.30		25,27	0.70	
Neurospora	(38)*	1.28	31	(38)*	0.72	30
Aspergillus	29	1.28		32	0.69	
Drosophila	20.9	0.48		17.6	0.28	
Xenopus	40	0.53	40	43	0.30	44
mammals	33-40	0.53	≲50	36-43	0.33	≲30
HeLa cells (human)	42,45	0.53		43,45	0.30	
E. coli	54	1.1	34	54	0.55	21

* total rRNA

discrete 5S RNA. The ribosomal RNAs in most species map in adjacent segments of the mtDNA, separated by a gap of 120-160 bp, and form parts of a single primary transcript (e.g. 8,28). This, however, is not the organization found in *Saccharomyces* (4,29), *Neurospora* (30) and *Tetrahymena* (20), where they are separated by gaps of 25, 21, and 10 kbp, respectively. In addition, the gene for the large rRNA is split in both *Neurospora* and some, (ω^+), *Saccharomyces* strains containing inserts of 2500 and 1050 bp. Both ribosomal RNAs appear to be transcribed from the same DNA strand in these instances also, but in *Saccharomyces* they do not constitute a single transcriptional unit. Ribosomal RNA genes thus exhibit certain features suggesting both conservation and divergence of functional organization superimposed on a much more prevalent and rapid divergence of base sequences. Similarly, Jakovcic, Casey and Rabinowitz (31), on the basis of studies on the mitochondrial leucyl tRNAs from rat, mouse, guinea pig, monkey, chicken and yeast, concluded that such mtDNA sequences appeared to be conserved to a smaller extent than ones for cytoplasmic rRNA, 5S RNA or hemoglobin mRNA. The mitochondrial complement of tRNAs may turn out to be significantly less (by two or more species) than the minimum required on the basis of Crick's wobble hypothesis; possible models to circumvent this difficulty have been discussed by Borst and Grivell (4). Mitochondrial polypeptide chains are known to be initiated with fMet (32,33) as are those in bacteria, but the tRNA responsible ($tRNA_F^{Met}$) exhibits significant differences from both the procaryotic and the eucaryotic species (34).

TABLE 5

AMINO ACID COMPOSITION (MOL %) OF SUBUNITS ENCODED IN MITOCHONDRIAL DNA

	Cytochrome Oxidase						Cytochrome b		
	Subunit I			Subunit II					
	A	B	C*	A	B	C*	D	E	F
Asx	7.76	6.45	7.25	8.15	8.55	7.28	8.6	8.3	7.5
Thr	5.12	4.80	7.71	4.89	4.15	7.80	5.3	4.7	7.6
Ser	7.22	10.09	6.69	8.08	9.36	9.78	5.7	8.3	5.9
Glx	5.85	4.28	4.70	9.19	9.00	9.71	6.6	6.1	3.5
Pro	4.32	6.84	5.91	4.32	7.33	6.25	4.9	4.9	5.9
Gly	12.0	10.2	9.26	14.8	6.81	5.05	6.5	7.3	6.6
Ala	7.82	7.53	8.05	7.03	4.90	4.56	7.6	7.5	7.2
Cys	1.41	N.D.	N.D.	1.93	N.D.	N.D.	N.D.	0.8	1.0
Val	6.26	7.14	6.19	5.38	7.71	4.79	9.4	6.7	4.7
Met	3.20	1.27	5.77	1.74	2.05	7.08	2.6	2.3	3.8
Ile	7.19	8.72	5.63	6.51	9.16	5.00	8.7	8.6	9.4
Leu	11.5	13.4	12.2	8.51	11.7	14.0	8.8	12.4	15.7
Tyr	3.46	4.18	3.91	2.38	4.89	5.06	3.4	4.1	4.0
Phe	6.28	8.25	7.51	3.79	5.38	3.70	8.4	6.4	6.2
Lys	4.35	1.51	2.60	7.20	2.50	3.34	4.4	2.9	2.7
His	2.13	2.56	3.75	2.11	2.64	3.46	2.9	2.4	3.1
Arg	2.29	2.81	2.63	2.51	3.07	3.12	3.9	3.6	2.2
Trp	2.08	N.D.	N.D.	1.46	N.D.	N.D.	N.D.	2.7	3.0

*Contains N-terminal f-Met

—— Differences between S. cerevisiae and N. crassa

--- Differences between S. cerevisiae and beef heart

S. cerevisiae	[A] Poyton and Schatz (35)	[D] Katan et al. (38)
N. crassa	[B] Sebald et al. (36)	[E] Weiss and Ziganke (39)
Beef heart	[C] Steffens and Buse (37)	[F] Von Jagow et al. (40)

Divergence of Polypeptide Sequences. While substantial base sequence divergence can still be compatible with conservation of protein sequence, a really convincing demonstration of evolutionary changes in mitochondrial DNA would be provided by an examination of the amino acid sequences of some of the mitochondrial polypeptide products described in Table 1. With one exception this has not yet been accomplished; however, highly significant differences become apparent already on the level of amino acid composition. Relevant data for cytochrome oxidase subunits I and II, and cytochrome *b* are summarized in Table 5. Divergences between yeast and

Neurospora are seen in four residues of cytochrome *b* and in nine and eight residues of subunits I and II respectively. Sequence divergences in subunit I from human and mouse mitochondria are also suggested by the electrophoretic data of Jeffreys and Craig (41). By virtue of the requirement for a set of nuclear genes specifying the other members of these complexes, any sequence divergence in the mitochondrial partner probably implies an accommodating change in the nuclear one as well.

<u>The Small Hydrophobic Subunit of Mitochondrial ATPase.</u> This polypeptide provides one of the most interesting, and at the same time most puzzling, examples of the interplay of evolutionary constraints and divergences. It is an ubiquitous constituent of energy transducing ATPases (to be equated with ATP synthetases), and similar molecular entities have been found in and purified from *E. coli*, spinach chloroplasts, and mitochondria from *Neurospora*, beef heart and yeast (40, and Tzagoloff, this volume). The complete amino acid sequence for the molecules from mitochondria and *E. coli* have been determined by Sebald, Wachter and their collaborators (42,43, and private communication from Dr. W. Sebald). The segment containing the residue covalently modified by DCCD, as well as two mutational sites each in *Neurospora* and *Saccharomyces*, all resulting in oligomycin resistance in these organisms, are shown in Scheme 1. In spite of the evident homologies of sequence and function, the origin of the two proteins differs: in *Saccharomyces* it is encoded in mtDNA and translated within the organelle, while in *Neurospora* it is the product of a nuclear gene. In the mammalian protein the absence of a formylated N-terminal methionine suggests an extramitochondrial origin. However, evidence to the contrary has been provided by the studies of Dianoux, Bof and Vignais on the analoguous protein from rat liver (44). Clearly a number of possible models can be suggested

SCHEME 1

SEQUENCE HOMOLOGIES OF DCCD-BINDING ATPASES (SEBALD, 1979)

```
                                                                      S         Y
                  5         15        25        35        45        55↑    65   ↑    75
N. crassa     Y----IAQAMVEVSKNLGMGSAAIGLTGAGIGIGLVFAALLNGVARNPALRGQLFSYAILGFAFVEAIGLFDLMVALMAKFT
                                                                             _
Beef heart          DIDTAAKFIGAGAATVGVAGSGAGIGTVFGSLIIGYARNPSLKCQLFSYAILGFALVEAMGLFCLMVAFLILFAM
                                                                             _
S. cerevisiae    f-MQLVLAAKYIGAGISTIGLLGAGIGIAIVFAALINGVSRNPSIKATVRPMAILGFALSEATGVFCLMVSFLLLFGV
                                                                          ↓  _    ↓
                                                                          F       S

E. coli        f-MGNLNMDLLYMAAAVMMGLAAIGAAIGIGILGGKFLQGAARQPDLIPLLRTQFFIVMGLVDAIPMIAVGLGLYVMFAVA
```

to account for these findings, all of them relevant to the problem of mitochondrial evolution (4). An additional observation bearing on a choice of models is that ATPases inhibited by DCCD are not restricted to the mitochondria but are also present in yeast plasma membranes. My collaborators J. Mc Donough and P. Jaynes are currently engaged in determining whether this protein contains a separate DCCD binding subunit analoguous to that described above (Scheme 1), or a homologous region as part of a larger polypeptide (45).

ORGANIZATION AND REGULATION OF A MITOCHONDRIAL GENE

In our own work we have concentrated our attention on the region of mtDNA involved in the specification of apocytochrome *b* of the respiratory chain. This segment has been located physically and genetically between the antibiotic resistance markers *oli1* and *oli2* (46) and as shown on the simplified map in Fig. 2 (4) encompasses a sizable portion ($\gtrsim$10%) of the total length of mtDNA centered at about 9 o'clock, or about 20 kbp from the origin provided by the single *Sal* site.

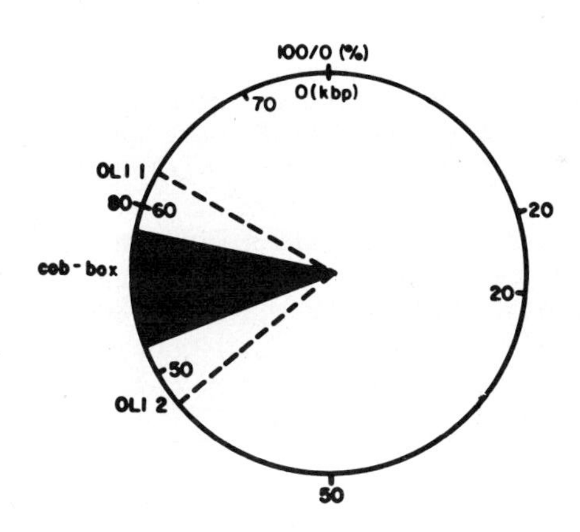

FIGURE 2. A simplified genetic map of *S. cerevisiae*

Properties of the *Cob-Box* Gene. We have isolated over 30 *mit* mutants in this region, all of which lack any mitochondrial cytochrome *b* and any measureable NADH:cytochrome *c* reductase activity. Most of these mutants exhibit a low but significant frequency of reversion to wild type and are thus due to lesions involving one, or only a few nucleotides. Six of them are relatively large deletions (Fig. 3). We have established their position on the genetic and physical (restriction endonuclease) map (Fig. 4) and we have investigated the proteins synthesized by their mitochondria *in vivo*. The principal conclusions from these studies (10,11,46), which are in complete agreement with those carried out simultaneously and independently by P.P. Slonimski and a group at Gif (47,48 and Slonimski, this volume), and by R. Schweyen and a group at Munich together with M. Solioz in Basel (49), working with even more extensive collections of mutants, are the following: i) The region variously referred to as *cob* or *box* is several times the size required for the specification of a polypeptide some 270 residues in length (M_R = 30,000). Our best current estimate (Figs. 3 and 4) based on the DNA segments missing in our deletion mutant PZ27, which lacks all sites in this region, and of *rho*⁻ mutant A12-4, which restores all our *mit*⁻ mutants is ~8500 bp. ii) Mutants in this region can be assigned to a minimum of eight genetically and physically distinct clusters, referred to here by the Gif nomenclature as *box1 - 8*. Precise localization has been achieved by a massive collaborative effort by investigators from all the various groups, conducted at Gif last fall and will be published elsewhere. iii) Three of the clusters (*box4, 1* and *6*) constitute *exons* in Gilbert's terminology (50) since they are responsible for the specification of the aminoacid sequence of mitochondrial cytochrome *b*, and in their aggregate form part of its structural gene: Some mutants in these clusters produce novel polypeptide products which can be unambiguously identified as prematurely terminated fragments of the wild type protein. As shown (Fig. 5) they increase in size in the order indicated and this fact together with other evidence strongly suggests that translation and hence transcription of the gene is in the direction *oli2* → *oli1* or from right to left in Figs. 3 and 4. Clusters *box5* and *8* may also belong to this class. Some exon mutants, especially those mapping in *box4* and *5* render the synthesis of cytochrome oxidase subunit I (Cox I) particularly susceptible to catabolite repression; one *box6* mutant (A104) leads to its overproduction. iv) Interspersed among these clusters are members of a second class, not involved directly in coding for cytochrome *b* mutations. These *introns* (*box3, 7* and *2*) produce sets of multiple new polypeptides not found in the wild type and frequently result in a selective and pronounced inhibition

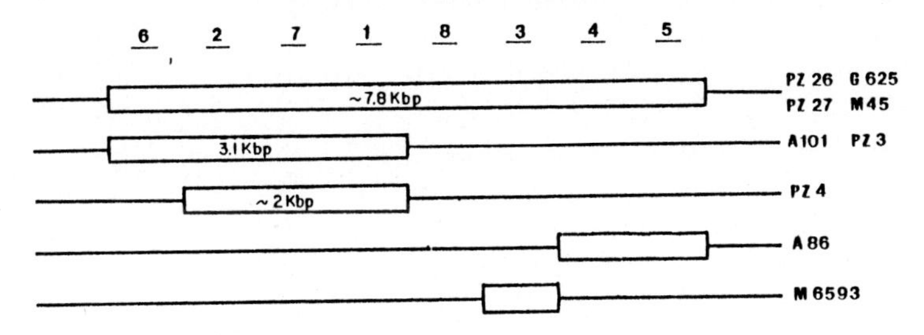

FIGURE 3. Deletion mutants in the *cob-box* region. The size of the deleted regions are based on restriction-endonuclease analysis (Fig. 4) while the retention of *cob* regions was analyzed by appropriate genetic crosses (10,11,46-48). The data for mutants G625, A86 and M6593 were obtained as part of a collaborative project with Drs. P.P. Slonimski and R. Schweyen.

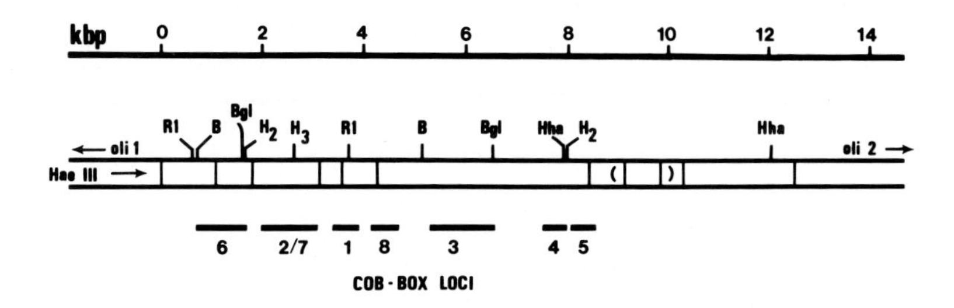

FIGURE 4. Combined physical and genetic[a] map of the *cob-box* region. Restriction endonuclease sites shown are: R1 - *Eco* R1; B - *Bam*H1; Bgl - *Bgl*II; H2 - *Hinc*II; H3 - *Hind*III; Hha - *Hha*I. [a]modified from (47)

of the expression of the synthesis of Cox I - and thereby of the activity of the enzyme under both repressing and non-repressing conditions.

Thus the sequence organization of the *cob-box* gene resembles that characteristic of eucaryotic and not of procaryotic genes. Because of the conditional nature of defects in mitochondrial biogenesis and function in *Saccharomyces* this organism - unlike other eucaryotes - permits ready identification of mutants in introns and renders them accessible to further study.

Nature of Intron Mutations. Lesions in these regions produce novel, characteristic polypeptide patterns which can be used not only as diagnostics for the various mutations and their assignment to sub-regions of the introns, but also as probes for the nature and consequences of the mutation itself. Presumably the mutant phenotype results from a defect in a processing step, required to convert a large transcript (pre-mRNA) into the active mRNA; that is, affecting the excision of the transcript of the intervening region from the pre-mRNA, the splicing of the resulting RNA segments to produce a continuous mature mRNA, or both. One of the most urgent questions is to what extent, if any, the novel polypeptides are specified by nucleotide sequences in adjacent exons, as well as by contiguous - or non contiguous - sequences in the intron. We are attempting to arrive at an answer by means of the following four approaches: i) Construction of double mutants (10) between all possible sets of *box* clusters in the hope of ascertaining the effects of a promoter-proximal mutation on a region "downstream". ii) Detailed fingerprint analysis (10) in the hope of ascertaining the source and interrelationships of the various polypeptides produced by intron mutants. iii) Construction of mutants introducing intron mutations in the *box2 - 7* region into a strain that lacks insert III in the *box3* region, in the hope of establishing whether this deficiency affects intron phenotype "downstream". iv) Probing the sequence organization within the gene and its presumed exons and introns by various means, including the malachite green technique for sorting restriction fragments by their AT/GC content (51). Very useful information concerning introns has already been obtained by our analysis of a number of double mutants (Scheme 2). The essential positive controls are provided by the introduction of a chain-terminating exon mutation "upstream" into an exon mutant, itself producing a larger fragment of apocytochrome *b*, e.g. *box4* into *box1*, *box5* into *box4* or *6*, *box4* into *box1* or *6*. In each of these cases the downstream exon phenotype is extinguished as predicted by the model for the structure of the *cob-box* region (Fig. 4). These results also bear on the relationship between *cob-box*

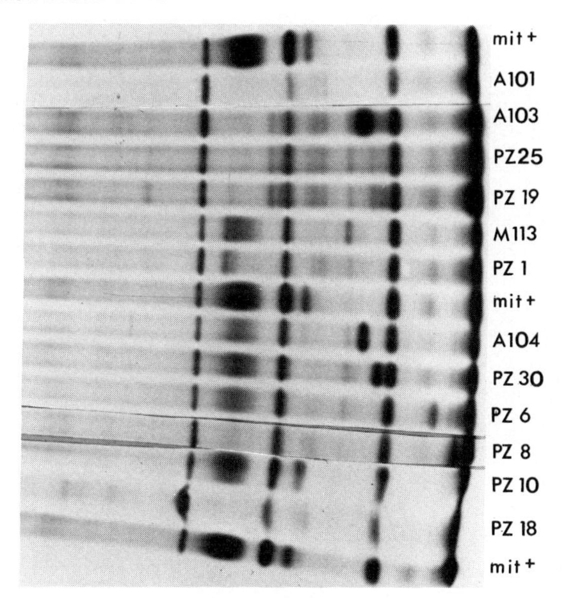

FIGURE 5. Polypeptides synthesized by *cob-box* mutants. The assignments are, in order, wild type (ρ^+), *box5* (PZ18-note altered p30); *box4* (PZ8-p15;PZ10-missense); *box1* (PZ6-p17); *box6* (A104); wild type; *box8* (PZ1); *box7* (PZ19,PZ25); *box2* (A103); deletion (A101); wild type.

GRID FOR POLYPEPTIDE PRODUCTS IN DOUBLE MUTANTS

(Direction of transcription and translation from left to right or top to bottom)

BOX (mutant)	5 (PZ18)	4 (EM25)	3 (PZ1)	8 (PZ24)	1 (PZ5)	7 (PZ16)	2 (A103)	6 (A104)
5 (PZ18)	NONE Cox I$^{-/+}$							
4 (EM25)	NONE Cox I$^{-/+}$	p15 Cox I$^{-/+}$						
3 (PZ1)		p15 Cox I^{-}	p45 Cox I$^{-/-}$					
8 (PZ24)				p15.5 Cox I^{+}				
1 (PZ5)		p15 Cox I^{-}	p45 Cox I^{-}		p18 Cox I^{+}			
7 (PZ16)	NONE Cox I$^{-/-}$		p45 Cox I/$^{-}$			p45,p14 p35,p23 Cox I$^{-/-}$		
2 (A103)	NONE Cox I/$^{-}$	p15 Cox I$^{-/-}$			p18 Cox I^{-}		p35 p25 Cox I$^{-/-}$	
6 (A104)	NONE Cox I^{+}	p15 Cox I$^{-/+}$	p45 Cox I/$^{-}$		p18 Cox I$^{+/+}$			p24 Cox I^{++}

SCHEME 2. The entries for Cox I refer to repressed (left) and derepressed (right) conditions.

mutations and oxidase expression; *box4* and *5* mutations greatly reduce the level of Cox I in cells grown on glucose (but not on galactose), while the downstream mutants (*box1* and *6*) have normal or elevated levels of Cox I, respectively (46). The oxidase phenotype of the *box4* mutant is therefore epistatic to that of the *box1* mutant, and that of the *box6* mutant appears to bear the same relation to that of the *box5* mutant.

Surprisingly, the placement of a chain-termination site upstream from an intron mutant, e.g. *box5* into *box7* or *box2*, *box4* into *box3* or *box2*, or *box1* into *box2* always generates a polypeptide pattern consisting of only the prematurely terminated fragment: The sets of novel polypeptides characteristic of intron mutants are absent from their double mutants. In contrast, their equally characteristic inhibition of expression of oxidase subunit I (Table 6) is retained in all their double mutants. Thus the double mutants exhibit a recombinant phenotype with a polypeptide due to the mutant exon and the Cox I defect due to the mutant intron. Clearly, in these exon-intron double mutants, the absence of this subunit has been dissociated from the presence of novel polypeptides and therefore can not be a consequence of, or linked to, their formation. It is likely, then, that this regulatory effect exerted by introns is at the level of transcription (Scheme 3). Furthermore, since intron mutations downstream do not interfere with the synthesis of a prematurely terminated fragment upstream, we conclude that processing at or near promoter-proximal introns must be independent of processing lesions further downstream. It appears, therefore, that processing events at different sites in the primary transcripts are independent, or, if not, are ordered in the direction of transcription.

Finally, we have constructed a mutant in which an intron has been combined with a chain-terminating event downstream from it. In this instance (*box3-box1*) the intron phenotype is epistatic to that of the exon. This result constitutes a further verification of the model in Scheme 3 that events controlled by intron regions proximal to the initiator are required for the propagation of the polypeptide chain further downstream, presumably by providing the locale for processing relevant portions of the primary transcript.

EXTENSION AND OUTLOOK

Principles of Mitochondrial Evolution. On the basis of the presentation so far, the inference appears warranted that mitochondrial evolution appears to have pursued a separate path strongly divergent from that of the procaryotic line. The constraints along this path appear to have been designed

TABLE 6

HOW TO DISTINGUISH INTRONS FROM EXONS

(COB-BOX REGION)

FEATURE IN REPR. MUTANTS	EXON	INTRON
POLYPEPTIDE(S) EXHIBITING NOVEL MOBILITY	YES	YES
COMPLEX PATTERN	NO	YES
HOMOLOGY[a] OF P30 TO W.T. APOCYTOCHROME <u>B</u>	VARIABLE	GREAT
HOMOLOGY[a] OF FRAGMENT POLYPEPTIDES TO W.T. <u>B</u>	GREAT	VARIABLE
ANTIBIOTIC RESISTANCE LOCI IN SAME SEGMENT	YES	NO
INHIBITION OF EXPRESSION OF CYTOCHROME OXIDASE ACTIVITY (AND ITS SUBUNIT I)[b]	NO	YES

[a] FINGERPRINTS WITH SEVERAL PROTEASES

[b] DEREPRESSED CONDITIONS

SCHEME 3

PROCESSING DEFECTS

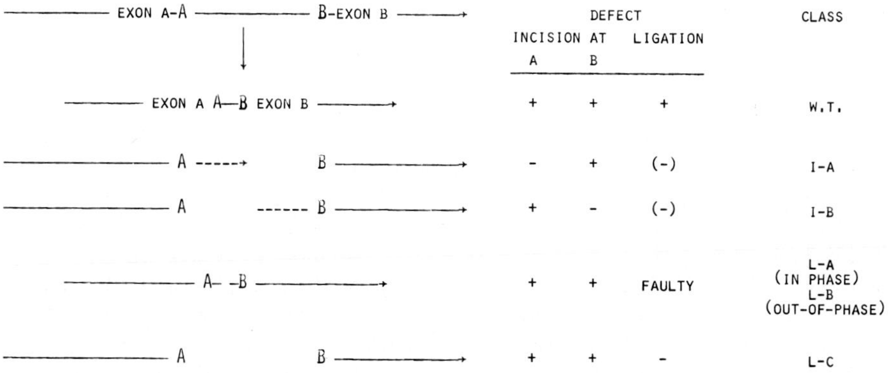

	DEFECT INCISION AT A	DEFECT INCISION AT B	DEFECT LIGATION	CLASS
——— EXON A-A ——— B-EXON B ——→ ↓ ——— EXON A A—B EXON B ——→	+	+	+	W.T.
——— A -----→ B ——→	-	+	(-)	I-A
——— A ------ B ——→	+	-	(-)	I-B
——— A– –B ——→	+	+	FAULTY	L-A (IN PHASE) L-B (OUT-OF-PHASE)
——— A B ——→	+	+	-	L-C

principally to insure extreme conservation of functions, while permitting the widest possible latitude, and rapid changes, in coding and aminoacid sequences. This combination is hardly the one expected of an evolutionary dead-end (4).

Sequence Organization in *Saccharomyces*. In at least one set of organisms (*Saccharomyces cerevisiae* and *carlsbergensis*) the sequence organization of mtDNA exhibits several highly unusual features: i) Genes appear to be flanked and separated by two classes of sequences exhibiting a relatively high degree of homology (4,21,52,Tzagoloff *et al.*, this volume): (a) extensive regions composed of AT-rich stretches ("spacers"?), a sub-set of which may be confined to precise locations (53, 53a); and (b), two types of G-C rich clusters; one of these appears to be a highly conserved palindromic sequence of 47 nucleotides containing closely spaced sites for HpaII and HaeIII restriction endonucleases (54, Tzagoloff *et al.*, this volume). ii) At least some of the genes exhibit a complex sequence organization (see above): not only do they contain exons interspersed among introns, and perhaps other non-coding sequences - reminiscent of the situation prevailing in eucaryotic genes - but, in certain strains, also contain within some of their introns additional insertion elements of unknown function (55, Grivell *et al.*, this volume).

How prevalent is this organizational scheme for other mitochondrial genes? There is suggestive evidence that it is obeyed in the case of another large gene, *oxi3* (4,56), containing the structural information for the large subunit of cytochrome oxidase. In one case that of *pho2 (oli1)* (cf. above) the evidence, based on DNA sequencing by Macino and Tzagoloff (57; cf. also Tzagoloff *et al.*, Grivell *et al.*, this volume) is compelling that this gene is *not* split. Thus one and the same mitochondrial genome exhibits both a simple and complex pattern, even for adjacent genes. It should prove extremely instructive to see whether this type of juxtaposition is general, or unique to *Saccharomyces*, perhaps because it represents an intermediate stage in the evolution of the mitochondrial genome - in which case we would predict its disappearance in the case of the small mtDNAs, e.g. in animals, or perhaps already even in *Aspergillus* or *Kluyveromyces*.

Possible Evolutionary Significance. The possible advantages of A-T rich sequences outside and between genes for rapid genomic evolution by favoring unequal cross-over events, involving quasi-homologous sequences, have been discussed by Bernardi and his collaborators (21,52) and criticized by Borst *et al.* (58). Another possibility is that such sequences external to structural genes proper are involved in RNA processing (58). A logical and plausible extension would be to propose that such sequences within genes (specifically at or close to exon-intron boundaries [59,60]) may be required for the processing events that transform primary transcripts into mRNAs.

Sequence homologies or quasi-homologies between introns may also provide a possible model for the stringent regulation of expression of the *oxi3* gene by certain intron mutants. For instance, if in such a mutant the intron or a portion thereof could be readily excised and reinserted, not only at the site of origin, but in the corresponding region in another intron, this would provide a mechanism for the propagation of a processing defect at a recombinational level. It remains to be seen whether this propagation is really as polar as it now appears or if one can discover the reciprocal event of the control of expression of cytochrome *b* by certain *oxi3* mutants.

Gilbert (50) and Darnell (61) have provided cogent arguments how having genes arranged "in pieces" might favor rapid and effective evolution of novel proteins. Even more relevant to our discussion is the suggestion advanced both by Doolittle (62) and Darnell (61) that the more complex gene organization found in eucaryotic chromosomes and viruses - and now in their mitochondria - may in fact be representative of the more ancient and primitive pattern. This hypothesis certainly provides a ready explanation of the "mixed" organization present in the mtDNA of *Saccharomyces*. In this connection, what are we to make of the observation that microfossils (called *Ramsaysphaera* [from South Africa] and *Isuasphaera* [from Greenland]) closely resembling contemporary asporogenous yeasts such as *Candida tropicalis* in size, shape and structural details have been recovered from some of the oldest sediments (3.4 and 3.8 x 10^6 years) on earth (63)?

Of great interest also is the suggestion by Blake (64) and by Darnell (61) that the exonic regions are to be equated to the integrally folded domain (or supersecondary) structures (65) found in many proteins. Blake proposes that active sites are to be found only at interfaces between the domains - and hence close to exon-intron boundaries. We would suggest that i) the existence of three (or more) sites for antibiotic resistance in the *cob-box* region (66,67 and Table 6) will find its explanation in terms of their being located precisely at such domain boundaries, cooperating in the formation of constellations of amino acid responsible for interaction with substrates - and inhibitors; and ii) that rapid evolutionary changes are permitted *outside* these restricted regions, so long as they fall within the rather loose constraints imposed by the requirements for domain formation. Both these predictions are susceptible to experimental verification.

ACKNOWLEDGEMENTS

We are greatly indebted to our collaborators N. Alexander, P. Anziano, S. Dhawale, D. Hanson, J. Johnson, D. Miller, M. Miller and S. Nixon for performing the experiments described in this communication and for permitting their publication. This research was supported by research grants GM 12228 (to H.R.M.) and GM 19607 (to P.S.P.) from the Institute of General Medical Sciences, National Institutes of Health. H.R.M. holds a Research Career Award from this Institute.

REFERENCES

1. Gillham, N.W., ed. (1978). "Organelle Heredity." Raven Press, New York.
2. Bandlow, W., Schweyen, R.J., Rolf, K., and Kaudewitz, F., eds. (1977). "Mitochondria 1977." de Gruyter, Berlin.
3. Bacila, M., Horecker, B.L., and Stoppani, A.O.M., eds. (1978). "Biochemistry and Genetics of Yeasts. Pure and Applied Aspects." Academic Press, New York.
4. Borst, P. and Grivell, L.A. (1978). Cell 15, 705.
5. Linnane, A.W. and Nagley, P. (1978). Arch. Biochem. Biophys. 187, 277; (1978) Plasmid 1, 324.
6. Buetow, D.E. and Wood, W.M. (1978). Subcell. Biochem. 5,1.
7. Mahler, H.R. and Raff, R.A. (1976). Intl. Rev. Cytolog. 43, 1.
8. Battey, J. and Clayton, D.A. (1978). Cell 14, 143.
9. Coruzzi, G., Trembath, M.K., and Tzagoloff, A. (1978). Eur. J. Biochem. 92, 279.
10. Hanson, D.K., Miller, D.H., Mahler, H.R., Alexander, N.J., and Perlman, P.S. (1979). J. Biol. Chem., in press.
11. Mahler, H.R., Hanson, D., Miller, D., Lin, C.C., Alexander, N.J., Vincent, R.D., and Perlman, P.S. (1978). In Ref. 3, pp. 513-547.
12. Dujon, B., Colson, A.M., and Slonimski, P.P. (1977). In Ref. 2, pp. 579-669.
13. Bechmann, H., Kruger, M., Boker, E., Bandlow, W., Schweyen, R.J., and Kaudewitz, F. (1977). Molec. gen. Genet. 155, 41.
14. Tzagoloff, A., Foury, F., and Akai, A. (1976). In "The Genetic Function of Mitochondrial DNA" (C. Saccone and A. M. Kroon, eds.), pp. 155-161. North-Holland, Amsterdam.
15. Borst, P. and Flavell, R.A. (1976). In "Handbook of Biochemistry and Molecular Biology" (D.G. Fasman, ed.), 3rd Ed., pp. 363-374. CRC Press, Cleveland.
16. Whittaker, R.H. (1969). Science 163, 150.
17. Ryan, R., Grant, D., Chiang, K-S., and Swift, H. (1978). Proc. Natl. Acad. Sci. USA 75, 3268.

18. Woese, C.R. and Bleyman, M.A. (1972). J. Molec. Evolution 1, 223.
19. Groot, G.S.P., Flavell, R.A., and Sanders, J.P.M. (1975). Biochim. Biophys. Acta 378, 186.
20. Goldbach, R.W., Arnberg, A.C., Van Bruggen, E.F.J., Defize, J., and Borst, P. (1977). Biochim. Biophys. Acta 477, 37; Goldbach, R.W., Bollen-De Boer, J.E., Van Bruggen, E.F.J., and Borst, P. (1978). Biochim. Biophys. Acta 521, 187.
21. Prunell, A. and Bernardi, G. (1977). J. Mol. Biol. 110, 53.
22. Sanders, J.P.M., Heyting, C., Verbeet, M.Ph., Meijlink, C.P.W., and Borst, P. (1977). Molec. gen. Genet. 157, 239.
23. Ramirez, J.L. and Dawid, I.B. (1978). J. Mol. Biol. 119, 133.
24. Buzzo, K., Fouts, D.L., and Wolstenholme, D.R. (1978). Proc. Natl. Acad. Sci. USA 75, 909.
25. Kroon, A.M., de Vos, W.M., and Bakker, H. (1978). Biochim. Biophys. Acta 519, 268; Hayashi, J-I., Yonekawa, H., Gotoh, O., Motohashi, J., and Tagashira, Y. (1978). Biochem. Biophys. Res. Comm. 81, 871.
26. Upholt, W.B. and Dawid, I.B. (1977). Cell 11, 571.
27. Brown, W.M., George, M.G., Jr., and Wilson, A.C. (1979). Proc. Natl. Acad. Sci. USA, in press.
28. Klukas, C.K. and Dawid, I.B. (1976). Cell 9, 615.
29. Levens, D. and Rabinowitz, M. (1979). This volume.
30. RajBhandhary, U.L., Heckman, J., Yin, S., Alzner-DeWeerd, B., and Ackermann, E. (1979). This volume.
31. Jakovcic, S., Casey, J., and Rabinowitz, M. (1975). Biochemistry 14, 2037.
32. Mahler, H.R. and Dawidowicz, K. (1973). Proc. Natl. Acad. Sci. USA 70, 111; Feldman, F. and Mahler, H.R. (1974). J. Biol. Chem. 249, 3702.
33. Bianchetti, R., Lucchini, G., Crosti, P., and Tortora, P. (1977). J. Biol. Chem. 252, 2519.
34. Heckman, J.E., Hecker, L.I., Schwartzbach, S.D., Barnett, W.E., Baumstark, B., and RajBhandary, U.L. (1978). Cell 13, 83.
35. Poyton, R.O. and Schatz, G. (1975). J. Biol. Chem. 250, 752.
36. Sebald, W., Machleidt, W., and Otto, J. (1973). Eur. J. Biochem. 38, 311.
37. Steffens, G. and Buse, G. (1976). Hoppe-Seyler's Z. Physiol. Chem. 357, 1125.

38. Katan, M.B., van Harten-Loosbroek, N., and Groot, G.S.P. (1976). Eur. J. Biochem. 70, 409.
39. Weiss, H. and Ziganke, B. (1978). Methods in Enzymology 53, 212.
40. Von Jagow, G., Schagger, H., Engel, W.D., Machleidt, W., and Machleidt, I. (1978). FEBS Lett. 91, 121.
41. Jeffreys, A.J. and Craig, I.W. (1977). FEBS Lett. 77, 151.
42. Wachter, E., Sebald, W., and Tzagoloff, A. (1977). In Ref. 2, pp. 441-449.
43. Sebald, W. and Wachter, E. (1978). In "29th Mosbacher Colloquium on 'Energy Conservation in Biological Membranes'" (G. Schafer and M. Klingenberg, eds.), in press. Springer-Verlag, Berlin.
44. Dianoux, A-C., Bof, M., and Vignais, P.V. (1978). Eur. J. Biochem. 88, 69.
45. Pick, U. and Racker, E. (1979). Biochemistry 18, 108.
46. Alexander, N.J., Vincent, R.D., Perlman, P.S., Miller, D.H., Hanson, D.K., and Mahler, H.R. (1979). J. Biol. Chem., in press.
47. Slonimski, P.P., Pajot, P., Jacq, C., Foucher, M., Perrodin, G., Kochko, A., and Lamouroux, A. (1978). In Ref. 3, pp. 339-368; Claisse, M.L., Spyridakis, A., Wambier-Kluppel, M.L., Pajot, P., and Slonimski, P.P. (1978). In Ref. 3, pp. 369-390.
48. Slonimski, P.P., Claisse, M.L., Foucher, M., Jacq, C., Kochko, A., Lamouroux, A., Pajot, P., Perrodin, G., Spyridakis, A., and Wambier-Kluppel, M.L. (1978). In Ref. 3, pp. 391-401.
49. Haid, A., Schweyen, R., Bechmann, H., Kaudewitz, F., Solioz, M., and Schatz, G. (1979). Eur. J. Biochem., in press.
50. Gilbert, W. (1978). Nature 271, 501.
51. Bünemann, H. and Müller, W. (1978). Nucleic Acids Res. 5, 1059.
52. Fonty, G., Goursot, R., Wilkie, D., and Bernardi, G, (1978). J. Mol. Biol. 119,213.
53. Bos, J.L., Van Kreijl, C.F., Ploegaert, F.H., Mol, J.N.M., and Borst, P. (1978). Nucleic Acids Res. 5, 4563.
53a. Sanders, J.R.M. and Borst, P. (1977). Molec. gen. Genet. 157, 263.
54. Cosson, J. and Tzagoloff, A. (1979). J. Biol. Chem. 254, 42.
55. Grivell, L.A. and Moorman, A.F.M. (1977). In Ref. 2, pp. 371-384.
56. Carignani, G., Dujardin, G., and Slonimski, P.P. (1979). Molec. gen. Genet. 167, 301.
57. Macino, G. and Tzagoloff, A. (1979). Proc. Natl. Acad. Sci. USA 76, 131.

58. Borst, P., Bos, J.L., Grivell, L.A., Groot, G.S.P., Heyting, C., Moorman, A.F.M., Sanders, J.P.M., Talen, J.L., Van Kreijl, C.F., and Van Ommen, G.J.B. (1977). In Ref. 2, pp. 213-254.
59. Catterall, J.F., O'Malley, B.W., Robertson, M.A., Staden, R., Tanaka, Y., and Brownlee, G.G. (1978). Nature 275, 510.
60. van den Berg, J., van Ooyen, A., Mantei, N., Schamböck, A., Grosveld, G., Flavell, R.A., and Weissmann, C. (1978). Nature 276, 37.
61. Darnell, J.E., Jr. (1978). Science 202, 1257.
62. Doolittle, W.F. (1978). Nature 272, 581.
63. Pflug, H.D. (1978). Naturwissenschaften 65, 611.
64. Blake, C.C.F. (1978). Nature 273, 267.
65. Yon, J.M. (1978). Biochimie 60, 581.
66. Colson, A-M. and Slonimski, P.P. (1979). Molec. gen. Genet. 167, 287.
67. Subik, J. and Takacsova, G. (1978). Molec. gen. Genet. 161, 99.

Restriction Enzyme Analysis of Mitochondrial DNA from Closely Related Species of *Paramecium*[1]

Donald J. Cummings, Arthur E. Pritchard
and Richard A. Maki

Department of Microbiology and Immunology,
University of Colorado Medical Center
Denver, Colorado 80262

ABSTRACT. Mitochondrial (mt) DNA from *Paramecium* is a linear molecule, 14 μm in length (1). Replication is initiated by closure at a unique end, followed by unidirectional synthesis via a lariat intermediate, terminating in a head-to-head linear dimer (1,2). The processing of the linear dimer to two semi-conservatively constituted (3) monomer length molecules can be blocked by growth in the presence of chloramphenicol (1,2). This allows us to identify readily the initiation segment of DNA by the use of appropriate restriction endonucleases. Several species of *Paramecium* exist and by microinjection of isolated mitochondria, Beale and Knowles (4) have shown that species 1, 5 or 7 mitochondria can function and replicate in the nuclear environment of each other, although to different extents, but that species 4 can neither support the growth of mitochondria from the other species nor serve as a successful donor for microinjection of its mitochondria into the other species.

In the present study, we have utilized restriction enzymes to 1) compare the mt DNA patterns obtained from the different species; 2) determine which fragment contains the DNA initiation sequences; 3) locate the rDNA genes; 4) identify interspecies DNA fragment homology by Southern transfer; and 5) measure the extent of interspecies homologies directly by filter hybridization. A preliminary report of some of these studies has already been published (5).

INTRODUCTION

Unlike other animal mitochondria, mt DNA from *Paramecium* is a linear molecule, 14 μm in length (1). In this regard, it is similar to the only other ciliate examined, Tetrahymena where the mt DNA is also linear, 15 μm in length (6).

[1]This work was supported by USPHS Grant GM 21948

ISBN 0-12-198780-9

The two ciliates do differ in their mode of replication, however (7). For Paramecium, replication is initiated by closure of a unique end and proceeds unidirectionally via a lariat intermediate terminating in a dimer length linear molecule (1,2). Representative electronmicrographs of these molecules are depicted in Figure 1. Replication intermediates can be enriched for by growth in the presence of ethidum bromide (lariats) or chloramphenicol (dimers). Schematically, this is illustrated in Figure 2.

There are several features of this scheme which warrant some discussion. Since chloramphenicol leads to the build-up of linear length dimer molecules, it would appear that the enzymes necessary for processing are synthesized on mitochondrial ribosomes. At the moment, there is no evidence regarding the enzyme necessary for the initiation of synthesis (step B). With regard to structure, the scheme indicates that the dimer molecule is a head-to-head configuration of two monomer length molecules. Snapback renaturation studies show that this is indeed the case (2). An interesting prediction of this arrangement is that endonuclease digestion of dimer molecules should yield a new fragment, twice the molecular weight of the corresponding monomer fragment at the initiation region of the molecule. This is one feature of mt DNA structure which will be examined here. Another aspect of structure which is of interest is the arrangement of the genes for ribosomal RNA. Except for yeast (9) and recently, Tetrahymena (10) these genes are close together on the mt genome.

Biologically, the function and biogenesis of Paramecium mitochondria has been studied using microinjection techniques. Knowles (8) and Beale and Knowles (4) isolated antibiotic resistant mitochondria from different species and injected them into homologous and heterologous cells sensitive to the antibiotic. They were able to show that mitochondria from species 1 and 5 were essentially interchangeable, but that while species 1 and 5 mitochondria could function in species 7, the reciprocal "cross" was not allowed. Mitochondria from species 4 could not function in the environment of the other species nor could species 4 serve as a host for heterologous mitochondria. Beale and Knowles concluded that the nucleus played some role in this "restriction" but it is not clear whether the mitochondria themselves possess certain properties which could account for these findings. Earlier, we (11) were able to show that hybrid cells constructed by microinjection contained mitochondrial DNA unique to the species which served as donor. Here we will present evidence on the extent of homology between the species as

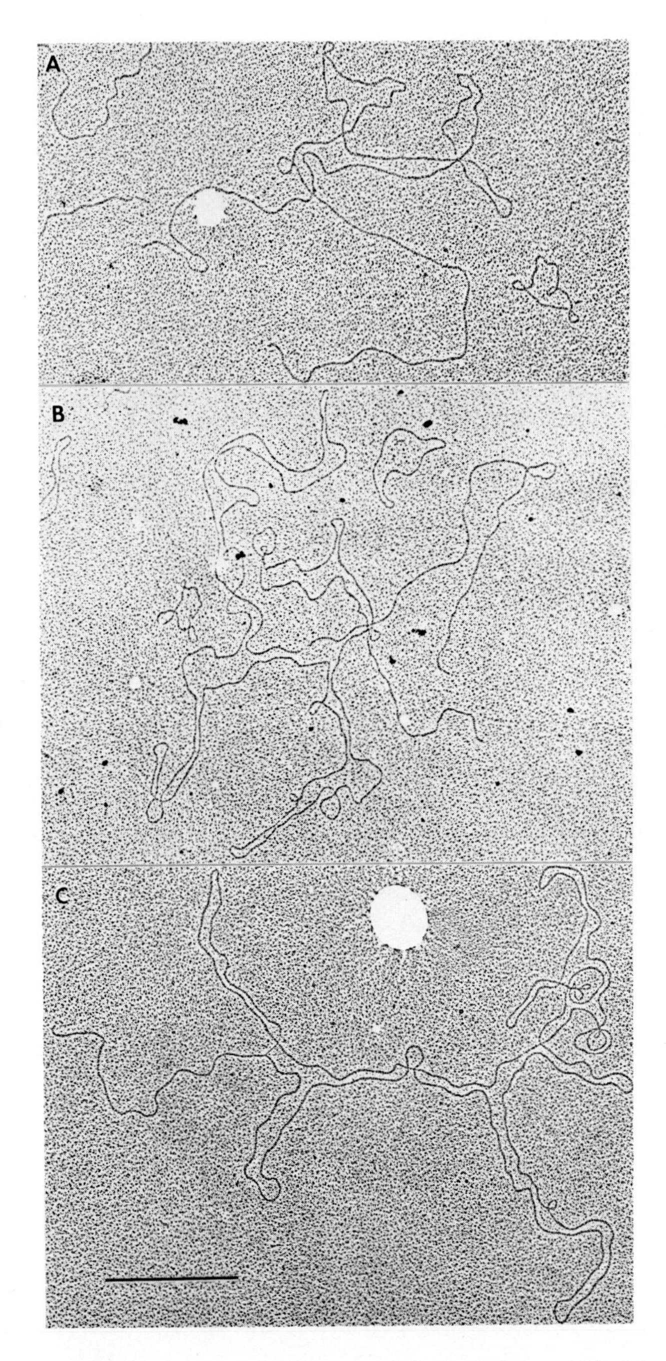

Figure 1. Electronmicrographs of A) Monomer; B) Dimer; and C) Lariat molecules.

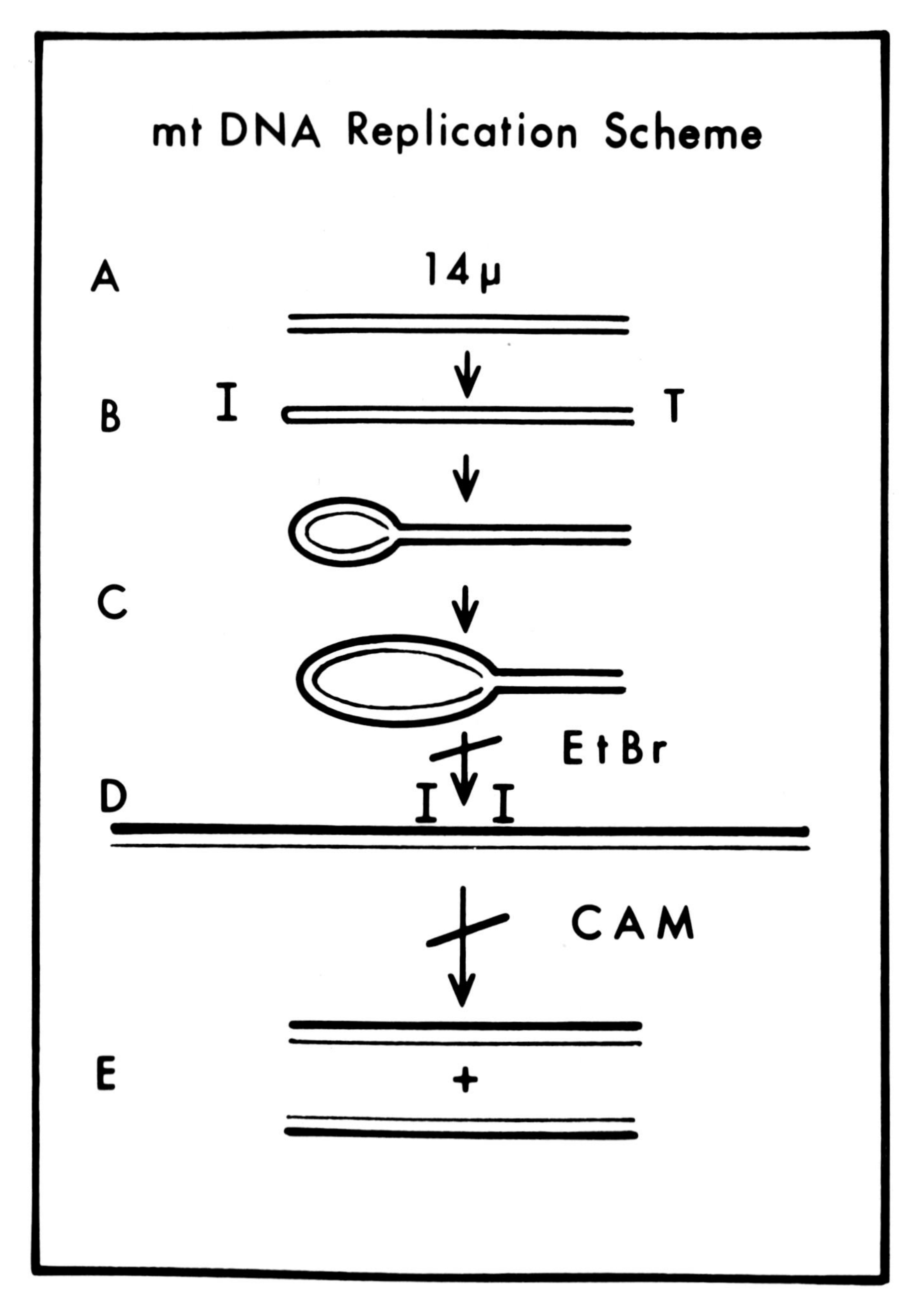

Figure 2. Schematic Representation of mt DNA Replication.

measured by both Southern blot techniques (12) and filter hybridization (13). These results are discussed in correlation with the DNA restriction fragments containing the genes for ribosomal RNA and DNA initiation.

MATERIALS AND METHODS

Isolation of Mitochondrial DNA. DNA was extracted from purified mitochondria and purified on CsCl density gradients as described by Goddard and Cummings (1,2).

Restriction Enzyme Analysis. Reaction conditions for endonuclease cleavage of mt DNA were as described by Green *et al*. (15). The DNA fragments were separated by agarose gel electrophoresis. Gels were photographed using a short-wave UV-transilluminator with a Polaroid camera equipped with a Wratten 23A gelatin filter and type 665 film.

Isolation of Ribosomal RNA. RNA was isolated directly from mitochondrial pellets or from purified ribosomes. Ribosomes were isolated using methods described by Tait and Knowles (14). The RNA was purified by one extraction with phenol followed by another extraction with chloroform-octanol (9:1).

Iodination of RNA and Hybridization to DNA. Mitochondrial rRNA was labelled with ^{125}I as described by Getz *et al*. (16). DNA fragments were transferred from agarose gels to nitrocellulose filters by methods described by Southern (12). Hybridization of the ^{125}I-labelled RNA to DNA on nitrocellulose filters was done as described by Bell *et al*. (17).

^{32}P-labelling and Hybridization to DNA. Mitochondrial DNA from species 1 and 4 was labelled with ^{32}P by nick translation (18). Transfer of DNA fragments to nitrocellulose filters and hybridization to ^{32}P-labelled DNA was done by modification of methods described by Southern (12) and Bukhari *et al*. (19).

RESULTS

Restriction Endonuclease Analysis. Earlier, we reported (11) that each species of *Paramecium* gave rise to a unique set of fragments after digestion with EcoRI or HaeII. This is illustrated for EcoRI in Figure 3. While previously we found (11) that different stocks of the same species yielded the same digestion fragments, here it can be noted that for one mutant at least, species 4E, the pattern is altered slightly. This is a mutant selected for Erythromycin resistance and was a gift from Dr. A. Adoutte. We show it here

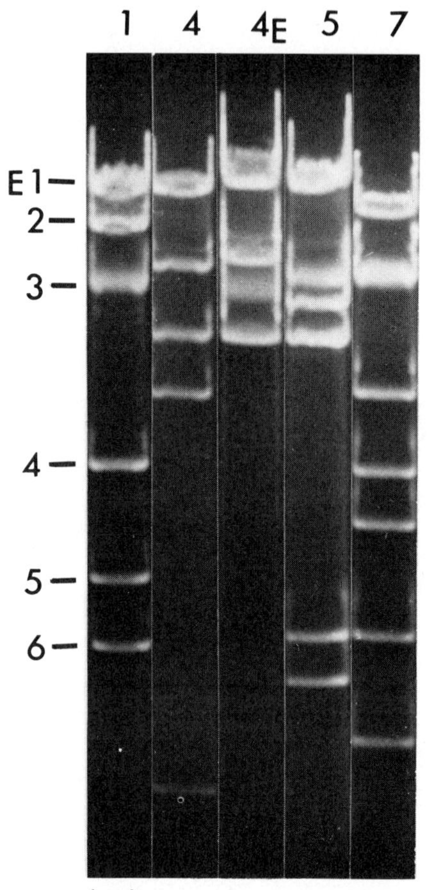

Figure 3. EcoRI Fragment Patterns. The E numbers refer to species 1 fragments.

mainly because the new fragment (E3) for species 4E contains the genes for ribosomal RNA (see later). The uniqueness of the fragment patterns for the different species is also illustrated for HindIII in Figure 4. Also included in this array is a dimer molecule of species 1 which contains a new fragment (I_1) of about 1.2×10^6 daltons which represents a dimer of the initiation region. A similar fragment can be discerned at about 3.5×10^6 in species 4 monomer preparation (I_4). We always observe these fragments in dimer molecules and can also see them in so-called monomer length molecule preparations but to different extents, as one might expect.

The disappearance and appearance of new fragments from digests of dimer length molecules is depicted in more detail in Figure 5. As can be seen, a new fragment does not always

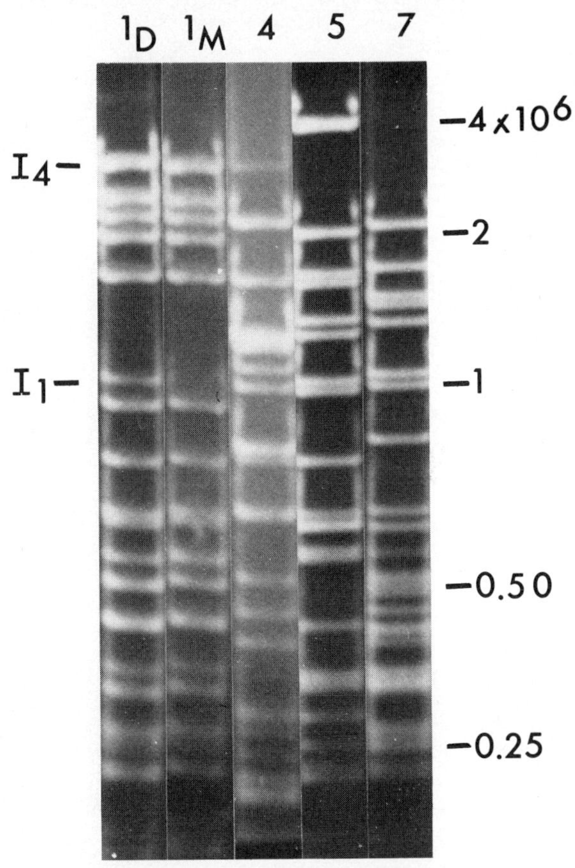

Figure 4. HindIII Fragment Patterns. 1_D refers to species 1 dimers and 1_M, monomers.

appear since it may well have the same molecular weight of a fragment not concerned with initiation. With EcoRI, species 1 and 7 mt DNA dimers do not yield a new fragment. With BamHI endonuclease, it is clear that species 1 dimers contain a new fragment at about $8x10^6$ daltons and lack a monomer fragment (B4) at about $4x10^6$ daltons. For species 7, no new fragment appeared but B2 appears to be more intense in the dimer preparation and B3 is absent. We have observed recently a new fragment for species 7 dimers after digestion with SalI. For both species 4 and 5, a new band appears in EcoRI dimer digests at about $5x10^6$ dalton molecular weight, without any apparent loss of a fragment. This fragment is twice the molecular weight of E3 from species 4 and E4 from species 5. Comparing dimer to monomer digests, there is an apparent decrease in the intensity of these two fragments suggesting that two fragments are comigrating. In fact, as we will see,

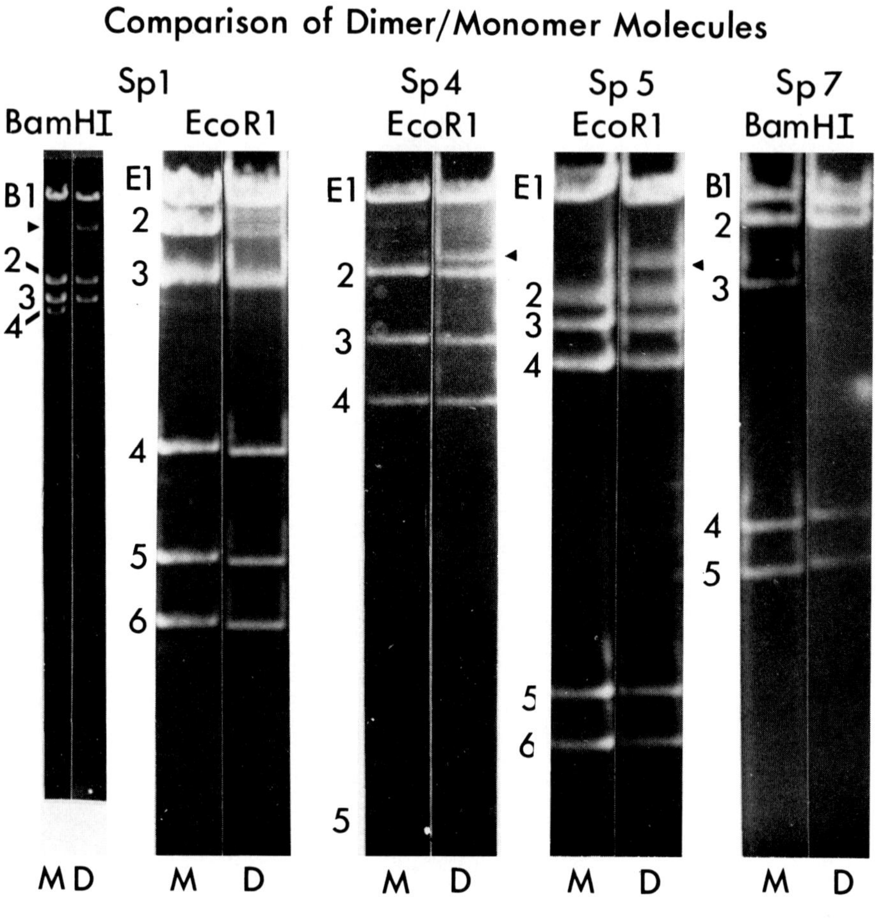

Figure 5. Dimer/Monomer Fragment Patterns

E3 from species 4 contains genes for ribosomal RNA. Although we have not yet constructed a map of species 4, we do expect to find comigrating DNA fragments since the 5 fragments detected in gels after EcoRI digestion, add up to only 21×10^6 daltons. The molecular weight of species 4 DNA as measured by electron microscopy is about 27×10^6 daltons (11). In any case, Figure 5 illustrates the ease with which the fragments containing the initiation sequences can be detected.

<u>Location of Ribosomal RNA Genes.</u> Ribosomal RNA was isolated from purified mitochondria from species 1 (14) and labelled *in vitro* with ^{125}I (16). Denatured DNA fragments from EcoRI digests were transferred from agarose gels to

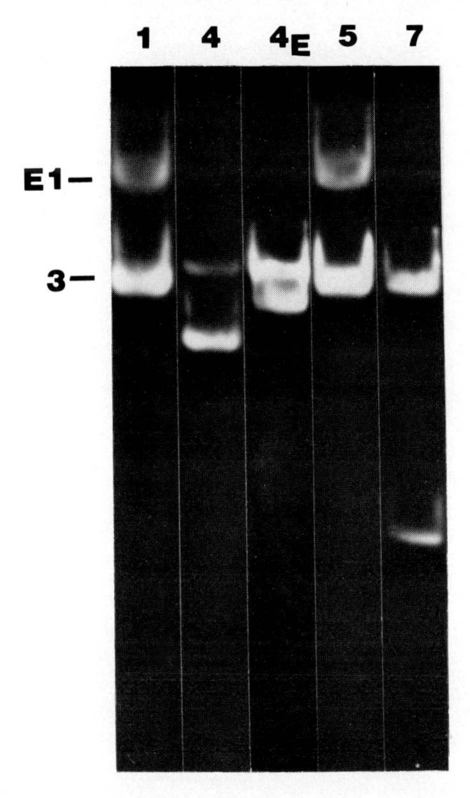

Figure 6. Southern Blot Autoradiograph of rRNA Hybridized to EcoRI Fragments. E1 and E3 refer to species 1 fragments.

nitrocellulose filters and hybridized with ^{125}I rRNA. Autoradiographs of the dried filters gave the results presented in Figure 6. For species 1 and 5, primarily one band, E3 for species 1 and E2 for species 5 was homologous with rRNA (refer to Figure 3).

Varying amounts of hybridization occurred with E1 for both species 1 and 5 and while further work is required to be certain, we attributed this to partial digestion. The molecular weight of the E3 fragment from species 1, or E2 from species 5 is about 4×10^6 or 6100 bp. Since the 20s and 14s rRNA subunits represent about 2800 bp and 1600 bp, respectively, it would appear that each gene subunit is represented once and the genes are close together. For both species 4

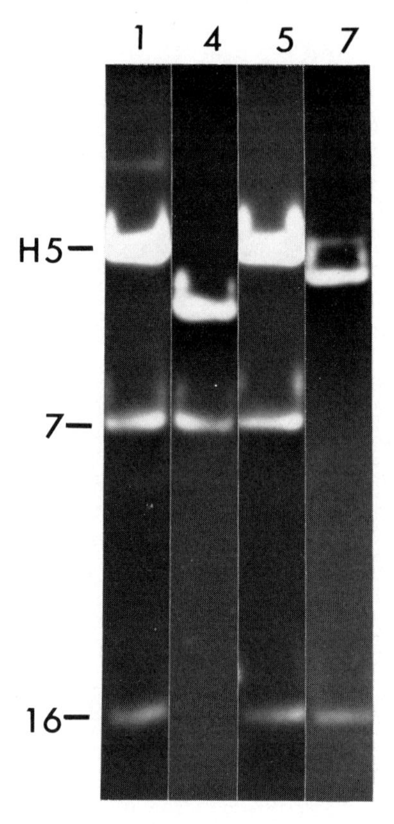

Figure 7. Southern Blot Autoradiograph of rRNA Hybridized to HindIII Fragments. H5, 7 and 16 refer to species 1 fragments (see Figure 4).

and 7, two DNA fragments were homologous: E2 and 3 for species 4, and E2 and 5 for species 7. The molecular weights of these fragments are $6.9x10^6$ and $5.2x10^6$ daltons for species 4 and 7, respectively. For the mutant of species 4 (Erythromycin-resistant), two fragments were homologous, E2 and 3, totalling about $7.6x10^6$ daltons. It may be of some interest that an Erythromycin resistance mutant has an altered arrangement of rRNA genes.

A clearer picture of the total molecular weights of DNA fragments homologous to rRNA emerged using HindIII digests (refer to Figure 4). These results are given in Figure 7, where it can be seen that species 1 and 5 contain 3 fragments totalling about $2.8x10^6$ daltons or 4500 bp, a value quite close to the equivalent number of bp required for the rRNA

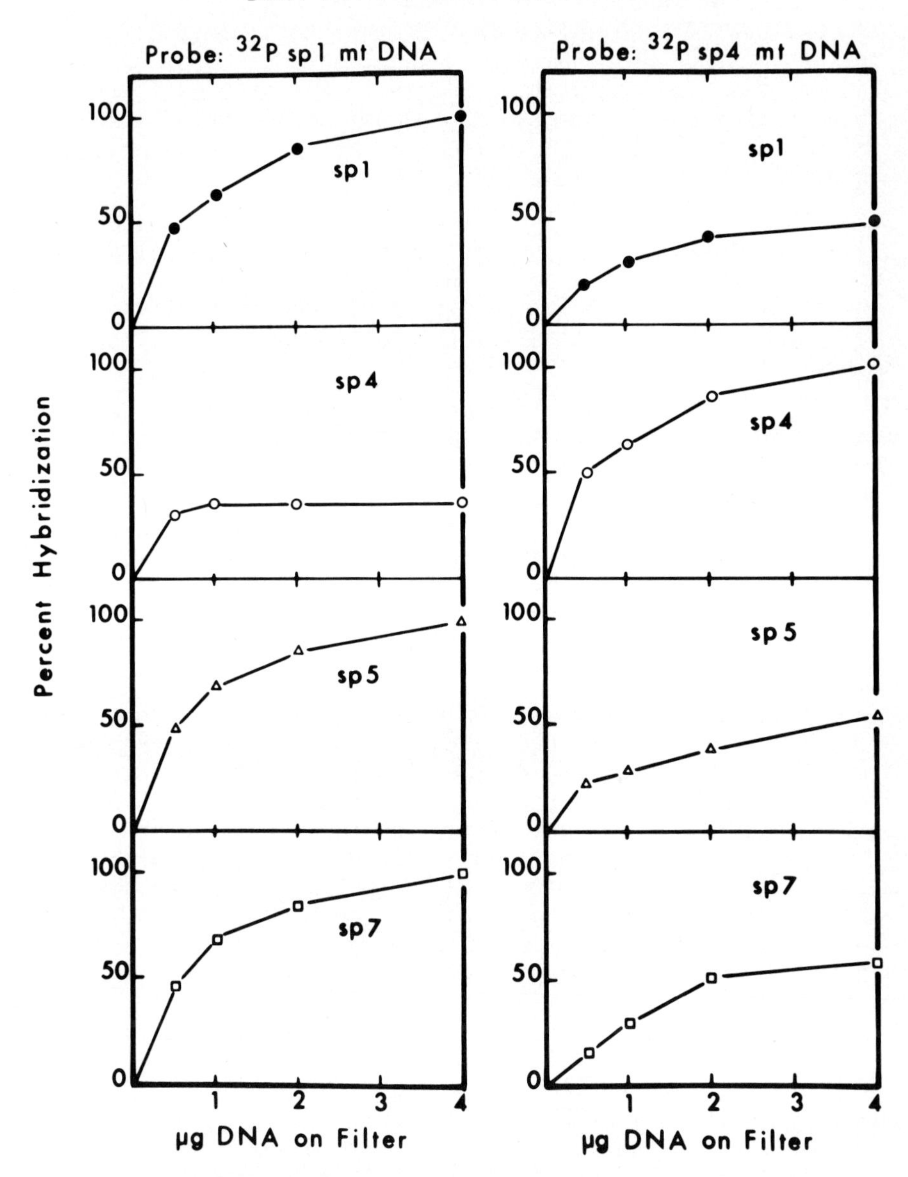

Figure 8. Filter Hybridization of Species 1 or Species 4 mt DNA to species 1, 4, 5 and 7 mt DNA.

genes. Occasionally, a trace amount of hybridization was noted to H2 of species 1 but this was not reproducible. Species 5 never gave hybridization to fragments other than

H4, H9 and H17. Species 4 had only two HindIII fragments homologous to rRNA. These fragments (H3 and H6; this numbering excludes the so-called I_4 fragment) total only 2.2×10^6 daltons or 3500 bp. However it is clear that H3 of species 4 has at least two comigrating species. For species 7, the three homologous fragments (H2, H3 and H19) total 3.4×10^6 daltons or about 5400 bp, quite similar to that number in the homologous EcoRI fragment (E2).

Interspecies Homology. As discussed earlier, mitochondria from the different species do not function equally well when injected into the other species. In order to determine whether the mitochondrial genome itself could be playing a role in this "restriction", interspecies homology was determined. This was done in two ways.

First, DNA-DNA homology of total genomes was measured by filter hydridization (13), using ^{32}P-labelled species 1 or species 4 DNA as probes. In Figure 8, it can be seen that species 1 probe is essentially 100% homologous to both species 5 and 7 but is only 40% homologous to species 4. The reciprocal challenge gave similar results; species 4 probe was homologous to species 1, 5 or 7 to only about 50%. Since the ribosomal RNA genes account for only about 10% of the total genome (4400 bp out of 43000 bp), the remainder of species 4 DNA is homologous to species 1, 5 or 7 to only 30-40%.

Similar results were obtained using Southern blot transfers (12). In some respects, this technique gave less specific information since the degree of homology to each band could not be specified; only homology itself was detected. On the other hand this technique allows us to determine which of the DNA fragments contain homologous sequences. In Figure 9, using ^{32}P-labelled species 1 DNA as probe, it can be seen that species 1 is homologous to all the EcoRI fragments of species 1, 5 or 7, as expected. However, only E1, 2 and 3 of species 4 contain homologous sequences; two of these, E2 and E3 contain the genes for ribosomal RNA. Species 4 E1 fragment contains about 40% of the total bp, thus accounting for the homology measured by filter hybridization. Essentially the same results for species 1 and 4 were obtained using ^{32}P-labelled species 4 mt DNA as probe (Figure 10). Here EcoRI fragments E1, E2 and E3 of species 1 were homologous, with only trace amounts detectable in the other fragments. Similarly for species 5, where E1, 2 and 4 contained the main homologous bands. For species 7, the only fragments which showed a high degree of homology to species 4 were E2, and E5, the rRNA genes. E1 also showed homology but to a lesser extent.

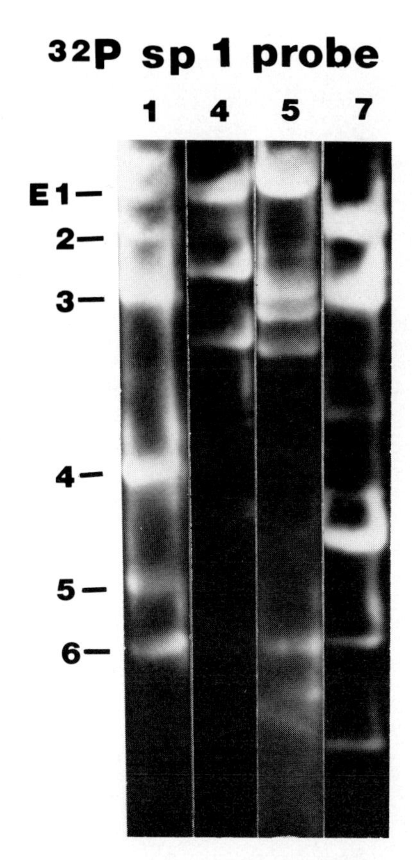

Figure 9. Southern Blot Autoradiograph of Species 1 mt DNA Hybridization to Species 1, 4, 5 and 7 EcoRI Fragments.

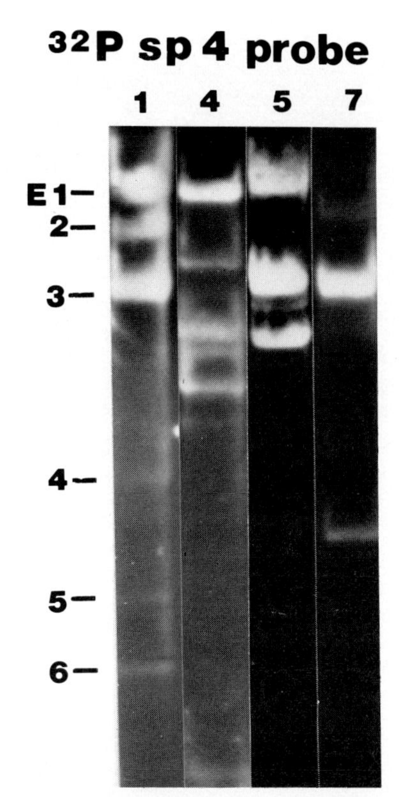

Figure 10. Southern Blot Autoradiograph of Species 4 mt DNA Hybridization to Species 1, 4, 5 and 7 EcoRI Fragments. The E numbers here and in Figure 9 refer to species 1 fragments.

Homology to HindIII Initiation Fragment of Species 1. Aside from the fragments containing the ribosomal RNA genes, the only other identifiable fragments showing homology with species 4 were those containing sequences in the initiation region: E2 of species 1 and E4 of species 5. Since the degree of homology is difficult to measure using total DNA in the Southern blot, we resorted to using a cloned ^{32}P labelled HindIII fragment identified as containing initiation sequence (see I_1 in Figure 4). These results are contained in Figure 11 where EcoRI fragments of species 1, 4, 5 and 7 were transferred to nitrocellulose and challenged with ^{32}P-labelled I_1. As expected, E2 of species 1 showed the greatest homology with this fragment. E4 of species 5, a fragment previously identified as containing initiation sequences,

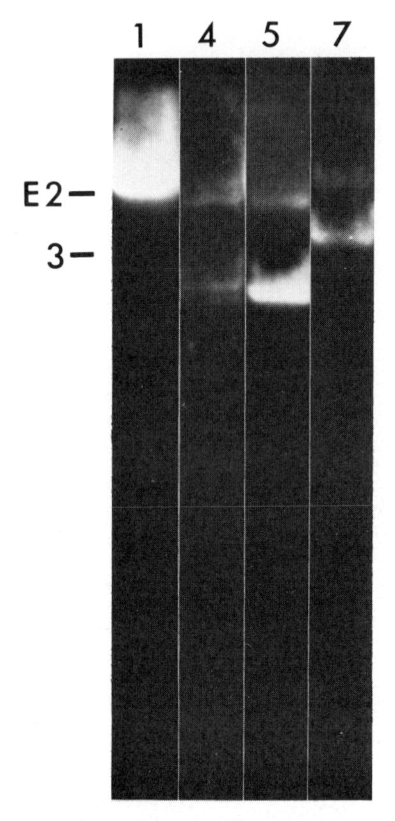

Figure 11. Southern Blot Autoradiograph of Species 1 initiation Fragment (I_1) Hybridized to Species 1, 4, 5 and 7 EcoRI fragments. E2 and E3 refer to species 1 fragments.

also showed good homology. Next in order of apparent homology was species 7 E2. E2 in species 7 appears to contain two comigrating fragments (see Figure 3), one the rRNA genes and the other the initiation fragment. Finally, species 4 showed the least homology, E3. Curiously, E3 fragments of species 1 showed some homology, although much less than the primary fragment E2. It is possible that some of the sequences contained in the 1.2×10^6 dalton, I_1 fragment are repeated elsewhere in the genome. The other fragment showing homology in species 4, 5 and 7 represents the small quantity of dimer length DNA present in the preparations.

DISCUSSION AND SUMMARY

We had two related goals in reporting these experiments. First, we wanted to present a fairly comprehensive picture of the replication mechanism of this linear mt DNA, and its consequences. One of the important features of this replication scheme is its value in determining which endonuclease fragment contains sequences necessary for initiation of DNA replication. This allows us to begin sequence analyses of the initiation region. It also gives us the opportunity to look for the enzymes required for the processing of the DNA since this is the target fragment for that processing. We have avoided discussion of the problems involved in completion of replication of this linear genome (20), but it is our hope that once we obtain the sequence for the initiation region as well as information on the enzymes involved that we will also gain some insight into this problem as well.

Second, we wished to detail some of the properties of this linear DNA with regard to rRNA genes and interspecies homology. As with many other organisms, rRNA genes appear to be among the most conserved. No differences were noted in the degree of homology of rRNA from species 1 with mt DNA from species 1, 4, 5 or 7. In general, the evidence suggests that each rRNA gene is represented once in the species examined. However, it is possible, at least for species 1 and 5, that some homology exists in the E1 fragment as well. The Eco RI map which we have constructed (21) indicated that E1 and E3, the fragment containing the major portion of the rRNA genes, are well separated from each other, with E3 being terminal. Since species 5 appears to be so similar to species 1, this is likely to be true for this species also. We have not yet constructed maps for species 4 or 7 so we can not specify how close together the two homologous Eco RI fragments are on the genome. Interspecies homology studies have provided some data which bear on the "restriction" observed in microinjection studies, at least with regard to species 4. This species is only about 50% homologous to the other species whether measured by Southern transfer or by filter hybridization. The main identifiable homologous fragments were those containing the genes for ribosomal RNA. In addition, very little homology was noted between the initiation fragment of species 1 and species 4. It is possible that this is a critical region for the establishment of one species of mitochondria within the background of another species. Sequence analyses should provide critical information on this point. In terms of size, the major homologies observed between all the species were for E1, the largest molecular weight Eco RI fragment. This fragment presumably contains genes for tRNA but we have no evidence for this as yet.

To summarize, studies on protist mt DNA structure may provide valuable insight into replication mechanisms with regard to evolution and biogenesis. In many respects, Paramecium mt DNA has similarities with Tetrahymena. The differences, however, suggest that the study of other ciliate mt DNA may be beneficial.

REFERENCES

1. Goddard, J.M., and Cummings, D.J. (1975). J. Mol. Biol. 97, 593.
2. Goddard, J.M., and Cummings, D.J. (1977). J. Mol. Biol. 109, 327.
3. Cummings, D.J. (1977). J. Mol. Biol. 117, 273.
4. Beale, G.H., and Knowles, J.K.C. (1976). Molec. gen. Genet. 143, 197.
5. Cummings, D.J., Maki, R.A., and Paroda, C.M. (1978). In "Alfred Benzon Symposium" 13.
6. Suyama, Y., and Miura, K. (1968). Proc. Natl. Acad. Sci. U.S. 60, 235.
7. Clegg, R.A., Borst, P., and Weijers, P.J. (1974). Biochim. Biophys. Acta 361, 277.
8. Knowles, J.K.C. (1973). Ph.D. Thesis, University of Edinburgh, Scotland.
9. Sanders, J.P.M., Heyting, C., and Borst, P. (1975). Biochem. Biophys. Res. Commun. 65, 699.
10. Goldbach, R.W., Borst, P., BollendeBoer, J.E., and VanBruggen, E.F.J. (1978). Biochim. Biophys. Acta 521, 169.
11. Maki, R.A., and Cummings, D.J. (1977). Plasmid 1, 106.
12. Southern, E.M. (1975). J. Mol. Biol. 98, 503.
13. Denhardt, D.T. (1966). Biochem. Biophys. Res. Commun. 23, 641.
14. Tait, A., and Knowles, J.K.C. (1977). J. Cell Biol. 73, 139.
15. Green, P.J., Betoch, M.C., Goodman, H.M., and Boyer, H.W. (1974). In "Methods in Molecular Biology" (ed., R.B. Wickner), Vol. 7, p. 87-105, Dekher, New York.
16. Getz, M.J., Attenburg, L.C., and Saunders, G.F. (1972). Biochim. Biophys. Acta 287, 485.
17. Bell, G.I., DeGennaro, L.J., Gelford, D.H., Bishop, R.J., Valenzuela, P., and Rutter, W.J. (1977). J. Biol. Chem. 252, 8118.
18. Maniatis, T., Jeffrey, A., and Kleid, D.G. (1975). Proc. Natl. Acad. Sci. U.S. 72, 1184.

19. Bukhari, A.I., Froshauer, S., and Botchan, M. (1976). Nature 264, 580.
20. Watson, J.D. (1972). Nature 239, 197.
21. Maki, R.A. (1978). Ph.D. Thesis, University of Colorado Medical Center, Denver, Colorado.

THE EXTRACHROMOSOMAL RIBOSOMAL RNA GENES IN TETRAHYMENA; STRUCTURE AND EVOLUTION[1]

J. Engberg and Nanni Din

Biochemical Institute B, University of Copenhagen,
DK-2200 Copenhagen N, Denmark

ABSTRACT The macronuclear rDNA from a number of strains within several species of *Tetrahymena* have been characterized. Restriction enzyme analysis revealed that individual strains all contained entirely homogenous populations of extrachromosomal rDNA. The linear rDNA molecules were shown to be giant palindromes, varying in size from 18 kilobases (kb) to 21 kb in different strains, and to contain two genes for rRNA separated by a centrally located spacer. The sizes of the centrally located spacer and of the genes were approximately the same in all strains investigated and altogether an extended structure homology was shown to be maintained throughout all the species. Nevertheless, differences both inside and outside the gene region could be detected when rDNA from different species were compared. The differences were most pronounced in the spacer regions. In two out of five different species (*T. pigmentosa* and *T. borealis*), interbreeding strains were found which exhibited different restriction pattern of their rDNA. These strains will be useful in investigating recombinational events in the rDNA during sexual crosses, but their demonstration questions the usefulness of comparing restriction patterns in estimating relatedness. Detailed restriction mapping of rDNA from strains within the *T. pigmentosa* group gave evidence of an intervening non-ribosomal RNA sequence of about 350 bp in size within the structural gene for 26S rRNA in some of the strains. In the *T. thermophila* group all strains investigated contained a similar-sized intervening sequence at exactly the same position within the 26S gene region. Restriction enzyme mapping of the similar-sized intervening sequences from these two *Tetrahymena* species showed differences in their sequence.

[1]This work was partly supported by the Danish Research Council for Natural Sciences (fellowship to N.D.) and by Nato Research Grant 1296.

ISBN 0-12-1

INTRODUCTION

In eukaryotes the nuclear genes that code for the common precursor molecules of the cytoplasmic ribosomal RNAs are present in several hundred copies per haploid genome and most, if not all, are either clustered in distinct regions of specific chromosomes, or, in some cell systems, as extrachromosomal amplified nucleolar units (see ref. 1 to 6). It has been shown by various techniques that these genes are tandemly arranged in repeating units each consisting of alternating transcribed regions and non-transcribed spacer regions. In a given cell system the actual gene sequences are virtually identical while a certain degree of size and sequence heterogeneity may exist in the non-transcribed spacer regions of a gene cluster (7,8). Between related species extensive spacer heterogeneity may exist (in transcribed as well as in non-transcribed spacer regions), while the gene regions are still apparently identical (9), but in more distantly related species also the gene regions may be demonstrably different, although they apparently evolve more slowly than the spacer region (10).

The ciliated protozoan *Tetrahymena* contains a germinal diploid micronucleus which maintains the genetic continuity of the organism, as well as a polyploid macronucleus, which is formed after conjugation from the zygotic nucleus. The macronucleus functions as a somatic nucleus directing most of the cell's transcriptional activity during vegetative growth, but is destroyed at the end of each sexual generation. The two nuclei differ drastically in their content of ribosomal RNA genes. The germinal micronucleus contains a single chromosomally integrated copy per haploid genome (11), while the somatic macronucleus contains about 600 extrachromosomal gene copies per haploid genome (about 1% of the total nuclear DNA, ref. 12). In the two different species of *Tetrahymena* used in the study of Karrer & Gall (13) and Engberg *et al* (12) the extrachromosomal rDNA was found to be linear DNA molecules each containing two copies of the rRNA genes arranged as inverted repeats about a central axis of symmetry (so-called palindromes). The size and structure of the rDNA from these two species were clearly different as revealed by restriction enzyme mapping. Each species, however, contained entirely homogeneous rDNA populations.

We have now examined the extrachromosomal rDNA from a large number of different *Tetrahymena* strains to provide insight in the degree of sequence divergence which exists in a specific gene among closely related lower eukaryotes.

METHODS

All *Tetrahymena* strains, with the exception of *T. pyriformis*, *T. thermophila* BIV and BVII were kindly provided by Ellen Simon of the University of Illinois at Urbana-Champaign. Stock designation and species affiliation for the strains used are given in ref. 14.

All strains were cultivated in a complex proteose peptone, yeast extract medium as previously described for the strain *T. pyriformis* GL (15). The rDNA was isolated in the form of snap-back molecules by the method of Klenow (16) or in its native form by a modification of the hot phenol extraction procedure developed for isolation of mitochondrial DNA (17). The products of restriction enzyme digestion were fractionated by electrophoresis on agarose slap gels (12). Restriction enzyme fragments of phage λ wt, λ d *rif* D18 or pBR were used as molecular weight markers. The fragments were visualized by fluorography after ethidium bromide staining before transfer onto cellulose nitrate filters (18). Filters were hybridized to *in vitro* labelled rRNA as described by Yao & Gall (11).

RESULTS

Restriction enzyme mapping of rDNA from different species

A structural comparison of the free rDNA molecules from different *Tetrahymena* species was initiated by digesting purified native rDNA molecules or snap-back molecules with different restriction enzymes followed by size determination of the restriction fragments produced by gel electrophoresis on agarose gels. The restriction fragments were further characterized by transfer to cellulose nitrate filters followed by hybridization to ^{32}P-labelled rRNA prepared from *T. thermophila*. This probe cross-hybridizes between 75% and 100% with rDNA from the other *Tetrahymena* species (19). Restriction maps were constructed using conventional methods involving limited and double digestions, the details of which will be published elsewhere. The *Hind* III, *Eco* RI and the *Bam* HI maps are shown in Fig. 1.

The maps are striking in their uniformity in the localization of a number of restriction sites. Hence, all the species except one have *Bam*, *Eco* and *Hind* sites located almost exactly 4 x 10^6 D from the center of the molecule. The only exception is *T. thermophila*, which lacks the *Eco* site, while the *Bam* and the *Hind* sites are present at a distance of 4.2 x 10^6 D from the center of the molecule. Further, a number

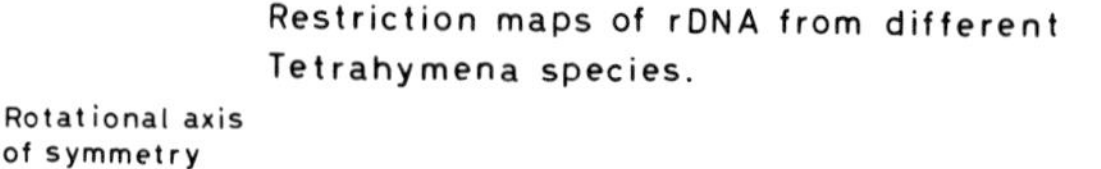

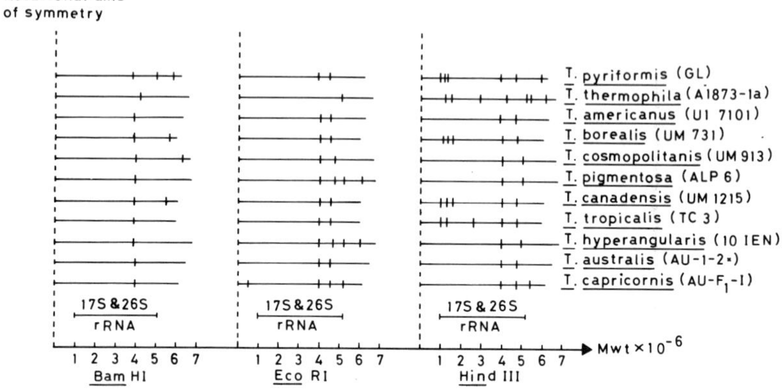

FIG. 1 shows the Bam HI, Eco RI and Hind III maps of rDNA molecules isolated from representative strains of eleven different Tetrahymena species. Note that the vertical lines which symbolize double-stranded DNA only show one half of the palindromic molecules.

of the species (T. thermophila, T. cosmopolitanis, T. pigmentosa and T. hyperangularis) all have an Eco RI site and a Hind III site located exactly 0.75 x 10^6 D and 1 x 10^6 D, respectively, distal to the 4.0 sites. In the rest of the species, a similar pair of Eco/Hind sites are located 0.55 x 10^6 and 0.80 x 10^6 D, respectively, from the 4.0 sites. If nearly identical localization of restriction sites may be taken to indicate that a very high degree of sequence conservation is maintained in a given DNA region, then, with respect to the region located between 4 x 10^6 and 5 x 10^6 D from the center of the rDNA molecules, only two classes of rDNA molecules exist in all the different Tetrahymena species.

However, when other regions of the rDNA molecules are compared, clear differences in the localization of restriction sites can easily be found. The differences are most pronounced in the outer non-coding region, as evidenced both by the Bam and Eco maps, but differences are also detected in the central non-coding region and in the coding region, as evidenced by the Hind maps. Thus, when the combined restriction maps are compared, it is clear that none of the species contain entirely identical rDNA. However, certain patterns exist, suggesting that some of the species are more closely related than others.

Comparison of rDNA from different strains within species

Five species were selected for a study of rDNA structure in interbreeding strains. This study was undertaken in order to investigate whether the differences detected when rDNA from different species were compared, reflected specific species characteristics. If this was the case, different strains from anyone species should contain identical rDNA molecules. If, on the contrary, different strains within a species contained different rDNA molecules, it must be concluded that genetic polymorphism with respect to rDNA structure exist in *Tetrahymena*. And in that case, any assessment of evolutionary distance among species on the basis of rDNA structure from just one representative strain from each species would be futile.

The five selected species were *T. thermophila*, *T. americanis*, *T. borealis*, *T. pigmentosa* and *T. australis*. Except for the strains from *T. thermophila* all the interbreeding strains originated from different natural geographical habitats. The *T. thermophila* strains used has been maintained separately for several years, but originated from one cross. They all had different mating types. From four of the species three different strains were examined, and from the fifth six different strains. From the restriction analyses of the rDNA from the strains of the four latter mentioned species it could be concluded that strains belonging to a given species contained identical rDNA, with the exception of a small size difference detected at the end of the rDNA from one of the *T. borealis* strains.

However, when rDNA from different strains of *T. pigmentosa* were examined clear differences were immediately seen (Fig. 2).

Two of the strains, IL 12 and UM 1286, contain identical rDNA, as judged from these restriction maps, while the rDNA from the rest of the strains clearly differ. The two strains containing the identical rDNA molecules do not belong to the same syngen (20). Thus, it must be concluded that in *T. pigmentosa*, and even in the separate and narrower older classification, syngen 6 and syngen 8, genetic polymorphism in the rDNA structure exist. We do not know whether the failure to demonstrate genetic polymorphism in the other species is accidental or whether it reflects a true difference between these species and *T. pigmentosa*.

Demonstration of intervening regions in the 26S coding region of different rDNA molecules

The differences detected in the restriction pattern in the gene region of the rDNA from the *T. pigmentosa* strains are suggestive of the presence of an intervening region of a

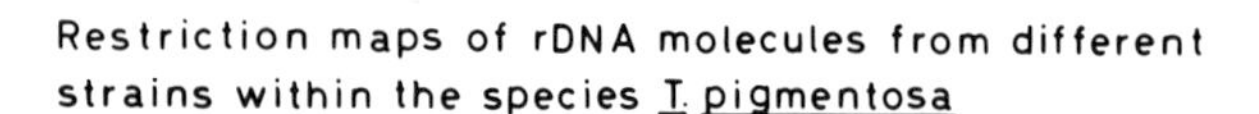

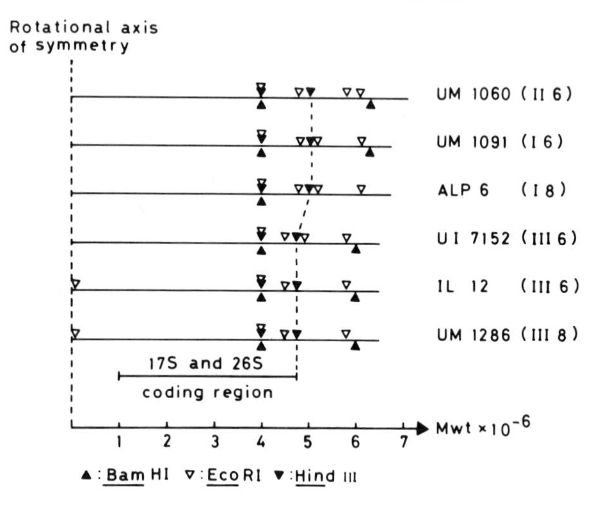

FIG. 2 shows the *Bam* HI, *Eco* RI and *Hind* III restriction maps of rDNA molecules from different strains of the *T. pigmentosa* group. Note that the vertical lines which represents double stranded DNA only show one half of the palindromic molecules.

size of 0.25×10^6 D in the rDNA of three of the strains, as first pointed out by Martha Wild and later demonstrated (Wild and Gall, in press) by restriction analyses and by visualization in the electron microscope of R-loops between rDNA and 26S rRNA. We have confirmed and extended these observations by similar techniques and by Berk and Sharp-type experiments (21) where S1 nuclease treated rRNA/rDNA hybrids were analysed by alkaline agarose gel electrophoresis. Thus, in three out of six strains the size of an intervening sequence has been determined by various techniques to be 375-400 basepairs (about 0.25×10^6 D) and its right end located about 950 bp (0.60×10^6 D) from the *Hind* site located 5.0×10^6 D from the center of the rDNA molecule. This *Hind* site coincides exactly with the end of the 26S coding region which is about 3.7 kb (2.5×10^6 D) in size.

The particular restriction site differences in the 26S coding region between the strains of *T. pigmentosa* harbouring the intervening region and the other *T. pigmentosa* strains are identical to the differences detected in the 26S coding region of the rDNA of the different species, which were described in the previous section. It was pointed out that the rDNA of

T. thermophila, T. cosmopolitanis and T. hyperangularis was very similar to the rDNA of T. pigmentosa with respect to the right end of the 26S coding region in terms of restriction sites and overall size (see Fig. 1). We, therefore, tested the possibility of finding intervening sequences in one of these other Tetrahymena species, viz., T. thermophila. R-loops between rDNA and 26S rRNA from T. thermophila were examined by electron microscopy and were found to be composed of two adjacent bubbles with a short double stranded DNA loop at their junction, exactly as found in the corresponding case of T. pigmentosa, strain ALP 6. Measurements of a number of T. thermophila R-loops gave the following figures for the size and localization, respectively, of the intervening sequence: 0.25×10^6 D and 0.6×10^6 D from the outer end of the 26S coding region, i.e. values identical to the ones determined in ALP 6. These findings were confirmed by Berk and Sharp-type experiments where measurements of size were done of S1 nuclease resistant R-loops formed between rDNA restriction fragments and 26S rRNA (results not shown).

DISCUSSION

The following basic observations have been made in the present work: 1) The individual populations of the extrachromosomal rDNA molecules isolated from many different Tetrahymena strains all consist of palindromic molecules which appear homogeneous with respect to size and base sequence. 2) Basic sequence similarities exist in the rDNA isolated from different species but each species can be unequivocally distinguished from the others by its restriction pattern resulting from the use of several restriction enzymes. 3) In the group of T. pigmentosa genetic polymorphism with respect to rDNA structure was clearly evident. 4) Intervening sequences of similar size and location within the 26S coding region was demonstrated in two different Tetrahymena species.

The absence of size and sequence heterogeneity in spacer as well as in gene region in individual extrachromosomal rDNA populations is most easily explained by assuming that these molecules are derived from a single chromosomally integrated rRNA gene copy (22). This is in accordance with the fact that the micronucleus in contrast to the macronucleus has been shown to contain one chromosomally integrated rDNA copy per haploid genome (11) and that rDNA amplification has been implied to take place at a specific stage of the macronuclear development (23,24).

However, the amicronucleate species T. pyriformis, that cannot go through a sexual cycle also contained a homogeneous population of extrachromosomal, palindromic rDNA molecules. Amicronucleate *Tetrahymena* strains are quite common (40% of the *Tetrahymena* found in nature are amicronucleate) but it is not known whether these strains are derived from micronuclear species or not. Available data show *T. pyriformis* to be evolutionary very distant from any of the known micronucleate species (20). Likewise, it is not possible to invoke a likely micronuclear ancestor for the *T. pyriformis* strain from the present restriction data.

The demonstration of the homogeneity of the individual rDNA populations, together with the uniqueness of the rDNA molecules isolated from different *Tetrahymena* strains, may be helpful for identification purposes. The use of restriction patterns for estimating relatedness may, however, be doubtful in view of the demonstration that strain differences exist within species. The demonstration of these strain differences, on the other hand, is very helpful for other studies involving genetic crosses.

The demonstration of intervening sequences of similar size and location within the 26S coding region in different *Tetrahymena* species raises the question whether these sequences are identical. Recent evidence from restriction analyses show that this is not the case, thereby implying that the actual size and location, rather than the primary sequence of the intervening sequences are remarkably evolutionary stable. The species of *Tetrahymena* have diverged an estimated one Myr ago (25). A similar conclusion in quite another system has recently been reached in the case of the β-globin gene from mouse and rabbit (26).

REFERENCES

1. Gall, J.G. (1969). Genetics (Suppl.) 61, 121-132.
2. Gall, J.G. (1974). Proc. Natl. Acad. Sci. USA 71, 3078-3081.
3. Busch, H. & Smetana, K. (1970). The Nucleolus, Academic Press, New York.
4. Birnstiel, M.L., Chipchase, M. & Speirs, I. (1971). Progr. Nucleic Acid Res. 11, 351-389.
5. Engberg, J., Christiansen, G. & Leick, V. (1974). Biochem. Biophys. Res. Commun. 59, 1356-1365.
6. Tobler, H. (1975). In: Biochemistry of Animal Development, vol. III, ed. Weber, R., pp. 91-143. Academic Press, New York.
7. Trendelenburg, M.F., Scheer, U., Zentgraf, H. & Franke, W.W. (1976). J. Mol. Biol. 108, 453-470.
8. Wellauer, P.K., Reeder, R.H., Dawid, I.R. & Brown, D.D. (1976). J. Mol. Biol. 105, 487-505.

9. Forsheit, A.B., Davidson, N. & Brown, D.D. (1974). J. Mol. Biol. 90, 301-314.
10. Sinclair, J.H. & Brown, D.D. (1971). Biochem. 10, 2761-2769.
11. Yao, M.-C. & Gall, J.G. (1977). Cell 12, 121-132.
12. Engberg, J., Andersson, P., Leick, V. & Collins, J. (1976). J. Mol. Biol. 104, 455-470.
13. Karrer, K.M. & Gall, J.G. (1976). J. Mol. Biol. 104, 421-453.
14. Engberg, J. and Din, Nanni. (1979). Proceedings of the 13th Alfred Benzon Symposium: Specific Eukaryotic Genes; Structural Organization and Function. Eds. Engberg, J., Klenow, H. and Leick, V. Munksgaard, Copenhagen. In press.
15. Engberg, J. & Pearlman, R.E. (1972). Eur. J. Biochem. 26, 393-400.
16. Klenow, H. (1979). Proceedings of the 13th Alfred Benzon Symposium: Specific Eukaryotic Genes: Structural Organization and Function. Eds. Engberg, J., Klenow, H. and Leick, V. Munksgaard, Copenhagen. In press.
17. Arnberg, A.C., Goldbach, R.W., Van Bruggen, E.F.J. & Borst, P. (1977). Biochim. Biophys. Acta 477, 51-69.
18. Southern, E. (1975). J. Mol. Biol. 98, 503-517.
19. Allen, S.L. & Gibson, I. (1973). Genetics of _Tetrahymena_. In: Biology of _Tetrahymena_, ed. A.M. Elliott. p. 363. Dowden, Hutchinson & Ross, Inc. Stroudsburg, Pennsylvania.
20. Nanney, D.L. & McCoy, I.W. (1976). Trans. Am. Microscop. Soc. 95, 664-682.
21. Berk, A.J., and Sharp, P.A. (1977). Cell 12, 721-732.
22. Collins, J., and Engberg, J. (1977). J. theor. Biol. 66, 573-582.
23. Yao, M.-C., Blackburn, E., and Gall, J. (1978). Cold Spring Harbour Quant. Biol. In press.
24. Pearlman, R.E., Anderson, P., Engberg, J. and Nilsson, J. (1979). Proceedings of the 13th Alfred Benzon Symposium: Specific Eukaryotic Genes; Structural Organization and Function. Eds. Engberg, J., Klenow, H. and Leick, V. Munksgaard, Copenhagen. In press.
25. Borden,D.,Miller,E.T.,Whitt,G.S. and Nanney,D.L. (1977) Evolution 31,91-102.
26. Berg van den,J.,Ooyen van,A.,Mantei,N.,Schamböck,A., Grosveld,G.,Flavell,R.A. and Weissman,C. (1978) Nature 276,37-44.

MOLECULAR HETEROGENEITY AMONG MITOCHONDRIAL DNAS FROM DIFFERENT MAIZE CYTOPLASMS[1]

C. S. Levings, III*, D. M. Shah[2], W. W. L. Hu**,
D. R. Pring***, and D. H. Timothy**

Departments of Genetics* and Crop Science**
North Carolina State University
Raleigh, North Carolina 27650

SEA, USDA***
Department of Plant Pathology
University of Florida
Gainesville, Florida 32611

ABSTRACT Maize mitochondrial DNAs (mtDNA) from three cytoplasmic sources, normal, S, and T, were isolated and examined by electron microscopy. The normal cytoplasm is male fertile while S and T carry the cytoplasmic male sterility trait. Covalently closed circular molecules of mtDNA were demonstrated. Maize mtDNA is composed of a heterogenous population of circular molecules. Several possible causes of intermolecular heterogeneity were discussed. Comparison of mtDNAs from normal, S, and T cytoplasms indicated that each type had a unique distribution of molecular sizes. These results suggested that substantial alterations in the mitochondrial genome are associated with the cytoplasmic male sterility trait.

INTRODUCTION

The physico-chemical properties of mitochondrial DNA (mtDNA) of higher plants are not fully understood. In animals, the entire informational content of the mtDNA is encoded on a single circular molecule of the size 5-6 μm (1). MtDNAs of higher plants have been isolated as circular

[1]This work was supported in part by grants from NSF (PCM76-09956 & DEB78-00538) & Pioneer Hi-Bred International, Inc.

[2]Present Address: Department of Microbiology, University of Chicago, Illinois 60637.

ISBN 0-12-198780-9

molecules; however, they have been reported to exist as a single molecular class (2) in some species and as multiple molecular classes in others (3, 4). The mitochondrial genome of higher plants appears to be the largest in nature; mtDNAs ranging in size from 70-165 x 10^6 have been described.

In maize, as well as in many other plant species, there is an extrachromosomally inherited trait termed cytoplasmic male sterility (cms). Plants carrying this trait do not produce viable pollen. Several unique sources of cms are recognized in maize which are distinguished on the basis of specific nuclear genes which restore pollen fertility. In a series of studies, we have demonstrated by restriction enzyme analyses that the mitochondrial DNAs from normal (fertile) and male sterile cytoplasms are distinctive (5, 6, 7). These results, in conjunction with several other types of studies, have furnished compelling evidence that the factors responsible for the cms trait are borne on the mitochondrial genome (see review 8).

The analysis of maize mtDNA from normal and male-sterile cytoplasms has revealed substantial differences. When restriction patterns from *Hin*dIII digests of mtDNA from normal and T cytoplasm were compared, distinctions in 17 bands were observed (7). The minimum molecular weight of mtDNA with normal and T cytoplasm were estimated to be 131- and 116 x 10^6, respectively, by sizing the fragments resulting from restriction endonuclease digestion. In another cms, the S cytoplasm, two plasmid-like DNAs with molecular weights of 3.45- and 4.10 x 10^6 have been found associated with the mitochondria (9). In addition, mitochondria from all maize cytoplasms studied contained a fast-migrating DNA, as demonstrated by gel electrophoresis, with a molecular weight of about 10^6. In this study, we have compared the mtDNA of normal and male-sterile cytoplasms by electron microscopy.

METHODS

Our methods for isolating and purifying maize mtDNA have been described previously (5, 6, 7). MtDNA was spread for electron microscopy by the formamide technique (4, 9, 10). Contour lengths of circles were determined by using a replica grating of 2160 lines/mm as an external standard. Molecular weights were determined with an internal standard, ϕX174 RF II DNA with a known size of 5375 base pairs (11). Molecular classes were identified by their means in those cases where discrete groups were apparent in the frequency distributions.

The sources of the mtDNAs were B37 x NC236, normal cytoplasm; B73 x Mo17, S cytoplasm; and B37 x NC236, T cytoplasm. Each mtDNA was isolated and spread at least twice;

only the combined data are presented. One hundred and four, 88, and 83 molecules were measured from normal, S and T cytoplasms, respectively. Forty-one molecules from the S cytoplasm were measured from a spreading without an internal standard. Molecules which were difficult to follow or of doubtful circularity were excluded.

RESULTS

Sarkosyl lysates of mitochondrial fractions were centrifuged to equilibrium in CsCl-ethidium bromide gradients. After centrifugation examination of the gradients with long wave UV light revealed a single fluorescent band. Electron microscopy demonstrated that this mtDNA band contained mostly linear and a few relaxed circular molecules; supercoiled molecules were not observed in this fraction. A second fraction was removed from the gradient about a centimeter below the single fluorescent band. This is the approximate position where covalently closed circular molecules would be expected to band. Electron microscopic examination showed that this fraction contained supercoiled and linear molecules as well as a few open-circular types. Approximately 5% of the molecules occurred as covalently closed circles. These results indicated that the band containing supercoiled molecules was present in subvisible amounts, therefore, subsequent observations were performed on mtDNA fractions drawn from this position on the gradient.

The mtDNA of maize from normal (fertile) cytoplasm was examined by electron microscopy. One hundred and four circular molecules which included both supercoiled and open-circular types were measured (Fig. 1, 2). It is evident from the distribution of contour lengths (Fig. 3) and molecular weights (Fig. 4) that maize mtDNA contains a heterogenous population of circular molecules. The largest molecular class had a mean contour length of 21 μm and a molecular weight of 45×10^6; this class contained 48% of the circular molecules. Twenty percent of the molecules were found in a class with a mean length of 15 μm and a weight of 33×10^6. Two classes of larger mtDNA molecules were observed, 30 μm (66×10^6) and 41 μm (91×10^6). These larger molecules may be dimers of the 14 and 21 μm classes. The remaining thirteen percent of the molecules range in size from 4 to 10 μm.

MtDNA from two cms cytoplasms, S and T, were also analyzed by electron microscopy. Molecular heterogeneity was observed among the mtDNAs of both S and T cytoplasms (Fig. 3, 4). From the S cytoplasm, 41% of the molecules were found in a single class with a mean contour length of

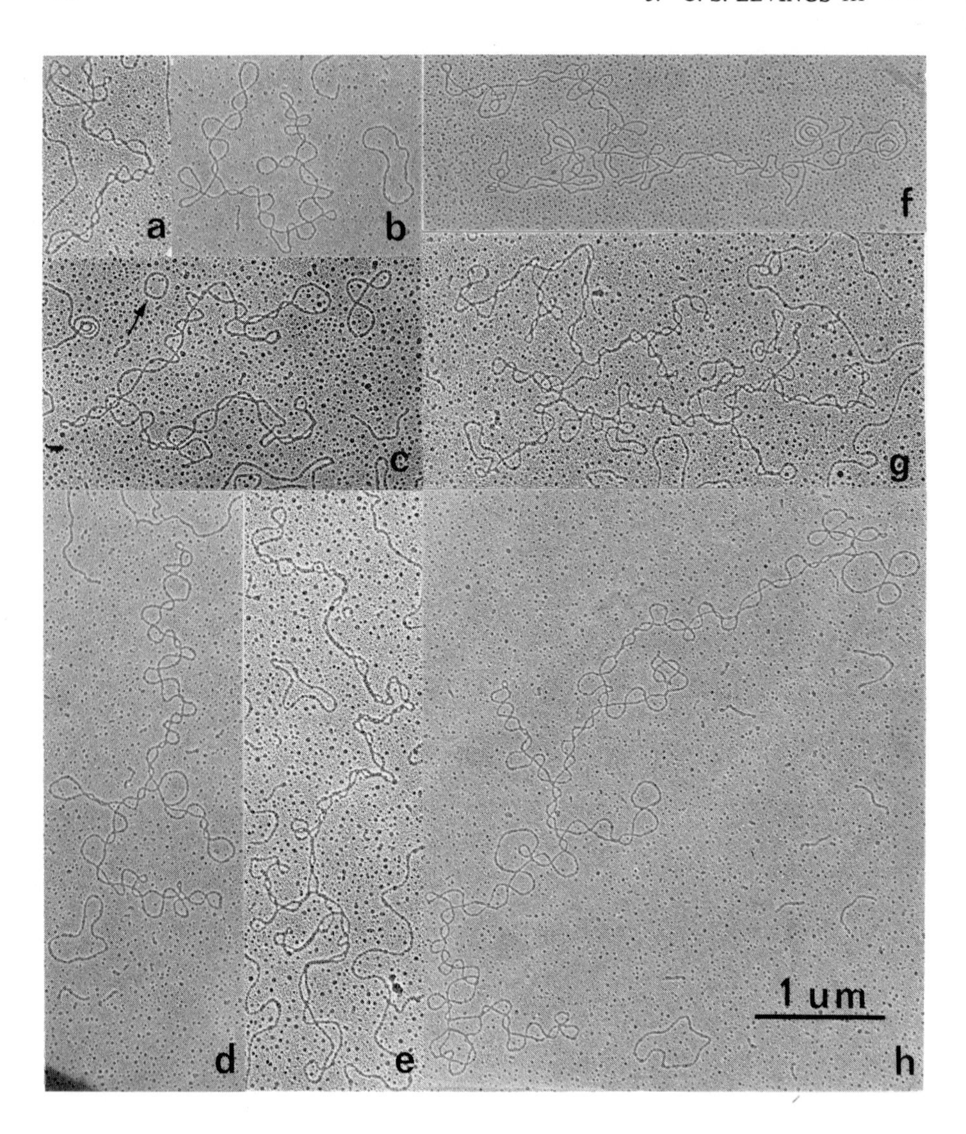

Figure 1. Supercoiled mitochondrial DNA molecules isolated from normal (N) and male-sterile maize, (S) and (T) cytoplasms. (a) 7 μm, N; (b) 11 μm S; (c) 15 μm, N, arrow indicates minicircle of 0.6 μm; (d) 17 μm, S; (e) 21 μm, N; (f) 25 μm, T; (g) 32 μm, N; (h) 36 μm, S. Small circles are ϕX174 RF II DNA.

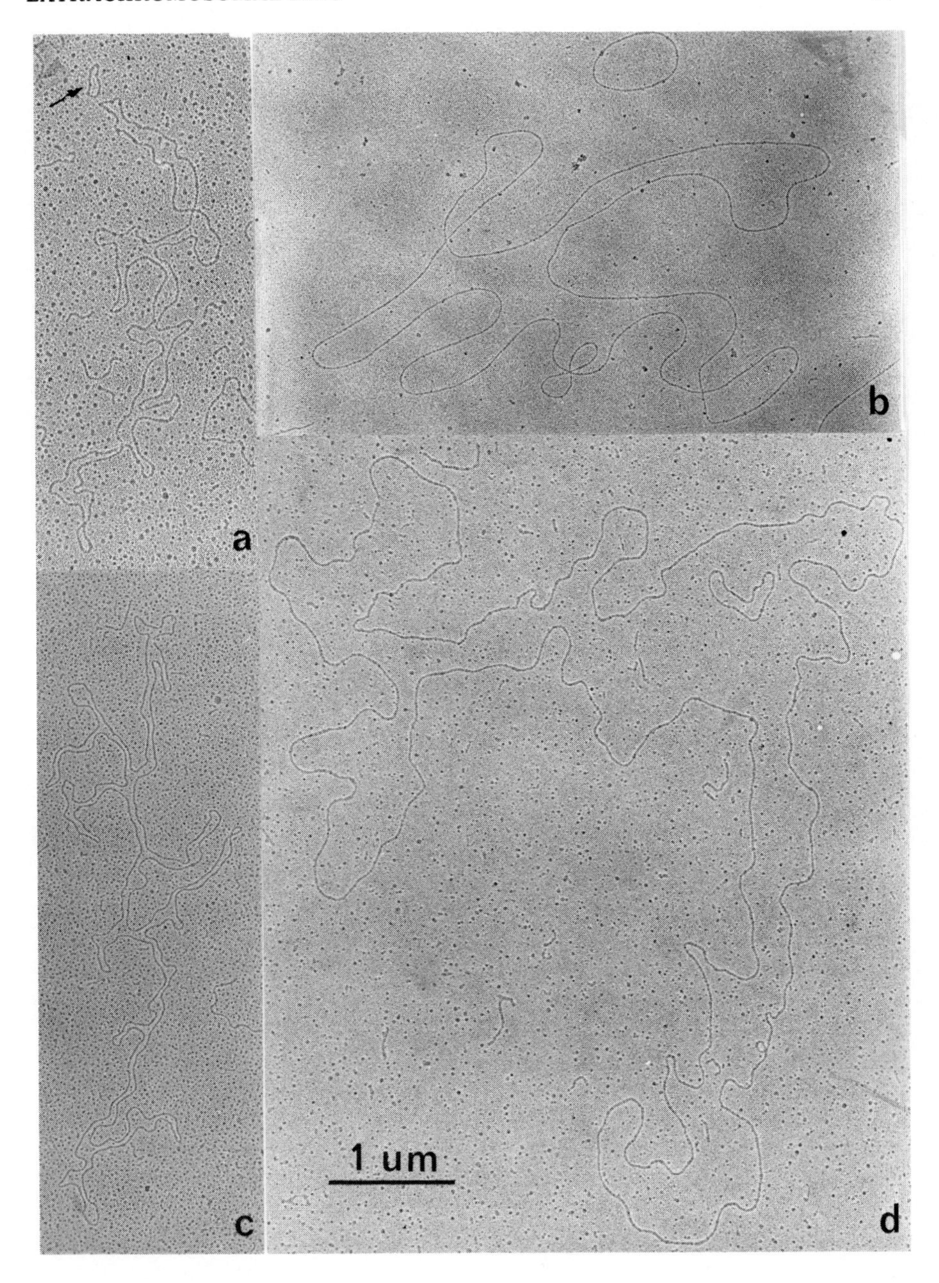

Figure 2. Relaxed circular mitochondrial DNA molecules isolated from normal (N) and male-sterile maize, (S) and (T) cytoplasms. (a) 14 µm, N, arrow indicates minicircle of 0.6 µm; (b) 22 µm, N; (c) 25 µm, T; (d) 37 µm, S. Small circles are ϕX174 RF II DNA.

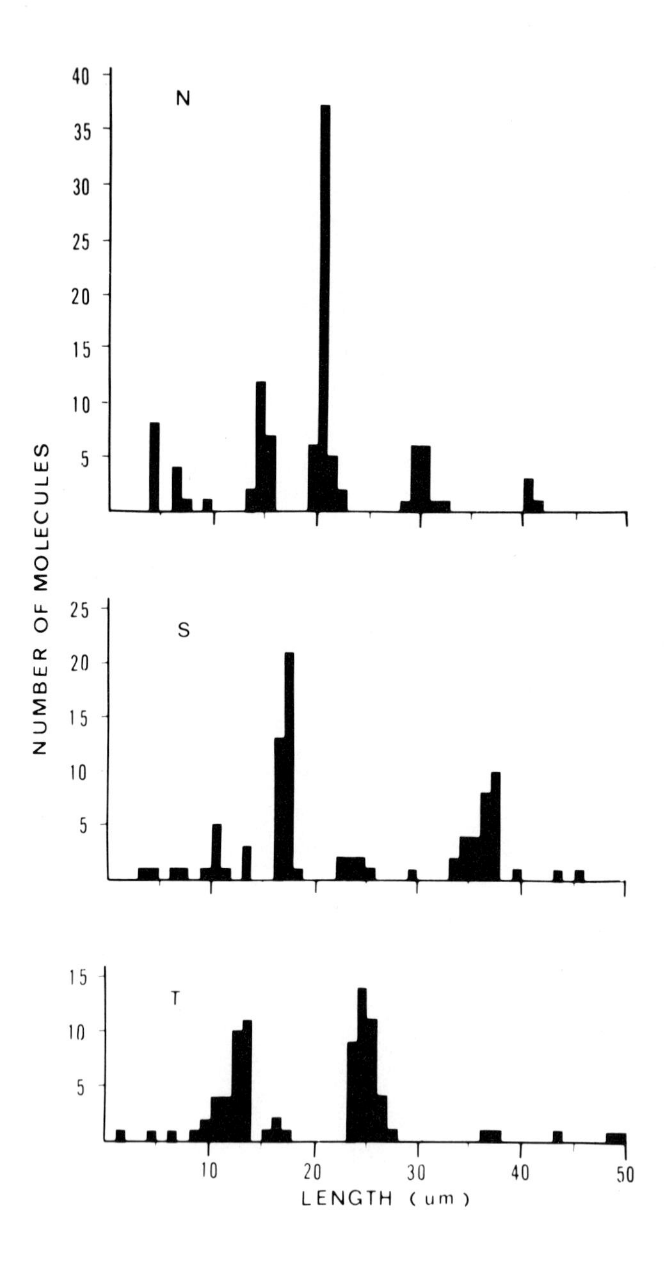

Figure 3. Frequency distribution of contour lengths of circular mitochondrial DNA molecules isolated from normal (N) and male-sterile maize, (S) and (T) cytoplasms.

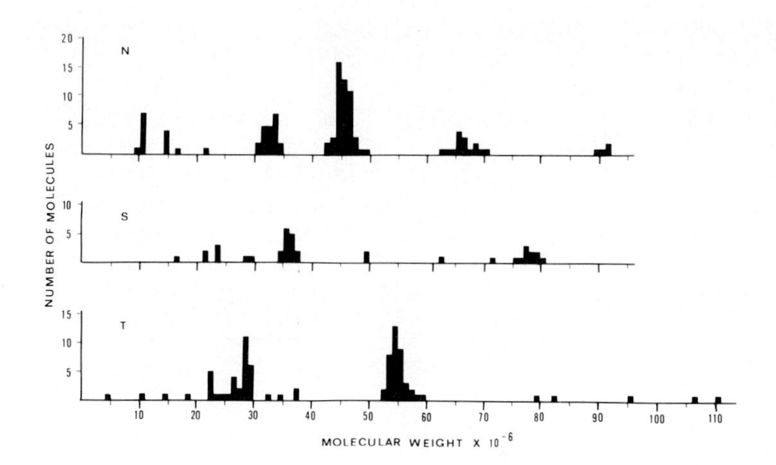

Figure 4. Frequency distribution of molecular weights of circular mitochondrial DNA molecules isolated from normal (N) and male-sterile maize, (S) and (T) cytoplasms.

17 μm and a molecular weight of 36 x 10^6; and 30% were in a 30 μm (78 x 10^6) group. The remaining molecular sizes occurred in low frequencies and ranged from 3 to 46 μm.

The S cytoplasm contains two plasmid-like DNAs (S-S and S-F) which have molecular weights of 4.10 and 3.45 x 10^6 (9). In earlier studies it appeared that these DNA species occurred as linear molecules rather than circles. To verify this, the internal standard, ϕX174, was omitted in one of our spreadings because the size of ϕX174 (3.55 x 10^6) is very similar to that of S-S and S-F DNAs. No circular molecules of the size 3.45 and 4.10 x 10^6 were observed, (Fig. 3, 4), therefore, these results indicate that the plasmid-like DNAs exist in a linear configuration.

The mtDNA from T cytoplasm had a major molecular class with a mean contour length of 25 μm and a molecular weight of 55 x 10^6 (Fig. 3, 4). Forty-seven percent of the circular mtDNA were found in this group. The second most common class contained 37% of the circular molecules and had a mean length of 12 μm and a molecular weight of 37 x 10^6. The remaining circles range in size from 4 to 49 μm and were present in low frequency.

Also observed in the mtDNA preparations from the normal, S and T cytoplasms were minicircles, molecules of less than one μm in length. Measurements from 150 circles revealed that these molecules had contour lengths of 0.6 μm and a molecular weight of 1.3 x 10^6. In the S cytoplasm exclusively, another class of minicircles was discovered which were 0.3 μm in length and had a molecular weight of 0.6 x 10^6. No effort was made to determine the frequency of minicircles.

In Table 1 we have presented the size and frequency of the two major molecular classes from the three cytoplasmic types. The dominant class for each cytoplasm comprises between 40-48% of the circular molecules. Significantly, the mean lengths of the various classes are different and range from 17 to 25 μm. The second most frequently occurring classes for the three cytoplasms are even more variable in length and frequency. It is clear from the data that each cytoplasm has a distinctive distribution of mtDNA molecules.

TABLE 1.

THE LENGTHS (μm) AND FREQUENCY (%) OF THE TWO MAJOR CLASSES OF MTDNA MOLECULES FROM THE NORMAL, S, AND T CYTOPLASMS OF MAIZE.

Cytoplasm	Most frequent class		Second most frequent class	
	μm	frequency %	μm	frequency %
Normal	21	48%	15	20%
S	17	40%	36	32%
T	25	47%	12	37%

DISCUSSION

From the data it is clear that the mtDNA of maize contains a variety of circular molecules which are heterogenous in length. Restriction enzyme fragment analysis estimated the size of mtDNA from normal (fertile) cytoplasm to be at least 131 x 10^6 daltons. None of the molecules which we observed are large enough to account for such a high molecular weight.

Molecular heterogeneity of mtDNAs appears to be common in higher plants. The mtDNA of soybeans contains seven classes of molecules which range in contour lengths from 5.9 to 29.9 μ (4). Quetier and Vedel (3) determined by restriction endonuclease analysis that the molecular weights of Virginia creeper, cucumber, wheat and potato mtDNAs ranged from 90 to 165 x 10^6. These values were much larger than the native molecular weight of 60 to 70 x 10^6 as measured by electron microscopy. The large discrepancy in the molecular weights led them to suggest that mtDNA molecules in these higher plants were a heterogenous population of circles with a unique length of 30 μ. In contrast, Kolodoner and Tewari (2) have studied mtDNA of pea by electron microscopy and renaturation kinetics; these studies failed to show any evidence of intermolecular heterogeneity. Thus, although molecular heterogeneity of mtDNAs is widespread in higher plants, it is not a universal phenomenon.

Although we have demonstrated covalently closed circular molecules in the mtDNA of maize, the frequency of these molecules was not large. The low yield of supercoiled molecules may simply result from isolation difficulties. Indeed, difficulties in isolating mtDNA in a supercoiled form have also been encountered in other eukaryotes, for example, yeast, *Neurospora* and *Euglena* (12, 13, 14). Even though it appears that the native configuration of maize mtDNA is that of a covalently closed circular molecule, it is possible that some linear forms exist. For instance, the plasmid-like DNAs associated with mtDNAs from the S cytoplasm of maize seem to be present only in the linear configuration.

Several possible explanations could account for the intermolecular heterogeneity present in maize mtDNA. The total informational content of maize mtDNA may be encoded in more than one molecule (3). Although multiple mitochondrial chromosomes have not been previously reported in higher plants, this possibility is compatible with our results. The presence of a heterogenous population of circular molecules is, nevertheless, not without precedent. MtDNA isolated from the wild type *Neurospora crassa* strain 5297 has been found to contain circular molecules heterogeneous in length. Circular mtDNA molecules from 0.5 to 7 μm were detected in addition to the main class of 19 μm circles (15). Recently linear mtDNA isolated from strain GL of *Tetrahymena pyriformis* has been reported to be heterogeneous in length ranging from 17 to 26 μm and without discrete classes (16).

An interesting but speculative alternative is suggested by the possibility that mitochondrial populations *per se* are not homogeneous. In that case, the different mitochondria might very well contain their own unique DNA of varying lengths. Currently, no evidence supports this hypothesis.

Recombinational events among mtDNA molecules might be responsible for the generation of different sizes of circular molecules. Since recombination is recognized among bacterial plasmids (17), serious consideration should be given to this mechanism in mtDNAs. Recombination among mtDNA molecules has been shown to occur in interhybrid somatic cells of animals (18). In addition, mtDNA molecules of different sizes may also result from incomplete or superfluous replication of the parental molecule.

Finally, contamination by alien DNAs from other organelles is an important concern. However, circular molecules described in the present study are believed to be of mitochondrial origin because supercoiled molecules of nuclear origin have not been demonstrated in higher plants and the chloroplast DNA of maize has been shown to consist of a homogenous population of circular molecules with a molecular weight of 85.4 x 10^6 (19).

In earlier studies, restriction enzyme analyses have shown that mtDNAs from normal, S and T cytoplasms are distinctive. The present electron microscopic studies of the mtDNAs from normal, S and T cytoplasms indicate that each type has a unique distribution of molecular classes. These findings suggest that substantial alterations in the mitochondrial genome are associated with the cms trait. The mechanism by which these changes were brought about is not known. In the S cytoplasm, the cms trait has been shown to be unstable, and in one case, the instability has been correlated with a change in the constitution of the plasmid-like DNAs (20, 21). Finally, it is tempting to speculate that events responsible for molecular heterogeneity are related to the origin of the cms trait.

ACKNOWLEDGMENTS

We thank Jane Mooneyhan for technical assistance. This has been a cooperative investigation of the Science and Education Administration, U. S. Department of Agriculture, and Institute of Food and Agricultural Sciences, University of Florida, and North Carolina Agricultural Research Service. Mention of a trademark name, proprietary product, or specific equipment does not constitute a guarantee or warranty by the U. S. Department of Agriculture and does not imply its approval to the exclusion of other products that may also be suitable.

REFERENCES

1. Borst, P. (1972). Ann. Rev. Biochem. 41, 333.
2. Kolodner, R. and Tewari, K. K. (1972). Proc. Natl. Acad. Sci. USA 69, 1830.
3. Quetier, F. and Vedel, F. (1977). Nature 268, 365.
4. Synenki, R. M., Levings, III, C. S., and Shah, D. M. (1978). Plant Physiol. 61, 460.
5. Levings, III, C. S., and Pring, D. R. (1976). Science 193, 158.

6. Levings, III, C. S., and Pring, D. R. (1976). Proc. 31st Annual Corn and Sorghum Research Conf., pp. 110-117.
7. Pring, D. R., and Levings, III, C. S. (1978). Genetics 89, 121.
8. Levings, III, C. S., and Pring, D. R. (1979). *In* "Physiological Genetics," (Scandalios, J. G., ed.), Academic Press, New York, (in press).
9. Pring, D. R., Levings, III, C. S., Hu, W. W. L., and Timothy, D. H. (1977). Proc. Natl. Acad. Sci. USA 74, 2904.
10. Davis, R. W., Simon, M., and Davidson, N. (1971). *In* "Methods in Enzymology" Vol. XXI, pp. 413-428. Academic Press, New York.
11. Sanger, F., Air, G. M., Barrell, B. G., Brown, N. L., Coulson, A. R., Fiddes, J. C., Hutchison, III, C. A., Slocombe, P. M., and Smith, M. (1977). Nature 265, 687.
12. Hollenberg, C. P., Borst, P., and Van Bruggen, E. F. J. (1970). Biochem. Biophys. Acta 209, 1.
13. Schafer, K. P., Bugge, G., Grandi, M., and Küntzel, H. (1971). Eur. J. Biochem. 21, 478.
14. Nass, M. M. K., Schori, L., Ben-shaul, Y. and Edelman, M. (1974). Biochem. Biophys. Acta 374, 283.
15. Aggsteribbe, E., Kroon, A. M., and Van Bruggen, E. F. J. (1972). Biochem. Biophys. Acta 269, 299.
16. Møller, J. K., Bak, A. L., Christiansen, C., Christiansen, G., and Stenderup, A. (1976). J. Bacteriol. 125, 398.
17. Clowes, R. C. (1972). Bacteriol. Rev. 36, 361.
18. Horak, I., Coon, H. G., and Dawid, I. B. (1974). Proc. Natl. Acad. Sci. USA 71, 1828.
19. Kolodner, R., and Tewari, K. K. (1975). Biochem. Biophys. Acta 402, 372.
20. Laughnan, J. R., and Gabay, S. J. (1975). *In* "International Maize Symposium: Genetics and Breeding," (Walden, D. B., ed.), John Wiley and Sons, Inc. (In press).
21. Levings, III, C. S., Pring, D. R., Conde, M. F., Laughnan, J. R., and Gabay-Laughnan, S. J. (1979). Maize Genet. Coop. N. L. (In press).

GENETIC ANALYSIS OF CHLOROPLAST DNA FUNCTION IN *CHLAMYDOMONAS*[1]

Nicholas W. Gillham, John E. Boynton, David M. Grant, Hurley S. Shepherd,[2] and Edwin A. Wurtz[3]

Departments of Botany and Zoology, Duke University, Durham, North Carolina 27706

ABSTRACT *Chlamydomonas* is the only plant whose chloroplast genome can be studied by recombination analysis. Our previous results show that chloroplast mutations resistant to antibacterial antibiotics define a group of 7 closely linked loci comprising a ribosomal region and cause the chloroplast ribosomes to become drug resistant. Recently we have isolated many chloroplast mutations defective in photosynthesis using the thymidine analog 5-fluorodeoxyuridine which increases specifically the yield of all known chloroplast mutant phenotypes. The 16 non-photosynthetic mutants characterized so far fall in 9 recombinationally distinct loci, one of which has now been mapped some distance from the ribosomal region. Mutations at 2 of these non-photosynthetic loci cause defects in chloroplast ribosome assembly; mutations at 3 loci result in the loss of chlorophyll-protein complex 1 from thylakoid membranes; and mutations at 3 other loci cause the disappearance of at least 2 of the 4 CF_1-ATPase polypeptides from the thylakoids. To orient our genetic map of the chloroplast genome with respect to the physical map made with restriction endonucleases, we are screening our collection of chloroplast mutants for deletions, insertions or other alterations in the chloroplast genome. Once such physical markers have been found, we will map them genetically with respect to mutants at the 16 chloroplast loci identified by recombination. Thus far we have observed several differences in restriction digest patterns of chloroplast DNA. One

[1]This work was supported by NIH grant GM-19427; Anna Fuller Foundation and NIH postdoctoral fellowship GM-05822 to DMG; NIH postdoctoral fellowship GM-05396 to EAW and NIH predoctoral traineeship GM-02007 to HSS.
[2]Present address: Department of Biology, University of California, La Jolla CA 92093.
[3]Present address: Department of Physiology and Biophysics, University of Illinois, Urbana IL 61801.

ISBN 0-12-198780-9

non-photosynthetic mutant stock contains deletions in two separate EcoRI fragments. Preliminary results indicate that this stock cotransmits both deletions in crosses with the maternal inheritance pattern characteristic of all known chloroplast mutations.

INTRODUCTION

The chloroplast genomes of all green algae and higher plants so far studied, with the exception of *Acetabularia*, are circles ranging in size from 40-60 μm, with the capacity to code for more than 250 proteins of M_r 20,000 (cf. 1). DNA-RNA hybridization has revealed that both chloroplast rRNAs and tRNAs are coded by the chloroplast genome in tobacco, maize, pea, *Euglena*, and *Chlamydomonas* (1-5). However, only a small proportion of the coding capacity of chloroplast DNA can be accounted for by the stable RNAs (cf. 1). The only polypeptide so far shown by physical mapping studies to be a product of chloroplast DNA is the large subunit of the CO_2-fixing enzyme ribulose-1,5-bisphosphate carboxylase (6,7).

Mutations have been used extensively to probe chloroplast DNA functions in *Chlamydomonas*, and to a lesser extent in higher plants (cf. 1,2). So far, however, *C. reinhardtii* is the only organism in which chloroplast genes can be mapped by recombination analysis. This unicellular green alga contains a single chloroplast and several mitochondria. The DNA molecules of both organelles have been characterized physically (cf. 1) and maps of both DNAs have been made using restriction enzymes (8,9 and Figs. 1 and 2). The chloroplast genome of *C. reinhardtii* consists of a 60 μm circle (10) present in ∼75 copies per cell while the mitochondrial genome is 4.7 μm in size and is present in 40-50 copies per cell (11). While nuclear genes segregate 2:2 in crosses, chloroplast genes are normally transmitted uniparentally by the maternal (*mt*$^+$) parent (Fig. 3). As discussed later in this paper, this pattern of inheritance reflects the maternal pattern of inheritance of chloroplast DNA in crosses. A few zygotes transmit chloroplast genes from both parents (biparental zygotes) or, rarely, only from the mating type minus (*mt*$^-$) parent (paternal zygotes). Putative mitochondrial mutations (12,13) are inherited in a biparental but non-Mendelian fashion (Fig. 3).

Mapping of chloroplast genes would verge on the impossible were it not for the fact that the frequencies of exceptional zygotes transmitting paternal chloroplast genes can be greatly increased by UV irradiation of *mt*$^+$ gametes prior to

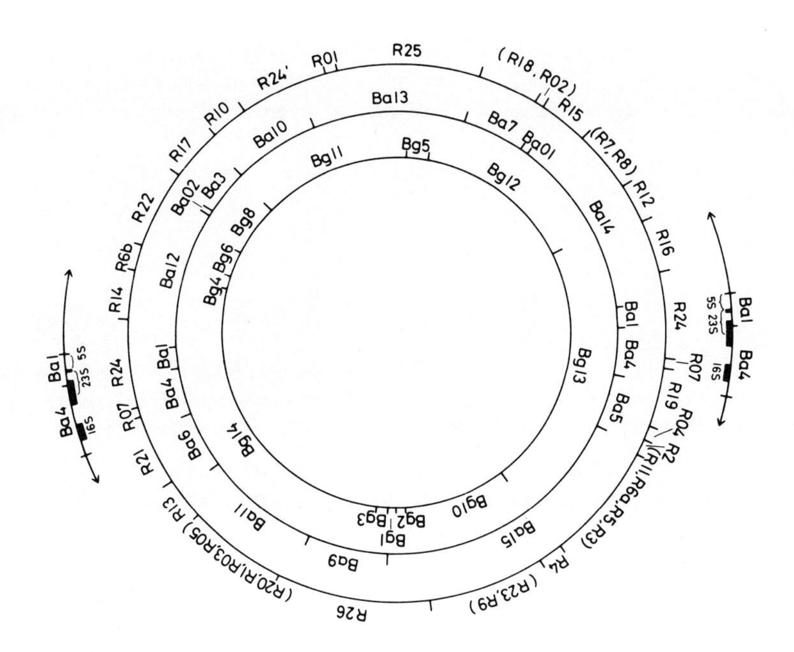

FIGURE 1. Restriction endonuclease map of C. reinhardtii chloroplast DNA. The 3 circles from the inside to the outside represent the BglII, BamHI and EcoRI restriction maps, respectively. The order of the fragments in parentheses has not been determined. The 2 inverted repeats containing the rRNA genes are indicated (from Rochaix,8).

FIGURE 2. Restriction endonuclease map of C. reinhardtii mitochondrial DNA. Mitochondrial DNA was prepared by method B of Ryan *et al.* (11) and the map determined by analysis of fragments produced by partial and double enzyme digests (9). The enzymes used were BamHI (▽), SalI (▼), and EcoRI (■). Although the map is linear as are the majority of isolated mitochondrial DNA molecules, the fact that supercoiled molecules of the same size are occasionally observed has led Grant and Chiang (9) to propose that the linear molecules result from site specific cleavage of these supercoiled molecules (from Grant and Chiang, 9).

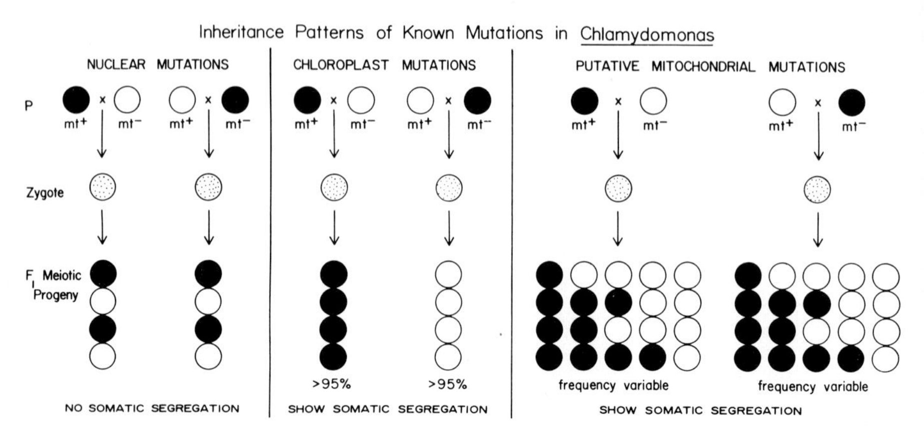

FIGURE 3. Patterns of inheritance of nuclear, chloroplast, and putative mitochondrial mutations in C. reinhardtii (from Gillham, 1).

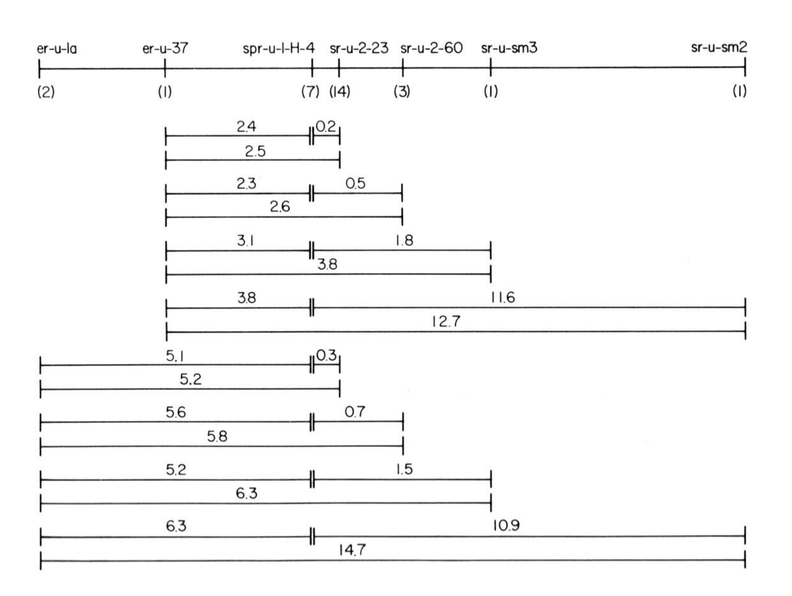

FIGURE 4. Map of chloroplast genes that mutate to antibiotic resistance in C. reinhardtii based on recombination in zygote clones. Numbers in parentheses below each gene symbol refer to numbers of alleles presently known at each locus. Gene symbols are as follows: er-u-la, er-u-37 = erythromycin resistance; spr-u-H-4 = spectinomycin or neamine/kanamycin resistance; sr-u-2-23, sr-u-2-60, sr-u-sm3, sr-u-sm2 = streptomycin resistance (from Harris et al., 22).

mating (14,15). Segregation and recombination of chloroplast genes have been studied among the progeny of biparental zygotes by pedigree analysis, by analysis of successive generations of zoospore progeny growing in liquid culture, and by zygote clone analysis (cf. 1,16,17,18). The relative merits of different methods for mapping chloroplast genes have been contrasted in several reviews (cf. 1,16,17,19). In the zygote clone analysis method used in our laboratory, 64 progeny clones are randomly selected from 50-100 biparental zygotes. Recombination frequencies are calculated as (recombinant progeny)/(total progeny) for pooled data from all biparental zygotes, and in this respect our method resembles the random diploid method used for mitochondrial genes in yeast (20). The few ($<1\%$) remaining heteroplasmic progeny are excluded from the calculations.

Until recently most of our efforts at mapping chloroplast genes have been confined to a series of chloroplast mutations resistant either to erythromycin, neamine, spectinomycin or streptomycin (16,21,22). These mutations confer antibiotic resistance on chloroplast ribosomes as shown by in vitro protein synthesis experiments employing synthetic polynucleotides and radioactive amino acids (21-25). Moreover, subunit exchange experiments have established that the ribosomal subunit affected by each mutation is the same as that affected by similar mutations in E. coli (cf. 23,24). To establish a map of these chloroplast mutations, we first tested all the known mutants with similar antibiotic resistant phenotypes for allelism. Somatic segregation of chloroplast genes in diploids as well as haploids has so far prevented the use of the classical dominance/complementation tests used in vegetative diploids for nuclear gene mutations to determine whether one or more gene functions are involved. Instead allelism must be judged entirely on the ability of two antibiotic resistant mutants with similar phenotype to produce recombinant progeny which are antibiotic sensitive. These experiments established the existence in the chloroplast genome of two loci for erythromycin resistance, one locus at which mutations can confer resistance to spectinomycin or to neamine and kanamycin, and four loci for streptomycin resistance (22). Until the primary gene product coded by each of these recombinationally distinct loci is identified we cannot say whether they are equivalent to functionally distinct genes defined by complementation tests. The order and relative positions of these 7 loci on the chloroplast genome have been established by a series of 8 different mapping crosses analyzed by recombination in zygote clones (22, Fig. 4). Although these crosses yielded a consistent order and remarkably similar map distances for the same intervals, we have as yet no idea of the relationship of map distance to physical distance.

Since antibiotic-resistant mutations represent a limited class of mutant phenotypes with which to probe the chloroplast genome, we decided to develop new techniques for the isolation of chloroplast mutations affecting chloroplast components other than the antibiotic resistance of chloroplast ribosomes. Many Mendelian mutations defective in photosynthesis have been isolated and characterized in *C. reinhardtii* because the alga is capable of respiring acetateas an energy source (cf. 26-28). Thus chloroplast mutations defective in photosynthesis were the most obvious class to be sought. In this paper we review our development of a new technique for the specific isolation of chloroplast mutations in *C. reinhardtii*. We then discuss the genetic and biochemical characterization of the numerous non-photosynthetic chloroplast mutations obtained for the first time using this technique. Finally, we consider experiments presently underway in our laboratory whose ultimate purpose is to correlate the genetic map of the chloroplast genome in *C. reinhardtii* with the physical map made recently in Rochaix (8) using restriction enzymes (Fig. 1).

RESULTS AND DISCUSSION

Specific Increase in Chloroplast Gene Mutations Following Growth of C. reinhardtii in 5-fluorodeoxyuridine (FdUrd): Before the kinds of selective methods used routinely in genetics can be employed efficiently to obtain desired classes of organelle mutations, a method for the specific induction or recovery of mutations in the organelle genome must be sought. This two-step approach is essential for two reasons. First the biogenesis of chloroplasts or mitochondria requires the participation of both the nuclear and organelle genomes. When mutations are selected which affect the phenotype of the organelle, they frequently turn out to be nuclear in origin. Hence the investigator must sort through large numbers of Mendelian mutations with no assurance of obtaining the desired mutations in the organelle genome. Second, mitochondria and chloroplasts contain several to many identical genomes and one might expect that most newly arising organelle mutations would never be expressed since the majority of the organelle genomes would carry wild type alleles of the mutation.

Some years ago Sager (29,30) reported that streptomycin was an effective mutagen for chloroplast genes in *Chlamydomonas* and the data presented were consistent with the interpretation that the antibiotic could induce chloroplast mutations to streptomycin resistance. More recently, Putrament *et al*. (31) demonstrated that Mn^{+2} was a powerful mutagen for mitochondrial genes in *Saccharomyces cerevisiae*. However, neither streptomycin nor Mn^{+2} have proved effective in our hands for the isolation of chloroplast gene mutations of *Chlamydomonas*.

Recently we reported that growth of *C. reinhardtii* in the thymidine analog FdUrd leads to a selective reduction in the amount of chloroplast DNA per cell (32, Fig. 5) and perturbs the normal maternal pattern of chloroplast gene trans-

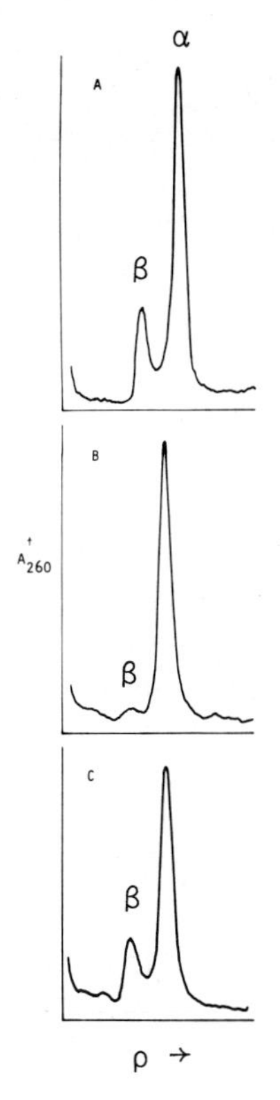

FIGURE 5. Tracings of CsCl equilibrium gradients of DNA from cells grown photosynthetically for 8 to 9 generations without FdUrd (A), with 0.5 mM FdUrd (B) and with 0.5 mM FdUrd + 0.5 mM thymidine (C). α = nuclear DNA; β = chloroplast DNA.

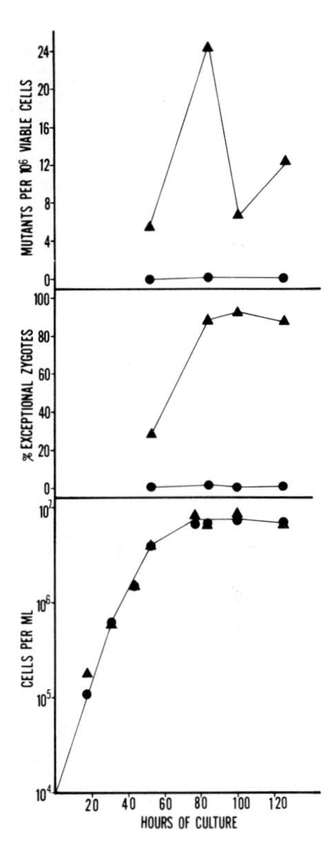

FIGURE 6. Effect of growth in FdUrd to different points in the cell cycle on the frequency of chloroplast mutations to spectinomycin resistance and the frequency of exceptional zygotes transmitting paternal chloroplast genes. ● = control; ▲ = 0.5 mM FdUrd (from Wurtz *et al*., 33).

mission in crosses (32). We have now found that growth of C. reinhardtii in the presence of FdUrd also increases at least 10 to 20 fold the frequency of cells expressing antibiotic resistant mutations in the chloroplast genome and in addition yields for the first time numerous chloroplast mutations defective in photosynthesis (Table 1 and 33). There is no increase in nuclear gene mutations with similar phenotypes under these conditions. The efficiency of recovery of chloroplast gene mutations rises as the FdUrd concentration increases from 0.1 to 1.0 mM. At higher concentrations of FdUrd, growth rates and mutant yields are reduced. We have also found that FdUrd has its strongest effect on chloroplast gene mutation frequency and on perturbing the pattern of inheritance of chloroplast genes about 24 hours after cells grown in the presence of the analog reach stationary phase (Fig. 6).

TABLE I

RECOVERY OF STREPTOMYCIN-RESISTANT AND NON-PHOTOSYNTHETIC ACETATE-REQUIRING MUTANTS FOLLOWING GROWTH OF CELLS OF C. REINHARDTII IN 0.1 mM FdUrd (see 33 for details).

Experiment	Treatment	Chloroplast mutations to streptomycin resistance/10^6 viable cells	Non-photosynthetic acetate-requiring mutants/10^6 viable cells	
			Chloroplast	Nuclear
I	Control	0.18	0	0.64
	FdUrd	3.29	0.82	0.56
II	Control	0	0	0.56
	FdUrd	6.31	0.69	0.69

Since labelled thymidine is known to be incorporated into chloroplast but not nuclear DNA in vegetative cells of Chlamydomonas, we have postulated that growth of cells in the presence of FdUrd causes the chloroplast to become thymidine starved by blocking the action of thymidylate synthetase in this organelle (32). This interpretation is consistent with the observed reduction in chloroplast DNA in FdUrd-grown cultures, its amelioration by equimolar concentrations of thymidine (Fig. 5), and the known mode of action of the analog in bacteria. If FdUrd is causing the chloroplast to become thymidine starved, the analog might promote the selective recovery of chloroplast mutations by a two-step process. First, FdUrd might act indirectly as a mutagen by causing thymidine starvation which is known to be mutagenic in bacteria (37,38), bacteriophage (39), and the mitochondrial genome of Saccharomyces cerevisiae (40). Second, FdUrd would promote the expression of chloroplast mutations by reducing the ploidy of the chloroplast genome and quite possibly causing damage to many of the remaining copies (41). Wiseman and Attardi (42,43) have found that ethidium bromide can be used to increase the yield of cytoplasmically transmitted mutations affecting mitochondrial function in mammalian cells in culture. The dye causes a reversible ten-fold reduction in the amount of mitochondrial DNA per cell, and may uncover spontaneous mitochondrial mutants that would otherwise not be expressed. Evidence that growth of Chlamydomonas in FdUrd may permit chloroplast mutations to be expressed that otherwise might be lost comes from our recent experiments in which the FdUrd-grown cells were subsequently mutagenized with the frameshift mutagen ICR-191. The number of spectinomycin resistant mutants, an indicator of chloroplast mutation frequency (33), increased 1800 fold when cultures grown in the absence of FdUrd were exposed to 20 μg/ml of ICR-191, but the increase was 57,000 fold when cultures grown in 0.5 mM FdUrd were treated with the mutagen. In contrast, growth of cells in FdUrd alone produced a 280 fold increase in the mutation rate to spectinomycin resistance over the untreated control.

Genetic and biochemical characterization of non-photosynthetic chloroplast mutants isolated following growth of cells in FdUrd: Until recently only three chloroplast mutations affecting photosynthetic parameters had been reported. One of these causes the loss of chlorophyll protein complex 1 (CP1) from the thylakoid membranes (44), the second alters the electrophoretic mobility of a different thylakoid membrane polypeptide (45), and the third appears to affect photophosphorylation (46,47). The two experiments summarized in Table I yielded 19 stable, non-photosynthetic chloroplast mutants following selection on arsenate plates (48). Of these 16 have now been characterized genetically (49).

Non-photosynthetic mutants found by tetrad analysis to be inherited uniparentally were tested for their ability to recombine in pairwise matings to yield photosynthetically competent progeny. All crosses were done at least twice, with between 50,000 and 100,000 zygotes plated each time. Assuming a minimum of 50,000 zygotes screened, an average of 70% germination, and 1% spontaneous biparental zygotes transmitting chloroplast markers from both parents, there should be at least 350 biparental zygotes in which recombination could occur. In crosses of antibiotic resistant mutants with similar phenotypes, progeny of 100 or fewer biparental zygotes are scored (22) for antibiotic sensitive recombinants. Thus a large number of zygotes can be analyzed conveniently in the case of the non-photosynthetic mutants since photosynthetically competent recombinants are positively selected on plates lacking acetate. With antibiotic resistant mutants, positive selection of antibiotic sensitive recombinants is not possible.

The 16 non-photosynthetic mutants characterized to date were found to fall in 9 distinct loci within which recombination was not detectable (Fig. 7). Photosynthetically competent progeny were produced in all crosses between mutants at different loci. These loci were designated ac-u-a through ac-u-i. Both mutants at the ac-u-b locus were isolated in a single experiment. Mutants at the ac-u-c locus were isolated in two separate experiments. The C1 mutant of Bennoun et al. (44) was found to map in the ac-u-i locus.

Experiments were carried out to ascertain whether any of the mutants were reverting to photosynthetic competence at a rate which would give spurious results in either the genetic or physiological analyses. Uniparentally inherited reversions or suppressors conferring wild type phenotype were found at frequencies of about 1 x 10^{-7}/mutant cell for specific mutants at most loci (Fig. 7). At the ac-u-d and ac-u-h loci Mendelian suppressor mutations were recovered at frequencies of around 1 x 10^{-5}/mutant cell, as is the case for nuclear mutants of similar phenotype (48).

Mutants at 7 of the 9 loci were found to have near wild type levels of ribulose bisphosphate carboxylase activity, photosystem II activity, and chlorophyll (49). Mutants in the 2 remaining loci, ac-u-d and ac-u-h, were deficient in both ribulose bisphosphate carboxylase and photosystem II activities, a syndrome characteristic of mutations affecting chloroplast ribosome assembly which are defective in chloroplast protein synthesis (48). Analysis of ribosomes from mutants at both loci on sucrose gradients showed that they lacked chloroplast ribosome monomers (70s) and accumulated 54s particles which may represent large subunits of the chloroplast ribosome (Fig. 8). SDS polyacrylamide gels of thylakoid membrane polypeptides isolated from these mutants revealed that CP1 and polypeptides 4.1-4.2,5,6,9,10 and 19 were drastically reduced (Fig. 9A). Representative mutants from the other loci had normal amounts of 70s chloroplast ribosomes (data not shown).

Isolated thylakoid membranes from mutants in 3 of the other 7 loci, ac-u-c, ac-u-g, and ac-u-i almost totally lack CP1 when analyzed on SDS polyacrylamide gels (Figs. 9B,9C). No other thylakoid membrane polypeptides are markedly affected in the membrane preparations of the 2 mutants at the ac-u-i locus (Fig. 9B). In the case of mutants at the ac-u-c and ac-u-g loci, we have also observed that polypeptides 4.1-4.2 vary in amount in different preparations (Fig. 9C). How-

Locus	Mutant	1-15	1-10	1-5	2-21	2-13	2-9	1-7	1-20	2-29	2-43	2-12	1-24	2-31	2-3	2-17	2-25	C1	Revertant frequency/ 10^7 cells
ac-u-a	1-15	-	+	+	+	+	+	+	+	+	+	+	+	+	+	+	+	+	0.2
ac-u-b	1-10		-	-	+	+	+	+	+	+	+	+	+	+	+	+	+	+	0.2
	1-5			-	+	+	+	+	+	+	+	+	+	+	+	+	+	nt	0.4
ac-u-c	2-21				-	-	-	-	-	-	-	+	+	+	+	+	+	+	0
	2-13					-	-	-	-	-	-	+	+	+	+	+	+	+	nt
	2-9						-	-	-	-	-	+	+	+	+	+	+	nt	0
	1-7							-	-	-	-	+	+	+	+	+	+	nt	nt
	1-20								-	-	-	+	+	+	+	+	+	nt	0
	2-29									-	-	+	+	+	+	+	+	nt	nt
	2-43										-	+	+	+	+	+	+	nt	nt
ac-u-d	2-12											-	+	+	+	+	+	+	98.0
ac-u-e	1-24												-	+	+	+	+	+	2.2
ac-u-f	2-31													-	+	+	+	+	0.33
ac-u-g	2-3														-	+	+	+	0.25
ac-u-h	2-17															-	+	+	122.0
ac-u-i	2-25																-	-	1.25
	C1																	-	nt

FIGURE 7. Recombination matrix of non-photosynthetic chloroplast mutants in *C. reinhardtii*. + = occurrence of photosynthetically competent recombinants; - = absence of photosynthetically competent recombinants; nt = not tested (from Shepherd *et al.*, 49).

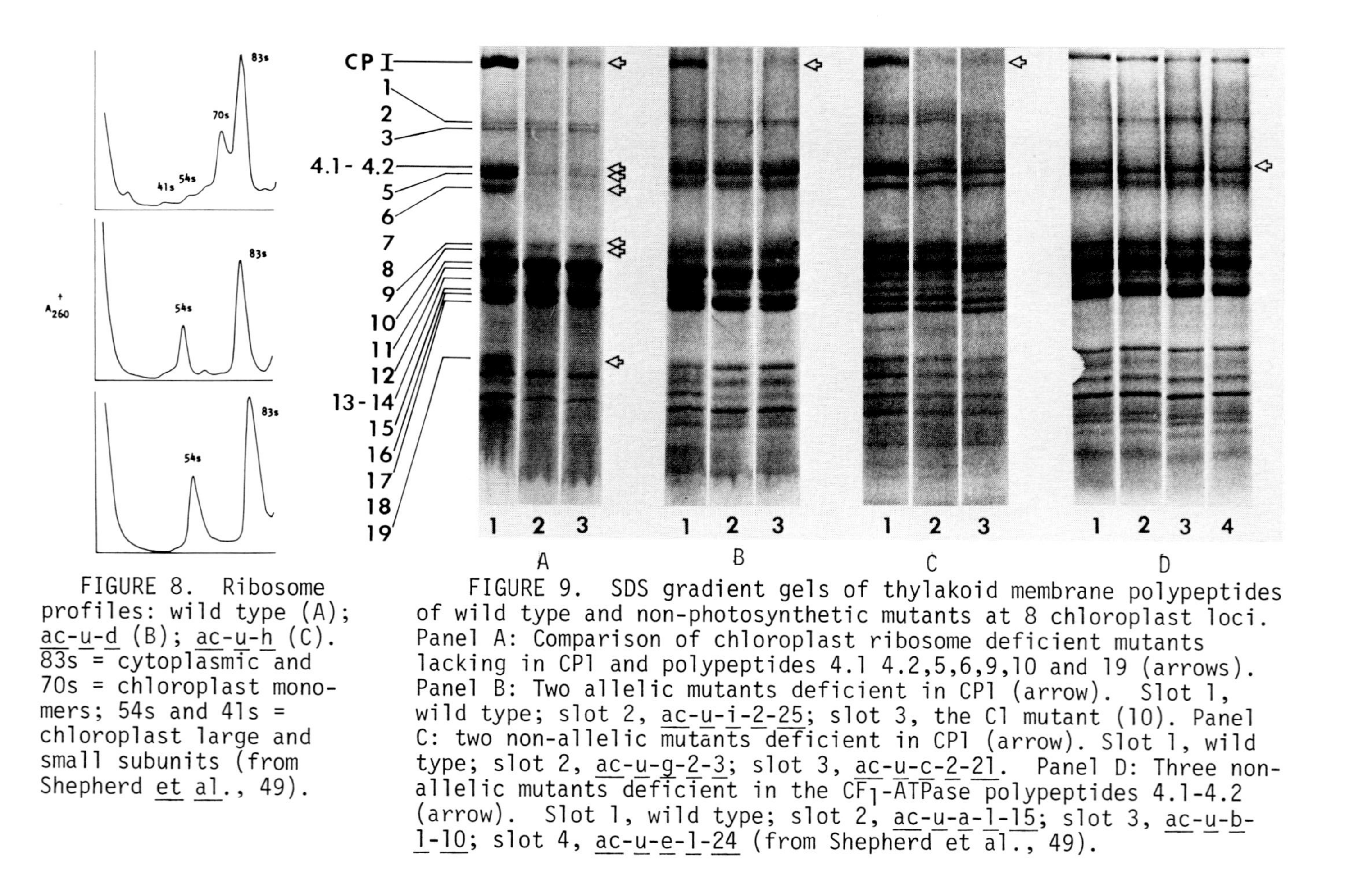

FIGURE 8. Ribosome profiles: wild type (A); ac-u-d (B); ac-u-h (C). 83s = cytoplasmic and 70s = chloroplast monomers; 54s and 41s = chloroplast large and small subunits (from Shepherd *et al*., 49).

FIGURE 9. SDS gradient gels of thylakoid membrane polypeptides of wild type and non-photosynthetic mutants at 8 chloroplast loci. Panel A: Comparison of chloroplast ribosome deficient mutants lacking in CP1 and polypeptides 4.1 4.2,5,6,9,10 and 19 (arrows). Panel B: Two allelic mutants deficient in CP1 (arrow). Slot 1, wild type; slot 2, ac-u-i-2-25; slot 3, the C1 mutant (10). Panel C: two non-allelic mutants deficient in CP1 (arrow). Slot 1, wild type; slot 2, ac-u-g-2-3; slot 3, ac-u-c-2-21. Panel D: Three non-allelic mutants deficient in the CF_1-ATPase polypeptides 4.1-4.2 (arrow). Slot 1, wild type; slot 2, ac-u-a-1-15; slot 3, ac-u-b-1-10; slot 4, ac-u-e-1-24 (from Shepherd *et al*., 49).

ever, Chua (personal communication) has independently analyzed thylakoid membrane polypeptides from the same ac-u-c and ac-u-g mutants shown in Fig. 9C and has confirmed that they lack specifically CP1 but have normal amounts of polypeptides 4.1-4.2. Since only about 5% of the total chlorophyll of thylakoid membranes is tightly associated with polypeptide 2 in the CP1 complex analyzed on SDS gels (50), we were not surprised to find that most of the mutants at these 3 loci had nearly normal levels of chlorophyll.

Thylakoid membranes isolated from mutants in the ac-u-a, ac-u-b, and ac-u-e loci are missing only polypeptides 4.1-4.2 (Fig. 9D) which are thought to be part of the CF_1-ATPase complex. The mutant at the remaining locus, ac-u-f, has no defects in any of the photosynthetic parameters yet analyzed.

The 9 chloroplast loci reported by Shepherd et al. (49) more than double the number of chloroplast loci defined by mutations in Chlamydomonas. Recombination analysis has shown that the ac-u-b locus is loosely linked to the chloroplast antibiotic loci (Fig. 10) as are the ac-u-a and ac-u-i loci (data not shown). Similar mapping studies with the other ac-u loci are now in progress.

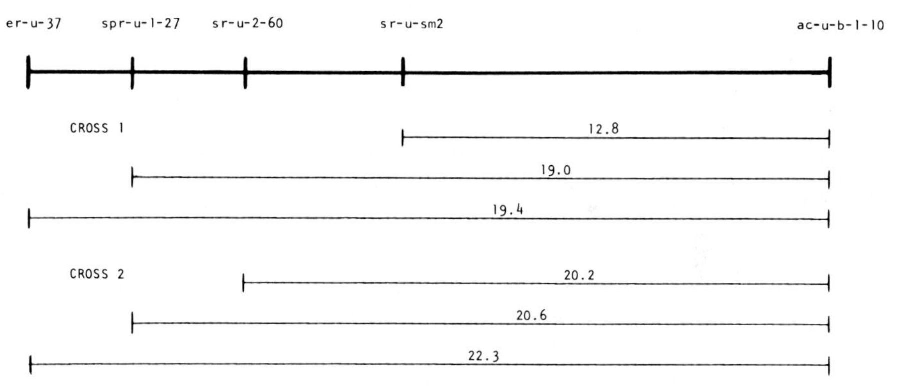

FIGURE 10. Position of the first non-photosynthetic chloroplast mutation (ac-u-b-1-10) mapped by zygote clone analysis with respect to 4 of the 7 antibiotic loci comprising the ribosome region. See Fig. 4 legend for symbols.

Characterization of these non-photosynthetic chloroplast mutations reaffirms the interplay between nuclear and chloroplast genes in the biogenesis of both chloroplast ribosomes and chloroplast lamellae. Mutants at the ac-u-d and ac-u-h chloroplast loci and several nuclear cr loci (48) accumulate what appear to be the large (54s) subunit of the chloroplast ribosome but are deficient in the small (41s) subunit. Thus, several loci in both the nuclear and chloroplast genomes may affect the synthesis of the small subunit of the chloroplast ribosome. The pattern of thylakoid membrane polypeptides of the ac-u-d and ac-u-h mutants is similar to that seen in wild type cells of *Chlamydomonas* following long-term grown in chloramphenicol, an inhibitor of chloroplast protein synthesis (51). In both cases, CP1 and polypeptides 4.1-4.2,5,6,9, 10 and 19 are affected (Fig. 9A) substantiating the fact that these lamellar polypeptides depend on chloroplast protein synthesis for their formation or incorporation into thylakoid membranes.

The results presented here show that at least 3 chloroplast loci specifically affect CP1, including locus ac-u-i which contains the C1 mutant described by Bennoun *et al.* (44, Fig. 9B). Both Mendelian and non-Mendelian mutations are now known to cause the loss of the CP1 complex and its principal component (polypeptide 2) from thylakoid membranes (44,52).

Mutants at the ac-u-a, ac-u-b, and ac-u-e loci lack polypeptides 4.1-4.2. Bennoun and Chua (52) have shown that these 2 polypeptides as well as polypeptide 8.1 are missing in thylakoid membranes of the Mendelian mutant F54 described by Sato *et al.* (53) as being defective in photophosphorylation. Since Sato *et al.* (53) reported that the mutant had an active, but non-latent, CF_1-ATPase, Bennoun and Chua (52) speculated that these missing polypeptides are associated with the membrane sector and interact with CF_1. Recently, however, Bennoun *et al.* (54) have reported that polypeptides 4.1-4.2 comigrate in SDS gels with the largest CF_1 subunits α and β , while the γ and ε subunits comigrate with polypeptides 8.1 and with one of a group of less well resolved low molecular weight polypeptides, respectively. On the basis of these recent findings, the ac-u-a, ac-u-b, and ac-u-e mutants are very likely blocked in the synthesis, assembly, or integration of chloroplast coupling factor CF_1 since they lack polypeptides 4.1-4.2. At least 3 subunits of CF_1 are known to be synthesized in isolated chloroplasts of pea (55) and spinach (56) and polypeptides 4.1-4.2 are known to be products of chloroplast protein synthesis in *Chlamydomonas* (51). If the CF_1 complex must be assembled before association with the thylakoid membrane, any defect which causes failure of proper assembly would result in the absence of CF_1 polypeptides from membrane preparations. The presence of these polypeptides in the cytoplasm might be detected using antibodies to the specific subunits.

The availability of a method which expedites the specific recovery of chloroplast mutants in *C. reinhardtii* will undoubtedly spur further research in this area. Bennoun *et al.* (54) have already made use of our method for the isolation of other new non-photosynthetic chloroplast mutations. The mutants obtained by these investigators were either affected in photosynthetic electron transport or defective in photophosphorylation. One mutant, FUD7, is deficient in photosystem II activity. This mutant has a membrane polypeptide pattern similar to that of the Mendelian mutant F34. Polypeptides 6 and 12 are missing, only traces of 5 are present, and 19 and 24 are found in reduced amounts. Three mutations (FUD2, FUD3, FUD25) are deficient in Photosystem I activity. One of these, FUD2, shows no alteration in membrane polypeptides and the membrane phenotypes of the other two mutants have yet to be reported. Whether any of these mutants are allelic with mutants at the ac-u-c, ac-u-g, or ac-u-i loci remains to be established. Bennoun *et al.* have also examined 9 chloroplast mutations defective in photophosphorylation and all of them are missing polypeptides 4.1- 4.2, and 8.1. These mutants are similar in phenotype to mutants mapping in the ac-u-a, ac-u-b, and ac-u-e loci and allelism tests between them as well as the photophosphorylation mutant isolated by Hudock and Togasaki (46,47) are clearly warranted.

Identification of Physical Markers in Chloroplast DNA with Restriction Endonucleases and Their Uses: A primary goal of our research is to be able to correlate the genetic map we have made of the chloroplast genome (Figs. 4,10) with the restriction endonuclease map published recently by Rochaix (Fig. 1, ref.8). In order to do this we have been systematically screening our collection of non-photosynthetic and antibiotic-dependent and -resistant chloroplast mutants for changes in chloroplast DNA restriction enzyme patterns. Such changes could arise because of sizable deletions or insertions or base pair changes at individual restriction sites. To determine whether a given mutant phenotype results from the observed physical change one can ask whether the mutation and the physical change revert together or whether the 2 recombine in crosses. If the mutation and the physical change corevert, there is good reason to believe they did so as the result of a single event. On the other hand, if the "reverted" strain still contains the physical change present in the original mutation, the so-called "revertant" may actually have resulted because of an intra- or intergenic suppressor mutation. Thus, mutations containing restriction site changes may be expected to revert while deletion mutations should not. If a given mutation and a physical change present in that strain recombine, this is also good evidence that they are unrelated genetically. However, if the 2 do not recombine, within the limits of resolution of the system, one can assume they are at the same locus. Even if further genetic analysis proves that the mutation and the physical change recombine at a low frequency and are not at precisely the same site within the locus, the position of the mutant is still fixed approximately.

We have now screened chloroplast DNA from over 30 chloroplast mutants with restriction endonuclease EcoRI (GAATTC) and HaeIII (GGCC) including all of the non-photosynthetic mutants shown in Fig. 7. Two of the non-photosynthetic mutant stocks exhibit differences in chloroplast DNA with respect to wild type following cleavage with EcoRI.

The mutant stock ac-u-g-2-3 contains 2 deletions, one of ~800 base pairs in fragment R21 and another of approximately the same size in fragment R19 (Fig. 11) as defined in the Rochaix map (Fig. 1). Comparison of the BamH1 (GGATCC) fragment pattern from the ac-u-g-2-3 mutant and wild type (data not shown) reveals that the mutant also has deletions of corresponding size in Bam5 and 6 (Fig.1), consistent with the EcoRI results. Fragment mobility shifts resulting from similarly sized deletions are also seen when chloroplast DNA from this strain is restricted with MspI (CCGG) and HindIII (AAGCTT) (Fig. 11). However, in these cases only one of the 2 deletions visualized with BamHI and EcoRI is apparent for reasons we do not yet understand. Chloroplast DNA from a double mutant recombinant obtained from a cross between ac-u-g-2-3 and spr-u-1-27-3, a chloroplast mutation to spectinomycin resistance, lacks both deletions. The fact that the recombinant is acetate-requiring and lacks both the R19 and R21 deletions shows that the non-photosynthetic mutation and the deletions are distinct from one another. We have also found that a uniparentally inherited "revertant" of ac-u-g-2-3 contains both deletions. Hence this "revertant" of ac-u-g-2-3 could either be a true same site reversion or could result from a chloroplast suppressor mutation.

Identification of the deletion in R19 is complicated by the fact that chloroplast DNA preparations from C. reinhardtii are often contaminated to some extent with mitochondrial DNA (8; Grant, unpublished) and that the largest EcoRI mitochondrial band comigrates with R19 (8,9). However, comparison of the stoichiometry of the bona fide chloroplast bands R17, R18 and R19 to that of the mitochondrial bands M1, M2 in the ac-u-g-2-3 EcoRI digest (Fig. 11) shows that a deletion is present in R19 of that strain. This problem is not encountered in the BamHI digests since none of the chloroplast and mitochondrial BamHI fragments comigrate.

The mutant stock ac-u-c-2-43 also contains a deletion of 1500-2000 base pairs in fragment R19 and Bam5. We do not yet know whether this deletion is related to the ac-u-c-2-43 mutation as revertants and recombinants have not been studied. So far none of the other nonphotosynthetic mutants have shown differences in fragment pat-

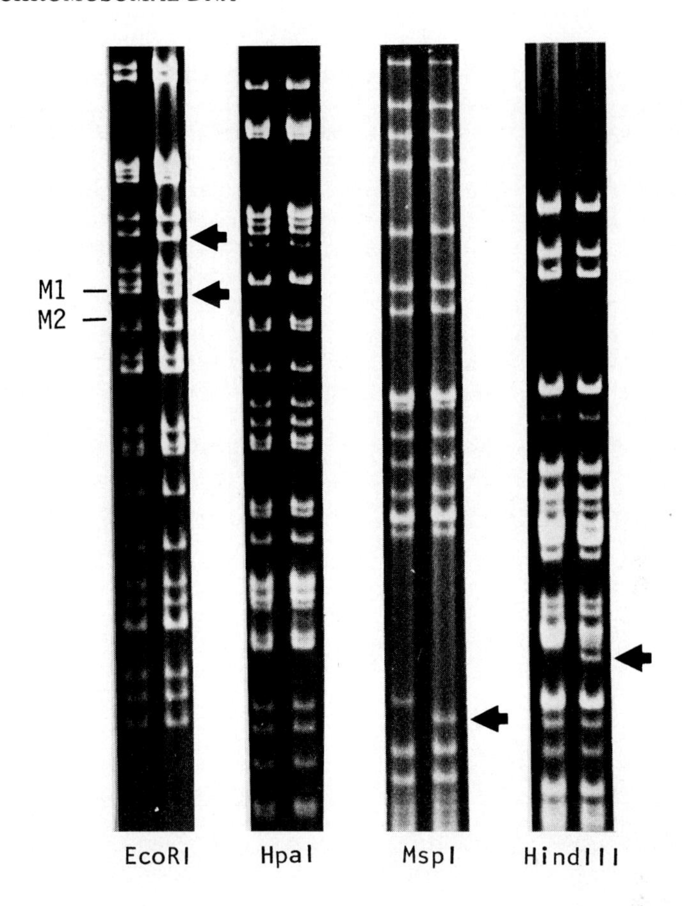

FIGURE 11. Digests of chloroplast DNA from photosynthetically competent (CW-15) mt+ (left) and ac-u-g-2-3 mt+ (right) strains. Late log cells were suspended in TEN (0.15 M NaCl,10 mM EDTA, 10 mM tris-HCl pH 8.0), lysed by SK + SDS (2% each), digested 48 hr at 4° with heat-treated Pronase (2 mg/ml) and extracted with PCIA (TEN saturated phenol:$CHCl_3$:isoamyl alcohol, 25:24:1). Nucleic acids were concentrated ~ 5 fold with 2-butanol, dialyzed vs TEN and centrifuged to equilibrium in NaI gradients containing EtBr. The chloroplast DNA band, identified by UV fluorescence, was precipitated with 1 vol EtOH, dissolved in 10 mM tris-HCl pH 7.5 and centrifuged in isokinetic gradients (5% top) in 10 mM tris-HCl pH 7.5 (SW-27 rotor, 25krpm, 4°, 19 hrs) to remove RNA. DNA was precipitated with 2 vol EtOH, dissolved in 10 mM tris-HCl pH 7.5 and stored at 4° over $CHCl_3$. 50 µl digestions contained 5-10 µg DNA and 10-20 units of enzyme. Fragments were separated on 35 cm 1% agarose gels at 1.4 V/cm for 41 hrs. Running buffer was 36 mM NaH_2PO_4, 1 mM EDTA, 40 mM tris-HCl pH 7.4. Arrows mark deletions.

terns of chloroplast DNA digested with either EcoRI or HaeIII with respect to wild type.

The availability of physical markers makes it possible to determine unambiguously the pattern of inheritance of chloroplast DNA in Chlamydomonas. This can then be related to the maternal inheritance pattern characteristic of chloroplast genes. We have now crossed a mt^+ stock containing the ac-u-g-2-3 mutation and the R19 and R21 deletions to a mt^- strain containing the erythromycin resistance marker er-u-37 and neither deletion. Chloroplast DNA from clones derived from each of the 4 meiotic products of 2 tetrads was isolated and digested with EcoRI and MspI (Fig. 12). The EcoRI digests

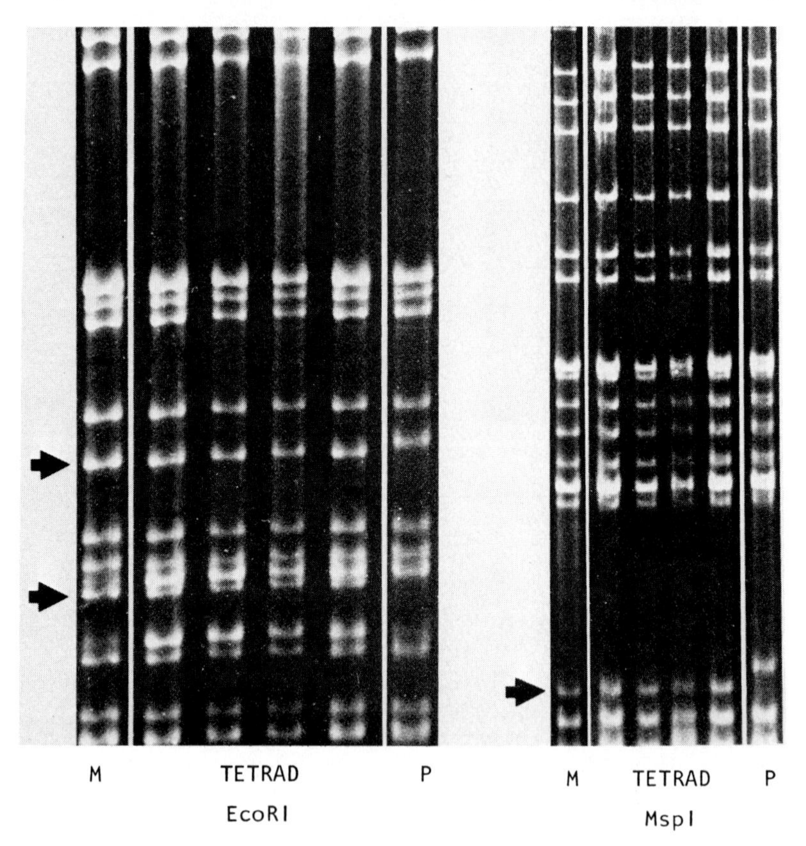

FIGURE 12. Inheritance of chloroplast DNA deletions in a maternal zygote from the cross ac-u-g-2-3 mt^+ x er-u-37 mt^- examined in both EcoR1 and MspI restriction digests. M = maternal (mt^+) and P = paternal (mt^-) parents. Tetrad = clones derived from the 4 meiotic products which segregated 2:2 for the nuclear mating type gene and carry both the ac-u-g-2-3 marker and the 2 deletions from the maternal parent. See Fig. 11 legend for methods. The EcoR1 digest was run for 70 hrs in order to resolve the 2 deletions better.

showed that in every case both the requirement for acetate and both deletions were cotransmitted to all progeny by the maternal parent while the paternal er-u-37 marker as well as the undeleted fragments were not. Since R19 and R21 are separated by about one-half of the chloroplast genome (Fig. 1), these results suggest that a large part, if not all, of the chloroplast genome is transmitted maternally. The single deletion visualized with MspI can also be seen to be maternally transmitted in this cross (Fig. 12). Obviously, further crosses must be done to substantiate these preliminary findings. These will include backcrosses of paternal (mt^-) progeny containing the deletions. In this regard we should note that Mets (57) has also obtained convincing evidence of maternal inheritance of chloroplast DNA in interspecific crosses between C. eugametos and C. moewusii which differ in chloroplast DNA restriction enzyme patterns. The results we have so far obtained are consistent with those published several years ago by Sager and Lane (58) which showed the maternal inheritance of density labelled chloroplast DNA in crosses of C. reinhardtii. Our experimental approach has the advantage of obviating the potential pool problems one may encounter using either density or radioactive labelling protocols which have been discussed in detail by Chiang (59). In addition our approach allows one to ascertain the pattern of inheritance of chloroplast DNA among the progeny of individual zygotes rather than populations of zygotes. Therefore, we can compare directly the inheritance of chloroplast DNA among the progeny of zygotes identified as being biparental, maternal or paternal on the basis of chloroplast marker segregation. We also hope to measure among the progeny of biparental zygotes the recombination frequencies of these deletions with respect to our genetic markers so that we can construct a genetic map that is congruent with the physical map.

REFERENCES

1. Gillham, N.W. (1978). "Organelle Heredity." Raven Press NY.
2. Gillham, N.W., Boynton, J.E., and Chua, N.-H. (1978). Curr. Topics Bioenerget. 8, 209.
3. Haff, L., and Bogorad, L. (1976). Biochem. 15, 4105.
4. Tewari, K.K., Kolodner, R., Chu, N.M., and Meeker, R.(1977). In"Nucleic Acids and Protein Synthesis in Plants" (L. Bogorad and J.H. Weil, eds.), pp. 15-36. Plenum Press NY.
5. Rochaix, J.D., and Malnoe, P. (1978). Cell 15, 661.
6. Gelvin, S., Heizmann, P., and Howell, S.H. (1977). Proc. Natl. Acad. Sci. USA 74, 3193.
7. Coen, D.M., Bedbrook, J.R., Bogorad, L., and Rich, A. (1977). Proc. Natl. Acad. Sci. USA 74, 5487.
8. Rochaix, J.D. (1978). J. Mol. Biol. 126, 597.

9. Grant, D.M., and Chiang, K.-S. (1977). Biophys. J. 17:284a.
10. Behn, W., and Herrmann, R.G. (1977). Mol. Gen. Genet. 157, 25.
11. Ryan, R., Grant, D., Chiang, K.-S., and Swift, H. (1978). Proc. Natl. Acad. Sci. USA 75, 3268.
12. Alexander, N.J., Gillham, N.W., and Boynton, J.E. (1974). Mol. Gen. Genet. 130, 275.
13. Wiseman, A., Gillham, N.W., and Boynton, J.E. (1977). Mol. Gen. Genet. 150, 109.
14. Sager, R., and Ramanis, Z. (1967). Proc. Nat. Acad. Sci. USA 58, 931.
15. Gillham, N.W., Boynton, J.E., and Lee, R.W. (1974). Genetics 78, 439.
16. Boynton, J.E., Gillham, N.W., Harris, E.H., Tingle, C.L., Swift, K. V-W., and Adams, G.M.W. (1976). In "Genetics and Biogenesis of Chloroplasts and Mitochondria" (T. Bücher, W. Neupert, W. Sebald, and S. Werner eds.) pp. 313-322. Elsevier/North-Holland Biomedical Press, Amsterdam.
17. Sager, R. (1977). Adv. Genet. 19, 287.
18. Forster, J.K., Grabowy, C.T., Harris, E.H., Boynton, J.E., Gillham, N.W., and Antonovics, J. (1979). In preparation.
19. Adams, G.M.W., Van Winkle-Swift, K., Gillham, N.W., and Boynton, J.E. (1976). In "The Genetics of Algae" (R.A. Lewin, ed.) pp. 69-118. Blackwell Scientific, Oxford.
20. Coen, D., Deutsch, J., Netter, P., Petrochilo, E., and Slonimski, P.P. (1970). Symp. Soc. Exp. Biol. 24, 449.
21. Conde, M.F., Boynton, J.E., Gillham, N.W., Harris, E.H., Tingle, C.L., and Wang, W.L. (1975). Mol. Gen. Genet. 140, 183.
22. Harris, E.H., Boynton, J.E., Gillham, N.W., Tingle, C.L. and Fox, S.B. (1977). Mol. Gen. Genet. 155, 249.
23. Schlanger, G., and Sager, R. (1974). Proc. Natl. Acad. Sci. USA 71, 1715.
24. Fox, S.B., Grabowy, C.T., Harris, E.H., Gillham, N.W., and Boynton, J.E. (1977). J. Cell Biol. 75, 307a.
25. Bartlett, S.G., Harris, E.H., Boynton, J.E., and Gillham, N.W. (1979). In preparation.
26. Levine, R.P. (1969). Annu. Rev. Plant Physiol. 20, 523.
27. Levine, R.P. (1971). In "Photosynthesis and Nitrogen Fixation" Part A (A. San Pietro, ed.), Methods in Enzymology, 23, 119. Academic Press NY.
28. Levine, R.P., and Goodenough, U.W. (1970). Annu. Rev. Genet. 4, 397.
29. Sager, R. (1962). Proc. Natl. Acad. Sci. USA 48, 2018.
30. Sager, R. (1972). "Cytoplasmic Genes and Organelles." Academic Press NY.
31. Putrament, A., Baranowska, H., and Prazmo, W. (1973). Mol. Gen. Genet. 126, 357.
32. Wurtz, E.A., Boynton, J.E., and Gillham, N.W. (1977). Proc. Natl. Acad. Sci. USA 74, 4552.

33. Wurtz, E.A., Sears, B.B., Rabert, D.K., Shepherd, H.S., Gillham, N.W., and Boynton, J.E. (1979). Mol. Gen. Genet. 170, 235.
34. Okazaki, R., and Kornberg, A. (1964). J. Biol. Chem. 239, 269.
35. Santi, D.V., and McHenry, C.S. (1972). Proc. Natl. Acad. Sci. USA 69, 1855.
36. Santi, D.V., McHenry, C.S., and Sommer, H. (1974). Biochem. 13, 471.
37. Coughlin, C.A., and Adelberg, E.A. (1956). Nature 178, 531.
38. Bresler, S., Mosevitsky, M., and Vyacheslavov, L. (1970). Nature 225, 764.
39. Smith, M.D., Green, R.R., Ripley, L.S., and Drake, J.W. (1973). Genetics 74, 393.
40. Barclay, B.J., and Little, J.G. (1978). Mol. Gen. Genet. 160, 33.
41. Nakayama, H., and Hanawalt, P. (1975). J. Bact. 121, 537.
42. Wiseman, A., and Attardi, G. (1978). Mol. Gen. Genet. 167, 51.
43. Wiseman, A., and Attardi, G. (1978). J. Cell Biol. 79, 321a.
44. Bennoun, P., Girard, J., and Chua, N.-H. (1977). Mol. Gen. Genet. 153, 343.
45. Chua, N.-H. (1976). In "Genetics and Biogenesis of Chloroplasts and Mitochondria" (Bücher, Th., Neupert, W., Sebald, W., and Werner, S., eds.) pp. 323-330. Elsevier/North-Holland Biomedical Press, Amsterdam.
46. Hudock, M.O., Togasaki, R.K., Lien, S., Hosek, M., and San Pietro, A. (1975). Plant Physiol. 56, s10.
47. Hudock, M.O., and Togasaki, R.K. (1974). J. Cell Biol. 63, 149a.
48. Harris, E.H., Boynton, J.E., and Gillham, N.W. (1974). J. Cell Biol. 63, 160.
49. Shepherd, H.S., Boynton, J.E., and Gillham, N.W. (1979). Proc. Natl. Acad. Sci. USA, 76, 1353.
50. Chua, N.-H., Matlin, K., and Bennoun, P. (1975). J. Cell Biol. 67, 361.
51. Chua, N.-H., and Gillham, N.W. (1977). J. Cell Biol. 74, 441.
52. Bennoun, P., and Chua, N.-H. (1976). "Genetics and Biogenesis of Chloroplasts and Mitochondria" (Bücher, Th., Neupert, W., Sebald, W., and Werner, S., eds.) pp. 33-39. Elsevier/North-Holland Biomedical Press, Amsterdam.
53. Sato, V.L., Levine, R.P., and Neumann, J. (1971). Biochim. Biophys. Acta 253, 437.

54. Bennoun, P., Masson, A., Piccioni, R., and Chua, N.-H. (1979). "Proceeding of the International Congress on Chloroplast Development." Elsevier/North-Holland Biomedical Press, in press.
55. Ellis, R.J. (1977). Biochim. Biophys. Acta 463, 185.
56. Mendiola-Morgenthaler, L.R., Morgenthaler, J.-J., and Price, C.A. (1976). FEBS Lett. 62, 96.
57. Mets, L. (1979). J. Supramolec. Struct. Supp. 3, 144.
58. Sager, R., and Lane, D. (1972). Proc. Natl. Acad. Sci. USA 69, 2410.
59. Chiang, K.-S. (1976). In "Genetics and Biogenesis of Chloroplasts and Mitochondria" (Th. Bücher, W. Neupert, W. Sebald, and S. Werner, eds.) pp. 305-312. Elsevier/ North-Holland Biomedical Press, Amsterdam.

METHYLATION AND RESTRICTION OF CHLOROPLAST DNA: THE MOLECULAR BASIS OF MATERNAL INHERITANCE IN CHLAMYDOMONAS[1]

Ruth Sager

Sidney Farber Cancer Institute
Boston, Massachusetts 02115

ABSTRACT A common feature of chloroplast DNA transmission in the sexual cycle of higher organisms is maternal inheritance. This paper describes the molecular basis of maternal inheritance in the simple eukaryote *Chlamydomonas*, and proposes that a similar mechanism may be present in higher organisms such as *Pelargonium* in which physical exclusion of male cytoplasm is insufficient to explain the observed patterns of inheritance. In *Chlamydomonas*, chloroplast DNA of maternal origin is methylated first in gametes and second in zygotes within the first six hours after zygote formation. The homologous DNA of paternal origin is not methylated in gametes and is degraded (i.e. restricted) just after zygote formation, before the two chloroplasts of male and female origin have fused within the common zygote cytoplasm. Two enzymes, a site-specific endonuclease and a methyl transferase which may be involved in the methylation and restriction processes are described and their potential roles in maternal inheritance are discussed.

INTRODUCTION

The maternal pattern of inheritance of chloroplast and mitochondrial genes is a fundamental property of the genetics of eukaryotic organisms. The first reports of chloroplast heredity were by Correns, who described strict maternal inheritance in *Mirabilis* (1) and other plants (2, 3) and by Baur, who described maternal, paternal and biparental patterns

[1]This work was supported by grants GM 22874 and ACS RD-30.

ISBN 0-12-198780-9

of transmission, all non-Mendelian, in the geranium plant, Pelargonium (4). In the subsequent years, examples of maternally inherited chloroplast traits have been widely reported in higher plants and the phenomenon is now firmly established (5).

Throughout the history of this field, maternal inheritance has been the hallmark by which chloroplast and mitochondrial genes were distinguished from nuclear genes. The mechanism of maternal inheritance in higher organisms was widely assumed to result from the virtual absence of cytoplasm in the male gamete at the time of fertilization. And similarly, the occasional transmission of chloroplast traits from the male parent was assumed to result from transmission of occasional male proplastids or mitochondria in fertilization. Despite this "conventional wisdom", no evidence was available to support it, and some of the complex breeding results e.g. with Oenothera and Pelargonium (5) cast doubt on the "physical exclusion" hypothesis as the sole mechanism of maternal inheritance (6).

Our discovery of maternal inheritance of chloroplast genes in Chlamydomonas (7) led us to a serious re-examination of the problem, since in this organism, both parents contribute equally to the zygote (5). In the sexual cycle of Chlamydomonas, pairs of morphologically identical haploid cells of opposite mating type, mt^+ and mt^-, fuse to form the diploid zygote which later undergoes meiosis and gives rise to 4 haploid zoospores, 2 mt^+ and 2 mt^-. Mating type is regulated by a nuclear gene or gene cluster. Chloroplast genes from the mt^+ parent are transmitted through the zygote to all 4 zoospores; the homologous chloroplast genes from the mt^- parent are not transmitted. This pattern of transmission is formally identical with the maternal inheritance of chloroplast genes in higher plants such as maize (8); and by analogy with the plant systems, we have considered the mt^+ parent as the female and the mt^- parent as the male. Exceptions to the rule of maternal inheritance occur with a frequency of about 0.1% in some strains of Chlamydomonas and up to 10-20% in other strains. In these exceptional zygotes, chloroplast genes are transmitted biparentally or rarely, from the paternal parent only (5).

The occurrence of maternal inheritance in Chlamydomonas, despite the equal contribution of cytoplasm from both parents, led us to seek an alternative to the physical exclusion hypothesis. Following the discovery of a high molecular weight species of DNA with a unique base composition in the chloroplast of *Chlamydomonas* (9, 10) we turned our attention to an examination of chloroplast DNA transmission in the sexual cycle of *Chlamydomonas*. Our thinking was much influenced by the process of modification and restriction of foreign DNA in bacteria, first described by Luria (11) and subsequently analyzed in an elegant series of studies by Werner Arber and coworkers (12). According to the modification-restriction hypothesis, foreign (e.g. phage) DNA entering a bacterial cell is *degraded* (i.e. restricted) by endonucleolytic attack at specific sequences on the DNA unless those sequences are protected by a secondary modification in DNA structure, later shown to be methylation. The methylating and restricting activities of *E. coli* B map at the same bacterial locus and were shown to be controlled by a pair of methylating and restricting enzymes (13).

Modification-restriction is an ideal mechanism for the destruction of one DNA genome in the presence of a homologous genome that was previously methylated to protect it from restriction. We therefore decided to search specifically for such a modification-restriction system in *Chlamydomonas*. For technical and other reasons, the search has been far more laborious and fraught with hazard than we could ever have predicted. The search has however been successful.

This paper summarizes our evidence that maternal inheritance of chloroplast DNA in *Chlamydomonas* is governed by a methylation-restriction system. The evidence for modification of chloroplast DNA from zygotes was first reported as a density shift in chloroplast DNA of maternal origin (14); and subsequently the density shift was shown to result from methylation (15). Our initial evidence for restriction was the disappearance from cesium chloride gradients of chloroplast DNA of male origin, using ^{14}N- and ^{15}N-DNAs as differential density prelabels of male and female DNA's (14). In subsequent studies, ^{3}H- and ^{14}C-adenine were used as differential prelabels (16) but serious pool problems were encountered; most recently we have used ^{3}H-thymidine to provide an unambigous tracer of the differential fates in the zygote of chloroplast DNAs from the two parents (15). We have also partially purified and characterized a site-specific endonuclease (17) and a methyl transferase (18) from *Chlamydomonas* and their properties will be summarized.

RESULTS

Methylation of chloroplast DNA. The bouyant density shift in chloroplast DNA as originally reported (14) is shown in Figure 1. A similar shift has been seen in numerous subsequent experiments in our laboratory with both analytical and preparative cesium chloride gradients. A recent example (15) in which prelabeling with [^{3}H]-thymidine was used to distinguish the parental origins of the DNA, is shown in Figure 2. This figure shows dramatically both the density shift in chloroplast DNA of maternal origin and the disappearance of the homologous DNA of paternal origin within 6 hours after zygote formation.

[^{3}H]-deoxycytidine and [^{3}H]-adenine were used as prelabels in separate experiments to look for methylation in vegetative, gametic, and zygotic DNAs (15). The [^{3}H]-adenine serves to label both adenine and guanine, and would identify methylated derivatives of either of these bases. The label from [^{3}H]-deoxycytidine is distributed into both cytosine and thymidine peaks, and would identify methylated cytosine.

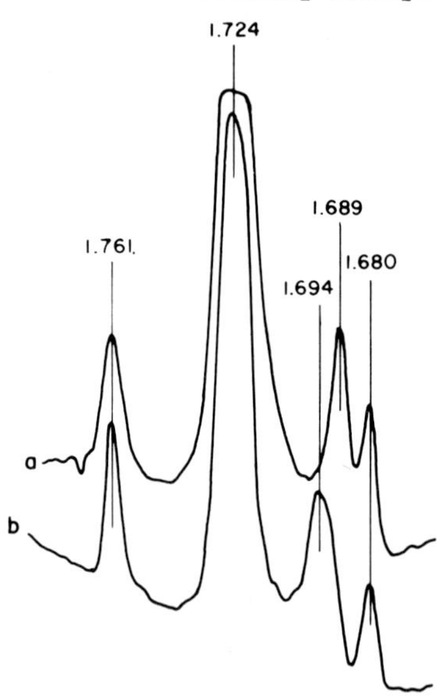

FIGURE 1. Microdensitometer tracings of UV-absorption bands of DNA after CsCl density equilibrium centrifugation. (a) DNA from 24-hr zygotes, from ^{14}N x ^{14}N cross, (b) gamete DNA from one of the parents. Outside peaks are markers: SP-15 DNA at 1.761 g/cm^3 and poly(dAT) at 1.680 g/cm^3. Nuclear DNA (overloaded) at 1.724 g/cm^3, chloroplast DNA at 1.694 g/cm^3 in (b), and at 1.689 g/cm^3 (shifted density in (a). (44,000 rpm, 20 hr, 25°). From (14).

A chromatographic method to separate the methylated from non-methylated bases (19, 20) was adapted to our system, using formic acid hydrolysis followed by high pressure liquid chromatography on an Aminex A6 column. The positions of the bases were established with the authentic bases and the absorptions measured under column conditions at pH 9.98 were normalized to molar absorptions empirically using known DNA standards, i.e. maize nuclear DNA containing 5-methyl C, and phage lambda DNA. By this method we established that both nuclear and chloroplast DNAs of Chlamydomonas were recovered from the column with molar ratios comparable to published values.

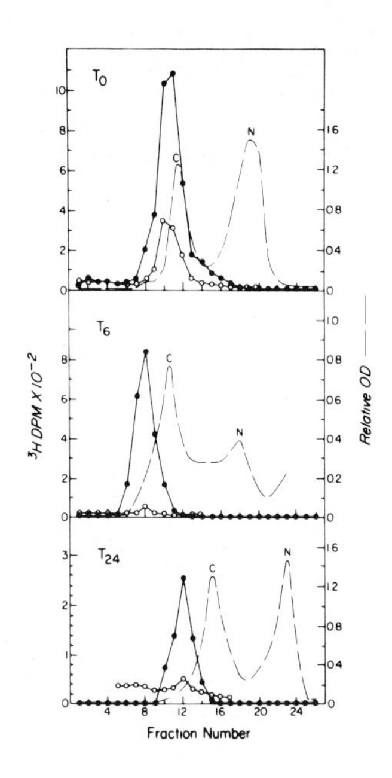

FIGURE 2. Bouyant density shift in zygotic chloroplast DNAs. Chloroplast DNA was extracted from zygotes of a pair of reciprocal crosses in which either strain 21gr (female) was prelabeled with [^{3}H]-thymidine and fused with unlabeled 5177D (male) DNA (●-●-●-●) or strain 5177D was labeled and fused with unlabeled 21gr DNA (o-o-o-o). Zygotes were sampled immediately after fusion (T_0), at 6 hours (T_6) and at 24 hours (T_{24}). Carrier cells were added, DNA extracted and fractionated in two successive CsCl gradients. OD tracing (----) of chloroplast (C) and residual nuclear (N) peaks and provides basis for estimating bouyant density of [^{3}H]-thymidine labeled maternal (●-●-●-●) and paternal (o-o-o-o) zygotic chloroplast DNA. From (15).

The results of numerous experiments to examine methylation of chloroplast DNA both published (15) and unpublished, are summarized here. In all these experiments, controls were performed to measure the mating efficiency, the viability of zygotes and extent of maternal inheritance. (The importance of high mating efficiency and zygote viability are obvious. The extent of maternal inheritance varies to some extent with the conditions, but most dramatically with the genetic background. The presence of even 5-10% biparental or paternal zygotes would be detectable in labeling experiments and would alter the pattern of maternal inheritance. In the studies reported here the frequencies of non-maternal zygotes were less than 1%.)

1) Using [^{3}H]-adenine as prelabel in either parent, no radioactivity is seen in the position of 6-methyl adenine, or in any position other than the adenine and guanine peaks.

2) Using [^{3}H]-thymidine as prelabel in either parent, no radioactivity is seen except in the thymidine peak.

3) Using [^{3}H]-deoxycytidine as prelabel, a peak is seen in the position of 5-methyl cytosine in the following DNAs:

a. Not in vegetative chloroplast DNA from either parent.
b. Not in gametic chloroplast DNA from the male parent.
c. Present as 2% of the total label in gametic chloroplast DNA from the female parent.
d. Present as 22% of the total label in chloroplast DNA extracted from 6 hr zygotes, when <u>female</u> cells were prelabeled; and as 4% of the total label when <u>male</u> cells were prelabeled.

An example of these results is shown in Figure 3 (further discussed below). The identification of 5-methyl cytosine in these gradients was further established by an independent method, using two-dimensional paper chromatography (21). (We thank Dr. Prem Reddy of this Institute for suggesting and performing the paper chromatography experiments.)

The identification of 5-methyl cytosine as responsible for the density shift in chloroplast DNA is based upon two lines of evidence.

1) The amount of 5-methyl cytosine in maternal chloroplast DNA increases from 2% in gametes to 22% (on radioisotope basis) in zygotes within 6 hours after zygote formation.

2) The extent of methylation of zygotic chloroplast DNA, estimated as 4-8 mol percent from absorption measurements of unlabeled DNA after high pressure liquid chromatography correlated well (22) with the observed shift in bouyant density of approximately 6-10 mg/cm^3 (15).

From these results we conclude that during formation of zygotes that show maternal inheritance, chloroplast DNA of maternal origin was methylated during the first 6 hours after zygote formation whereas the homologous DNA of paternal origin was not methylated but was degraded. The extent of degradation was virtually total as judged from thymidine labeling (see below).

Restriction of chloroplast DNA of male origin. In 1972 (14) we showed that chloroplast DNA of male origin disappeared from CsCl gradients of total extracted DNA within 6 hours after zygote formation. In these experiments, reciprocal crosses were examined in which one of the two parents was differentially prelabeled by growth in $^{15}N\text{-}NH_4Cl$. Genetic controls were performed to determine the extent of maternal inheritance which in this instance was less than 1%.

In unpublished experiments carried out at the same time, we followed the fate of chloroplast DNAs in one to six day old zygotes. We found that with increasing time after zygote formation, label of paternal origin which had initially disappeared from our gradients began to reappear and that concurrently the percent label from the maternal parent decreased. Clearly further DNA synthesis was occurring, either replication or repair or perhaps both, and labeled precursors of both male and female origin were being drawn from a precursor pool. It was evident from these results that examination of germinating zygotes or zoospores several days after zygote formation to assess the distribution of the initial gametic DNAs would be meaningless. What was required was a quantitative assessment of the events occurring immediately after zygote formation. After all we had already seen the qualitative loss of chloroplast DNA of male origin, and density shift of the DNA of female origin, in support of the modification-restriction hypothesis.

With the availability of high pressure liquid chromatography it became possible to re-examine the early events after zygote formation using radioactivity instead of ^{15}N for distinguishing the two parental DNAs. Preliminary experiments with [^{3}H]- and [^{14}C]-adenine and indicated a serious pool problem, not surprising in view of RNA turnover (23). Following the demonstration that chloroplast DNA but not nuclear DNA could incorporate thymidine (24), we used both [^{3}H]thymidine and [^{3}H]deoxycytidine in separate experiments, to follow the early fates of chloroplast DNAs from the two parents in the first few hours after zygote formation (15).

As shown in Figure 3, the dramatic increase in the 5-methyl cytosine peak was seen when the maternal parent prelabeled with [^{3}H]-deoxycytidine (Fig. 3A) but not when the

3A.

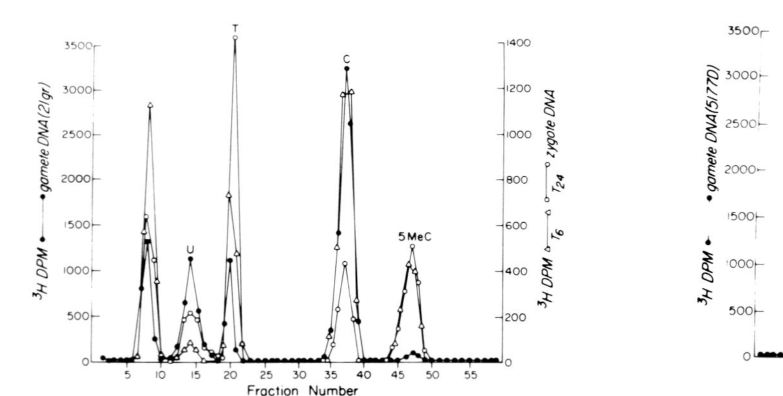

3B.

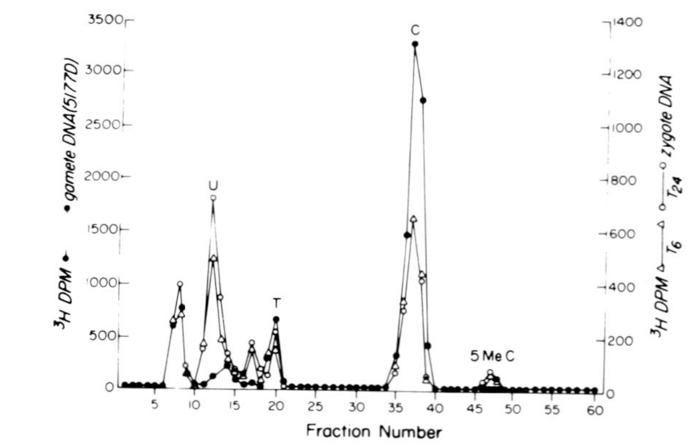

FIGURE 3. Chloroplast DNAs from gametes and zygotes in which one parent was prelabeled with deoxycytidine. One strain was prelabeled with [G-^{3}H]deoxycytindine (50 μCi/ml), gametes were prepared and fused with unlabeled gametes of the other strain, and zygotes were sampled at 6 and 24 hr. Chloroplast DNAs were purified and hydrolyzed with formic acid, and bases were separated on HPLC Aminex A-6 column. Positions of marker bases are shown: U, uracil; T, thymine; C, cytosine; 5MeC, 5-methyl cytosine. Peak at left is void volume, containing [^{3}H] deoxyribose. (A) Strain 21gr (female) labeled. (B) Strain 5177D (male) labeled. From (15).

paternal parent was prelabeled (Fig. 3B). The small 5-methyl cytosine peaks seen in Fig. 3B probably result from incorporation of [^{3}H]-deoxycytidine from a precursor pool into maternal DNA after zygote formation by either replication or repair synthesis. Further evidence for DNA synthesis between 6 and 24 hours comes from the relative decrease in cytosine and increase in thymidine labeling (Fig. 3A). In view of the total loss of thymidine of paternal origin seen in Figure 2, we think that the residual cytosine in Fig. 3B from 6 and 24 hour zygotes represents resynthesis occurring in maternal DNA utilizing labeled deoxycytidine from the precursor pool.

The results of experiments using [^{3}H]-thymidine as prelabel (c.f. Figure 2) have shown the disappearance of label coming from the paternal parent in gradients of total and of purified chloroplast DNA within the first six hours after zygote formation. These results show quantitatively and unambiguously the destruction of chloroplast DNA of male origin in parallel with the loss of chloroplast genetic markers from the male parent in the same experiments.

<u>Restriction endonuclease from Chlamydomonas</u>. A specific search was instituted to look for a restriction enzyme that might be responsible for the degradation of male chloroplast DNA. Enzyme isolation procedures and assay systems developed by Dr. Richard Roberts in his highly successful isolation of restriction endonucleases from bacteria (25) were applied to <u>Chlamydomonas</u> extracts (17). A site-specific endonuclease was recovered from vegetative cells of <u>Chlamydomonas</u> (<u>mt</u>$^+$) and purified as far as possible without loss of activity. The enzyme preparation cleaved viral DNA from adenovirus-2 with the production of discrete fragments that formed bands upon electrophoresis in an agarose gel, suggesting that it might be making double strand cuts. However, electron microscopy of cleavage fragments revealed gaps, i.e. long single-stranded regions within duplex fragments.

The single strand nicking ability of the enzyme preparation and the relative frequency of recognition sites was indicated by studies with SV40 DNA (26). The <u>Chlamydomonas</u> enzyme preparation converted covalently closed, twisted circular SV40 molecules first to open circles and then to full-length linear molecules. When the linear molecules were further digested with EcoRl, which cuts at only a single site in SV40 DNA, the resulting fragments produced some 20-30 discrete bands after electrophoresis in agarose.

From these and other data, we inferred (17) that the <u>Chlamydomonas</u> enzyme makes single-stranded nicks and gaps at specific sites in DNA and that the eventual appearance of discrete fragments results from the presence of overlapping gaps on complementary DNA strands. Studies of site-specificity with synthetic polynucleotides showed that the enzyme cleaved only at sites containing T, but 5'termini present in cleaved and gapped adeno-2 DNA included about 25% G as well as 66% T residues. To account for these results, we proposed that the enzyme attacks initially at a site containing T but that a second cut, leading to excision of a single stranded fragment, might occur at a site with different specificity as diagrammed in Figure 4.

Although further studies are required to purify the enzyme and to establish its mode of action, these results clearly demonstrate its site-specific nicking and gapping ability and lead us to view its biological potentialities for recombination and repair with great interest. Whether this enzyme is responsible for degradation of zygotic chloroplast DNA in maternal inheritance is as yet unknown.

Methyl transferase from *Chlamydomonas*. We have isolated a methyl transferase from vegetative cells of *Chlamydomonas* (18) and characterized its properties at a 300-fold level of purification. As yet, further purification has led to inactivation. The native enzyme catalyzes the transfer of methyl groups from S-adenosylmethionine to the 5 position of deoxycytidine in DNA. Native DNA is a 10-fold better substrate than denatured DNA. Kinetic analysis showed that the enzyme obeys a random sequential mechanism. The molecular weight is estimated at 55,000-58,000 from glycerol gradients and gel filtration experiments.

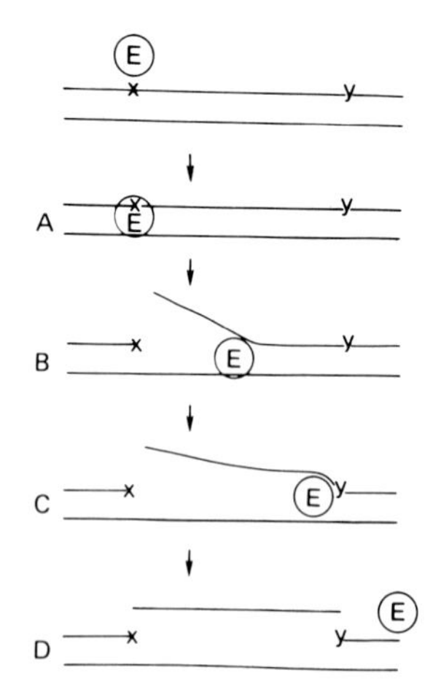

FIGURE 4. Model for the action of the *Chlamydomonas* endonuclease. x = site of initial endonuclease cleavage (incision site). y = site of second cleavage (excision site). E = enzyme. From (17).

The extent of saturation with methyl groups depends upon the species from which the DNA was obtained. Both prokaryotic and eukaryotic DNAs can be methylated by the enzyme, although the prokaryotic DNAs do not contain 5-methyl cytosine in nature. Site-specificity of methylation was studied by nearest neighbor analysis of *in vitro* methylated DNAs from *Micrococcus luteus* and *Chlamydomonas* (nuclear). With *M. luteus*, T-$\overset{*}{C}$-A and T-$\overset{*}{C}$-G were equally methylated, whereas with *Chlamydomonas* DNA, T-$\overset{*}{C}$-A was the major product (60%) followed by T-$\overset{*}{C}$-G (30%) and T-$\overset{*}{C}$-C (10%).

A different approach was used by methylating phage øX174 RF DNA, and digesting with the restriction endonuclease HaeIII. Twenty-five methylation sites were identified and their distribution was nonrandom. With respect to the T-$\overset{*}{C}$-Pu sequence revealed by nearest neighbor analysis, the number of sites recognized by the enzyme is much fewer than the number of T-C-Pu sequences present. In a computer study of sequences around T-C-Pu no general sequences were found which could account for the frequencies of observed methylation by the enzyme.

Further studies are in progress to identify the binding site and methylation site of the enzyme. However, it is already evident that the site-specificity is very high. Chloroplast DNA is methylated only to about 0.16% of the total cytosine residues by the enzyme. This level of activity could account for the gametic methylation of maternal chloroplast DNA, but not for the extensive methylation seen in the zygotes.

DISCUSSION

The results obtained in our laboratory over the past several years and summarized in this paper provide consistent and unambiguous evidence for the methylation of chloroplast DNA of female origin and concomitant degradation of the homologous male DNA as the molecular basis of the genetic phenomenon of maternal inheritance of chloroplast genes.

Studies of the differential fates of chloroplast DNAs in *Chlamydomonas* zygotes have also been carried out by Chiang (27, 28, 29) and it is important to consider why he has reached conclusions totally at variance with ours. In Chiang's initial studies of chloroplast DNA transmission in zygotes (27), he used differential prelabeling with [^{3}H]- and [^{14}C]-adenine, and examined zygote DNA from mature zygotes and from zoospores obtained after zygote germination. He concluded that chloroplast DNAs from both parents were conserved in the zygote and physically preserved in the progeny (i.e. zoospores).

In more recent studies, (28, 29) Chiang has conceded that pool problems with adenine misled him, and he has retracted his erroneous conclusion concerning conservation. However, in the new studies in which [^{3}H]-thymidine was used to label one parent in a pair of reciprocal crosses with unlabeled cells of the other parental strain, he again examined zygote DNA at very late times in zygote development, i.e. from germinating zygotes and from zoospores (28). He found that most of the label from both parents was gone by this time, and concluded from these studies that rather than being totally and equally conserved, the chloroplast DNA of both parents was largely and equally degraded.

Aside from pool problems, there are two serious criticisms to be raised that apply to all Chiang's studies. First, the conclusions are based on the unsubstantiated and unlikely expectation that neither replication nor repair of chloroplast DNA occurs during the many days of zygote maturation. As we have now shown (15) substantial changes occur already within the first six hours after gametic fusion. Second, no genetic data were provided to demonstrate whether the zygotes formed in these crosses showed maternal, paternal, or biparental inheritance or some mixture of all three. Indeed, it was not even shown whether the zygotes were viable.

Viability is of real concern, since in Chiang's early work (30, 31) an erroneous conclusion was drawn because neither viability tests or genetic controls were performed in conjunction with his physical studies of DNA. He found that only one round of semiconservative replication occurred between the time of gametic fusion and appearance of eight zoospores, the equivalent of three cell divisions. On this basis it was proposed that in the normal cell cycle gametes contained twice the nuclear DNA content of zoospores. The explanation for this strange result has now been provided by the work of Tan and Hastings (32) who have shown that the zoospores with reduced DNA content were inviable.

Turning now to our enzyme studies, we have described an endonuclease (17) and a methyl transferase (18) from *Chlamydomonas*, and their potential roles in maternal inheritance merit discussion.

With respect to methylation, there appear to be two steps: the low level methylation of chloroplast DNA in female but not in male gametes, and the high level of methylation of the same DNA of female origin in zygotes just after zygote formation. The site-specificity of the methyl transferase that we have described represents the right order of magnitude for the gametic methylation events, but cannot account as such for the high level of zygotic methylation. The two processes may be related, however, in an indirect way: the gametic methylation may signal or initiate the subsequent zygotic methylation.

It seems likely that a second methyl transferase may be induced at the time of zygote formation, and be localized within the chloroplast of female origin. In wild type cells, chloroplast fusion occurs some 5-6 hours after gametic fusion, providing a period of time during which the chloroplast compartments are separate but within a common cytoplasm. Thus, methylation of chloroplast DNA of female origin within its compartments could be completed before fusion with the chloroplast of male origin in which its DNA has already been degraded (33, 34).

With respect to restriction, here too one may predict the action of two enzymes: a site-specific endonuclease followed by a relatively non-specific exonuclease. This combination has been shown to determine the rapid breakdown of phage DNA seen in bacterial restriction systems (35). The nicking and gapping enzyme we have described could provide the initial breakage in this process, although no direct evidence has yet been developed on this point. The rate of ultimate breakdown however could depend upon the properties of the second enzyme, and could vary from one strain to another. This concept of a second enzyme with different activity levels from one strain to another could provide an explanation for certain discrepancies in the genetic results of experiments from our laboratory and from the laboratory of Gillham and Boynton (36).

For example, their strains show a much higher frequency of spontaneous exceptions to the rule of maternal inheritance, i.e. biparental zygotes, than ours. Secondly, their strains show a much higher frequency of recombination of chloroplast genes during zygote maturation in biparental zygotes than do ours (37). The difference is also shown by the rarity in their strains of zoospores from biparental zygotes containing all the parental chloroplast markers compared to the high frequency of these heterozygotes in our material (37). All these differences could be explained by a slow breakdown of male chloroplast DNA in the Gillham and Boynton strains, leading to marker rescue by the recombination of large fragments of male DNA into female DNA in the zygote; whereas, in our strains, there is rapid degradation of the male DNA (e.g. Figure 2) and recombination in zygotes rarely occurs.

The enzymology of methylation and restriction in zygotes is clearly of continuing interest to us. The use of mutants with altered patterns of maternal inheritance (38) as well as inhibitors (33) that affect methylation and restriction should help in the further analysis. It is noteworthy that from an evolutionary point of view, the enzymatic mechanisms of maternal inheritance in *Chlamydomonas* must have long preceded the physical exclusion mechanisms that play a role in higher organisms. Thus, the phenomenon of maternal inheritance of organelle genomes must have appeared very early in protist evolution and been retained ever since. We still do not understand the evolutionary advantage of maternal inheritance, but it is evident that this advantage exists, and is a fundamental feature of organelle heredity in eukaryotes. The recent discovery of a high rate of evolutionary change in mitochondrial DNA sequences (39) raises the possibility that rapid evolution is favored by clonal inheritance.

The recent work of Tilney-Bassett (6) with *Pelargonium* has provided new evidence on the inheritance of chloroplast genes in this plant and has established the influence not only of environmental factors but especially of nuclear genes upon the proportion of maternal, biparental, and paternal transmission. I previously proposed (6) that the results with *Pelargonium* could be interpreted in terms of a modification-restriction mechanism similar to that operating in *Chlamydomonas*. With our new finding of methylation of chloroplast DNA, it may be possible to test this hypothesis directly with *Pelargonium* or *Oenothera*. Applications of the methylation-restriction mechanism to "selective silencing" of DNA in other organisms have also been proposed (40). With the availability of new sensitive methods to identify methylated bases in DNA, it may now be possible to look for methylation in certain of these systems.

REFERENCES

1. Correns, C. (1909). Z. Vererbungslehre 1, 291-329.
2. Correns, C. (1909). Z. Vererbungslehre 2, 331-340.
3. Correns, C. (1937)."Nicht Mendelnde Vererbung" (F. von Wettstein, ed.), Borntraeger, Berlin.
4. Baur, E. (1909). Z. Vererbungslehre 1, 330-351.
5. Sager, R. (1972). "Cytoplasmic Genes and Organelles." Academic Press, New York.
6. Sager, R. pp. 252-267, and Tilney-Bassett, R.A.E. pp. 268-308. (1975). In "Genetics and Biogenesis of Mitochondria and Chloroplasts" (C.W. Birky, Jr., P.S. Perlman, and T.J. Byers, eds.), Ohio State University Press, Columbus Ohio.
7. Sager, R. (1954). Proc. Natl. Acad. Sci. USA 40, 356-363.
8. Rhoades, M.M. (1946). Cold Spring Harbor Symp. Quant. Biol. 11, 202-207.
9. Chun, E.H.L., Vaughan, M.H. and Rich, A. (1963). J. Mol. Biol. 7, 130-141.
10. Sager, R., and Ishida, M.R. (1963). Proc. Natl. Acad. Sci. USA 50, 725-730.

11. Luria, S.E. (1953). Cold Spring Harbor Symp. Quant. Biol. 18, 237.
12. Arber, W. (1974). Proj. Nucleic Acid Res. Mol. Biol. 14, 1-35.
13. Arber, W., and Linn, S. (1969). Annu. Rev. Biochem. 38, 647.
14. Sager, R., and Lane, D. (1972). Proc. Natl. Acad. Sci. USA 69, 2410-2413.
15. Burton, W.G., Grabowy, C.T., and Sager, R. (1979). Proc. Natl. Acad. Sci. USA 76, 1390-1394.
16. Schlanger, G. and Sager, R. (1974). J. Cell Biol. 63, 301a.
17. Burton, W.G., Roberts, R.J., Myers, P.A. and Sager, R. (1977). Proc. Natl. Acad. Sci. USA 74, 2687-2691.
18. Sano, H., and Sager, R. (1979). Manuscript submitted.
19. Singhal, R.P. (1972). Arch. Biochem. Biophys. 152, 800-810.
20. Singhal, R.P. (1974). Separation and Purification Methods 3, 339-398.
21. Flanegan, J.B., and Greenberg, G.R. (1977). J. Biol. Chem. 252, 3019-3027.
22. Kirt, J.T.O. (1967). J. Mol. Biol. 28, 171-172.
23. Siersma, P.W., and Chiang, K.-S. (1971). J. Mol. Biol. 58, 167-185.
24. Swinton, D.C., and Hanawalt, P.C. (1972). J. Cell Biol. 54, 592-597.
25. Roberts, R.J. (1976). Crit. Rev. Biochem. 3, 123-164.
26. Burton, W.G., and Sager, R., unpublished.
27. Chiang, K.-S. (1968). Proc. Natl. Sci. USA 60, 194.
28. Chiang, K.-S. (1975). In "Les cycles cellulaires et leur blocage chez plusiers protistes" (M. Lefort-Tran and R. Valencia, eds.), pp. 147-158. Editions du Centre National de la Recherche Scientifique, Paris.
29. Chiang, K.-S. (1976). In "Genetics and Biogenesis of Chloroplast and Mitochondria" (T. Bucher, W. Neupert, W. Sebald, and S. Werner, eds.), pp. 305-312. North Holland, Amsterdam.
30. Chiang, K.-S. (1965). Meiotic DNA Replication Mechanism in Chlamydomonas Reinhardi. Ph.D. Dissertation, Princeton University.
31. Chiang, K.-S., and Sueoka, N. (1967). J. Cell Physiol. 70, 89 (Suppl. 1).
32. Tan, C.K., and Hastings, P.J. (1977). Molec. gen. Genet. 152, 311-317.
33. Sager, R., and Ramanis, Z. (1973). Theor. Appl. Genet. 43, 101-108.
34. Sager, R. (1977). Adv. Genet. 19, 287-340.
35. Meselson, M., Yuun, R., and Heywood, J. (1972). Annu. Rev. Biochem. 41, 447-465.

36. Gillham, N.W. (1978). "Organelle Heredity." Raven Press, New York.
37. Sager, R., and Ramanis, Z. (1976). Genetics 83, 303-321.
38. Sager, R., and Ramanis, Z. (1974). Proc. Natl. Acad. Sci. USA 71, 4698-4702.
39. Brown, W.M., George, M., and Wilson, A.C. (1979). Proc. Natl. Acad. Sci. USA 76, 1967-1971.
40. Sager, R., and Kitchin, R. (1975). Science 189, 426-433.

GENES ON THE MAIZE CHLOROPLAST CHROMOSOME[1]

Lawrence Bogorad, Gerhard Link, Lee McIntosh
, and Setsuko O. Jolly

Department of Biology, Harvard University
Cambridge, Massachusetts 02138

ABSTRACT In carlier work, genes for a few chloroplast ribosomal proteins were located in the nuclear and plastid genomes of the alga *Chlamydomonas reinhardii* by a combination of transmission genetic analyses and biochemical methods.

Another approach has been taken to map genes on the *Zea mays* chloroplast chromosome. By physical methods genes were located for 15S, 23S and 5S chloroplast ribosomal RNAs, the large subunit (LS) of ribulose-1,5 bisphosphate carboxylase (RuBPcase) and for a 32,000 dalton protein of the photosynthetic membrane (Photogene 32). The LS RuBPcase gene is now shown to be 1600 base pairs long and uninterrupted. A 4,350 base pair-long maize chloroplast DNA sequence cloned in *E. coli* as part of chimeric plasmid carries the gene for all of the LS RuBPcase mRNA as well as for a 1400 base pair-long uninterrupted sequence constituting part of a 2200 nucleotide-long RNA; these two genes, which are expressed differently in chloroplasts of two cell types in maize leaves, lie on the maize chloroplast chromosome head-to-head. Their 3' ends are separated by 330 base pairs of DNA which appears to be untranscribed.

[1]This work was supported in part by research grants, from the National Institute of General Medical Sciences, and from the National Science Foundation as well as from the Maria Moors Cabot Foundation of Harvard University.

ISBN 0-12-198780-9

INTRODUCTION

Mature chloroplasts of green plants conduct photosynthesis and other biochemical functions essential for the life of the cell and the organism. Plastids of higher plants and some mutant algae are immature in dark-grown organisms. They mature into chloroplasts only upon illumination. The array of differentiated forms of plastids is greatest in higher plants. As examples: proplastids are the primordial plastids in meristems; chromoplasts, e.g. in ripe tomato skins, contain crystals of carotenoids; amyloplasts are plastids for starch storage. Each different specialization is an opportunity for studying functions of extrachromosomal DNA in the plastid, mechanisms of intergenomic integration in eukaryotic cells, and molecular devices for the control of gene expression.

Structural genes for a few chloroplast ribosomal proteins (rProt) have been mapped by transmission genetics in the alga *Chlamydomonas reinhardii*. More recently structural genes for two chloroplast proteins and for chloroplast rRNAs in *Zea mays* chloroplast DNA have been identified and mapped by physical methods. The latter methods are applicable to a wider range of species.

A few years ago we searched for structural genes for chloroplast rProts, by isolating a group of an erythromycin-resistant mutants of the alga *C. reinhardii* and studying the patterns of transmission of the resistance traits and the properties of ribosomes and rProts of wild-type and erythromycin-resistant strains (1,2). Among the nine examined initially, one mutation proved to be transmitted uniparentally and was designated ery-U 1a. This mutation resulted in an alteration in chloroplast rProt LC4 (2,3) and has since been mapped to a locus on the chloroplast genome by Boynton *et al* (4).

Several Mendelianly transmitted erythromycin-resistant mutations have also been identified. The best studied of these, ery-M 1a, b, c, and d, all map to one place, about 15 map units from the centromere, on nuclear linkage group XI (5). The same chloroplast rProt, LC6, is altered in all four ery-M1 mutuants; a total of three different alterations of LC6 have been distinguished among mutant strains examined (5). The conclusion that the locus ery-M1 on linkage group XI is the site of the structural gene for LC6 was further strongly supported by the analysis of chloroplast rProts of vegetative diploids constructed between wild-type and ery-M1 mutants of *C. reinhardii* (6). Each heterozygous diploid contained both the wild-type and mutant forms of LC6. Ribosomes from such diploids generally contain 60-70 percent of the wild-type form of protein LC6 and 30-40 percent of the altered form of this rProt. (Three additional mutants that map to the ery-M1 locus have also been identified (7).)

Two other groups of Mendelianly transmitted erythromycin-resistant mutations, both unlinked to ery-M1, have been studied. Mutants of class ery-M2 have a normal LC6 rProt but another protein of the large subunit of the chloroplast ribosome is altered (2); five mutants have been isolated and the mutation has been mapped to a position 15-16 map units from the centromere of an as yet unidentified nuclear linkage group (7). Mutants in the group ery-M3 map to a third linkage group (also unidentified) about 30 map units from the centromere (7).

The identification of nuclear genes for chloroplast rProts has been extended by isolating reverants of ery-M1b to erythromycin sensitivity (8). One gene for reversion to erythromycin sensitivity, es101, is located on the same linkage group as ery-M1; it is about 7 map units from the centromere on linkage group XI lying between the loci pf-2 and ery-M1. Among the other suppressor mutations are two unlinked to ery-M1. These and other reports on research into chloroplast ribosomal protein genes have been reviewed recently by Gillham (9).

Genes for rProts of the large ribosomal subunit, discussed above in *C. reinhardii*, are scattered among several nuclear linkage groups plus the chloroplast chromosome. Mechanisms by which the production of chloroplast rProts in the nuclear-cytoplasmic and organelle compartments are integrated are yet to be studied--let alone understood. To begin to explain this and other similar cases of intergenomic integration in eukaryotic cells we need to know about the organization and expression of genes in the organelle genome. Transmission mapping of genes in the chloroplast genome is limited to *C. reinhardii* at the present time. Although uniparental transmission of plastid genes is well known in a number of plants, *C. reinhardii* is the only organism in which chloroplast genes can be mapped in this way and, for practical reasons, mapping of nuclear genes for chloroplast components is also limited to *Chlaymdomonas* at present. Consequently, recent developments in molecular biology and molecular genetics, especially the advent of recombinant DNA technology, have been important in permitting the identification and mapping of chloroplast genes. We have employed methods of *in vitro* genetics to locate genes in the chloroplast genome of *Zea mays*. The current status of this work is presented here.

Thus far, maize genes for 5S, 23S and 16S rRNAs, the large

subunit (LS) of ribulose-1,5-bisphosphate (RuBPcase)--a key enzyme of the photosynthetic carbon reduction cycle, the gene for a 2.2 kilobase (KB) transcript present in maize cpRNA and the gene for a 32,000 dalton chloroplast photosynthetic membrane (thylakoid) protein (Photogene 32) have been identified and mapped on the maize chloroplast chromosome.

RESULTS

A physical map of Zea mays chloroplast DNA. The maize chloroplast chromosome is an 85 $\times 10^6$ dalton circle (10,11). All of the 11 recognition sites for the endonuclease Sal I as well as some sequences recognized by Eco-RI and Bam HI have been located on the chromosome (Fig. 1) (12). As a result of this work two inverted repeated segments were identified. Furthermore, the observation that the Sal I-generated fragments sum to 85 x 10^6 daltons indicates that all of the 30-60 chromosomes present in each of the 40-60 chloroplasts each cell are the same! (One fragment was found to occur in 80 percent of the chromosomes. This could be an experimental error or, in fact, 80 percent of the chromosomes could in fact be different from the rest. This variation was found in the 6.3 KB fragment Sal G.)

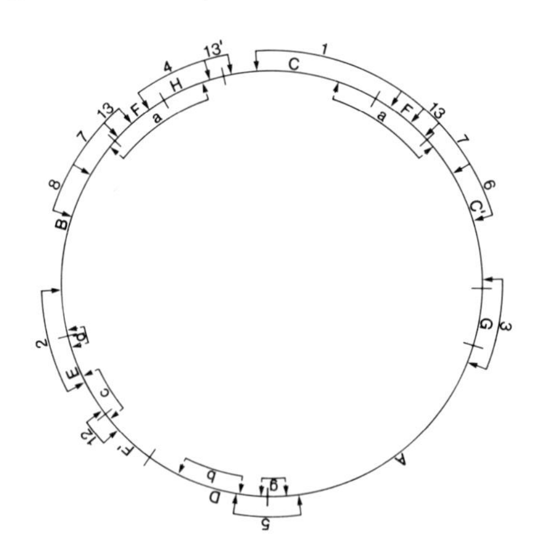

FIGURE 1. A map of recognition sites on the maize chloroplast chromosome for endonuclease Sal. I (straight lines across the solid line), Bam HI (arrows pointing into the circle) and Eco RI (arrows pointing outward). Fragments produced by Sal. I are designated by capital letters, those produced by Eco RI by lower case letters, those by Bam HI by arabic numerals [Adapted from (12).]

Genes for chloroplast rRnas. Restriction fragments of chloroplast DNA (cpDNA) have been cloned in *E. coli* using various bacterial plasmid vehicles. The genes for 5S, 23S, and 16S chloroplast rRNAs have been located on the largest fragment produced by Eco RI digestion, Eco *a*. Two copies of Eco RI fragment *a* are present per chloroplast chromosome; one on each inverted repeat. Thus there are two copies of each of these rRNA genes per chromosome.

The arrangement of the 5S, 23S and 16S genes on the chromosome, as well as the orientation of two sets of rRNA genes with respect to one another, has been established by mapping of restriction endonuclease recognition sites; by hybridization of total or separated *in vitro* ^{32}P-labelled rRNAs to restriction fragments; by DNA-RNA heteroduplex mapping; and by R-loop analyses (13).

The 16 and 23S sequences in each unit are separated by a 2.1 KB pair spacer and the 5S rDNA is close to that for the 23S rDNA. The two rDNA units on the circular maize chloroplast cpDNA molecule are separated by 18.5 KB in one direction and by 106.1 KB in the other direction; they have an inverted orientation with respect to each other (Fig. 2).

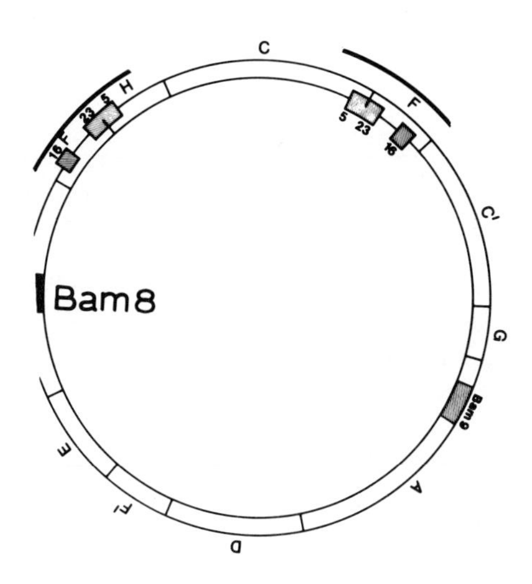

FIGURE 2. The two concentric circles represent the two DNA strands of the maize chloroplast chromosome. Sal I recognition sites are shown by lines connecting the two circles. Locations are shown of genes for 55, 23S and 16S rRNAs as well as of fragment Bam 9 (containing the gene for LS RuBPcase and part of the 2.2 KB gene) and the part of fragment Bam 8 containing Photogene 32.

The gene for the large subunit of ribulose-1, 5 bisphosphate carboxylase. RuBPcase is an important and abundant enzyme in photosynthetic organisms. RuBPcase catalyzes the production of 2 molecules of 3-phosphoglyceric acid from one molecule of CO_2 and one molecule of ribulose-1, 5bisphosphate. The phosphoglyceric acid is subsequently reduced to form phosphoglyceraldehyde, the first sugar produced in photosynthesis. The plant RuBPcase is comprised of two different sized subunits. In maize the larger of these is a 52,000 dalton polypeptide. At the time this work was begun there was evidence from its uniparental transmission in tobacco that the large subunit (LS) might be the product of a chloroplast gene and that the small subunit of the enzyme was most likely a nuclear gene product because it is transmitted in the normal Mendelian manner (14).

Maize DNA sequence Bam 9 cloned after insertion into the bacterial plasmid RSF 1030 directs the synthesis *in vitro* of an LS-sized polypeptide (15) that can be precipitated by an antibody produced by rabbits immunized with RuBPcase. The linked *in vitro* transcription-translation system used to transform the information from DNA into a polypeptide was comprised of *E. coli* RNA polymerase (for transcription) with a rabbit reticulocyte lysate (for translation). ^{35}S-methionine was included to label the polypeptide products. Proteolytic fragments of ^{35}S-methionine-labeled **LS** isolated from maize plants correspond to those obtained by digestion of the immunoprecipitated *in vitro*-synthesized polypeptides.

The DNA sequence Bam 9 is 4.3 KB pairs long. Since the minimum coding length for a polypeptide of 52,000 daltons is about 1500-1600 KB pairs, Bam 9 could contain 2 copies of the LS genes. A map of recognition sites for Bgl II, Eco RI and Pst I i.e. a restriction map, was first constructed. Then Bam 9 was digested with various restriction endonucleases and the ability of each digest to direct the synthesis of the full 52,000 dalton LS polypeptide was studied (16). From the results of these experiements and a knowledge of the restriction map, it was possible to determine that a single copy of the structural gene for LS lies in a 2.5 KB pair-long portion of Bam 9 delimited by a recognition site for Bgl II at one end and a site for Sma I at the other end (16). One copy of this DNA sequence is present in each maize chloroplast chromosome.

We have taken advantage of the availability of cloned Bam 9 (pZmc 37) and of a new clone, pZmc 3711, containing the 2.7 KB pair-long Bgl-Bgl unit within Bam 9 cloned in RSF 1030, to study the structure of the LS gene in detail. Since, as best we could judge the Bgl-Bgl fragment contains only (or virtually only) the LS gene we could use unfractionated chloroplast RNA in the nuclease S1 protection method of Berk and Sharp (17) to compare the mRNA for LS with the DNA sequence from which it is transcribed, i.e. to measure the size of the mRNA, by finding the amount of DNA protected by hybridization, and to determine whether the mRNA is colinear with an uninterrupted or a split DNA sequence. In these experiments pZmc 37 digested with Bam I or pZmc 3711 digested with Bgl II was incubated with total chloroplast RNA under conditions favoring RNA-DNA hybridization. After digestion with nuclease S1 and separation of surviving DNA fragments on alkaline agarose gels by electrophoresis, the DNA was transferred to nitrocellulose filter sheets (18) and located by hybridization with ^{32}P-RNA transcribed by _E. coli_ RNA polymerase from the 2.7 KB Bgl-Bgl fragment recovered from pZmc 3711, i.e. by hybridization with radioactive copy RNA (cRNA). As shown in Figures 3 and 4, a 1.6 KB uninterrupted DNA sequence was found. This shows that the functional transcript for LS is colinear with a 1.6 KB uninterrupted DNA sequence. If there are any sequences in the gene which are absent from the mRNA, they must be very near one or both ends.

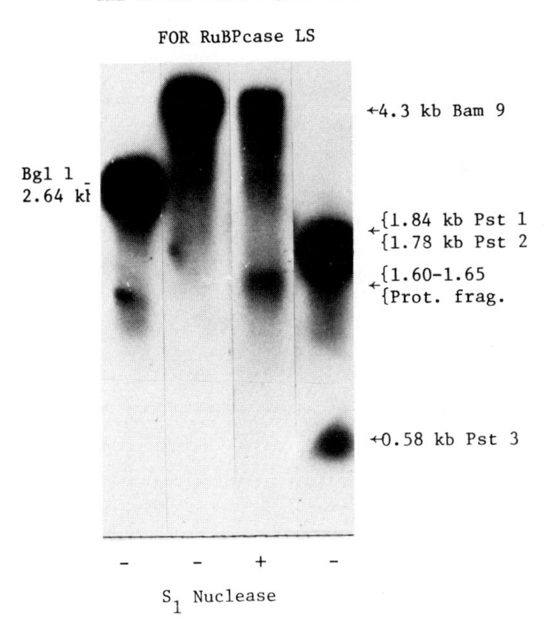

FIGURE 3. An autoradiograph showing hybridization of ^{32}P-cRNA of the Bgl-Bgl 2.7 KB portion of Bam 9 to DNA fragments transferred to a nitrocellulose filter sheet after electrophoresis on an alkaline agarose gel. Left: the

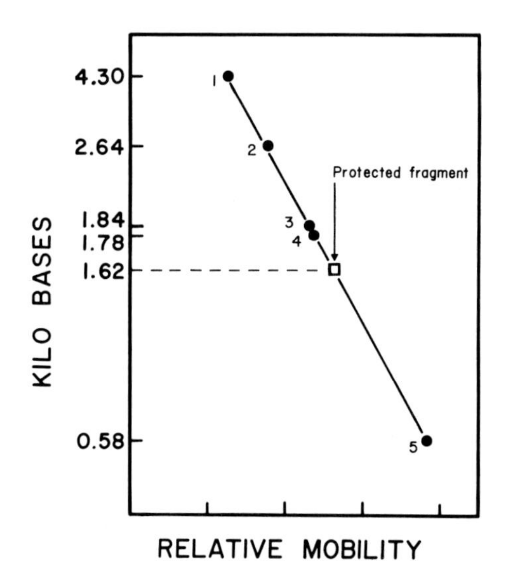

FIGURE 4. A plot of sizes of DNA markers vs relative electrophoretic mobility to determine the size of the protected fragment in Figure 3.

Bgl-Bgl fragment. Second track: Bam 9. Third track: Bam 9 digested with nuclease S1 showing the locating of the 1.6 KB DNA sequence protected from nuclease digestion by hybridization with cpDNA. Right: the Bgl-Bgl fragment shown in the left-and right-hand tracks were included as size standards (see Fig. 4). The material at the 4.3 KB position in track 3 was not digested by S1 because of DNA-DNA hybridization. Some partially digested sequences are also visible.

The conclusion that the LS gene is not interrupted by non-coding sequences is supported by the observation (15,16) that the cloned cpDNA sequence Bam 9 directs the synthesis of the LS polypeptide in the *E. coli*-rabbit reticulocyte linked transcription-translation system, which is unlikely to be capable of processing transcripts of the natural chloroplast gene. LS is the first structural gene for a cpDNA-coded protein to be identified, mapped and measured.

In the subsequent experiments, the 2.7 KB Bgl-Bgl fragment (from pZmc 3711) or the 4.3 KB Bam 9 fragment was digested with the endonuclease Pst I, which yields three major fragments (plus a very small one derived from one end only of Bam 9). An unresolved mixture of these fragments was hybridized with total cpRNA, treated with nuclease S1, electrophoresed in agarose, and transferred to nitrocellulose filter paper. Specific ^{32}P-cRNA probes from the right-, left-hand or center portion of the cpDNA fragment were used to determine from which of the three major Pst-generated sequences were used to determine each DNA fragment that survived S1 digestion (because of hybridization with RNA) was derived.

These specific probes were prepared by first isolating, either by agarose gel electrophoresis or sucrose density centrifugation, individual DNA sequences produced from Bam 9 or from the Bgl-Bgl insert by digestion with Pst I or Eco RI. ^{32}P-cRNA was prepared separately from each DNA sequence and used as a hybridization probe to permit assignment of each surviving DNA sequence to a position on the cloned cpDNA sequence and, consequently, on the chloroplast chromosome. It has been concluded from these experiments that the LS gene commences 0.04 KB to the left of Eco RI site B (Fig. 5) and continues without interruption to a site 0.06 KB just before (i.e. to the left) of Bgl site B (19).

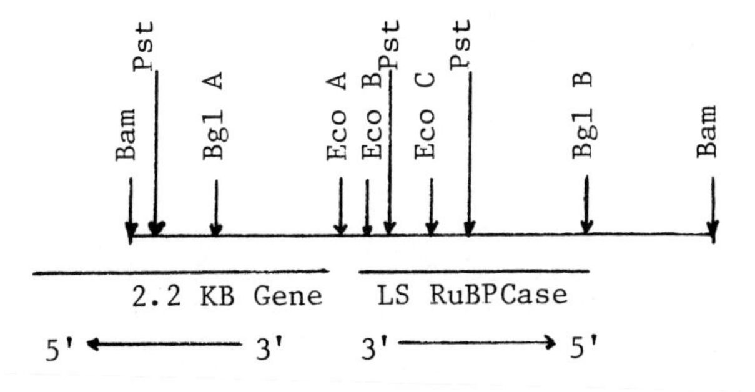

FIGURE 5. A diagram of the location of LS RuBPcase and 2.2 KB RNA genes on Bam 9.

To determine the direction of transcription of the LS gene, the Bgl-Bgl fragment from pZmc 3711 was digested with Pst I and the 3' ends of the resulting stretches of DNA were labeled. After hybridization with total cpRNA and S1 treatment, the surviving DNA fragments were separated electrophoretically on agarose. By determining which fragments retain the ^{32}P label. The conclusion that the LS gene is transcribed from left to right as drawn in Figure 5 has been reached by determining which fragments retain ^{32}P at their 3' end (19).

Availability of the cloned LS gene and knowledge of its location on the cloned fragment Bam 9 have been exploited to study the expression of this gene in chloroplasts of mesophyll and bundle sheath cells, two types of cells in the leaves of maize which differ in their carbon dioxide fixing activities. This plant and others of its photosynthetic type, i.e. C_4 plants, first fix carbon dioxide into oxalacetate by a phosphoenolpyruvate carboxylase catalyzed reaction in mesophyll plastids. Malate or asparate derived from the oxalacetate is transported to bundle sheath cells where it is decarboxylated and the carbon dioxide is refixed, this time into phosphoglyceric acid by RuBPcase. Mesophyll cells lack the latter enzyme.

RNA from maize bundle sheath cells directs the synthesis of the LS RuBPcase in a cell-free rabbit reticulocyte translation system; mesophyll cell RNA does not. Within the limits of the detection no mRNA for LS could be detected in mesophyll cell RNA (labeled with ^{32}P in vitro after isolation) by hybridization with the 2.7 KB Bgl-Bgl sequence of Bam 9 (produced by digesting pZmc 37 with BgII, separating the fragments produced by agarose gel electrophoresis, and transferring the DNA to nitrocellulose filter sheets). On the other hand, mRNA for LS was readily detected in bundle sheath RNA in the same assay (20). Furthermore, hybridization of in vitro labeled Bgl-Bgl DNA with DNA from mesophyll and from bundle sheath cells showed the LS gene to be present in the chloroplast chromosomes of both cell types (20).

The 2.2 KB RNA gene. In the course of studying differences between RNAs of bundle sheath and mesophyll cells, it was discovered that some RNA from both types of chloroplasts hybridized to the 4.3 KB Bam 9 DNA but only that from bundle sheath cells hybridized to the smaller (2.7 KB) Bgl-Bgl or (2.5 KB) Sma-Bgl fragment of Bam 9. We concluded that Bam 9 in fact contains coding sequences for two transcripts present in cpRNA (20).

Mesophyll cell RNA contains a little if any of a transcript that hybridizes to the Sma-Bgl fragment but it does contain RNA that hybridizes to some other portion of Bam 9--a region outside of the LS gene. In the course of analyzing transcripts in cpRNA which protect portions of Bam 9 from S1 digestion--for the work described above in examing the LS gene in detail--a 1.4 KB sequence running to the left end of Bam 9 (Figure 5) was found to be protected. Since this 1.4 KB stretch could represent only part of the gene, we determined the length of this transcript by an independent method. Chloroplast RNA was denatured by glyoxylation, separated electrophoretically, and transferred to DBM paper using the method of Alevine _et al_. (21). Nick-translated DNA derived from a fragment within the 1.4 KB protected sequence, serving as a probe, hybridized to a 2.2 KB RNA (19). Again using the 3' labeling and RNA-DNA hybridization-S1 protection method we have determined that this gene is transcribed in the opposite direction from the LS gene and that a 0.33 KB DNA sequence intervenes between the 3' ends of these two genes (19). At least one binding site for maize RNA polymerase (22) has been identified in this 0.33 KB region (L. McIntosh, S. O. Jolly and L. Bogorad, unpublished).

Photogene 32. When germinated and grown in darkness (on seed reserves) plastids in maize seedlings develop to the characteristic chlorophyll-less "etioplast" stage. Some but not all thylakoid membrane proteins of mature, photosynthetically functional chloroplasts are present in etioplasts. Chlorophyll synthesis commences within a few minutes after illumination begins and thylakoid proteins required for acquisition of photosynthetic competence are subsequently made and set into place (23,24). Isolated maize plastids produce a 34,500 dalton polypeptide (26) which appears to be processed to the 32,000 dalton form. The latter is prominent among the several thylakoid polypeptides that accumulate during photo-induced plastid development (24).

A 34,500 dalton polypeptide is also a conspicious product of cell-free protein synthesis (in a rabbit reticulocyte lysate) directed by cpRNA from green or greening maize but is absent (or virtually so) when etioplast cpRNA is used. RNAs were isolated from etioplasts, greening plastids, or mature chloroplasts, end-labeled with ^{32}P-_in vitro_, and hybridized with endonuclease-digested cpDNA separated by agarose gel electrophoresis and transferred to nitrocellulose filters. A species of RNA present in mature or maturing plastids hybridizes to cpDNA Bam fragment 8 but is lacking (or virtually so) from RNA isolated from etioplasts (25).

The location of Bam fragment 8 on the maize chloroplast chromosome is shown in Figure 2. A 2.2 KB Bam-Eco Ri portion of the 4.3 KB-long Bam 8 that has been inserted into pBR 322 and cloned in E. coli has been found in preliminary experiments to direct the synthesis in a linked transcription-translation system of a polypeptide whose proteolytic products are like those of the 32,000 dalton polypeptide of membrane polypeptides. (L. McIntosh and L. Bogorad, unpublished).

A 1.3 KB DNA sequence within the cloned Bam-Eco R1 DNA is protected against digestion with nuclease S1 by hybridization with cpRNA. Thus the mRNA for this polypeptide, like that for LS RuBPcase, is colinear with an uninterrupted cpRNA sequence. A binding site for chloroplast RNA polymerase has been roughly located outside of this apparent coding sequence. (L. McIntosh, S. O. Jolly, G. Link and L. Bogorad, unpublished).

DISCUSSION

Two regions of the maize chloroplast chromosome have now been analyzed in detail. The duplicated sequence containing rRNA genes was described earlier (13). The second region, initially identified because it includes the gene for the first protein mapped on any chloroplast chromosome: LS RuBPcase (15,16), contains the entire LS gene plus part of at least one additional gene.

The research reported has shown that based on its colinearity with cpRNA, the maize LS gene is uninterrputed by non-coding sequences. The functional transcript is 1.6 KB long, i.e. just about the minimum DNA sequence length needed to code for a 52,000 dalton protein. No precursor RNA of greater length has been detected by separating a cpRNA mixture electrophoretically and probing with a 0.58 KB DNA sequence from the middle of the gene. If there is a larger precursor, it would seem more likely to be extended at the 3' than the 5' end of the transcript. The 3' end of the LS gene is just 0.33 KB from the 3' end of the gene for the 2.2 KB transcript adjacent to the 5' end of the gene. (We do not know at this time what the lowest limit of detection may be nor have we examined cpRNAs prepared from variously differentiated plastids.)

The genes which have been mapped on the maize chloroplast chromosome fit into a few expression classes. First, transfer of dark-grown maize seedlings to light results in a quantitative increase in the transcription of chloroplast rRNAs without significant effect on the rate of incorporation of ^{32}P-phosphate into cytoplasmic rRNAs (23). Second, light controls the expression of Photogene 32 more absolutely (25).

Third, the expression of the maize gene for LS RuBPcase is somehow differently regulated in mesophyll and bundle sheath cells of maize; it is not transcribed in the former but is in the latter (20). Finally, the 2.2 KB transcript of the gene partially located on the cloned cpDNA sequence Bam 9, and separated from the LS gene by 0.33 KB of DNA, is expressed in both mesophyll cells which lack LS mRNA as well as in bundle sheath cell plastids which have LS mRNA (19).

The techniques which have been employed to date for physically locating genes on the maize chloroplast chromosome will find future use in locating the 20-30 tRNA genes plus the perhaps 50 or so additional protein structural genes which this chromosome may contain. The availability of cloned genes of different developmental classes as well as purified chloroplast DNA-dependent polymerase when joined with techniques for determining DNA sequences should open the way toward understanding DNA structures and mechanisms for the control of gene expression in this system. With other students of extrachromosomal DNA, we look forward to using such knowledge to design experiments which will reveal the mechanisms by which the expression of nuclear and chloroplast genes are coordinated for the production of multimeric chloroplast structures such as ribosomes, enzymes and thylakoids (27).

ACKNOWLEDGMENTS

We are indebted to Mrs. Brigitte Link, Mrs. Linda Scrafford-Wolff and Miss Susan Adam for their skilled technical assistance which permitted this work to be done. We are also deeply indebted to our former colleagues: John R. Bedbrook, Donald M. Coen and Richard D. Kolodner whose pioneering work on the maize chloroplast chromosome formed an essential foundation for our new findings; Jeffrey N. Davidson, Maureen R. Hanson and Laurens J. Mets whose deep analyses of chloroplast ribosomal protein genetics provided important insights into genome interactions; Harriet Jane Smith and Warwick Bottomley whose early work on the maize chloroplast DNA-dependent-RNA polymerase opened the way to another facet of chloroplast molecular biology.

REFERENCES

1. Mets, L., and Bogorad, L. (1971). Science 174,707.
2. Mets, L., and Bogorad, L. (1972) Proc. Natl. Acad. Sci. U.S.A. 69,3779.
3. Hanson, M. R., Davidson, J. N., Mets, L. J., and Bogorad, L. (1974) Molec. Gen. Genet. 132,105.
4. Boynton, J. E., Gillham, N. W., Harris, E. H., Tingle, C. L., Van Winkle-Swift, K., and Adams, G. M. (1976).

In "Genetics and Biogenesis of Chloroplasts and Mitochondria" T. H. Bucher, W. Neupret, and S. Werner, (eds.), pp. 312-322. North Holland Press, Amsterdam.
5. Davidson, J. N., Hanson, M. R., and Bogorad, L. (1974). Molec. Gen. Genet. 132,119.
6. Hanson, M. R., and Bogorad, L. (1977). Molec. Gen. Genet. 153,271.
7. Davidson, J. N., Hanson, M. R., and Bogorad, L. (1978). Genetics 89,281.
8. Davidson, J. N., and Bogorad, L. (1977). Molec. Gen. Genet. 157,39.
9. Gillham, N. W. (1978). "Organelle Heredity" Raven Press New York.
10. Kolodner, R. D., and Tewari, K. K. (1975). Biochem. Biophys. Acta 402,372.
11. Hermann, R. G., Bohnert, H-J., Kowallik, K. D., and Schmidtt, J. M. (1975). Biochem. Biophys. Acta 378,305.
]2. Bedbrook, J. R., and Bogorad, L. (1976). Proc. Natl. Acad. Sci. U.S.A. 73,4309.
13. Bedbrook, J. R., Koldner, R., and Bogorad, L. (1977). Cell 11,739.
14. Kung, S. D. (1977). Ann. Rev. Plant Physiol. 28,401.
15. Coen, D. M., Bedbrook, J. R., Bogorad, L., and Rich, A. (1977). Proc. Natl. Acad. Sci. U.S.A. Vol. 74, 5487.
16. Bedbrook, J. R., Coen, D. M., Beaton, A. R., Bogorad, L., and Rich, A. (1979). J. Biol. Chem. 254,905.
17. Berk, A. T., and Sharp, P. A. (1977). Cell 12,721.
18. Southern, A. M. (1975). J. Mol. Biol. 98,503.
19. Link, G., and Bogorad, L., in preparation.
20. Link, G., Coen, D. M., and Bogorad, L. (1978). Cell 15,725.
21. Alewine, J. C., Kemp, D. J., and Stark, G. R. (1977). Proc. Natl. Acad. Sci. U.S.A. 74,5350.
22. Smith, H. J., and Bogorad, L. (1974). Proc. Natl. Acad. Sci. U.S.A. 71,4839.
23. Bogorad, L. (1968). In, "Control Mechanisms in Developmental Processes" The 26th Symposium of the Soc. for Devel. Biol. pp. 1-31. Academic Press, New York.
24. Grebanier, A. A., Steinback, K. E., and Bogorad, L. (1979). Plant Physiol. 63,436.
25. Bedbrook, J. R., Link, G., Coen, D. M., Bogorad, L. and Rich, A. (1978). Proc. Natl. Acad. Sci. U.S.A. 75,3060.
26. Grebanier, A. A., Coen, D. M., Rich, A., and Bogorad, L. (1978). J. Cell. Biol. 78,734.
27. Bogorad, L. (1975). Science 188,891.

CHLOROPLAST DNA OF *EUGLENA GRACILIS* GENE MAPPING AND SELECTIVE IN VITRO TRANSCRIPTION OF THE RIBOSOMAL RNA REGION [1]

Richard B. Hallick, Keith E. Rushlow, Emil M. Orozco, Jr., Gary L. Stiegler and Patrick W. Gray[2]

Department of Chemistry, University of Colorado, Boulder, Colorado 80309

ABSTRACT. Purified chloroplast 5S, 16S, and 23S rRNAs and tRNAs were labeled *in vitro* and hybridized by the Southern method to ctDNA restriction fragments and cloned ctDNA fragments. Transfer RNA coding loci were found scattered throughout the genome, as well as in the rRNA coding region. 5S, 16S and 23S rRNAs and tRNA(s) all hybridized to the same strand of a 5.6 kbp DNA, repeated three times in tandem on the chloroplast genome. The RNA product transcribed *in vitro* by *Euglena* chloroplast RNA polymerase in a purified, transcriptionally active complex with ctDNA has been characterized as largely a selective transcript of the rRNA coding region. In preliminary experiments, sequence homology between *Euglena* chloroplast DNA and a fragment coding for the large subunit of RuDP carboxylase in *C. reinhardii* has been mapped.

INTRODUCTION

The chloroplast DNA of *Euglena gracilis* is a circular, duplex molecule of 130-140 kbp (1,2). The major RNA transcripts of this genome are the 16S and 23S ribosomal RNAs (3,4), 5S rRNA (5), and approximately 25 tRNAs (6,7). These stable, abundant RNAs account for nearly one fourth of the 60-65 kb of chloroplast DNA transcribed into RNA in light grown *Euglena* (8,11). Among the remaining transcripts are presumably mRNAs for chloroplast coded polypeptides, such as

[1]This work was supported by NIH Grant GM21351. R.B.H. is recipient of NIH Research Career Development Award K04-GM 00372.

[2]Present address: Department of Biochemistry and Biophysics, University of California at San Francisco, San Francisco, California 94143.

ISBN 0-12-198780-9

the large subunit of ribulose-1,5-bisphosphate carboxylase, which is encoded in chloroplast DNA in other organisms (9,10).

One of our major goals has been to understand the mechanisms involved in the control of chloroplast transcription in *Euglena*. In studies on chloroplast transcription mechanisms, a detailed knowledge of chloroplast DNA structure and gene location is essential. We have previously described a restriction endonuclease cleavage map of the *Euglena* chloroplast genome (2). To this map we can now add 28 sites on the circular map, and additional detailed mapping in the rRNA coding region. The positions of the 5S rRNA, and several tRNA coding regions have been identified, and a more detailed description of the rRNA coding region has been obtained. With this restriction map as reference, we have been able to locate and characterize the RNA transcripts made *in vitro* from a previously described (12,13) transcriptionally active chromosome purified from *Euglena* chloroplasts.

RESULTS

Restriction Endonuclease Cleavage Map of Euglena Chloroplast DNA. A composite restriction endonuclease cleavage map of *Euglena* chloroplast DNA is shown in Figure 1. Several features have been added to our previously published map. The four fragments from KpnI and six fragments from PvuII digestion have been positioned. Also shown in Figure 1 is a nearly complete EcoRl cleavage site map. The nomenclature and size estimates for these fragments has been described (14). There are at least 28 different EcoRI fragments, designated EcoA-Z (including J1 and J2), and Eco aa, ranging in size from EcoA (21 kbp) to Eco aa (0.5 kbp). We have constructed and characterized a series of recombinant DNAs between different chloroplast EcoRl fragments and the plasmid vector pMB9. Several plasmids have been particularly useful in the gene mapping studies described below. There have been three independent reports that a 5.6 kbp DNA segment, which codes for both 16S and 23S chloroplast rRNA, is repeated three times in tandem on the chloroplast genome (2,15,16). The location of the rRNA coding region is shown in Figure 1.

Chloroplast tRNA Genes. For mapping *Euglena* chloroplast tRNA coding loci, total chloroplast tRNAs were isolated from purified chloroplasts. Following mild digestion with snake venom phosphodiesterase to activate the 3'-ends, the tRNAs were radioactively labeled with [α^{32}P]CTP by the action of *E. coli* tRNA nucleotidyl transferase. [^{32}P]tRNAs were

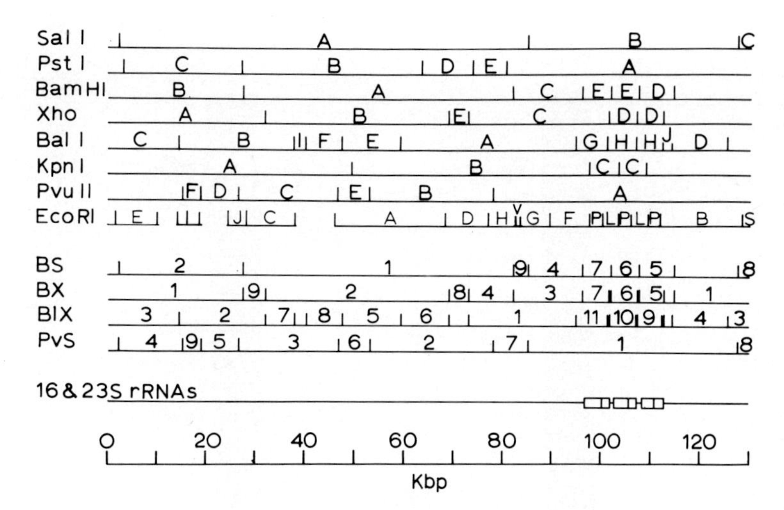

FIGURE 1. Restriction Endonuclease Cleavage map of Euglena Chloroplast DNA. The linear map is derived by opening the circular map in the middle of fragment SalC. Nomenclature and sizes of many fragments have been reported (2,14). BS, BX, B1X and PS are double digestion products from the enzyme combinations BamHI-SalI, BamHI-XhoI, BalI-XhoI, and PvuII-SalI, respectively (2).

hybridized to chloroplast DNA restriction fragments by the method of Southern (17). Products from the following enzyme digestions and multiple digestions were analyzed for tRNA hybridization: EcoRI, EcoRI-SalI, EcoRI-BamHI, BamHI-SalI-XhoI, BalI-XhoI, PvuII, and PvuII-SalI. In figure 2, results from tRNA hybridization to PvuII-SalI, and to BamHI-SalI-XhoI digestion products are shown. Strong hybridization is obtained to the fragments SalB, PvuB, D, E, and F, and PS-4 in lane 1, and to fragments XhoB, BS-2,4,9, and the identical BX-5,6,7 in lane 2. In some experiments, weak but distinct hybridization was also observed, such as with PvuC in lane 1. Weak hybridization could be due to a low concentration of a particular tRNA species in our [^{32}P]tRNA, to a low efficiency of labeling *in vitro*, or both.

In constructing a composite tRNA map from the results of all hybridization experiments, we started with areas of strong hybridization, and subtracted areas from overlapping fragments with either no or weak hybridization. This is

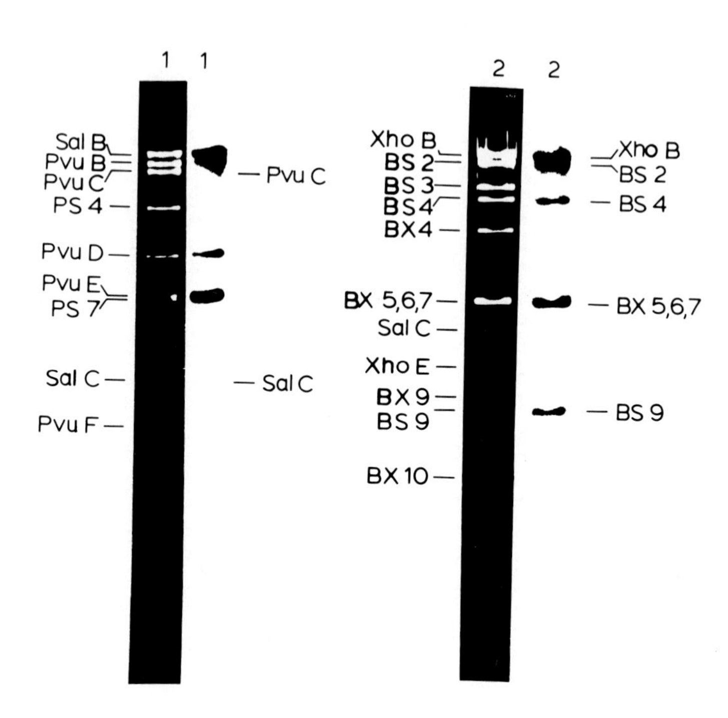

FIGURE 2. Hybridization of Euglena chloroplast tRNAs to chloroplast DNA restriction fragments. Digests of ct DNA are 1. PvuII-SalI and 2. BamHI-SalI-XhoI. The map positions of the resulting products are shown in Figure 1. Both the ethidium-bromide stained gel (left) and autoradiogram (right) are shown for each lane.

diametrically illustrated in Figure 3. For example, with the EcoRI, EcoRI-BamHI, and EcoRI-SaI fragments of chloroplast DNA, strong hybridization was observed to EcoA. PvuE, which is within EcoA, also gives strong hybridization (Figure 2). Two fragments which each partially overlap EcoA, PvuC and B1X6, only hybridized weakly to tRNA. The composite map showing all the areas of strong, weak, and no hybridization is shown in Figure 3.

The most precise mapping of tRNA genes has been possible in EcoRI fragments EcoG and EcoP, where the tRNA genes have been positioned to within $\leq$ 0.7 kbb. In these cases recombinant plasmids were available for the tRNA hybridization experiments. The tRNA genes in EcoP are described below.

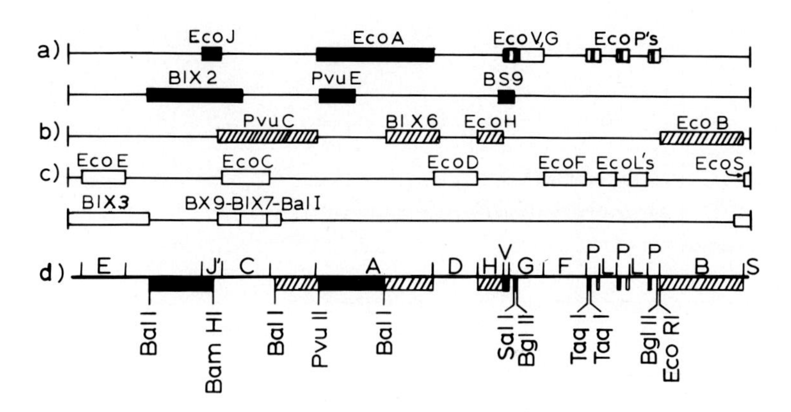

FIGURE 3. Location of Regions on the Euglena chloroplast DNA which hybridize with [^{32}P]chloroplast tRNA. Shown are areas of a) strong hybridization (filed boxes), b) weak hybridization (dashed boxes), c) no hybridization (open boxes), and d) a composite map. The fragments are designated according to the nomenclature in Figure 1.

Gene Organization in the Chloroplast rRNA Gene Region. Hybridization of [^{32}P]tRNA to EcoP, and to other restriction nuclease fragments containing rRNA genes, such as BX5,6, and 7 (Figure 2) was observed in all experiments. A recombinant plasmid, pPG50, containing the chloroplast DNA fragments EcoP and Ecoaa, and the plasmid vector pMB9, was chosen for detailed mapping of the tRNA genes in the rRNA coding region. EcoP is known to contain the entire 16S rRNA gene, an intergenic spacer, and part of the 23S rRNA gene (2,5,15,16). A restriction nuclease map of EcoP with the location of TaqI, XbaI, HincII, HindIII, and BglIII cleavage sites is shown in Figure 4. [^{32}P]tRNAs, as well as [^{125}I]16S rRNA and [^{125}I]-23S rRNAs were hybridized by the method of Southern (17) to pPG50 DNA, which had been digested with TaqI, EcoRI-TaqI, and/or HindIII-TaqI. The results are summarized diagramatically in Figure 4.

The tRNA coding region maps in the intergenic spacer region between the 16S and 23S rRNA genes. It is possible that a second tRNA coding region is present in EcoP, preceeding the 16S rRNA gene. Weak hybridization with [^{32}P]tRNA is consistently obtained with DNA fragments containing sequences to the right of the BGlII site in EcoP. We also find that the start of the 16S rRNA gene is >300bp from the end of

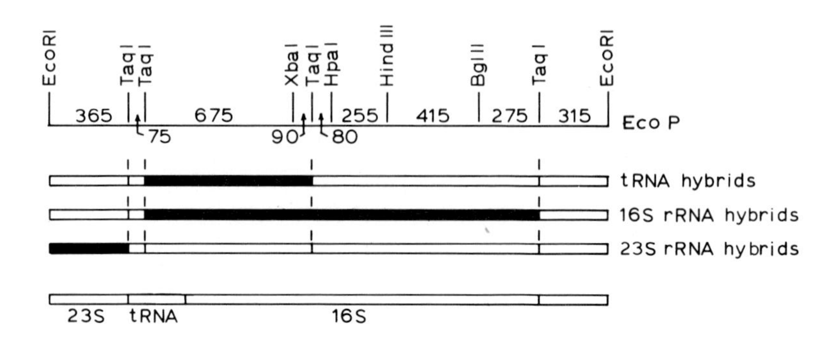

FIGURE 4. Restriction Endonuclease cleavage map of EcoP. The cleavage sites and resulting fragment sizes in bp for the enzymes BglII, HincII, HindIII, TaqI, and Xbal are shown. Fragment size estimates are based on electrophoretic mobility in 3.5% acrylamide gels with HaeIII digested ØX174 DNA fragments as size markers. Regions hybridizing with chloroplast tRNA, 16S rRNA, 23S rRNA and the resulting composite map are shown.

EcoP. Recently the 5S rRNA genes have been located at the distal end of the 23S rRNA gene in each 5.6kbp rRNA repeat (5). A summary of the gene mapping data for the rRNA coding region is shown in Figure 5. The 23S rRNA gene is bounded by the tRNA gene(s) of the intergenic spacer, and the 5S rRNA gene, located at the BamHI cleavage site. The 5S, 16S, and 23S rRNAs, and the tRNA genes are all located with a region of approximately 4.7-5 kbp, only slightly larger than the size of the transcripts. The clustering of these genes may be due to the fact that they are all part of a single transcription unit. If these genes are all from a single operon, all of the RNA species would be transcribed from the same DNA or another EcoP strand. To test this, pPG50 or another EcoP plasmid, pPG2 (5), linearized by digestion with SalI, were denatured. The DNA strands were separated by electrophoresis, and hybridized to chloroplast RNAs by the Southern method. [^{125}I]16S and 23S rRNAs, and [^{32}P]tRNAs all hybridized exclusively to the electrophoretically slower strand of pPG50 or pPG2.

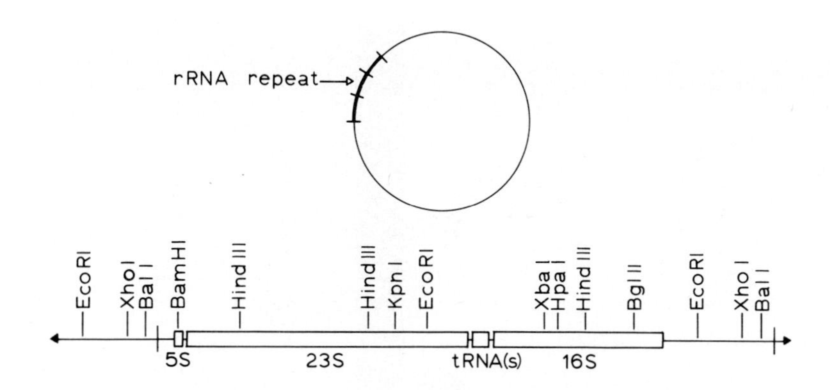

FIGURE 5. The location of rRNA and tRNA genes in the rRNA repeat. The middle DNA sequence of three identical, tandem repeats is illustrated.

In a related experiment, [^{125}I]5S rRNA was found to hybridize to the same strand as 16S and 23S rRNAs of a recombinant plasmid containing the BamE fragment.

The above experiments are consistent with the clustering or rRNA and tRNA genes on a single operon. However it will be necessary to identify a common precursor of these RNAs to confirm this hypothesis.

Selective in vitro transcription of the Euglena Chloroplast rRNA genes. An important approach to understanding mechanisms involved in the control of RNA synthesis is to obtain a purified *in vitro* transcription system that duplicates the RNA transcripts of the intact cell. To this end we have been studying the properties of a highly purified complex of chloroplast RNA polymerase and chloroplast DNA, termed a "transcriptionally active chromosome" or "TAC", isolated directly from *Euglena* chloroplasts (12,13). In the TAC, all RNA synthesis is from the endogenous chloroplast DNA template.

To test if the RNA product represents a selective transcript of certain regions of the chloroplast genome, the transcript was labeled with [α^{32}P]CTP *in vitro*, and hybridized by the Southern method to restriction fragments of chloroplast DNA. Hybridization to EcoRI fragments is shown in Figure 6. Strong hybridization is obtained with only 4 of the 28 EcoRI fragments, EcoB,F,L and P. EcoF,L, and P contain the rRNA coding region (Figure 1). EcoB has been reported to have partial sequence homology with EcoP (E. Stutz, personal communication). In similar experiments (18),

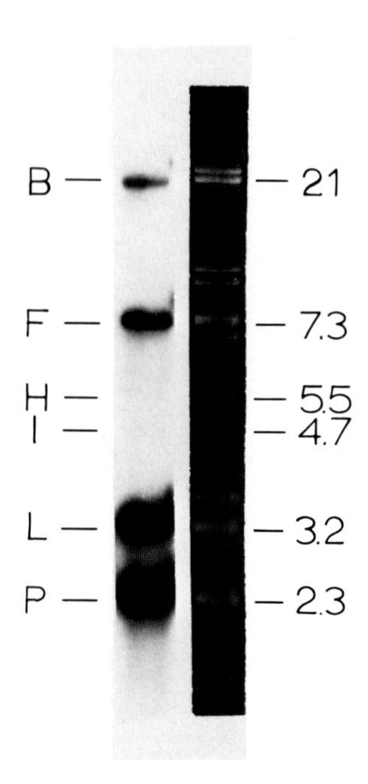

FIGURE 6. Hybridization of RNA synthesized by the TAC *in vitro* to EcoRI fragments of chloroplast DNA. Right, ethidium bromide stained gel. Left, autoradiogram.

in vitro RNA was found to hybridize predominately with fragments BalG and H, XhoD, and BamD and E, all from the rRNA region. Therefore the TAC product is largely a selective transcript of the rRNA coding region. The transcript must be coded at both ends of the rRNA repeat, since both EcoP, which contains the 16S rRNA gene, and EcoF and L, which contains the 5S gene and most of the 23S gene, gave similar hybridization results.

Two lines of evidence are consistent with the conclusion that RNA synthesis occurs only from the sense strand of the DNA. First, when excess non-radioactive chloroplast DNA is added to a Southern hybridization of *in vitro* RNA, essentially all of the hybridization is competed. Second, both RNA synthesized *in vitro*, and RNA isolated from chloroplasts hybridizes to the same DNA strand of the rRNA coding region. In these strand separation, Southern hybridization experiments, 16S rRNA, 23S rRNA, 5S rRNA, tRNA, and the TAC RNA product all hybridize exclusively to the electrophoretically slower strand after strand separation in agarose gels. From these experiments we conclude that the predominant RNA product of the TAC *in vitro* is a selective transcript of the rRNA genes.

We have previously reported that RNA chains are iniated *in vitro* (12). One of the most interesting properties of the TAC is that initiation and elongation of RNA are insensitive to heparin. As shown in Figure 7, initiation of RNA, as measured by the incorporation of [γ^{32}P]ATP into RNA, is inhibited only 20-30% in the presence of 1mg/ml heparin. Elongation of RNA chains, as measured by the incorporation of [^{3}H]UMP into RNA, was also inhibited only 20-30% in the presence of 1-5mg/ml heparin. We also find a dramatic increase in the size of RNA made in the presence of heparin. The lack of inhibition by heparin may be an indication that chloroplast RNA polymerase may not be released from the DNA following the completion of an RNA chain. Following termination the RNA polymerase is possibly translocated along the DNA, and subsequently reinitiates on the same DNA molecule.

Since RNA transcription occurs selectively from the rRNA genes, we reasoned that it might be possible to interrupt *in vitro* RNA synthesis by treating the TAC with restriction endonucleases that cut in the rRNA region. TAC was treated with four different enzymes, HindIII, EcoRI, KpnI, and BamHI, each of which cut once in each rRNA repeat (Figure 5). RNA synthesis was then assayed in the enzyme-treated TAC samples. The results are presented in Table 1.

TABLE I
EFFECT OF RESTRICTION ENDONUCLEASE TREATMENT ON TAC RNA SYNTHESIS ACTIVITY

Enzyme	RNA Synthesis (% Control)	Template Length[a] (% Control)
HindIII	19.3	15
EcoRI	41.2	48
KpnI	56.4	54
BamHI	85.0	98

[a]Assuming that a full length template of approximately 4.8 kbp runs from the beginning of the 16S rRNA gene at or near the TaqI site to the end of the 5S rRNA gene, this value is calculated as the distance from the beginning of the 16S gene to the appropriate cleavage site.

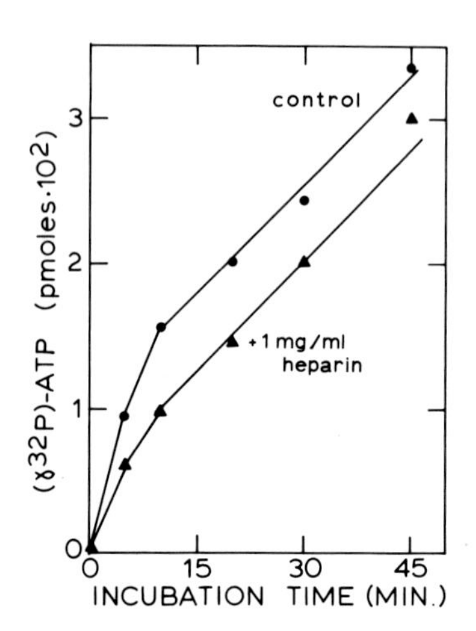

FIGURE 7. Effect of heparin on the initiation of RNA synthesis. RNA synthesis with $[\gamma^{32}P]$ATP was determined as previously described (12). Reactions were run in the presence and absence of 1mg/ml heparin.

In each case RNA synthesis is inhibited by prior treatment with a restriction endonuclease. The extent of inhibition correlates well with the distance from the presumed site of the beginning of RNA synthesis near the end of the 16S gene to the cleavage site. One possible interpretation of this result is that cleavage sites serve as RNA chain terminators for RNA synthesis.

To study the nature and selectivity of the RNA synthesis initiation reaction, in vitro synthesized RNA, labeled at the 5'-end with $[\gamma^{32}P]$ATP was prepared for nuclease digestion experiments. The rationale is that digestion of a selectively initiated, $[\gamma^{32}P]$cRNA with RNAseTI or RNAseA would result in only discrete oligonucleotides. In this experiment the products of digestion of 5'-$[\gamma^{32}P]$RNA with various nucleases were separated on a 30% polyacrylamide/7M urea gel, and detected by autoradiography. The results are shown in Figure 8.

As seen in Figure 8 (lanes 3,4,9) the 5'-end of the RNA is resistant to digestion by either RNAseTI or RNAseA. Most of the product remains at the origin, under electrophoresis conditions in which 5S rRNA moves well into the gel. These nuclease conditions normally are sufficient for a limit

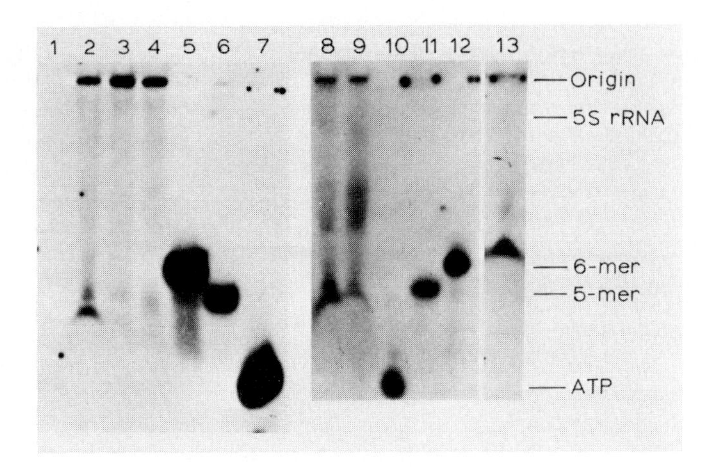

FIGURE 8. 5'-end analysis of RNA initiated by the TAC *in vitro*. 5'-[γ^{32}P]RNA was digested under the following conditions: Lane I, 0.1M KOH for 45 min. at 90°; lanes 2 and 8, RNAseTI, at an enzyme/substrate of 1/10 for 30 min. at 50° in 20mM NaCitrate, pH4.5, 1mM EDTA, 7M urea; lanes 3 and 9, RNAseTI, at an enzyme/substrate of 1/10 for 30 min. at 37° in 10mM Tris-HCl, pH7.4, 1mM EDTA; lane 4, Pancreatic RNAseA, same conditions as lanes 3 and 9; lane 13, SI-nuclease for 2 hr. at 50° in 0.03M NaAcetate, 0.3M NaCl, 3mM $ZnCl_2$, 10ug/ml denatured calf thymus DNA (pH 4.5). The following electrophoresis markers were used: lanes 7 and 10, [γ^{32}P]ATP; lanes 6 and 11, a synthetic, 5'-[α^{32}P]deoxypentanucleotide; lanes 5 and 12, a synthetic 5'-[α^{32}P]deoxyhexanucleotide. The oligonucleotides were generously provided by Dr. M. H. Caruthers. An additional marker, *E. coli* 5S rRNA was run on the same gel, and detected by fluorescence under UV-illumination after staining with ethidium bromide. All samples were heated at 50° in the presence of 5M urea prior to electrophoresis.

digestion of RNA (19). In parallel control experiments 16S, 23S, and 5S rRNAs from *E. coli*, and a 5'[γ^{32}P] *in vitro* RNA prepared with *E. coli* RNA polymerase were completely degraded by RNAseTI and RNAseA. The RNA product of the TAC was completely digested by KOH (lane 4; the product pppAp ran off the gel), and completely resistant to pronase digestion. (not shown).

When 5'-[γ^{32}P] in vitro RNA was treated with RNAseTI at 50° in 7M urea (Figure 8, lanes 2 and 8), most of the products still remained at the origin following electrophoresis, but two discrete oligonucleotides, with mobilities similar to that of the deoxypentanucleotide, are apparent. The resistance to nuclease digestion is most likely due to RNA secondary structure involving the 5'-end of the RNA. If there is extensive secondary structure, the RNA should also be resistant to SI-nuclease digestion. As shown in Figure 8 (lane 13) the product of an exhaustive SI-nuclease digestion is a discrete oligonucleotide, with electrophoretic mobility less than that of a deoxyhexanucleotide. These SI-nuclease digestions results are also consistent with the RNA initiated in vitro having a discrete 5'-end, which is protected from nuclease digestion by a stable secondary structure.

In summarizing the properties of the RNA synthesis reaction of the TAC, we find that the RNA is a selective transcript of the chloroplast rRNA genes. The RNA is initiated in vitro. The 5'-end of this RNA is protected from nuclease digestion by secondary structure, but can be converted to discrete 5'-oligonucleotides by exhaustive digestion. The TAC seems to represent an ideal model system for the study of mechanisms involved in the control of RNA synthesis.

Gene for the Large Subunit of RuDP Carboxylase. Recently the gene for the large subunit of Ribulose-1,5-bisphosphate carboxylase was located on *Chlamydomonas reinhardii* chloroplast DNA (10). The LS-gene is found on a 5.6kbp EcoRI fragment. We reasoned that there might be DNA sequence homology between the LS-genes of *Euglena* and *C. reinhardii*, and furthermore, that preliminary evidence on the LS-gene location in *Euglena* might be obtained by heterologous hybridization experiments with the *Chlamydomonas* 5.6kbp fragment. A recombinant plasmid, pCR34, containing the 5.6kbp EcoRI fragment with the LS-gene, was generously provided to us by Dr. J. Rochaix. A [^{32}P]cRNA transcript of denatured pCR34 DNA was made with *E. coli* RNA polymerase, and [α^{32}P]cTP as a substrate. The cRNA was then hybridized by the method of Southern to EcoRI restriction nuclease fragments of Euglena chloroplast DNA. The results are shown in Figure 9. With hybridization conditions allowing some DNA-RNA mismatching, the cRNA hybridized to fragment EcoC. In other experiments (not shown) the cRNA hybridized to fragment PvuC from a PvuII digest, and PvuC from a PvuII-KpnI double digest. We have also constructed a recombinant plasmid with EcoC. EcoC was cloned into the EcoRI site of the vector pMB9. The cRNA from pCR34 hybridizes to the EcoC fragment of this plasmid. EcoC contains a single XhoI site (Figure 1). The cRNA hybridizes

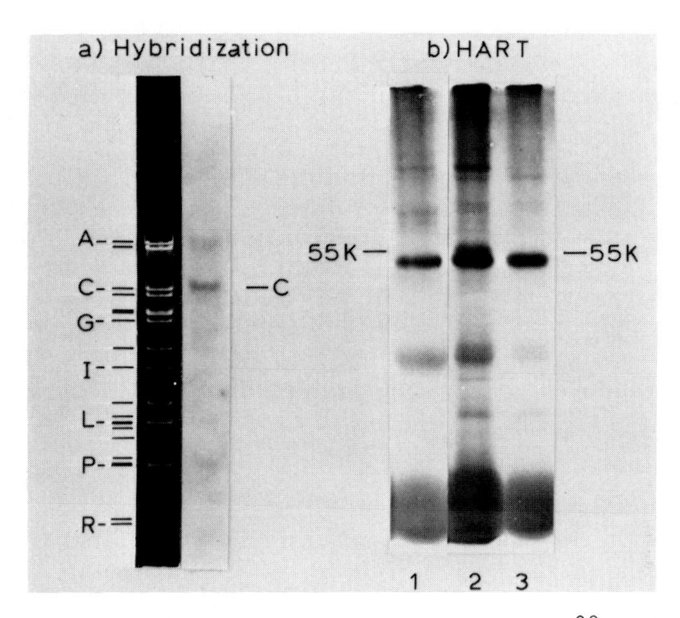

FIGURE 9. a) Southern hybridization of [^{32}P]cRNA transcript of pCR34 DNA and EcoRI fragments of Euglena chloroplast DNA. Hybridization in 6X SSC, 15% formamide was at 44° for 48 hr. Both the ethidium bromide-stained, 0.7% agarose gel and the resulting autoradiogram are shown. b) Hybrid arrest translation of Euglena chloroplast mRNA in a rabbit reticulocyte lysate. An autoradiogram of [^{32}S]polypeptides, separated in SDS-PAGE is illustrated. Lane 1, endogenous polypetides synthesized in reticulocyte system. Lane 2, Euglena chloroplast RNA was hybridized with excess pVK52 for 10' at 55° in 8mM Hepes, pH6.8, 1.0 M NaCl, 5mM EDTA, and 1% SDS, isolated by ethanol precipitation, and then translated in the reticulocyte lysate. pVK52, a recombinant plasmid containing the Euglena chloroplast BamE fragment (which codes for rRNA, was generously provided by Dr. J.R.Y. Rawson. Lane 3, chloroplast RNA was hybridized with pCR34 DNA as described above, and then translated in vitro.

only to the larger XhoI fragment of EcoC. From these experiments we conclude that there is homology between pCR34 DNA and a region in the EcoC fragment of Euglena chloroplast DNA. No regions of homology could be located outside the EcoC fragment.

A functional test for homology between pCR34 coding sequences and Euglena mRNA is possible with hybrid arrest translation experiments. The major transcript of Euglena chloroplast RNA in a rabbit reticulocyte lysate is a polypeptide of MW 55,000 which comigrates with authentic LS-polypeptide. It has previously been reported that the LS polypeptide is the major translation product of Euglena chloroplast RNA in a wheat germ translation system (20). The production of the MW 55,000 polypeptide in the reticulocyte lysate is not inhibited by hybridization of the chloroplast RNA with a plasmid DNA containing the chloroplast rRNA genes (Figure 9, lane 2). However translation of this polypeptide is completely abolished by hybridization to pCR34 DNA. Therefore there must be homology between the 5.6kbp EcoRI fragment of C. reinhardii chloroplast DNA containing the LS-gene and a Euglena chloroplast mRNA. Since pCR34 was only found to be homologous to the EcoC fragment in Euglena chloroplast DNA, the most likely conclusion is that the mRNA is encoded on EcoC. Our current working hypothesis is that the LS-gene in Euglena is located on the large XhoI fragment of EcoC. We plan to follow up these preliminary observations with more rigorous experiments.

REFERENCES

1. Manning, J.E., and Richards, O.C. (1972). Biochim. Biophys. Acta 259, 295.
2. Gray, P.W., and Hallick, R.B. (1978). Biochemistry 17, 284.
3. Scott, N.S., and Smillie, R.N. (1967). Biochim. Biophys. Res. Commun. 28, 598.
4. Stutz, E., and Rawson, J.R. (1970). Biochim. Biophys. Acta 209, 16.
5. Gray, P.W., and Hallick, R.B. (1979). Biochem. in press.
6. McCrea, J.M., and Hershberger, C.L. (1976). Nucleic Acids Res. 3, 2005.
7. Schwartzbach, S.D., Hecker, L.I., and Barrett, W.E. (1976). Proc. Natl. Acad. Sci. U.S.A. 73, 1984.
8. Chelm, B.K., Gray, P.W., and Hallick, R.B. (1978). Biochem. 17, 4239.
9. Coen, D.M., Bedbrook, J.R., Bogorad, L., and Rich, A. (1977). Proc. Natl. Acad. Sci. U.S.A. 74, 5487.
10. Rochaix, J.-D., and Malnoe, P. (1978). In "Chloroplast Development" (G. Akoyunoglou and J.H. Argyroudi-Akoyunoglou, eds.), pp. 581-586. Elsevier/North Holland Biomedical Press, Amsterdam.
11. Chelm, B.K., Hallick, R.B., and Gray, P.W. (1979). Proc. Natl. Acad. Sci. U.S.A., in press.
12. Hallick, R.B., Lipper, C., Richards, O.C., and Rutter, W.J. (1976). Biochem. 15, 3039.

13. Hallick, R.B., and Lipper, C. (1976). In "Molecular Mechanisms in the Control of Gene Expression" (D.P. Nierlich, W.J. Rutter, and C.F. Fox, eds.), pp. 355-360. Academic Press, New York.
14. Gray, P.W., and Hallick, R.B. (1977). Biochem. 16, 1665.
15. Rawson, J.R.Y., Kushner, S.R., Vapnek, D., Alton, N.K., and Boerma, C.L. (1978). Gene 3, 191.
16. Jenni, B., and Stutz, E. (1978). Eur. J. Biochem. 88, 127.
17. Southern, E.M. (1975). J. Mol. Biol. 98, 503.
18. Hallick, R.B., Gray, P.W., Chelm, B.K., Rushlow, K.E., and Orozco, E.M., Jr., (1978). In "Chloroplast Development", op. cit.
19. Barrell, B.G., (1971). In "Procedures in Nucleic Acid Research, Vol. 2" (G.L. Cantoni and D.R. Davies, eds.), pp. 751-779, Harper and Row, New York.
20. Sagher, D., Grosfeld, H., and Edelman, M. (1976). Proc. Natl. Acad. Sci. U.S.A. 73, 722.

THE MU PARADOX: EXCISION VERSUS REPLICATION[1]

Hajra Khatoon[2], George Chaconas, Michael DuBow and Ahmad I. Bukhari

Cold Spring Harbor Laboratory
Cold Spring Harbor, New York 11724

ABSTRACT Bacteriophage Mu DNA, as well as other procaryotic transposable elements, can be excised from the host chromosome at a low frequency. This excision is mostly imprecise; precise excision resulting in the restoration of the wild type sequences is seen only in a minority of cases. However, rapid transposition of Mu DNA to different sites on the E. coli chromosome during the Mu lytic growth does not involve excision of Mu DNA. The most frequent event appears to be replication of Mu DNA and transposition of a Mu DNA copy to a new site. Thus, a mechanism for Mu excision exists and yet, this mechanism is not used for rapid transposition of Mu DNA. To resolve this paradox, we propose that the A gene protein of Mu recognizes the ends of the Mu prophage. If replication functions are then provided, Mu DNA replicates and is transposed. If replication is blocked, the action of the A protein leaves the prophage in a state susceptible to excision.

It is now well recognized that the temperate bacteriophage Mu resembles procaroytic transposable elements in all essential respects (1). The transposable elements are natural constituents of the bacterial chromosomes and of extrachromosomal elements, and are not found to exist autonomously. Like the transposable elements, Mu is always associated with the host DNA. Even mature Mu DNA has heterogeneous host sequences at both ends (2). The structure of Mu DNA is depicted in Figure 1.

Both Mu and transposable elements can be excised from the host DNA at a low frequency (3, 4, 5). The excision can

[1]This work was supported by grants from the National Science Foundation and the National Institutes of Health. GC holds a postdoctoral fellowship from Medical Research Council of Canada. A.I.B. holds a Career Development Award of the National Institutes of Health.

[2]Present address: Department of Microbiology, University of Karachi, Karachi 32 Pakistan.

ISBN 0-12-198780-9

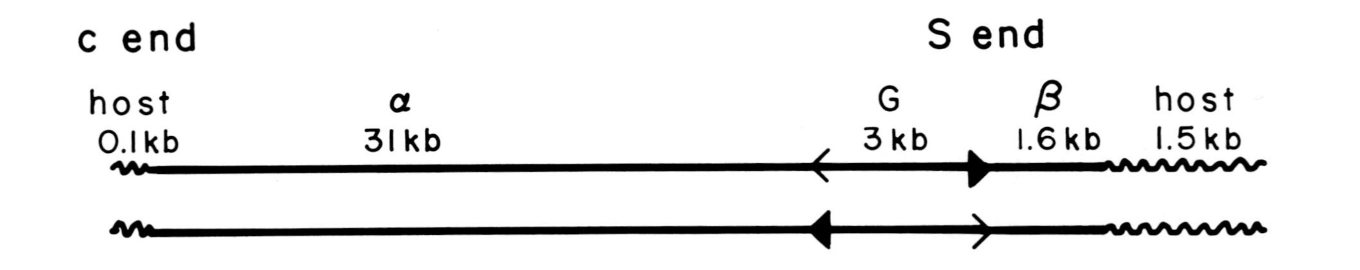

FIGURE 1. Structure of Mu DNA.

Mu DNA extracted from mature phage is a linear duplex of about 37 kilobases (see references 17 and 18). There are two characteristic features of Mu DNA. Both ends of Mu DNA are attached to host sequences picked up randomly from the host chromosome during growth. The left end (the c end) contains 50-100 base pairs (bp) of the host DNA, whereas the right end (the S end) carries about 1500 host bp, ranging from 500 bp to 3200 bp. The other feature is the invertible segment G which can flip-flop and is found in two different orientations in certain conditions (18). α, G, β are segments of Mu DNA and their sizes are indicated in kilobases. It should be noted that insertion of Mu into the host DNA amounts to transposition since Mu leaves one set of host sequences and inserts itself into another set. This shedding of host sequences during transposition has been hypothesized to result from replication events that begin specifically at the Mu ends (1, 8).

be precise, indicating that the ends of the elements can be recognized. However, the precise excision is less frequent than the imprecise excision. Most excision events do not lead to the restoration of the wild type host sequences at the site of insertion.

We have proposed that transposition of Mu DNA, and of other transposable elements, does not proceed via excision, but rather the mechanism of transposition involves replication of the element such that one copy of the element is left at the original site (6, 7, 8, 9). This proposal was based on experiments by Ljungquist and Bukhari (6) that showed that during transposition and replication of Mu DNA no excision of prophage DNA could be seen. Since Mu is similar to other transposable elements, it is reasonable to assume that they also follow a similar mode of transposition. The proposal that Mu and transposable elements move by mechanisms that have elements in common has received strong support from the observation that both Mu and transposable elements cause small duplications of host sequences at the site of insertion. Like the transposon Tn3 and the insertion element γδ, Mu generates a duplication of 5 base pairs at the site of insertion (10, 11, 12, 13) (R. Kahmann and D. Kamp, in preparation). The idea that both Mu and transposable elements may move by a mechanism that involves replication is thus gaining rapid acceptance (14, 15).

The replication transposition hypothesis raises a paradox. We know that a mechanism of Mu excision exists and yet it is not used for transposition. One way to resolve the paradox would be to consider excision as irrelevant to the process of transposition. We propose a different hypothesis here, that the excision process has elements in common with the transposition process and that excision in effect is abortive transposition. To develop the basis for this hypothesis, we will discuss here our recent results on excision and replication-transposition of Mu DNA.

EXCISION OF MU DNA

We have studied the excision of Mu DNA from the *Z* gene of the *lac* operon in *E. coli*. Normally, Mu insertions are completely stable. We showed earlier that Mu DNA excision at a low frequency can be seen if Mu contains the *X* mutation (3). Most of the spontaneous *X* mutants have turned out to be insertions in the *B* gene region of the Mu genome located near the left end of Mu DNA. The mutations directly tested thus far do not complement the *B* gene mutations. It is not clear whether all of the insertions are in the *B* gene, a locus whose gene product is apparently required for replication.

What is clear is that all X mutations inactivate the replication functions of Mu. The insertions in the B gene not only eliminate the B gene function but they also inactivate the kil function of Mu (16). Thus, the net result of X mutations is that the prophage cannot replicate its DNA and cannot kill the host cells.

Originally the X mutations were all shown to be caused by the insertion of the 800 base pair element IS1 (17, 18). We have now found that these mutations can also be caused by the insertion of other IS elements such as IS5 and IS2. In 43 independently isolated X mutants, 29 were found to have insertions of IS1, 4 had IS5 and 4 had IS2. No obvious insertion was seen in 6 X mutants by restriction endonuclease cleavage analysis and the nature of lesions in these mutants remains uncharacterized.

Excision Products. It was reported earlier that excision of Mu from the lacZ gene is mostly imprecise. That is, Z^-Y^+ revertants are more frequent than Z^+Y^+ revertants in our assay system. The Z^+Y^+ revertants resulting from precise excision of Mu DNA are detected as Lac^+ colonies, whereas Z^-Y^+ revertants are detected as colonies that can use the sugar melibiose as a carbon source at 41°C. At this temperature melibiose is transported into the cells by the $lacY^+$ permease only. The frequency of Z^+Y^+ revertants is 10^{-6} to 10^{-8} per cell, whereas the frequency of Z^-Y^+ revertants is 10 to 100-fold higher. In more than 99% of the Z^+Y^+ revertants Mu DNA is completely eliminated from the cells. That is, precise excision is not followed by integration of Mu DNA elsewhere. In about 80% of the Mel^+ revertants Mu DNA is also lost. The rest of the $melibiose^+$ revertants arise because of rearrangements in Mu DNA (see below).

Since excision of Mu DNA is a low frequency event and Mu DNA is generally lost from the cells, it is not possible to examine the state of Mu DNA after excision. However, we can examine the host DNA left behind after excision. We have isolated $melibiose^+$ (that is, Z^-Y^+) revertants from Mu X mutants located at 6 different sites on the lacZ gene. These Mel^+ revertants fall in 2 major groups. Those which can further revert to lac^+ and those which cannot revert to lac^+. Many Mel^+ revertants isolated from different X mutants were characterized in detail by fine genetic mapping and by DNA-DNA hybridization. For genetic mapping, the Z^-Y^+ revertants were crossed with a set of 27 deletions ending at different points in the Z gene and with several point mutations located close to the sites of Mu X insertions. For DNA-DNA hybridization the total DNA was extracted from the revertants and cleaved with the endonuclease BalI which gives a characteristic pattern for the X mutants (19). The DNA fragments were blotted by the Southern procedure after electrophoresis and hybridized to ^{32}P-labelled Mu DNA as described earlier (2).

These studies showed that the non-revertible Mel^+ colonies fall into two subgroups; 1) those that have no Mu DNA left and have a deletion in the Z gene, 2) those that have a deletion in the Z gene and a partial deletion on Mu. The revertible Mel^+ also fall into two subgroups; 1) those which still retain Mu DNA, and 2) those which retain no Mu DNA. The last class is the most frequent, comprising more than 50% of all Mel^+ revertants isolated.

Fig. 2 shows fine genetic mapping of some of the Mel^+ revertants isolated from the Mu insertion 8357. This insertion is located near the operator-promoter end of the Z gene (3). Except for three representatives, the major class of Mel^+ revertants (those which could revert to Lac^+) was excluded from this study. The figure shows that a large number of deletions start approximately from the point of Mu insertion and extend either to the right (revertant nos. 1, 7, 8, 12 and 13) or to the left (revertant nos. 2, 3). Other deletions span both sides of the Mu prophage (revertant nos. 5, 9, 11, 14, 15). No large deletion was detected in Mel^+ revertant no. 4 and revertant nos. 6 and 10. Revertants 6 and 10 have no Mu DNA left in them and can revert to Lac^+, whereas no. 4 still retains Mu DNA and can revert to Lac^+.

Those Mel^+ revertants that can further revert to Lac^+ and have no detectable Mu DNA have probably the remnants of 5 base pair duplication left behind. This point is being checked by determining the appropriate nucleotide sequences.

Mechanism of Excision. The original study on Mu excision suggested the involvement of Mu functions in the excision process (3). Mutations in the A gene of Mu were found to cause a reduction in the excision events. We have now confirmed that A gene function is required for Mu excision. We constructed Mu X A amber mutants and found that these mutants can be excised only in the presence of amber suppressors. Mu X B amber mutants on the other hand could revert in the presence or absence of the suppressors. A secondary Mu prophage inhibits the excision process. This effect is specific to Mu immunity since the Mu X mutants can be excised in the presence of a heteroimmune D108 phage. One explanation of the inhibitory effect of a second Mu prophage is that the A gene expression is leaky in the Mu cts X mutants and this leakiness is prevented by the secondary Mu prophage. The recovery of the revertants is also affected by the recA mutations. The Z^+Y^+ revertants are

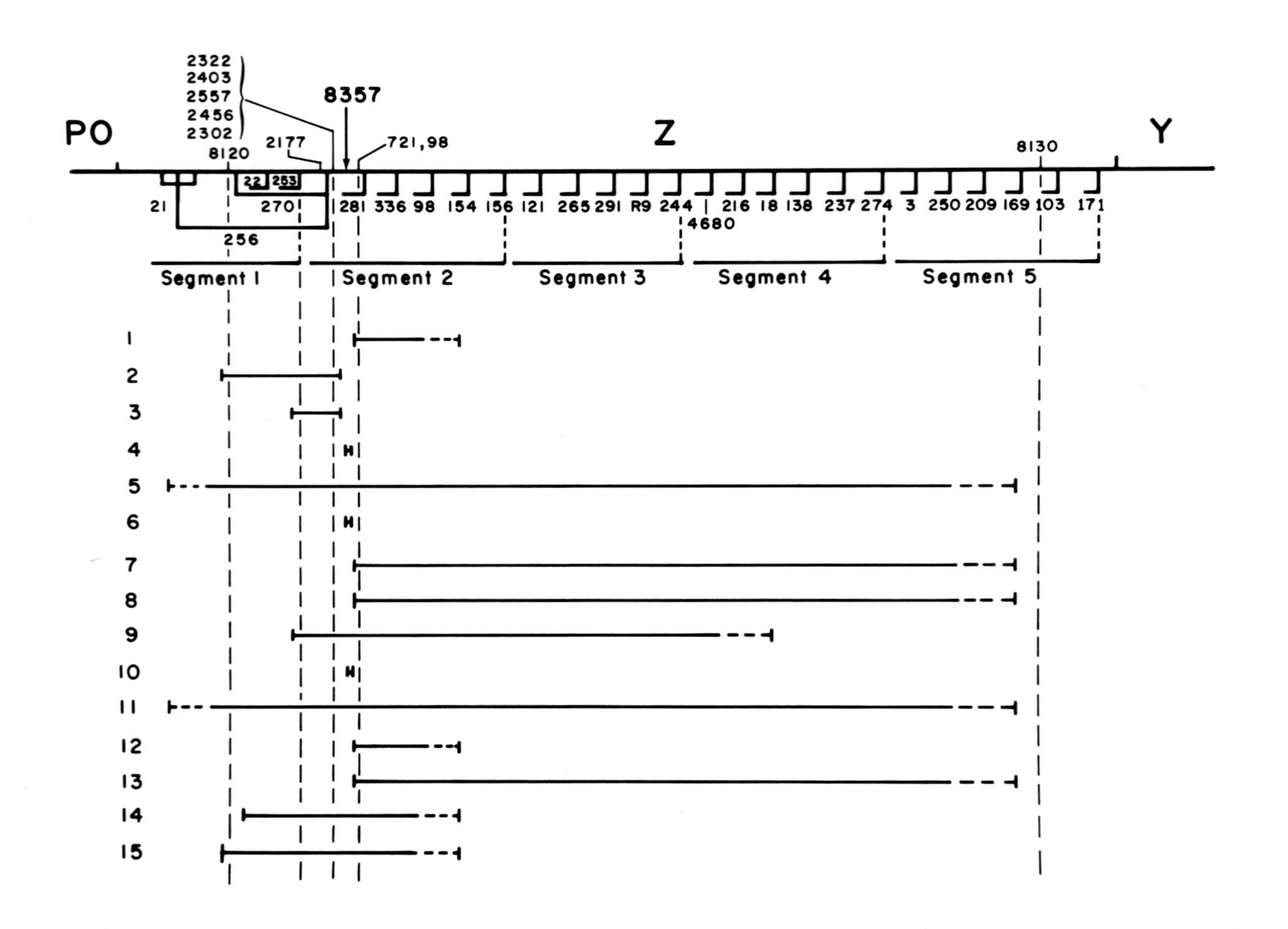

PO
Z
Y
8357
2322
2403
2557
2456
2302
8120
2177
721,98
8130
21
22
253
270
256
281 336 98 154 156 121 265 291 R9 244 216 18 138 237 274 3 250 209 169 103 171
4680
Segment 1
Segment 2
Segment 3
Segment 4
Segment 5
1
2
3
4
5
6
7
8
9
10
11
12
13
14
15

FIGURE 2. Analysis of Z^-Y^+ (Mel^+) revertants derived from Mu insertion 8357.

Several X mutants of the 8357 Mu insertion were isolated (3). Mel^+ revertants were isolated by plating 0.1 ml of overnight culture at minimal medium plates with melibiose as the sole carbon source at 41°C. Many Mel^+ revertants which could not revert to Lac^+ (from Z^-Y^+ to Z^+Y^+) were selected for genetic mapping. The revertants are identified by numbers from 1 to 15. In the figure, only three revertants (4, 6 and 10) could revert to Lac^+.

The lac genes in all Z^-Y^+ revertants were carried on an F' pro^+ lac episome. The mapping was done by conjugating the revertants with the appropriate strains carrying deletions or point mutations in the Z gene. The deletion end points are shown below the Z gene line and the deletions are indicated by numbers. The Z gene was arbitrarily divided into five segments for various mapping purposes. The numbers above the Z gene line shows the point mutations used for mapping studies. The deletions and point mutations were from D. Zipser's collection. The horizontal lines corresponding to the Mel^+ revertant numbers correspond to the deletions in these revertants. The dashed lines at the end of the horizontal line indicate that the precise end point of the deletion has not been determined. The dashed vertical lines represent some landmark point mutations used for mapping.

Similar analyses were done with Mel^+ revertants isolated from the Mu X mutants located at different sites in the Z gene (Khatoon and Bukhari, to be submitted to Genetics).

not seen in a recA$^-$ background and the Z^-Y^+ revertants are reduced in frequency by 3 to 10-fold.

The requirement for the A gene function indicates that the A protein of Mu is involved in the recognition of the ends of prophage Mu DNA.

REPLICATION AND TRANSPOSITION OF MU DNA

The hypothesis that integration of Mu DNA (transposition of Mu DNA from one set of host sequences to another) involves replication of Mu DNA stemmed from the studies which showed that there is no efficient excision of Mu DNA upon prophage induction (6). The Mu prophage DNA apparently persists at its original site and yet it replicates and many copies of Mu DNA are integrated at different sites on the host genome. This conclusion is consistent with the finding that Mu DNA is not converted to a covalently closed circular form upon infection of host cells (20). If linear Mu DNA were efficiently converted to a circular form by recombination at the ends, then the same process would be expected to occur upon induction of a prophage, resulting in excision of a circle from the host chromosome.

Recent evidence continues to sustain the replication-integration model for Mu DNA transposition. We have used small plasmids containing either the complete Mu genome or different parts of the Mu genome to study the process of Mu replication and transposition. The advantage of these plasmids is that all forms of the plasmids (supercoiled, open circles, and linear forms) can be readily separated from the bulk host DNA on low percentage agarose gels. Thus, we can examine the fate of the plasmids containing Mu after prophage induction. We have found no evidence for the excision of Mu DNA from the plasmids after Mu induction.

Construction of the Plasmids. The plasmids used in the studies on Mu replication were derived from the 9.09 kilobase plasmid pSC101 which contains a gene for tetracycline resistance (21). This plasmid is present in four to six copies per cell. M. Casadaban at Stanford University isolated a Mu insertion in this plasmid. He also inserted a segment encoding kanamycin resistance (kan) derived from another plasmid into the invertible G segment of Mu DNA on the plasmid. We mapped Casadaban's Mu insertion in pSC101 and found that it is located between the XhoI and PvuII restriction sites. We deleted the middle part of the Mu prophages by digesting the DNA with the enzyme PstI. There is no PstI site on pSC101 DNA but the enzyme cuts Mu DNA at 1.7 kb from the left end and 8 kb from the right end. Thus,

PstI digestion followed by recircularization and ligation removes about 27 kb of Mu DNA. These plasmids are referred to here as Mutiny plasmids. We have subsequently isolated several Mu insertions in pSC101, and these insertions are being further manipulated.

Behavior of Mu Containing Plasmids During the Mu Lytic Cycle. When strains carrying pSC101::Mu cts (temperature inducible mutant of Mu) are shifted to high temperature to induce the prophage, both the supercoiled and the relaxed forms of the plasmid disappear over time. This can be readily seen on 0.3% agarose gels. The plasmid is not cut or nicked and Mu DNA is not excised in any detectable manner from the plasmid; instead, the plasmid containing Mu begins to associate with the host DNA by 20 minutes or so after induction. By the onset of lysis, almost all the plasmid DNA can be found to be associated with the host DNA. This has been shown by hybridization of the DNA bands to ^{32}P-labelled pSC101 DNA and to ^{32}P-labelled Mu DNA following Southern blotting of the agarose gels. The same association with the host DNA can be seen with the Mutiny plasmids if a normal Mu phage is provided as a helper. For reasons not understood at present, the lytic cycle of Mu spans a longer time in the presence of Mutiny plasmids and thus, disappearance of the plasmids into the host chromosome takes a proportionally longer period of time. Our preliminary results indicate that the plasmid DNA association with the host chromosome is both non-covalent and covalent in nature. The noncovalent association appears to be the result of a topological trapping of the plasmids by the host DNA during transposition. Further analysis of this association would be necessary for understanding the molecular mechanism of transposition.

Using Mutiny kan plasmids we have shown that the ends of Mu are sufficient for transposition, if Mu functions are provided in trans. Both Mutiny plasmids with and without kan also give rise to cointegrate structures in which pSC101 is integrated in the host DNA. This integration apparently represents the covalent association of the plasmids with the host DNA. The structure of the cointegrate forms is under investigation. It is reasonable to assume that we will find the plasmid flanked by two mini Mu's (internally deleted Mu's). This type of structure has been reported for cointegrates generated by several other transposition elements (22, 23). R. Harshey in our laboratory has shown that, as expected, the mini Mu's can also be packaged in the presence of a helper phage.

We have further determined the extent to which mini Mu's replicate if provided with all the necessary functions.

To do this, we relied upon a method developed by us for measuring the kinetics of Mu DNA synthesis. We have shown that when Mu amp (Mu containing a gene for β-lactamase) is induced, the amount of β-lactamase increases and the enzyme synthesis apparently parallels Mu DNA synthesis. To study replication of mini Mu's, we first recombined the amp gene from Mu amp of Leach and Symonds (24) into the mini Mu. Subsequent induction of mini Mu amp in the presence of helper phage showed that mini Mu replicates about 10-fold. This means that the ends of bacteriophage Mu are sufficient for Mu DNA replication.

EXCISION AS ABORTIVE TRANSPOSITION

The above discussion shows that the ends of Mu DNA exhibit three important properties. 1) They can be recognized for excision of Mu DNA, 2) they are used for transposition to different host sequences 3) they are used for replication of Mu DNA and may well define sites for initiation of DNA replication.

We know that the A gene of Mu is involved in the first two processes. Chances are that it is also required for the third process, since no Mu DNA replication can be detected in A^- mutants (25).

To resolve the paradox that Mu DNA can be excised from the host DNA, albeit at a low frequency, and yet no excision can be seen when Mu is undergoing replication-transposition, we propose the following hypothesis (26). The A gene protein recognizes the Mu ends and initiates a process that would culminate in the replication-transposition of Mu DNA if all the necessary replication functions are provided. If Mu replication is blocked, as is the case with the Mu X mutants, then the interaction of A protein and the Mu ends can leave Mu DNA in a state which is susceptible to excision. This excision is not efficient and we propose that it requires replication of the host chromosome. Thus, one strand of Mu DNA may be cut out, leaving a nick or a gap. This nick or a gap must be repaired. Upon replication of the chromosome, one daughter chromosome will be without Mu DNA, whereas the other will retain Mu DNA. The deletions would arise when the repair of the chromosome after strand excision is faulty. The recA protein would be involved in keeping together the DNA complex during and after strand excision, so that it can be repaired properly. The reason imprecise excisions are more frequent than precise excisions then, is that DNA repair will only occasionally restore the wild type sequence that existed before the insertion occurred.

REFERENCES

1. Bukhari, A.I. (1976). Ann. Rev. Genetics 10, 389.
2. Bukhari, A.I., Froshauer, S., and Botchan, M. (1976). Nature 264, 580.
3. Bukhari, A.I. (1975). J. Mol. Biol. 96, 87.
4. Berg, D.E. (1977). In "DNA Insertion Elements, Plasmids and Episomes" (A.I. Bukhari, J. Shapiro and S. Adhya, eds.), pp. 205. Cold Spring Harbor Laboratory, New York.
5. Botstein, D., and Kleckner, N. (1977). In "DNA Insertion Elements, Plasmids and Episomes" (A.I. Bukhari, J. Shapiro and S. Adhya, eds.), pp. 185. Cold Spring Harbor Laboratory, New York.
6. Ljungquist, E., and Bukhari, A.I. (1977). Proc. Nat. Acad. Sci. USA 74, 3143.
7. Bukhari, A.I. (1977). Brookhaven Symp. of Biol. vol. 29, pp. 218.
8. Bukhari, A.I., and Ljungquist, E. (1978). In "Microbiology" (D. Schlessinger, ed.) American Society of Microbiology, Washington, D.C. pp. 52.
9. Bukhari, A.I., Ljungquist, E., DeBruijn, F., and Khatoon, H., (1977). In "DNA Insertion Elements, Plasmids and Episomes" (A.I. Bukhari, J. Shapiro and S. Adhya, eds.), pp. 249. Cold Spring Harbor Laboratory, New York.
10. Allet, B. (1979). Cell 16, 123.
11. Ohtsubo, H., Ohmori, H., and Ohtsubo, E. (1978). Cold Spring Harbor Symp. Quant. Biol. Vol. 43, pp. 1269.
12. Cohen, S.N., Casadaban, M.J., Chou, J., and Tu, C.P.T. (1978). Cold Spring Harbor Symp. Quant. Biol. Vol. 43, pp. 1247.
13. Reed, R.R., Young, R.A., Steitz, J.A., Grindley, N.D.F. and Guyer, M.S. (1979). Proc. Nat. Acad. Sci. USA. In press.
14. Grindley, N., and Sherratt, D. (1978). Cold Spring Harbor Symp. Quant. Biol. Vol. 43, pp. 1257.
15. Shapiro, J. A. (1979). Proc. Nat. Acad. Sci. USA 76, 1933.
16. van de Putte, P., Westmaas, G., Giphart, M., and Wijffelman, C. (1977). In "DNA Insertion Elements, Plasmids and Episomes" (A.I. Bukhari, J. Shapiro and S. Adhya, eds.), pp. 287. Cold Spring Harbor Laboratory, New York.
17. Bukhari, A.I., and Taylor, A.L. (1975). Proc. Nat. Acad. Sci. USA 72, 4399.

18. Chow, L.T., and Bukhari, A.I. (1977). In "DNA Insertion Elements, Plasmids and Episomes" (A.I. Bukhari, J. Shapiro and S. Adhya, eds.), pp. 295. Cold Spring Harbor Laboratory, New York.
19. Khatoon, H., and Bukhari, A.I. (1978). J. Bacteriol. 136, 423.
20. Ljungquist, E., and Bukhari, A.I. (1979). J. Mol. Biol. In press.
21. Cohen, S.N., and Chang, A.C.Y. (1973). Proc. Nat. Acad. Sci. USA 70, 1293.
22. Gill, R., Heffron, F., Dougan, G. and Falkow, S. (1978). J. Bact. 136, 742.
23. Meyer, R., Boch, G. and Shapiro, J. (1979). Mol. Gen. Genetics 171, 7.
24. Leach,D., and Symonds, N. (1979). Mol. Gen. Genetics. 172, 179.
25. Wijffelman, C., Gassler, M., Stevens, W.J., and van de Putte, P. (1974). Mol. Gen. Genetics. 131, 85.
26. Ljungquist, E., Khatoon, H., DuBow, M., Ambrosio, L., DeBruijn, F., Bukhari, A.I. (1978). Cold Spring Harbor Symp. Quant. Biol. vol. 43, pp. 1151.

INTEGRATION OF TRANSPOSABLE DNA ELEMENTS: ANALYSIS BY DNA SEQUENCING[1]

Nigel D. F. Grindley

Department of Biological Sciences, University of Pittsburgh, Pittsburgh, PA 15260

ABSTRACT. Recent DNA sequence analyses that have lead to some insights into recombinational events involving transposable DNA elements are reviewed here. They show that, with the exception of phage Mu, all elements examined, even the simplest insertion sequences (IS) have terminal sequences which are inverted repeats (sometimes imperfect) of each other. Integrations and deletions are absolutely site specific for the element involving the same (terminal) nucleotide at each end. Transposition of an element to a new site results in duplication of a short (generally a 5 or 9 base) sequence that pre-existed at the integration site: one copy of this sequence occurs at each end of the inserted element. The second copy of this DNA segment is not carried in by the transposing element, but rather is apparently copied from the sequence at the integration site during the insertion event. Models to explain these results are summarized. The sequencing studies also show that several elements are either related or contain common recognition sites close to their ends.

INTRODUCTION

Over the last decade a large body of research has underlined the importance of a novel recombinational mechanism in the evolution of bacterial genomes (1-5). Termed specialized, site-specific or non-homologous, this recombination requires little or no homology between the participating DNA molecules and it occurs in cells that are deficient in homologous

[1]Research in the author's laboratory is supported by a grant (GM25227) from the National Institutes of Health

ISBN 0-12-198780-9

(general) recombination. These genetic additions or rearrangements involve specific DNA segments called transposable elements. Such elements can move from one site to new genetic locations in the cell either within the same or to an alternative genome. This transposition is a precise event. It occurs at specific sites on the element, namely the ends, and mutations that result from insertions within a gene can revert indicating that there is generally no loss of DNA at the integration site. Other rearrangements promoted by transposable elements include deletions and inversions and the fusion of two replicons to form cointegrates (6,7). All these rearrangements occur in the absence of the host *recA* function.

There are two major categories of transposable elements: the insertion sequences (IS) and the transposons (Tn). In addition, certain bacteriophage chromosomes that integrate into the genome of their host can be considered as transposable elements (8,5).

The IS elements are rather small ranging in size from the 768 base pair (b.p.) IS1 (refs 9,10), to about 1500 b.p. They carry no known genetic markers and are usually detected as the causative agents of strong polar mutations in bacterial operons, or by heteroduplex analysis. DNA sequencing has recently shown for at least two IS elements (IS1 and IS2) that the ends are short, 20-40 b.p., imperfect inverted repeat sequences (see below).

The transposons from 2500 b.p. to several thousand b.p. in size, are more complex elements. They carry known genetic markers such as resistance to antibiotics (1,3) and heavy metal ions (11), production of enterotoxins (12), ability to ferment lactose (13) or degrade aromatic hydrocarbons (14). Structurally, transposons consist of a unique region bounded by a repeated sequence. This repeated sequence can vary between 35 b.p. and 1500 b.p. and generally occurs as an inverted repeat. In certain instances the repeated sequence has been identified as a particular IS element, for example Tn9 and Tn1681 are flanked by IS1 (refs. 15 and 12). In other cases, for example Tn10, the inverted repeat has been shown to act as an independent insertion element (16). It seems probable that almost any piece of DNA that is flanked by two copies of an IS-like element, either in opposite or identical orientation, will act as a transposable unit. Transposons that now have only short inverted repeat sequences as termini may have evolved from such structures by suffering a deletion of most of one of the repeated IS elements leaving only its distal end to provide a recognition site for transpositions. The ability of IS-like elements to promote integration of intervening non-IS DNA into new sites probably represents the most important role of the elements in the evolution of bacterial genomes.

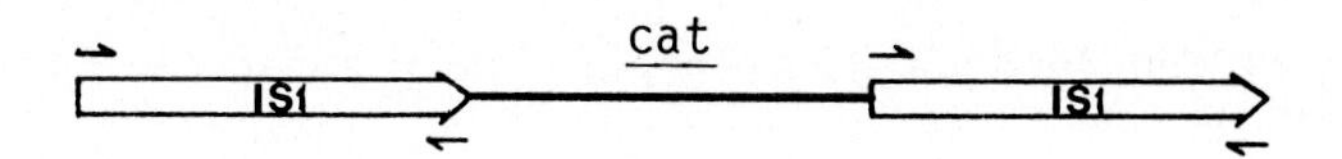

FIGURE 1. The Structure of Tn9.
The chloramphenicol resistance gene, cat, is flanked by two copies of IS1 in the same orientation (15). Because IS1 contains terminal inverted repeats (⟶), the extreme termini of Tn9 are also inverted relative to each other.

The chromosomes of bacteriophages λ and Mu can integrate into host DNA to form lysogens. Both these phage genomes have been considered as transposable elements and have served as models for the integration and transposition processes (5,8). As will be seen below, the mechanism of λ integration does not appear to typify that of other transposable elements while that of Mu may well do so. In the following discussion the term transposable element is taken to include Mu but to exclude λ.

SEQUENCE ANALYSIS OF TRANSPOSITIONS

Recombination mediated by transposable elements involves the formation of new DNA junctions between the ends of an element and sequences at the insertion site. During the last year a number of researchers have published the results of DNA sequence analyses of these sites for a variety of different elements (17-29; E. Auerswald and H. Schaller, personal communication; R. Musso, personal communication; D. Kamp and R. Kahmann personal communication). Four general observations have been made.

1. With the exception of Mu, both ends of any one of the analysed elements are almost identical consisting of imperfectly repeated sequences in inverted orientation. One consequence of this is that for the one transposon, Tn9, whose sequence is bounded by direct repeats of an IS element (IS1), the extreme ends consist of short inverted repeats (see Figure 1). While most elements have different terminal sequences from each other, there are some striking similarities (see below).

2. The new junctions formed always involve exactly the same base pair of the element. This applies both to transpositions (17-29) and to deletions (9). This presumably reflects a precise enzymatic recognition of the terminal sequences.

3. Insertion of a transposable element at a new site results in duplication of a short segment that pre-existed as a

single copy at the target site. One copy of this repeated sequence occurs in the same orientation adjacent to each terminus of the integrated element. The size of the segment is dependent on the element transposed. Integrations of IS1 (17,18,20) and Tn9 (19), Tn5 (E. Auerswald and M. Schaller, personal communication), Tn10 (22) and Tn903 (21) create 9 base duplications while those of IS2 (23,24, R. Musso, personal communication), Tn3 (25,29), γδ (27), a transposable sequence on pSC101 (26), and bacteriophage Mu (28, D. Kamp and R. Kahmann, personal communication) result in 5 base duplications. Although integration of an element always results in duplication of a specific length of host DNA, the sequence of this segment varies from one insertion site to another. Recent results suggest that insertions of IS4 into a single site in the E. coli galT gene (30) may result in an 11 b.p. flanking repetition (P. Starlinger, personal communication). If this finding is confirmed, IS4 will represent a new class among transposable elements, all of which so far create 9 b.p. or 5 b.p. repeats.

4. There appears to be no conservation of the sequence of the 9 or 5 base repeated segment associated with each element. Nor, in general, is there any significant homology between the ends of an element and the regions adjoining its integration site, or between the different integration sites of an individual element, although independent insertions at a single site have been observed (20,22,30). While the basis for specificity in the selection of a particular integration site is not yet understood, it appears not to reside simply in DNA homologies between the site and the ends of the element (19,27).

MODELS FOR TRANSPOSITION

The variation in sequence of the 9 or 5 base repetitions that result from transposition of an element suggest that these sequences alone do not carry the signals for recognition of the target site. These duplications, however, have implications for the mechanism of integration. When first observed, two possible origins for the repetitions were postulated (17,18). Either the transposing element carried the second copy, perhaps as an attachment sequence, or the repetition was made by copying the pre-existing target site sequence during the insertion process.

Integration of bacteriophage λ at its chromosomal attachment site provides a precedent for the first process. Insertion of DNA occurs by recombination between the 15 b.p. "core" regions common to both phage and bacterial attachment sites (31). This results in a direct repeat of the 15 b.p.

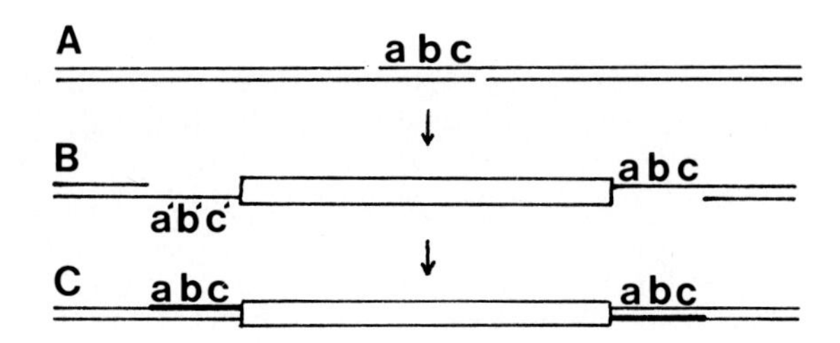

FIGURE 2. Duplication of a host sequence during transposition.
A. Staggered nicks are made at the target site.
B. The transposable element is inserted into the nicked site (no mechanism is implied)
C. The single stranded regions are filled in (heavy lines) resulting in a short host sequence (abc) being repeated at each end of the element.

sequence, one copy occurring at each end of the integrated prophage genome. Occasionally λ integrates into the bacterial chromosome at secondary sites which have only partial homology to the 15 b.p. core sequence. This results in non-identical sequences at the ends of the prophage, each segment being a hybrid of the phage and bacterial segments (32). A number of recent observations with Tn9 (19) Tn10 (22), γδ (27) and Tn3 (25) argue strongly that transposable elements integrate by a mechanism distinct from that of λ.

1. An element can transpose from a single location to many different sites. At each site an exact but different repetition is created (19,22,25,27).

2. An element can transpose to a single location from two different sites, each presumably associated with different flanking repetitions (22).

3. Even when an element is presented with a target DNA that contains an identical sequence to that flanking the "donor" element, subsequent transpositions show no preference for this sequence (19). Thus the element carries no memory of its original site and adjoining repeated sequence.

These observations indicate that the second copy of the repeated sequence is not brought in by the transposing element nor does the sequence adjoining the "donor" element play a role in the selection of an insertion site. These studies (19,22,27) also indicated that the flanking repetition is not required for subsequent transposition. This shows that recombination between the repeated sequences to generate a circular form of the element (as is the case in excision of a λ prophage) is not part of the normal transposition process. That such a mechanism may be used to generate revertants of insertion mutations remains to be tested.

All results, then, suggest that in contrast to the 15 b.p. segments flanking the λ prophage, those of 9 or 5 b.p. adjoining other transposable elements are generated by copying the single sequence present at the insertion site. The simplest and thus most favored model is that insertion occurs between staggered nicks in the target DNA. Subsequent filling in of the resulting single stranded regions would result in creation of a direct repeat of the original sequence between the nicks (see Figure 2).

A third possibility, that duplication occurs as an independent event before insertion of the element, has always appeared to us to be improbable. Recent analysis of γδ transposition to pBR322 (27,33), in fact indicates that a pre-existing five base duplication is not a strongly preferred site for integration. At least one such site is present in a non-essential region of pBR322 (nucleotides 3231-3240, G. Sutcliffe, personal communication) and insertion of γδ at this site would not affect the genes for tetracycline or penicillin resistance. Yet insertions in the vicinity of this site occur no more frequently than insertions that generate new 5 base duplications.

Staggered nicks at an element insertion site could be generated by host- or element-specified enzymes. The fact that transposition of several apparently unrelated elements resulted in the same size of host sequence duplication was used to support the idea that a single host enzyme might be used by each group of elements. As yet there is no evidence to support this speculation. At least some, and probably all, transposable elements are known to encode protein(s) required for transposition (34,35), although the specific role played by such proteins is not yet known. It seems likely that at least one function of an element-encoded transposition protein would be to recognize and make the precise cut (or direct the cutting) at the ends of the element.

Recently, two detailed molecular models for transposition have been described (36,37). In addition to explaining duplication of a host sequence during transposition, these models are consistent with other recA-independent genetic rearrangements associated with transposable elements such as the formation of deletions and replicon cointegrates. Both result in transposition of an element without excision or loss from its original site, as is suggested for Mu (38,39). Grindley and Sherratt (36) propose that transposition is initiated by transferring a single strand of one end of an element to a new site; the second end is only transposed after replication through the element is completed. In contrast, Shapiro (37) suggests that opposing single strands of both ends are transferred to the target site before any replication occurs. This results in recombinant structures as obligatory intermediates that have to be resolved by a second site-specific exchange to give the final transposition. For transposition from one replicon to another this intermediate is a fusion of the two replicons with two copies of the element, one at each new joint. The formation of such a structure, although possible in our model (36) is not a required intermediate in transposition. One particularly interesting result obtained recently with Mu indicates that both ends of an element are required for events that apparently only involve transfer of a single end, that is, the formation of deletions (39). This suggests that a specific interaction between the two ends (the inverted repeat sequences) of an element is required for promoting all end-specific recombination events. Such cooperativity, while predicted by Shapiro's model (37), is not an intrinsic part of our model (36) although it is not excluded.

RELATIONSHIPS AMONG TRANSPOSABLE ELEMENTS

The terminal sequences of four elements that are associated with 9 b.p. flanking repetitions and five that are associated with 5 b.p. repetitions have been determined in various laboratories (9, 17-29). These sequences, shown in Table I, indicate some interesting homologies. The most striking similarities exist between 3 of the 5 elements that create 5 b.p. duplications: Tn3, $\gamma\delta$ and an IS-like element on pSC101. These three elements have inverted repeats of 38,35 and 36 (imperfect) base pairs respectively. These terminal inverted repeats are highly (>80%) homologous. The longest contiguous stretch of bases in common is ACGAAAAC at positions 21-28. Remarkably, the first seven bases of this stretch occur at exactly the same position in the c arm of Mu and are present in the SE arm at positions 12-18. (A degenerate sequence, CGAAA, occurs in the "prototype" position, 22-26, in the SE arm). The conservation of this sequence could indicate that it is a required recognition site for these elements, or simply that they may have a common evolutionary origin. Significantly, perhaps, the fifth element that duplicates a 5 b.p. sequence upon transposition, IS2, does not contain this oligonucleotide; nor are there any other striking homologies between IS2 and the rest of this group of elements.

Among the elements with 9 base adjoining repetitions, significant homologies exist between IS1/Tn9, Tn10 and Tn903 but the fourth, Tn5, appears to be unrelated. Eleven of the 15 bases at positions 19-33 of the left arm of IS1 are found, shifted one place to the left, in Tn903. Overlapping this region, 9 of the bases between 15-25 at the left end of IS1 are

TABLE I

TERMINAL SEQUENCES OF TRANSPOSABLE ELEMENTS[a]

A. Those Associated with a 9 b.p. Repeated Sequence

Element	End		Sequence
IS1; Tn9	Left	5'	GGTGATGCTGCCAACTTACTGATTTAGTGTATGATGGTGT
	Right	5'	...A...ACT........T....AGTGTTTTATGTTCAGA
Tn10	Left	5'	CTGATGAATCCCCTAATGATTTTTATCAAAATCATTAAGT
	Right	5'	......................GGTA.............
Tn903	Both Ends	5'	GGCTTTGTTGAATAAATCAGATTTCGGGTAAGTCTCCCCC
Tn5	Both Ends	5'	CTGACTCTATACACAAGTAGCGTCCTGAACGGAACCTTTC

B. Those Associated with a 5 b.p. Repeated Sequence

Element	End		Sequence
IS2	Left	5'	TGGATTTGCCCC-TATATTTCCAGACATCTGTTATCACTT
	Right	5'	.A..C.G.....C.GA..C........A.CAA........
γδ	γ-End	5	GGGGTTTGAGGGCCAATGGAACGAAAACGTACGTTTATGG
	δ-End	5'	................................AAGGA
pSC101	End 1	5'	GGGGTTTGAGGTCCAACCGTACGAAAACGTACGGTAAGAG
	End 2	5'	....C.....G....TG.A.............T..GTGA
Tn3	Left	5	GGGGTCTGACGCTCAGTGGAACGAAAACTCACGTTAAGGG
	Right	5'	......................................CA
Mu	c-End	5'	TGCATTGATTCACTTGAAGTACGAAAAAAACCGGGAGGAC
	S-End	5'	TGAAGCGGCGCACGAAAAACGCGAAAGCGTTTCACGATAA

[a]5'-Terminal sequences of each end of an element are shown. Dots in place of letters indicate regions of the second end that are identical to the first end. In part A regions of homology discussed in the text are indicated either by underlining, or by crosses or dots above the sequences. In part B regions of pSC101 and Tn3 that are identical to γδ are underlined; the sequence ACGAAAA found also in Mu is shown in large type.

found in positions 12-22 of Tn10. Closer to the ends, the sequence TGATGAATCC is found at positions 2-11 of Tn10. Parts of this sequence occur in Tn903 at 6-13 (TGTTGAATaa); in the left arm of IS (TGATGctgCC), and in the right arm of IS1 (TaATGAcTCC) at positions 3-12. Again, it is unclear whether these homologies, absent in Tn5, indicate common recognition sites or common evolutionary origin or merely coincidence.

REFERENCES

1. Cohen, S.N. (1976). Nature 263, 731.
2. Starlinger, P., and Saedler, H. (1976). Current Topics Microbiol. Immunol. 75, 111.
3. Kleckner, N. (1977). Cell 11, 11.
4. Nevers, P., and Saedler, H. (1977). Nature 268, 109.
5. Bukhari, A.I., Shapiro, J.A., and Adhya, S., eds. (1977). DNA Insertion Elements, Plasmids and Episomes. Cold Spring Harbor Laboratory, New York.
6. Gill, R., Heffron, F., Dougan, G., and Falkow, S. (1978). J. Bacteriol. 136, 742.
7. Shapiro, J.A. and MacHattie, L.A. (1978). Cold Spring Harbor Symp. Quant. Biol. 43, In press.
8. Bukhari, A.I. (1976). Ann. Rev. Genet. 10, 389.
9. Ohtsubo, H., and Ohtsubo, E. (1978). Proc. Natl. Acad. Sci. USA 75, 615.
10. Johnsrud, L. (1978). Molec. Gen. Genet. 169, 213.
11. Stanisich, V.A., Bennett, P.M., and Richmond, M.H. (1977). J. Bacteriol. 129, 1227.
12. So, M., Heffron, F., and McCarthy, B.J. (1979). Nature 277, 453.
13. Cornelis, G., Ghosal, D., and Saedler, H. (1978). Molec. Gen. Genet. 160, 215.
14. Jacoby, G.A., Rogers, J.E., Jacob, A.E., and Hedges, R.W. (1978). Nature 274, 179.
15. MacHattie, L.A., and Jackowski, J.B., (1977). In ref. 5, pp. 219-228.
16. Sharp, P.A., Cohen, S.N. and Davidson, N. (1973). J. Mol. Biol. 75, 235.
17. Calos, M.P., Johnsrud, L. and Miller, J.H. (1978). Cell 13, 411.
18. Grindley, N.D.F. (1978). Cell 13, 419.
19. Johnsrud, L., Calos, M.P., and Miller, J.H. (1978). Cell 15, 1209.
20. Kuhn, S., Fritz, H.-J., and Starlinger, P. (1979). Molec. Gen. Genet. 167, 235.
21. Oka, A., Nomura, N., Sugimoto, K., Sugisaki, H., and Takanami, M. (1978). Nature 276, 845.
22. Kleckner, N. (1979). Cell 16, 711.

23. Rosenberg, M., Court, D., Shimatake, H., Brady, C., and Wulff, D.L. (1978). Nature 272, 414.
24. Ghosal, D., Sommer, H., and Saedler, H. (1979). Nucleic Acids Res. 6, 1111.
25. Ohtsubo, H., Ohmori, H., and Ohtsubo, E. (1978). Cold Spring Harbor Symp. Quant. Biol. 43, In press.
26. Ravetch, J.V., Ohsumi, M., Model, P., Vovis, G.F., Fischhoff, D., and Zinder, N.D. (1979). Proc. Natl. Acad. Sci. USA 76, 2195.
27. Reed, R.R., Young R.A., Steitz, J.A., Grindley, N.D.F. and Guyer, M.S. (1979). Proc. Natl. Acad. Sci. USA, In press.
28. Allet, B. (1979). Cell 16, 123.
29. Takeya, T., Nomiyama, H., Miyoshi, J., Shimada, K., and Takagi, Y. (1979). Nucleic Acids Res. 6, 1831.
30. Pfeifer, D., Habermann, P., and Kubai-Maroni, D. (1977). In ref. 5, pp. 31-36
31. Landy, A., and Ross, W. (1977). Science 197, 1147.
32. Bidwell, K., and Landy, A. (1979). Cell. 16, 397.
33. Guyer, M. (1979). J. Mol. Biol. 126, 347.
34. Heffron, F., and Bedinger, P., Champoux, J.J., and Falkow, S. (1977). Proc. Natl. Acad. Sci. USA 74, 702.
35. Meyer R., Boch, G., and Shapiro, J.A. (1979). Molec. Gen. Genet. 171, 7.
36. Grindley, N.D.F., and Sherratt, D.J. (1978). Cold Spring Harbor Symp. Quant. Biol. 43, In press.
37. Shapiro, J.A. (1979). Proc. Natl. Acad. Sci. USA 76, 1933.
38. Ljungquist, E., and Bukhari, A.I. (1977). Proc. Natl. Acad. Sci. USA 74, 3143.
39. Faelen, M., and Toussaint, A. (1978). J. Bacteriol. 136, 477.

IN VITRO REPLICATION OF ADENOVIRUS DNA

Thomas J. Kelly, Jr. and Mark D. Challberg

Department of Microbiology
Johns Hopkins University School of Medicine
Baltimore, Maryland 21205

ABSTRACT We have characterized a soluble enzyme system from adenovirus-infected cells that is capable of replicating exogenously added adenovirus DNA *in vitro*. Maximal DNA synthesis is observed when DNA-protein complex, isolated from purified virions, is added as template. Under these conditions DNA replication in the *in vitro* system closely resembles adenovirus DNA replication *in vivo*. Replication begins at or near either terminus of the template, and daughter strand synthesis proceeds exclusively in the 5' to 3' direction. Thus, the r daughter strand is synthesized from right to left on the conventional map of the adenovirus genome, and the l daughter strand is synthesized from left to right. The ends of the *in vitro* product are tightly associated with protein. In contrast, when deproteinized adenovirus DNA is added to the *in vitro* system, the limited DNA synthesis that is observed appears to be due to a repair-like reaction. In particular, synthesis begins at many sites within the template, and the synthetic product consists largely of short DNA chains that are covalently linked to template DNA strands. These results suggest that the specificity of initiation of *in vitro* replication depends upon the integrity of the 55K terminal protein attached to the 5' ends of the DNA strands of adenovirus DNA-protein complex.

INTRODUCTION

Little is known about the molecular mechanisms of DNA replication in eukaryotic cells. By analogy with recent experience in prokaryotic systems, it seems likely that analysis of the replication of the genomes of DNA viruses will provide a useful approach to this problem. It also seems clear that a complete description of the mechanisms of eukaryotic DNA replication will require the development and subsequent analysis of *in vitro* systems that carry out the replication of DNA templates of defined structure. We have

ISBN 0-12-198780-9

recently described a soluble enzyme system that is capable of replicating exogenously added adenovirus (Ad) DNA *in vitro* (1). The system consists of an extract derived from the nuclei of adenovirus-infected cells and utilizes a DNA-protein complex isolated from purified Ad virions as template. The latter consists of duplex adenovirus DNA molecules whose 5' termini are covalently linked to a protein with a molecular weight of 55,000 (2-6). Several lines of evidence indicate that DNA replication in this *in vitro* system closely resembles adenovirus DNA replication *in vivo* (see reference 7 for a review of *in vivo* replication). First, replication is dependent on factors that are specific to virus-infected cells; extracts prepared from uninfected cells are inactive. Second, during the course of the *in vitro* reaction branched DNA molecules with structural features identical to those of *in vivo* replication intermediates (8) are formed. Finally, the product of *in vitro* replication consists principally of long adenovirus DNA strands that are hydrogen-bonded, but not covalently linked, to the input DNA template. Many of the progeny DNA strands approximate full length.

In this report we present additional evidence that DNA replication in the *in vitro* system faithfully mimics adenovirus DNA replication *in vivo*. In particular, we show that replication begins at or near either terminus of the template and that daughter strands grow from their 5' termini toward their 3' termini. Our data further suggest that the specificity of initiation of *in vitro* replication depends upon the integrity of the 55K protein attached to the 5' termini of the template DNA strands. Finally, we present preliminary data suggesting that the ends of newly synthesized adenovirus DNA molecules are tightly bound to protein.

METHODS

<u>Conditions for *In Vitro* Synthesis.</u> Nuclear extracts and DNA templates were prepared as previously described (1). The standard reaction mixture for *in vitro* DNA synthesis contained 50 mM HEPES, pH 7.5, 5 mM $MgCl_2$, 0.5 mM dithiothreitol, 50 μM each dGTP, dATP, dTTP and dCTP, 2 mM ATP, 150 ng of Ad5 DNA-protein complex and 25 μl of nuclear extract, in a volume of 0.1 ml. Incubations were carried out at 37°C. For radioactive labeling of the *in vitro* product, the concentration of the four deoxynucleoside triphosphates was decreased to 20 μm each, and [α-^{32}P]deoxyguanosine triphosphate was added to a final specific activity of 1-10 Ci/mmole. To isolate the product of the reaction

following incubation, EDTA was added to a final concentration of 25 mM, and the reaction mixture was incubated with either 1 mg/ml Pronase (Calbiochem) in the presence of 0.6% SDS, or 0.1 mg/ml Proteinase K (EM Biochemicals) at 37°C for 2 hr. The resulting solution was extracted with phenol and dialyzed, first against 10 mM Tris-HCl, pH 8.0, 1.0 M NaCl, 1 mM EDTA, and then against 10 mM Tris-HCl, pH 8.0, 1 mM EDTA. In specific cases (see below) the *in vitro* product was isolated from reaction mixtures without the use of proteolytic enzymes by the following procedure: the product was collected by ethanol precipitation, redissolved in 10 mM Tris-HCl, pH 8.0, 1 mM EDTA, 1% SDS, 1% β-mercaptoethanol, and incubated at 60°C for 15 min. The DNA product was then separated from proteins and lower molecular weight substances by gel filtration on Sepharose 2B (Pharmacia) in a buffer containing 50 mM NaCl, 10 mM Tris-HCl, pH 8.0, 1 mM EDTA, 0.1% SDS. After pooling fractions containing DNA, the SDS was removed by extraction with isobutanol.

Restriction Enzyme Digestion and Gel Electrophoresis. DNA synthesized *in vitro* was digested with the Hpa I or Hind III restriction enzyme in 25 μl reaction mixtures containing 20 mM Tris-HCl, pH 7.5, 5 mM β-mercaptoethanol, 7 mM $MgCl_2$, 100 μg/ml gelatin and 5 units of enzyme (obtained from New England Biolabs). After incubation at 37°C for 2 hr, 2.5 μl of 0.4 M EDTA was added, and the reaction product was precipitated with ethanol. The precipitate was dissolved in 10 μl of 10 mM Tris-HCl, pH 8.0, 1 mM EDTA, 20% sucrose, 0.01% bromphenol blue and electrophoresed through a 1.4% agarose slab gel in a buffer containing 40 mM Tris-HCl, pH 7.5, 5 mM Na acetate, 1 mM EDTA.

To quantitate the radioactivity incorporated into the Hind III restriction fragments (Figure 1), the gel was dried and the fragments were located by autoradiography. Each fragment was cut from the dried gel and assayed for radioactivity in a liquid scintillation counter. The specific activity of a given fragment was calculated from the equation: Specific activity=$C/(N \cdot Xg)$, where C is the radioactivity incorporated into the fragment, N is a number of nucleotides in the fragment, and Xg is the mole fraction of guanine residues in the fragment. The mole fraction of guanine residues in each fragment was determined in an experiment in which Ad5 DNAs uniformly labeled *in vivo* with $[^{32}P]$orthophosphate and $[^{3}H]$ thymidine were mixed and digested to completion with the Hind III enzyme. After fractionation by agarose gel electrophoresis the ^{3}H and ^{32}P radioactivity in each fragment was determined. The mole fraction of guanine residues in a fragment, Xg, was calculated from the equation: $Xg=0.5-(X_t \cdot H)/P$, where X_t is the mole fraction of thymine residues in the Ad5 genome (0.215), H is the fraction of ^3H radioactivity in the fragment, and P is the fraction of ^{32}P radioactivity in the fragment.

Separation of the complementary strands of restriction fragments was accomplished by gel electrophoresis according to the method of Hayward (9).

RESULTS

Origin and Direction of Daughter Strand Synthesis. To localize the origin(s) of *in vitro* DNA synthesis within the Ad5 genome, we analyzed the distribution of radioactivity incorporated into the *in vitro* product during the course of the reaction. A standard reaction mixture containing [α-^{32}P] dGTP was sampled after incubation for 5, 10, or 30 min, and the radioactive products were digested with the Hind III restriction endonuclease. The resulting fragments were separated by agarose gel electrophoresis and assayed for radioactivity. As indicated in Figure 1, the terminal restriction fragments of the product were preferentially labeled at early times (5 min and 10 min). By 30 min the distribution of incorporated radioactivity in the product was almost uniform. Similar results were obtained when the products were digested with the Hpa I restriction enzyme (data not shown). The results of these experiments demonstrate that DNA synthesis *in vitro*, like that *in vivo*, begins at or near the termini of the template. This general conclusion is also supported by the results of a detailed electron microscopic analysis of *in vitro* replication intermediates which will be presented elsewhere.

The distribution of radioactivity between the separated strands of the terminal Hpa I restriction fragments of the *in vitro* product was also determined (Figure 2). After 10 min of incubation the distribution of radioactivity between the two strands of the terminal restriction fragments was highly asymmetric. Essentially all of the radioactivity incorporated into the Hpa I restriction fragment derived from the right end of the genome (fragment D) was recovered in the r strand. The radioactivity incorporated into the fragment derived from the left end of the genome (fragment E) was recovered in the l strand. After 60 min of incubation the incorporated radioactivity was still distributed unequally between the two strands of the terminal fragments, but the asymmetry was less marked. Since, by definition, the r strand has its 5' terminus at the right end of the genome, and the l strand has its 5' terminus at the left end of the genome, these results indicate that synthesis of both daughter strands starts at or near their 5' termini and proceeds towards their 3' termini.

Association of Protein with the Termini of Ad DNA Molecules Synthesized *In Vitro*. We have recently obtained evidence suggesting that the termini of adenovirus DNA molecules synthesized in the *in vitro* system are tightly bound to protein. The radioactive DNA product of a standard *in vitro* reaction was heated to 60°C for 15 min in 1% SDS, 1% β-mercaptoethanol and then separated from free proteins by gel filtration on Sepharose 2B in the presence of 0.1% SDS. After removal of the detergent, the product was digested with the Hpa I restriction enzyme, and the resulting fragments were analyzed by agarose gel electrophoresis.

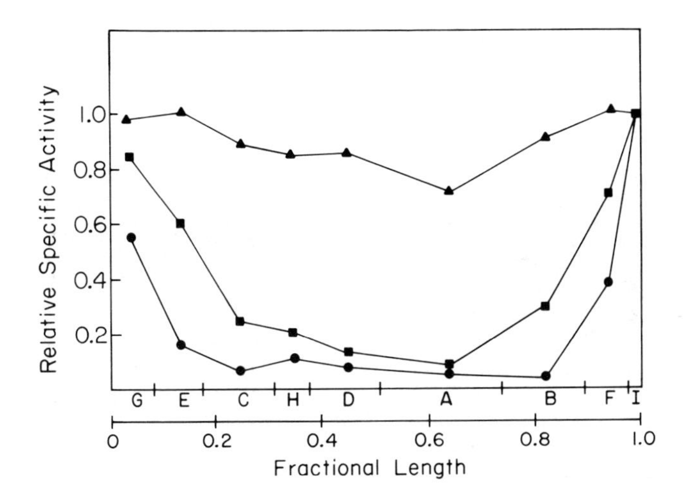

Figure 1. Distribution of incorporated radioactivity in the *in vitro* product as a function of time of incubation. A standard *in vitro* reaction mixture containing [α-^{32}P]dGTP was sampled after incubation for 5 min (●), 10 min (■), or 30 min (▲). The DNA product in each sample was digested with the Hind III restriction enzyme, and the specific radioactivity of each restriction fragment was determined as described in Methods. For purposes of comparison the specific activities of the fragments in all three samples were normalized to the specific activity of fragment I. The Hind III cleavage map of Ad5 DNA is taken from Sambrook, et al. (11).

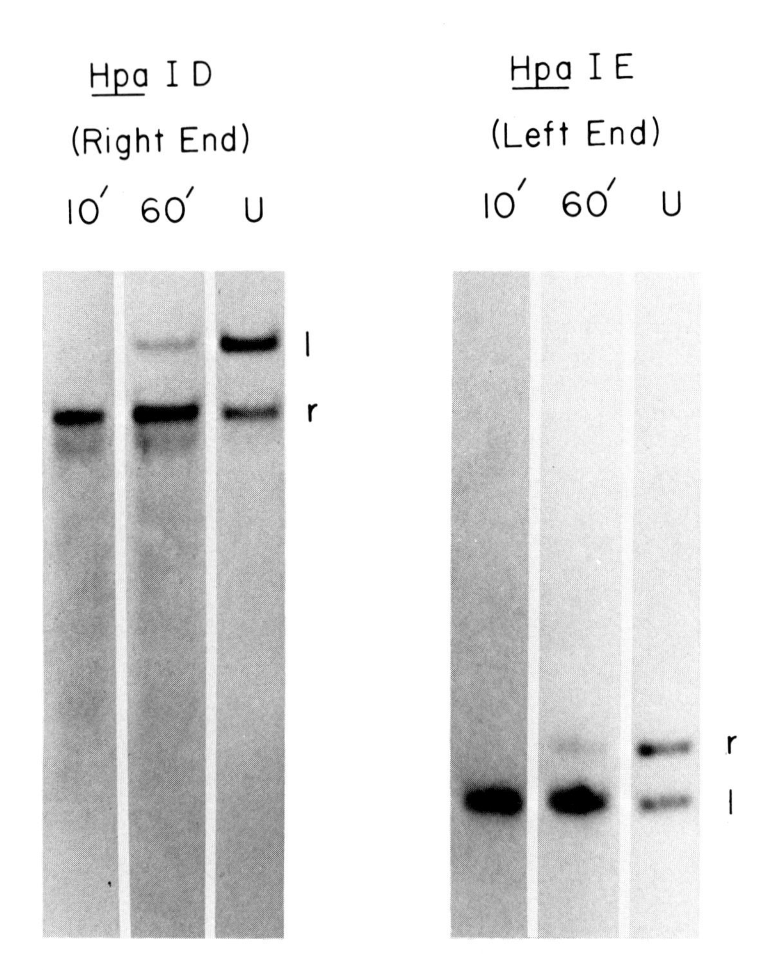

Figure 2. Distribution of radioactivity between the separated strands of terminal restriction fragments of the *in vitro* product. A standard reaction mixture containing [α-^{32}P]dGTP was sampled after incubation for 10 min or 60 min. The DNA in each sample was digested with the Hpa I restriction enzyme, and the two terminal fragments (D and E) were isolated by agarose gel electrophoresis. The complementary strands of each fragment were separated by electrophoresis through a 1% agarose gel according to the method of Hayward (9). Similar strand separations were performed on the terminal Hpa I fragments of uniformly ^{32}P-labeled Ad5 DNA (lanes marked U). The relative electrophoretic mobilities of the r and l strands of the terminal Hpa I fragments of adenovirus DNA were determined by J. Flint (personal communication).

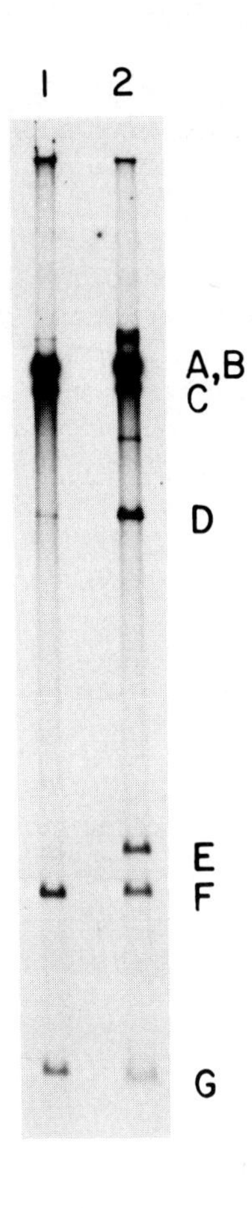

Figure 3. Agarose gel electrophoresis of Hpa I digests of the *in vitro* product. The radioactive DNA product synthesized during a 60 min incubation was purified from the reaction mixture by a procedure that did not involve the use of proteolytic enzymes (see Methods). An aliquot of the purified product was digested with the Hpa I restriction enzyme and electrophoresed through a 1.4% agarose gel (lane 1). A second aliquot was digested with the Hpa I enzyme and then treated with Proteinase K (10 μg/ml for 10 min at 37°) prior to electrophoresis (lane 2). The Hpa I cleavage map of Ad5 is given in reference 10.

As shown in figure 3, the bands corresponding to the internal fragments of the *in vitro* product were recovered in good yield, but the bands corresponding to the two terminal fragments (D and E) were virtually absent. In addition a significant fraction of the radioactivity was retained at the top of the gel. In contrast, when the restriction fragments were treated with Proteinase K prior to electrophoresis, the fraction of radioactivity retained at the top of the gel decreased, and both terminal fragments were recovered at the expected positions. This result strongly suggests that a protein(s) associated with the ends of the *in vitro* product is responsible for the anomalous behavior of the terminal restriction fragments on agarose gels. The obvious possibility that this protein(s) is related to the 55K protein known to be covalently attached to the 5' ends of mature adenovirus DNA strands is currently under investigation.

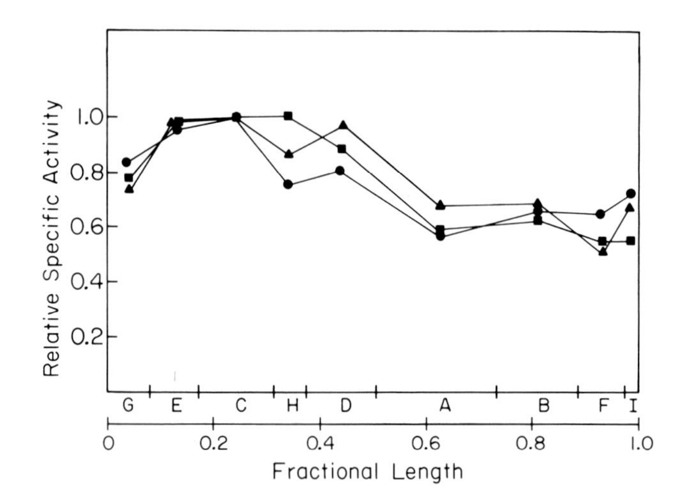

Figure 4. Distribution of radioactivity in the *in vitro* product synthesized with deproteinized Ad5 DNA as template. A reaction mixture containing [α-^{32}P]dGTP and deproteinized Ad5 DNA in place of Ad5 DNA-protein complex was sampled after incubation for 5 min (●), 10 min (■), or 30 min (▲). The DNA product in each sample was digested with the Hind III restriction enzyme, and the specific radioactivity of each fragment was determined as described in Methods. The specific activities of the fragments in all three samples were normalized to the specific activity of fragment C.

<u>DNA Synthesis with Deproteinized Adenovirus DNA as Template</u>. We have previously reported (1) that deproteinized adenovirus DNA is significantly less effective than adenovirus DNA-protein complex in stimulating DNA synthesis by infected-cell extracts. The extent of incorporation of radioactive precursors into DNA in reaction mixtures containing deproteinized Ad5 DNA as template is generally only 20-25 percent of that observed in standard reaction mixtures containing Ad5 DNA-protein complex. The experiments presented in figures 4 and 5 were carried out to determine whether this quantitative difference in the efficiency of the two templates is a reflection of a fundamental difference in

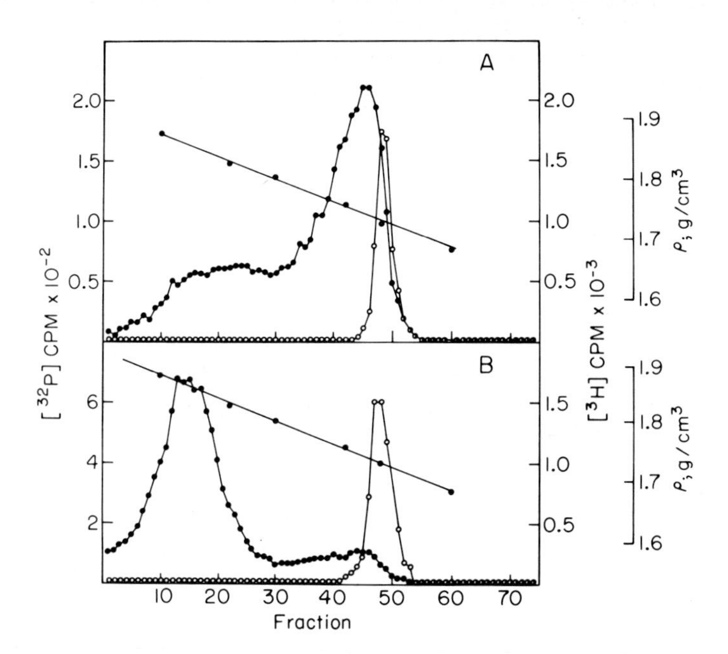

Figure 5. Isopycnic centrifugation of DNA synthesized *in vitro* in the presence of 5-bromodeoxyuridine triphosphate. Reaction mixtures containing [α-^{32}P]dGTP, 100 μM 5-BrdUTP (in place of dTTP) and either deproteinized Ad5 DNA (A) or Ad5 DNA-protein complex (B) as template were incubated for 2 hr. After deproteinization, the synthetic products were mixed with marker [^{3}H]Ad5 DNA, denatured, and analyzed by isopycnic centrifugation in neutral CsCl gradients as previously described (1). Assuming a G+C content of 57% for Ad5 DNA, the expected density for Ad5 DNA strands fully substituted with 5-BrdU is approximately 1.82g/cm^3 (12). (●), Product [^{32}P]DNA; (○), Marker [^{3}H]DNA.

reaction mechanism. Figure 4 shows the distribution of radioactivity in the *in vitro* product synthesized in response to deproteinized Ad5 DNA. In contrast to the results obtained when Ad5 DNA-protein complex was employed as template (Figure 1), the radioactivity incorporated during the first 5 min of the reaction was distributed almost uniformly throughout the genome. Furthermore, the observed distribution did not change upon longer incubation (10 min and 30 min). These findings indicate that the DNA synthesis that occurs in response to deproteinized Ad5 DNA can start at many sites within the template. Figure 5 shows the results of experiments in which deproteinized Ad5 DNA or Ad5 DNA-protein complex was added to reaction mixtures containing 5-bromodeoxyuridine triphosphate in place of dTTP. The DNA products synthesized in these reaction mixtures were denatured in alkali and analyzed by equilibrium centrifugation in CsCl gradients. As reported previously (1) the bulk of the product synthesized with Ad5 DNA-protein complex as template banded near the density expected for Ad5 DNA strands fully substituted with bromouracil. This result is consistent with the hypothesis that the extract can initiate DNA synthesis *de novo* on this template. In contrast, most of the product synthesized with deproteinized Ad5 DNA as template banded at densities only slightly greater than that of the unsubstituted marker DNA, indicating that it consists largely of short DNA chains that are covalently linked to the template DNA. We conclude that the limited DNA synthesis observed when deproteinized Ad5 DNA is added to the *in vitro* system consists predominately of a repair-like reaction which presumably occurs at nicks or single-stranded gaps in the template.

DISCUSSION

We have previously shown that extracts of adenovirus infected cells can support the replication of exogenously added adenovirus DNA *in vitro* (1). In this report we describe some of the characteristics of the initial phase of the *in vitro* replication reaction. Our data indicate that DNA synthesis begins at or near either terminus of the template and that daughter strand synthesis proceeds exclusively in the 5' to 3' direction. Thus, the origin of r strand synthesis is located at the right end of the adenovirus genome as convenionally drawn, and the origin of l strand synthesis is located at the left end. These findings confirm the close correspondence between *in vitro* and *in vivo* replication suggested by previous experiments (1). Our data are consistent with the hypothesis that *de novo* initiation of adenovirus DNA chains occurs in the *in vitro* system. However, direct confirmation of this possibility must await more detailed analysis of the early events in *in vitro* replication.

Little is known about the mechanism of initiation of adenovirus DNA replication *in vivo*. Since the adenovirus genome replicates as a linear DNA molecule, an understanding of the initiation process depends in a fundamental way upon an understanding of the mechanism responsible for priming the synthesis of the sequences at the extreme 5' terminus of daughter strands. It has been suggested (6) that the 55K adenovirus terminal protein may serve as the primer for daughter strand synthesis. In light of this hypothesis it is interesting that the ends of the DNA product of the *in vitro* replication reaction are complexed with protein. The linkage between DNA and protein in this complex must be quite stable since it is resistant to treatment with SDS and β-mercaptoethanol at 60°C. Further studies will be required to determine whether the protein component of the complex is related to the 55K adenovirus terminal protein and whether it is, in fact, covalently attached to the 5' ends of nascent daughter strands.

The data that we have presented suggest that the 55K terminal protein attached to the termini of the parental strands may also play an important role in the initiation process. Adenovirus DNA-protein complex is clearly more effective than deproteinized adenovirus DNA in stimulating DNA synthesis *in vitro*. Furthermore, the limited DNA synthesis that does occur in the presence of deproteinized adenovirus DNA does not resemble *in vivo* replication; in particular, synthesis can begin at many sites within the template, and the synthetic products consist largely, if not entirely, of short DNA chains that are covalently attached to template DNA strands. We believe that the terminal protein on the parental DNA serves to facilitate initiation of daughter strand synthesis at the molecular termini, perhaps via specific interactions with replication proteins present in the infected-cell extract. When the terminal protein is removed, enzymes in the extract can carry out repair synthesis at available 3'-OH termini within the template, but initiation of replication at the termini does not occur at a significant rate.

ACKNOWLEDGEMENTS

We thank Michael Stern for expert technical assistance and Stephen Desiderio and Gary Ketner for helpful discussions. This work was supported by a U.S.P.H.S. research grant from the National Cancer Institute.

REFERENCES

1. Challberg, M.D., and Kelly, T.J., Jr. (1979). Proc. Nat. Acad. Sci, USA 76, 655.
2. Robinson, A.J., Younghusband, H.B., and Bellett, A.J.D. (1973). Virology 56, 54.
3. Sharp, P.A., Moore, C., and Haverty, J.L. (1976). Virology 75, 442.
4. Carusi, E.A. (1977). Virology 76, 380.
5. Padmanabhan, R. and Padmanabhan, R.V. (1977). Biochem. Biophys. Res. Commun. 75, 955.
6. Rekosh, D.M.K., Russell, W.C., Bellett, A.J.D., and Robinson, A.J. (1977). Cell 11, 283.
7. Winnacker, E.L. (1978). Cell 14, 761.
8. Lechner, R.L. and Kelly, T.J., Jr. (1977). Cell 12, 1007.
9. Hayward, G.S. (1972). Virology 49, 342.
10. Mulder, C., Arrand, J.R., Delius, H., Kelly, W., Pettersson, U., Roberts, R.J. and Sharp, P.A. (1974). Cold Spring Harbor Symp. Quant. Biol. 39, 397.
11. Sambrook, J., Williams, J., Sharp, P.A. and Grodzicker, T. (1975). J. Mol. Biol. 97, 369.
12. Baldwin, R.L. and Shooter, E.M. (1963). J. Mol. Biol. 7, 511.

SELECTIVE CONSERVATION OF THE ORIGINS OF DNA REPLICATION (GENOMIC TERMINI) IN SPONTANEOUS DELETION MUTANTS OF THE PARVOVIRUS MVM[1]

Emanuel A. Faust and David C. Ward

Departments of Human Genetics and Molecular Biophysics and Biochemistry, Yale University School of Medicine, New Haven, Connecticut 06510

ABSTRACT We have isolated spontaneously arising deletion mutants of the parvovirus MVM after both a single high multiplicity passage and after serial undiluted passage. These deletion mutants, which possess 0.1-0.7 genome equivalents of DNA, contain two distinct types of viral DNA, designated type I D-DNA and type II D-DNA. Type I D-DNAs are predominantly single-stranded, "recombinant" molecules which have selectively conserved the self-complementary sequences derived from both genomic termini. Type II D-DNAs are double-stranded hairpin molecules which contain viral sequences that map entirely at the 5' end of the genome, between map coordinates 85.0 and 100.

Virtually all of the wt genome sequence is found in the total, heterogeneous, population of type I D-DNAs isolated after a single high multiplicity passage. Although the extent and position of the deletions in individual molecules vary significantly, sequences which map between coordinates 47.3 and 87.1 are clearly under-represented. The shortest molecules in the population lack ∿90% of the wt genome sequence and consist almost exclusively of sequences derived from within 5.0 map units (∿250 nucleotides) at both ends of the viral genome. These miniature recombinant molecules are selectively amplified during serial undiluted passage and are therefore believed to contain all of the critical recognition sites necessary for the replication of MVM DNA.

The type II DNAs, in sharp contrast to the type I genomes, are gradually lost from the total D-DNA population during serial undiluted passage (from ∿50% at passage 1 to <5% at passage 10). This suggests that type

[1]This work was supported by United States Public Health Service Grants GM-20124 and CA-16038

ISBN 0-12-198780-9

II D-DNA molecules are not competent for DNA replication and that they may arise as the result of fatal replication errors.

Deletion mutants of the type described here for MVM should be valuable generally as aids to future studies on parvovirus DNA replication, transcription and cell-virus interactions.

INTRODUCTION

The Parvoviridae are a family of animal and insect viruses characterized as having linear, single-stranded DNA genomes of 1.2-2.2 x 10^6 daltons (1). These viruses, which replicate within the nucleus of infected cells, provide a useful system for the study of both DNA replication and transcription in eukaryotes. Two groups of mammalian parvoviruses are recognized. The *Adeno-associated* virus (AAV) group are helper dependent in that they require a coinfecting Adenovirus to initiate AAV replication (2). They also encapsidate both plus and minus strands of DNA in separate virions (3,4). The *Parvovirus* group are autonomous viruses which package a unique strand of DNA (5,6) which is complementary in sequence to the viral transcription products (7,8).

The genome of Minute Virus of Mice (MVM), an autonomous parvovirus (AP), consists of ∿5000 nucleotide residues (6) with short, self-complementary sequences which form specialized duplex structures at each of the genome termini (9). *In vivo* studies of replicating molecules of the AP's H-1 and MVM as well as the helper-dependent AAV indicate that these duplex regions are likely to be intimately involved in the initiation of viral DNA replication (10-15). Direct sequence analysis at the 3' terminus of virion DNA obtained from four antigenically distinct AP's, including MVM (16), has shown that the first 115 nucleotide residues exist in a predominantly base-paired, hairpin configuration and that this region shares common features also found at the origin of DNA replication in several other animal and bacterial virus genomes (17-21).

In an attempt to define precisely the regions of the viral genome which are indispensable for DNA replication, spontaneously occurring deletion mutants of MVM have been isolated and the structure and sequence organization of the deleted viral genomes have been determined. These studies provide evidence that the critical sites for MVM DNA replication lie entirely within 200-300 nucleotides encompassing the self-complementary sequences at both ends of the viral genome.

MATERIALS AND METHODS

Cells and Virus. The virus strain, MVM (T) was used throughout (22). The virus was propagated in Ehlich ascites (EA) cells (23) and assayed for infectivity and haemagglutination (HA) as described previously (22, 24).

Incomplete virions were produced either during a single-cycle, high-multiplicity infection (MOI=10 PFU/cell) with wt virus or after multiple rounds of undiluted passage. Serial undiluted passage was carried out as follows. In the first passage, 10^7 cells were infected with wt virus at a multiplicity of 5 PFU/cell and the infection was allowed to proceed for 24 hours. The cells were harvested by centrifugation, the medium was decanted and cell pellets were resuspended in 1-2 ml of sterile TE 8.7 buffer (0.05 M Tris-HCl, 1.0 mM EDTA, pH 8.7) and dispersed by vortexing. The cells were then lysed by three cycles of freeze-thawing. Approximately 0.1 ml of this cell lysate, containing 3000 HA u was used to infect 10^7 cells in the second passage. Adsorption was carried out at 37° for 60 minutes at 10^7 cells/ml in 1.0 ml of phosphate-buffered saline containing 5 mM each of $CaCl_2$ and $MgCl_2$. Following adsorption, the cells were diluted to 2-3 x 10^5 cells/ml in D medium (Dulbecco modification of Eagle minimal medium), supplemented with 10% fetal calf serum. Infected cells were again harvested at 24 hours post-infection. Subsequent passages were carried out in a similar fashion.

Purification of Incomplete Virus Particles. Generally, full and empty virions were purified as described previously (24), but the procedure was modified to provide a better separation of defective virus particles. The protocol, which involves repeated cycles of equilibrium centrifugation in CsCl gradients, will be described in detail elsewhere.

Isolation of D-DNA from Purified Virions. Purified virions were precipitated from CsCl solution by the addition of 7 volumes of methanol (24). The virus pellets were dissolved in 0.1 M NaOH and the DNA purified by centrifugation in alkaline sucrose gradients (6). Centrifugation was carried out in an SW41 rotor at 38K for 24 hours at 4°C. The DNA was recovered from these gradients, dialyzed exhaustively against TE 7.4 buffer (0.01 M Tris-HCl, 0.2 mM EDTA), and concentrated by evaporation in a stream of nitrogen.

Enzymes and Enzyme Assay Conditions. *E. coli* DNA polymerase I (large fragment lacking 5' to 3' exonuclease activity) was obtained from Boehringer Mannheim Corp. Bacteriophage λ exonuclease (one unit gives 10 nanomoles acid-soluble nucleotides in 30 minutes at 37°) was the generous gift of C. Radding. Nuclease S_1 from *Aspergillus oryzae* was obtained from Sigma Chemical Co. Mung bean nuclease (specific activity 1.4 x 10^6 units/mg) was from P-L Biochemicals, Inc. Restriction endonucleases Mbo I from *Moraxella bovis*, Hinf I from *Haemophilus influenzae* R_f, Hind II and Hind III from *Haemophilus influenzae* R_d and HhaI from *Haemophilus haemolyticus* were obtained from New England Biolabs and Boehringer Mannheim.

The *in vitro* synthesis of wt RF DNA and D-RF DNA using *E. coli* DNA polymerase was done as previously described (6). The reaction and assay conditions used for S_1 nuclease were the same as those reported by Bourguignon *et al*. (6), except that dCTP was not included in the reaction mixture. Mung bean nuclease digestions were performed as described by Chow and Ward (9). Restriction endonuclease reaction mixtures contained in a total volume of 0.05-0.1 ml: 0.05 M NaCl, 0.005 M $MgCl_2$, 0.01 M Tris-HCl pH 7.4, 0.005 M dithiothreitol and 1-2 units of enzyme. Incubation was at 37° C for 30 minutes and the reactions were stopped by the addition of EDTA to a final concentration of 0.01 M. λ exonuclease digestion mixtures contained 67 mM glycine-KOH buffer pH 9.6, 4 mM $MgCl_2$, 3 mM dithiothreitol, 0.5 mM EDTA, 0.1-1.0 μg DNA and 0.2-0.5 units of exonuclease. The reaction was stopped by adjusting the reaction mixture to 0.01 M with respect to EDTA and then extracting the mixture with phenol:chloroform (1:1).

Hydroxylapatite chromatography. DNA samples were adsorbed to a 0.5 ml column of hydroxylapatite (Bio-Rad Laboratories, HTP grade) and then eluted with a 30 ml linear gradient of sodium phosphate buffer, pH 7.2 (0.05 M - 0.30 M). Step-wise elutions of single- and double-stranded DNA fractions were performed using 1.0 ml steps of 175 mM and 300 mM phosphate buffer respectively.

Benzoylated DEAE-Cellulose Chromatography. DNA samples were adsorbed to a 0.5 ml column of BD-cellulose (cellex-BD 100-200 mesh, Bio-Rad Laboratories) and eluted step-wise with solutions of TE buffer, pH 7.4 containing either 0.3 M NaCl, 1.0 M NaCl or 1.0 M NaCl plus 1.8% caffeine (Sigma Chemical Co.).

Electron Microscopy. The formamide modification of the Kleinschmidt technique described by Davis *et al*. (25) was used throughout. In general the procedure used was similar to that described by Singer and Rhode (13).

Gel Electrophoresis. 4% polyacrylamide slab gels (40 cm) with a 3.5% stacking gel were run in Jeppeson buffer (26). Electrophoresis was carried out at 0.5-1.0 ma/cm for 18-24 hours. Gradient gels were formed as described by Jeppeson (26). For autoradiography, gels were dried under vacuum and exposed to Kodak XR-5 double-sided X-ray film. Gels were prepared for fluorography by the method of Bonner and Laskey (27) and exposed to presensitized film (28) at -70° C.

RESULTS

Isolation of MVM Virions Containing Incomplete Genomes. Infectious "full" virions of MVM (ρ=1.41 gm/ml) are routinely resolved from "empty" virions (ρ=1.32 gm/ml) by centrifugation to equilibrium in CsCl density gradients. In previous studies (24), it was noted that these gradients contained a population of virions banding at intermediate densities which, although labeled with [^{3}H] thymidine, lacked infectivity. We report here an analysis of the structure and sequence organization of the deleted viral genomes found in these non-infectious particles after they have been obtained free of both "full" and "empty" virions by repeated banding in CsCl gradients. Typically, defective virions band as a heterogeneous population with densities in CsCl ranging from approximately 1.33 gm/ml to 1.38 gm/ml. This distribution is seen by labeling *in vivo* with [^{3}H] thymidine (^{3}H-TdR) or inorganic $^{32}PO_4$, by measuring the U.V. adsorption at 260 nm and also by measuring the HA activity. These defective virions can constitute a significant percentage of the total DNA-containing particles isolated from MVM-infected EA cells [$\sim$50% after a single high multiplicity passage (HMP)].

The average size of the D-DNA produced during a single HMP was estimated initially by sedimenting purified defective virions in alkaline sucrose gradients. Under these conditions, the DNA appears relatively heterogeneous in size and has an average sedimentation coefficient of 11-12 S. Thus, D-DNA has an average single-stranded molecular length of approximately 2000 nucleotides, i.e., roughly forty percent of the size of the intact viral genome (15 S, 5000 nucleotides). Additional measurements of the average size of this DNA population, which includes EM contour length measurements and sedimentation analysis in the presence of DNA markers of known molecular weight, have confirmed this initial size estimate. It should be noted in addition that the size of the DNA varies with particle density; particles which band at a greater density in CsCl contain larger DNA molecules than those which band at a lesser density (data not shown).

Defective virions of the related parvoviruses Lu III and AAV have been reported to have similar characteristics (29,30). All further analyses of D-DNA were performed on the material which was pooled following the sedimentation of purified defective virions in preparative alkaline sucrose gradients.

Resolution of Two Types of D-DNA. The existence of two subpopulations of D-DNA was demonstrated initially by chromatography on hydroxylapatite and by digestion of the D-DNA with the single-strand-specific nuclease S_1. The latter experiments were performed by mixing ^{32}P-labeled D-DNA and [^3H] thymidine-labeled wt virion DNA and subjecting the mixture to digestion with nuclease S_1. As expected, roughly seven to ten percent of the wt DNA in this analysis remains S_1-resistant (Figure 1). This is in agreement with the results obtained by Bourguignon *et al.* (6) and Chow and Ward (9) who demonstrated that this low level of S_1 resistance is due to the presence of two small duplex regions which comprise the hairpin termini of the viral genome. In contrast, the D-DNA exhibits a much greater degree of S_1 resistance (∿55%).

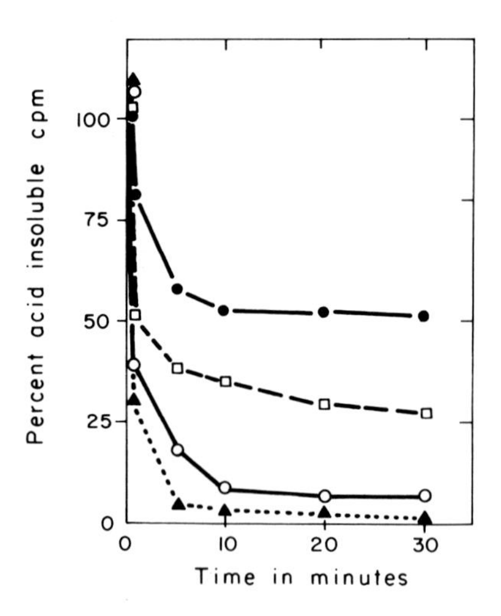

FIGURE 1. Kinetics of S_1 endonuclease digestion of wt MVM DNA (-O-) and incomplete virion DNA, before (-●-) and after (-□-) heat denaturation and quick-cooling. Pancreatic deoxyribonuclease digestion of incomplete virion DNA (-▲-).

Moreover, a significant portion of the D-DNA remains S_1-resistant even following heat denaturation and quick-cooling, suggesting in addition, that a significant portion of the D-DNA renatures spontaneously. Virtually all of the ^{32}P-labeled material in the D-DNA preparation was degraded by pancreatic deoxyribonuclease, as expected.

For chromatography on hydroxylapatite, ^{32}P-labeled D-DNA was adsorbed to the column together with [^{3}H] thymidine-labeled single-stranded and double-stranded positional markers and the column was developed with a sodium phosphate gradient. As can be seen in Figure 2, the D-DNA is resolved into two fractions. We conclude on the basis of these results and the results of Figure 1, that the D-DNA population consists of single-stranded as well as double-stranded molecules and that these two subpopulations are present in roughly equal amounts. This distribution has been observed for several independent D-DNA preparations, derived from virus stocks passaged a single time at a high MOI. The remainder of the paper is devoted to the analysis of the single-stranded (type I D-DNA) and double-stranded (type II D-DNA) fractions which were obtained following preparative fractionation of the total D-DNA population by hydroxylapatite (HAP) column chromatography.

Structure of Type II D-DNA. Type II D-DNA was subjected to a variety of denaturation and renaturation conditions and the treated DNA samples were then reanalyzed on HAP. Samples

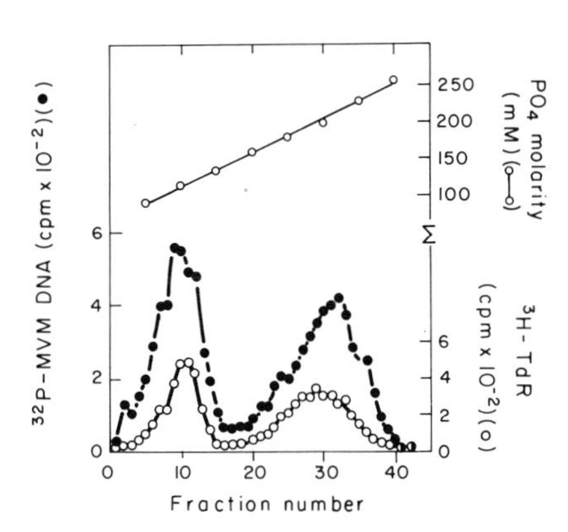

FIGURE 2. Hydroxylapatite chromatography of ^{32}P-labeled MVM D-DNA (-●-) in the presence of a mixture of wt MVM virion DNA and SV_{40} RFII DNA (-O-), both labeled with ^{3}H-thymidine.

treated in the following way were compared: (a) re-chromatography of native type II D-DNA, (b) denaturation at 96° C for 5 minutes and then quick-cooled in a 4° C ice-water bath, (c) digestion with mung bean nuclease followed by treatment as in b, (d) treatment as in c followed by incubation at 68° C to a $Cot_{1/2}$ value of approximately 10^{-2} and (e) treatment as in d but with a 20-fold molar excess of wt virion DNA added prior to denaturation. The following conclusions were drawn from these analyses (data not shown). First, approximately 85 percent of the D-DNA renatures spontaneously; second, this property of the DNA can be overcome for the most part by digestion with a single-strand-specific nuclease prior to denaturation; third, the spontaneously renaturable properties of type II D-DNA can be attributed to a covalent association of the complimentary strands via relatively large single-stranded loops; and fourth, the vast majority of the type II D-DNA is virus-specified. These conclusions are supported by the electron microscopic observations and restriction enzyme analyses outlined below.

Unfractionated D-DNA preparations containing both type I and type II molecules were examined in the electron microscope. The molecules observed were predominantly linear and exhibited a broad range of contour lengths (0.1-0.5 μ, avg. 0.25, n=65). For single-stranded DNA this corresponds to an average molecular length of ∿1800 nucleotides (13). A significant proportion of the molecules observed, about 30%, appear to consist of double-stranded stems and single-stranded loops (Figure 3). These "panhandle" structures are also heterogeneous in size. In addition, there is considerable variation in the size of both the loops and the stems in these molecules. For example, based on contour length measurements, the loops range in size from 300 to 1000 nucleotides. DNA samples treated with S_1 nuclease contain a much lower proportion of molecules judged to have "panhandle" structures, than do untreated samples. Circular structures lacking stems are not observed. The size distribution of the S_1-resistant molecules is similar to that seen for the stems in "panhandle" structures. Finally, only about 2 percent of the molecules examined appear as cross-over structures in which both ends of the molecules are free, making it unlikely that these could be mistaken for loop structures.

<u>Restriction Enzyme Analysis of Type II D-DNA</u>. Because of the predominantly double-stranded nature of type II D-DNAs, the sequence complexity of these molecules was examined directly after digestion with the restriction endonucleases Hinf I and Mbo I. The physical maps for these enzymes are given in Figure 4. This analysis revealed that type II

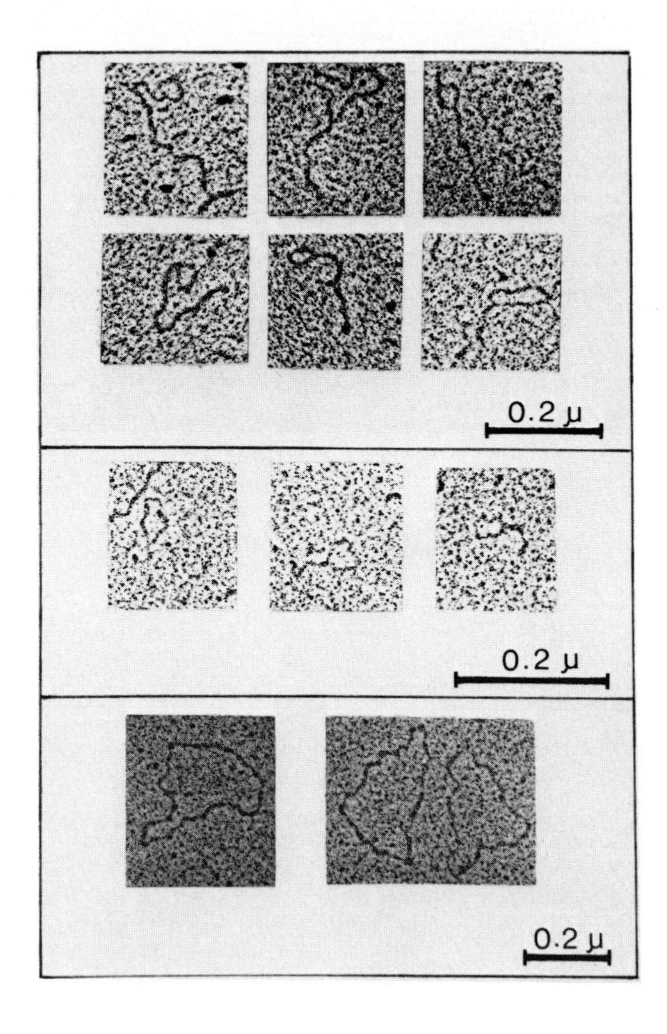

FIGURE 3. Electron micrograph of type II D-DNA molecules exhibiting a "panhandle" configuration. Molecules shown in the lower two panels are single-stranded circular ΦX174 DNA.

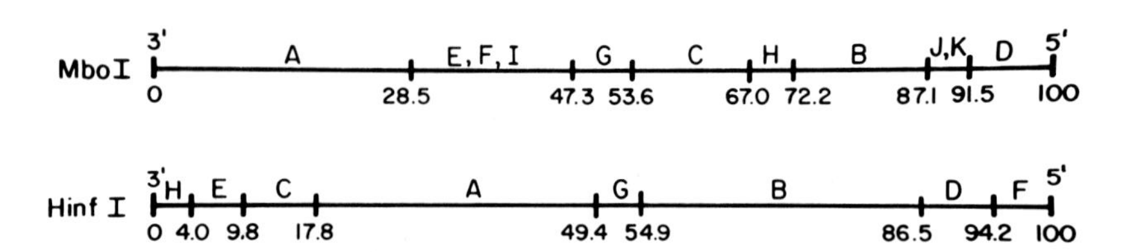

FIGURE 4. Cleavage maps of MVM RF DNA. The percentage values shown refer to a genome size of 5000 nucleotides. Mbo I fragments E (390 bp), F (390 bp) and I (190 bp) map within the boundaries shown, but their precise order is unknown. The same is true for Mbo I fragments J (175 bp) and K (80 bp). A 100 bp Hinf I fragment (I) has also not been mapped definitively but it is known to lie between positions 17.8 and 94.2. This fragment has been omitted from the map shown.

D-DNAs consist of sequences which are derived almost entirely from within approximately 15 map units at the 5' end of the viral genome. For example (see Figure 5), an Mbo I digest of type II D-DNA contains only the 5' terminal fragments 91.5/100 (D), J, and K, which comprise 9.0, 4.0 and 2.0 genome map units respectively. (Although Mbo I fragment K is not seen in Figure 5, this fragment has been detected in other analyses). There is also a trace of Mbo I fragment 72.2/87.1 (B) which maps just inboard of the terminal Mbo I fragments D, J and K. In contrast, Mbo I fragments mapping between 0/0 and 0/72 are not observed. After digestion of type II D-DNA with Hinf I, only two major bands are observed. These may correspond either to fragments C or D and E or F, since these sets of fragments are not resolved on the gels presented in Figure 5. Nevertheless, since Hinf I fragments 86.5/94.2 (D) and 94.2/100 (F) map together at the 5' end (see Figure 4) and since we do not detect either the 3' terminal Hinf I fragment 0/4 (H) or the internal fragments 17.8/49.4 (A), 49.4/54.9 (G) and 54.9/86.5 (B), this result is considered to be in support of the conclusion derived from the Mbo I digestion data.

<u>Analysis of Type I D-DNA Structure</u>. The wt MVM genome contains hairpin duplexes at both genome termini which can be detected as nuclease-resistant fragments following digestion

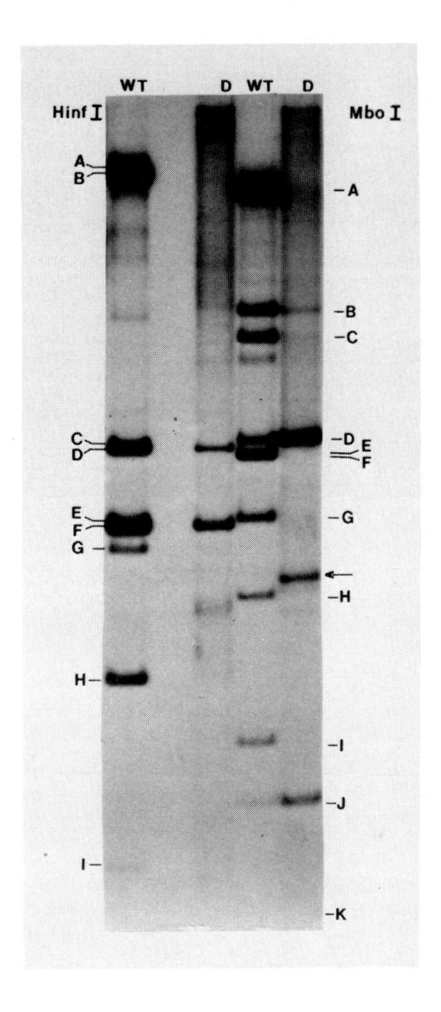

FIGURE 5. Restriction enzyme analysis of type II D-DNAs. Molecules labeled *in vivo* with ^{32}P-phosphate were cleaved with the restriction endonucleases Hinf I and Mbo I. *In vitro* synthesized wt RF was cleaved with the same enzymes and analyzed in parallel. Restriction enzyme fragments were resolved by electrophoresis in a 3.5 - 7.5% polyacrylamide gradient slab gel. Arrow denotes the 5' terminal fragment in "foldback" configuration (15).

with single-strand-specific nucleases (6,9). The smaller (115 nucleotide) hairpin fragment constitutes the 3' terminus of the viral genome. The larger (130 base pair) fragment constitutes the hairpin structure derived from the 5' terminus. In order to determine whether these small duplex regions are present in type I D-DNA, a sample of the DNA was treated with the single-strand-specific mung bean nuclease and the digestion products were analyzed by electrophoresis in an 8 percent polyacrylamide gel in parallel with a sample of wt virion DNA (Figure 6). The digestion of type I D-DNA clearly yields nuclease-resistant fragments. One of these comigrates precisely with the 115 nucleotide fragment seen in digests of wt DNA. The other nuclease-resistant band is seen as a doublet and migrates distinctly slower than the 130 base pair fragment found in digests of wt DNA. Using Hha I restriction fragments of ΦX174 RF as size makers, we have estimated these larger fragments to be between 140 and 145 base pairs in length. Although the basis for the apparent difference in size between these fragments and those derived from the 5' terminus of wt DNA is not known several possibilities can be suggested (see Discussion). Comparing the band intensities of nuclease-resistant fragments which are obtained in the case of both the wt and defective genomes, and taking into

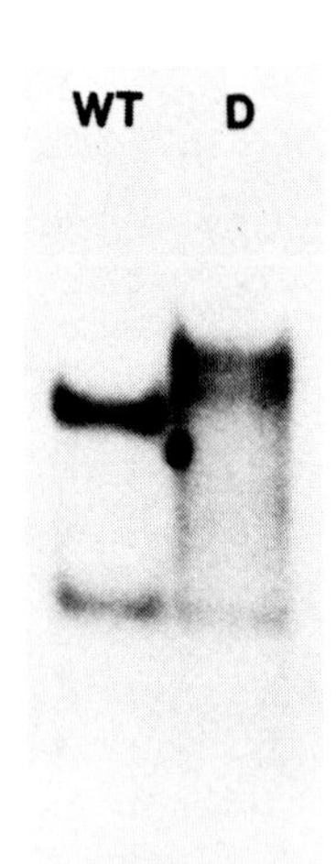

FIGURE 6. Polyacrylamide slab gel electrophoresis of MVM wt DNA and type I D-DNA, both labeled with ^{3}H-thymidine, after digestion with mung bean nuclease. Nuclease-resistant digestion products were analyzed in an 8% polyacrylamide gel. Electrophoresis was for 6 hours at 40 ma. The gel was dried and processed for fluorography.

account the smaller size of the D-DNA molecules, we have estimated that these small duplex regions are contained in a majority and possibly all of the molecules in the D-DNA population.

In order to examine the sequence organization within type I genomes, the DNA was first converted *in vitro* to a double-stranded form suitable for restriction enzyme analysis. Type I D-DNA molecules possess a 3' terminal hairpin structure that appears to be similar, if not identical, to that found in the wt genome. These molecules should be capable, therefore, of self-primed DNA synthesis (6), which leads to a duplex DNA in which the plus and minus strands are covalently linked. Thus, *E. coli* DNA polymerase I (lacking 5' to 3' exonuclease activity) was used to convert type I D-DNA into a duplex form (designated D-RF DNA) *in vitro*. The overall kinetics of this reaction are similar to that obtained when wt DNA is used as a template (data not shown). When compared with the cleavage patterns of wt RF DNA, certain Hinf I and Mbo I restriction fragments found in D-RF DNA digests were judged on a relative basis to be present in less than molar amounts (data not shown). These include Mbo I fragments G,C, H and B, which map contiguously on the viral genome between positions 47.3 and 87.1. Examination of a parallel Hinf I digest confirms that the deletions occur predominantly in this region. Accordingly, in a Hinf I digest of D-RF DNA, Hinf I fragments A,G and B are severely underrepresented. In contrast, restriction fragments which represent terminal portions of the viral genome, for example the 3' terminal Hinf I fragment 0/4 (H) and the 3' terminal Mbo I fragment 0/28.5 (A), were relatively well conserved. Also seen in relatively good yield were the 5' terminal Mbo I and Hinf I fragments D and J and D and F respectively. Thus, restriction enzyme digests of the total D-RF DNA population exhibit a gradient of band intensities which, in terms of the physical map, decreases toward the internal regions of the genome and increases toward the molecular termini. Essentially identical results were obtained in a separate restriction enzyme analysis involving hybrids formed between type I D-DNA and purified, ^{32}P-labeled (+) strands (data not shown). While an overall deletion of wt nucleotide sequence can be discerned in the total D-RF population by these techniques, further analyses were necessary in order to determine the sequence organization in individual molecules. Two experiments which address this problem are described below.

Type I D-DNA was fractionated according to size as indicated in Figure 7 (inset) and the sequence organization of these individual D-DNA populations was then examined by restriction enzyme analysis after conversion of the D-DNA to

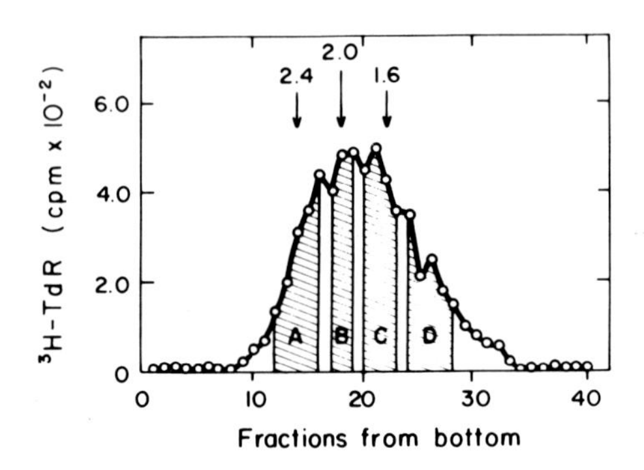

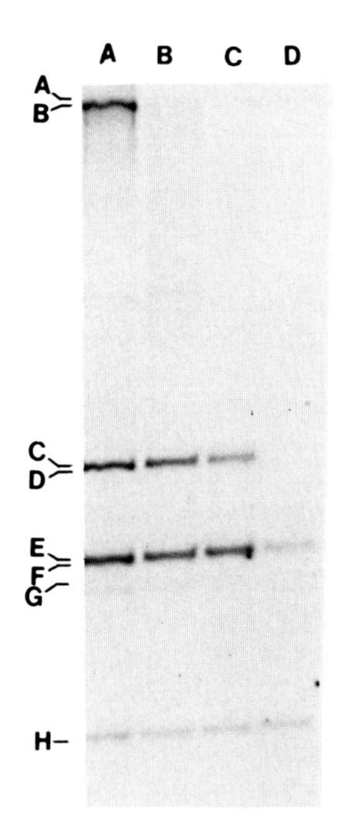

FIGURE 7. Restriction enzyme analysis of type I D-DNA fractionated on the basis of size. Various size classes of D-DNA were pooled from an alkaline sucrose gradient (as shown). After dialysis, the DNA was converted to D-RF *in vitro* and cleaved with restriction endonucleases. The digests were then analyzed by electrophoresis in 4% polyacrylamide slab gels.

the D-RF DNA form. It is evident that the sequence complexity of the D-DNA decreases as the DNA becomes smaller (see Figure 7). Moreover, the decrease in complexity derives from a progressive loss of DNA sequences which map at various internal positions in the viral genome. Accordingly, as the DNA becomes smaller, there is a rapid disappearance of the internal Hinf I fragments A,G and B, which map between 17.8 and 86.5. Indeed, the smallest DNA population appears to consist almost entirely of the 3' terminal Hinf I fragment 0/4 (H) and the 5' terminal Hinf I fragment 94.2/100 (F). Since the Hinf I fragments 4.0/9.8 (E) and 94.2/100 (F) are not resolved on the gels shown in Figure 7 it is possible that both of these fragments are present. However, we conclude that the majority, if not all, of this band represents the 5' terminal Hinf I fragment (F) on the basis of a parallel analysis of the D-RF DNA after cleavage with Mbo I (data not shown). In this study, a similar reduction in the sequence complexity was observed as the size of the D-DNA decreased. The increase in the relative molar abundance of the 5' terminal Mbo I fragment 91.5/100 (D) observed in the D-DNA from pools B and C can be rationalized only if the majority of the Hinf I (E/F) band represents the 5' terminal Hinf I fragment (F).

Proof that type I D-DNAs are true recombinants, with sequences from both genomic termini present together on individual molecules, was obtained in the following experiment. Restriction fragments representing regions within the 5' half and the 3' half of the viral genome were purified and these were then hybridized separately to samples of type I-D-DNA. The hybrids were purified on hydroxylapatite and BD cellulose columns; and the sequence organization of the type I D-DNA molecules selected by hybridization was examined after conversion _in vitro_ to D-RF DNA and digestion with Hinf I and Mbo I. The results of this analysis (not shown) demonstrated that D-DNA selected by hybridization to the 5' half of the viral genome contained the 3' terminal Mbo I fragment 0/28.5 (A) and the 3' terminal Hinf I fragment 0/4 (H). Conversely, D-DNA selected by hybridization to the 3' half of the viral genome contains the 5' terminal Mbo I fragment 91.5/100 (D). These results taken together with the data presented in Figure 7 suggest that individual molecules of type I D-DNA are recombinant genomes which contain sequences derived from both genomic termini although the extent and position of the deletions in individual molecules vary.

Changes in Type I and Type II D-DNAs during Serial Undiluted Passage. The incomplete genomes arising after several undiluted passages of a mixture of wild type and incomplete virions were found to differ in several respects from the single-passage D-DNA already described. Whereas the single-passage yield of D-DNA consists of the type I and type II subpopulations in roughly equal amounts, we have observed a progressive loss of the type II molecules during successive cycles of infection. Accordingly, at passage number ten, the D-DNA population consists almost entirely (95% or more) of type I structures as determined by HAP chromatography (data not shown). A second major change concerns the average size of the D-DNA. Although still extensively heterogeneous, the D-DNA obtained after the seventh undiluted passage is on the average only about 1200 nucleotides in length, as compared with an original size of 2000 nucleotides (data not shown). In order to derive some information about the possible significance of these changes, we have also examined the sequence complexity of the type I D-DNA molecules as a function of passage. As shown by restriction enzyme analysis, there is a significant reduction in the overall sequence complexity in the type I population which is related to passage number. Moreover, this reduction in sequence complexity derives from a progressive loss of internal regions of the viral genome and is accompanied by an enrichment of the sequences which map closer to both genomic termini. Thus, in digests of late passage D-RF DNA we have observed a relative increase in the band intensities of the terminal Mbo I fragments D,J and A as compared with the internal fragments E,F,G,H,C and B (Figure 8). This trend is confirmed by the Hinf I digestion pattern of late passage D-DNA, since we see a decrease in intensity of the internal fragments A,G and B whereas the 3' terminal Hinf I fragment 0/4 (H) is relatively well conserved. It is also apparent from the data presented in Figure 8, that there are a number of bands in these gel profiles which do not correspond to restriction fragments found in digests of wt RF DNA. Moreover, these also are not present in digests of single-passage D-RF DNA. (For example, compare lane a with lanes b and c in Figure 8.) Thus, after a number of high multiplicity passages, the defective genomes of MVM may acquire cellular DNA or undergo extensive sequence rearrangements. The possible implications of these findings are presented in the Discussion, below.

DISCUSSION

In the present study, we have obtained information relating to the structure and sequence organization of the incomplete genomes of the parvovirus, MVM. Based on a variety of experimental observations, we have concluded that the incomplete genomes exist as two structurally distinct populations, which can be easily distinguished on the basis of their degree of secondary structure.

Type I D-DNAs are a heterogeneous collection of predominantly single-stranded, recombinant molecules differing in size and sequence content. As a whole, this D-DNA population contains virtually all of the sequences found in wt virion

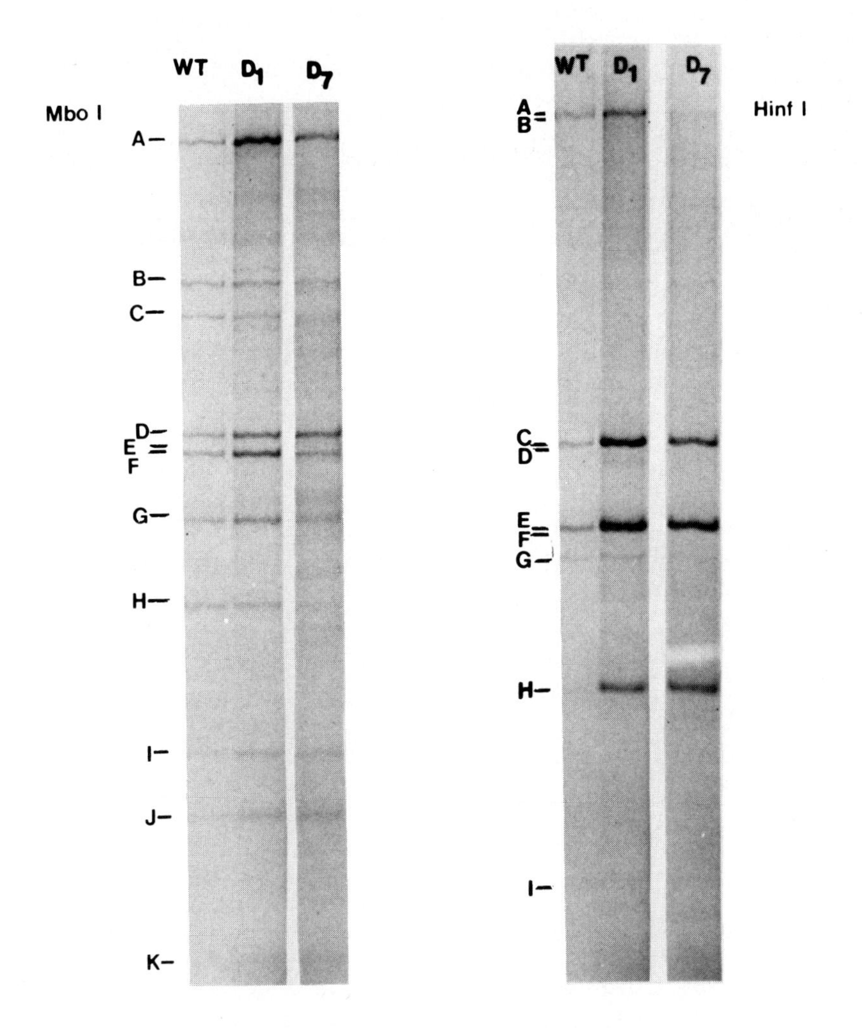

FIGURE 8. Restriction enzyme analysis of D-RF DNA prepared from type I D-DNA obtained after one (D 1) and seven (D 7) serial undiluted passages.

DNA, although certain regions are clearly underrepresented. Significantly, the smallest molecules in the D-DNA population (pool D, Figure 7) lack approximately ninety percent of the wt genome-sequence, and appear to consist exclusively of sequences which map within each of the terminal Hinf I restriction fragments, regions which include only about 200 and 300 nucleotides at the 3' and 5' terminus of the viral genome respectively. Moreover, based on studies in which incomplete genomes were selected by hybridization to specific portions of the viral genome, we have concluded that both genomic termini are present on individual molecules as part of a covalently continuous structure.

As in the case of wt DNA, type I genomes contain short, double-helical regions at their molecular termini. Although the 3' terminal hairpin in these molecules appears to be identical to that found in wt DNA, the self-complementary region at the 5' terminus appears modified in the type I genomes. The basis for this difference is not known. However, it is interesting to note that the incomplete genomes of H-1 virus acquire tandemly repeated insertions of unknown origin which appear to map in the vicinity of the self-complementary region at the 5' terminus of H-1 RF-DNA (31). Insertions at the 5' end in H-1 D-DNA (and possibly in MVM D-DNA) may represent tandem duplications of certain critical regions involved in the replication of parvovirus genomes. Alternatively, the apparent difference in the size of the 5' terminal hairpin duplex in D-DNA could reflect the presence of a 5' terminal DNA-protein complex similar to those which have been observed for other viral DNAs (32).

At the present time we do not know whether the type I genomes contain only single deletions, or whether individual molecules may also exhibit multiple deletions, sequence inversions, reiterations or even insertions of host-cell DNA. These kinds of changes have been described in some detail for the deleted genomes of SV_{40} where they have been detected by altered restriction enzyme cleavage patterns of the DNA (33). It is of interest to note therefore that restriction enzyme digests of MVM D-RF DNA contain fragments in addition to those seen in digests of wt RF DNA. This is particularly evident in digests performed with D-RF DNA derived from late passage D-DNA samples (for example, compare the gel profiles in Figure 8). However, in order to decide whether this DNA is viral or cellular in origin, and if cellular, whether it is covalently linked to viral sequences, will require an examination of cloned populations of type I D-DNA.

In contrast to the type I structures, type II molecules are hairpins, ranging in size up to 1000 base pairs in length and consist of sequences which map exclusively within approximately 0.2 fractional lengths at the 5' terminus of the viral genome. These molecules may arise by a mechanism similar to the one proposed originally by Daniell to explain the aberrant adenovirus genomes (34). Random nicks could occur internally in the viral (-) strand, possibly during strand displacement synthesis or on parental molecules. The 5' terminal portion of the nicked molecules would then contain a free 3' end capable of folding back and forming weak, transiently stable base-paired regions at a variety of sites on the template strand. In this configuration, the 3' end would be suitable as a primer and could be extended by DNA polymerase, thus leading to the formation of "looped" hairpin structures. Both the variability in nick sites and a variable degree of looping-over would account for the size heterogeneity of the type II molecules themselves and for the variability in the sizes of the loops and stems of which these molecules are constructed. Theoretically, hairpin molecules generated in this way could be as much as twice the molecular length of the viral genome. However, the largest type II DNA molecules seen are only about 50-60 percent genome size and so nicking may occur more frequently within the 5' half of the genome and less often toward the 3' end. Alternately, the nick sites may in fact be random and the distribution of type II molecules we have observed could be due to selective packaging of shorter molecules. The total absence of sequences from the 3' half of the MVM genome in type II D-DNA molecules is consistent with the previous suggestion (35) that encapsidation of MVM DNA is initiated at the 5' end of the virion DNA.

These studies on the deleted genomes of MVM were initiated in order to delineate required cis-acting functions, such as the origin for MVM DNA replication. A variety of independent observations have demonstrated the importance of the self-complementary sequences located at the termini of parvovirus genomes in the initiation of viral DNA replication. Important events such as site-specific nicking and "hairpin transfer" almost certainly occur at or near these specialized regions. The results obtained here provide additional support for this general view. For example, we have estimated that ∿30% of the type I D-DNA molecules from a single high multiplicity passage lack 85-90% of the genetic sequence of wt MVM DNA (see pool D, Figure 7). Significantly, the regions which are conserved in these molecules map almost entirely within 200 nucleotides at both genomic termini. Furthermore, type I D-DNAs arising during serial undiluted passage are enriched for shorter molecules in which there is an increased loss of internal wt sequence but a conservation of the self-complementary sequences from both genomic termini.

We have taken this to mean that type I genomes are part of an intracellular pool of actively replicating DNA at least in the presence of wt virus, and therefore that the selective conservation of the terminal self-complementary sequences in these molecules is of direct functional significance for the process of DNA replication. For these reasons, we have concluded that the critical recognition sites involved in MVM DNA replication lie entirely within a region, about 200 nucleotides in length, at either end of the viral genome. It is interesting to note in this regard that the 3' terminal 175 nucleotides of the DNA of the rodent parvoviruses MVM, H-1, H-3 and KRV have almost the identical nucleotide sequence (16).

In addition, we have noted the absence of type II genomes from late passage virus stocks. This observation lends itself to several possible interpretations. The first of these is that type II genomes do not take part in DNA replication, but arise as dead-end replication errors. The absence from these molecules of sequences mapping at the 3' terminus supports this view. In this event, the disappearance of type II genomes would be expected to be gradual, especially if these molecules arise from the type I genomes, as well as wt DNA, at each passage. This gradual disappearance of type II genomes is in fact observed. However, we would expect similar results if these molecules simply replicate at a much slower rate in comparison with the type I genomes. Clearly, this question can best be resolved by examining intracellular pools of replicating DNA.

The availability of a variety of deletion mutants, such as the ones described here for MVM, should be useful a) in the study of MVM transcription and mRNA processing, b) as probes for examining the activity of proteins involved in MVM DNA replication and c) as potential cloning vehicles for propagating foreign genetic elements in mammalian cells. For these reasons, elucidation of the structure of cloned populations of type I MVM genomes is of considerable interest.

ACKNOWLEDGEMENTS

We thank Dr. Steward Millward for generously providing temporary laboratory facilities to one of us (E.A.F.) during the initial stages of this study. We also thank L. Guluzian for his expect guidance in the spreading of DNA for electron microscopy and for preparing the electron micrographs. The assistance of D. Brooker and H. McGuire in the production of MVM is gratefully acknowledged. We also thank C. Cole for his gift of SV_{40} DNA and C. Hours for providing ΦX174 DNA.

E.A.F. was a recipient of a Canadian Medical Research Council Postdoctoral Fellowship.

REFERENCES

1. Berns, K.I., and Hauswirth, W.W. (1978). In "Replication of Mammalian Parvoviruses" (D.C. Ward and P. Tattersall, eds.), pp. 12-32. Cold Spring Harbor Laboratory.
2. Rose, J.A. (1974). In "Comprehensive Virology" (H. Fraenkel-Conrat and R.R. Wagner, eds.), v. 3, pp. 1-61. Plenum Press, New York.
3. Rose, J.A., Berns, K.I., Hoggan, M.D., and Koczot, F.J. (1969). Proc. Nat. Acad. Sci. USA 64, 863.
4. Berns, K.I., and Adler, S. (1972). J. Virol. 9, 394.
5. Salzman, L.A., and Redler, B. (1974). J. Virol. 14, 434.
6. Bourguignon, G.J., Tattersall, P., and Ward, D.C. (1976). J. Virol. 20, 290.
7. Green, M.R., Lebowitz, R.M., and Roeder, R.G. Cell, in press.
8. Dadachanji, D.K., and Ward, D.C. Unpublished results.
9. Chow, M.B., and Ward, D.C. (1978). In "Replication of Mammalian Parvoviruses" (D.C. Ward and P. Tattersall, eds.), pp. 205-217. Cold Spring Harbor Laboratory.
10. Hauswirth, W.W., and Berns, K.I. (1977). Virology 78, 488.
11. Hauswirth, W.W., and Berns, K.I. (1978). In "Replication of Mammalian Parvoviruses" (D.C. Ward and P. Tattersall, eds.), pp. 257-267. Cold Spring Harbor Laboratory.
12. Rhode III, S.L. (1977). J. Virol. 21, 694.
13. Singer, I.I. and Rhode III, S.L. (1977). J. Virol. 21, 713.
14. Straus, S.E., Sebring, E.D., and Rose, J.A. (1978). In "Replication of Mammalian Parvoviruses" (D.C. Ward and P. Tattersall, eds.), pp. 243-255. Cold Spring Harbor Laboratory.
15. Ward, D.C., and Dadachanji, D.K. (1978). In "Replication of Mammalian Parvoviruses" (D.C. Ward and P. Tattersall, eds.), pp. 297-313. Cold Spring Harbor Laboratory.
16. Astell, C.R., Smith, M., Chow, M.B., and Ward, D.C. (1979). Cell, in press.
17. Fife, K.H., Berns, K.I., and Murray, K. (1977). Virology 78, 475.
18. Langeveld, S.A., Van Mansfeld, A.D.M., Baas, P.D., Janoz, H.S., van Arkel, G.A., and Weisbeek, P.J. (1978). Nature 271, 417.
19. Sherrat, D. (1978). Nature 271, 404.
20. Soeda, E., Kimura, G., and Miura, K. (1978). Proc. Nat. Acad. Sci. USA 75, 162.
21. Subramanian, K.N., Dhar, R., and Weissman, S.M. (1977).

J. Biol. Chem. 252, 355.

22. Tattersall, P. (1972). J. Virol. 10, 586.
23. Van Venrooij, W.J., Henshaw, E.C., and Hirsch, C.A. (1970). J. Biol. Chem. 245, 5947.
24. Tattersall, P., Cawte, P.J., Shatkin, A.J., and Ward, D.C. (1976). J. Virol. 20, 273.
25. Davis, R.W., Simon, M., and Davidson, N. (1971). In "Methods in Enzymology" (L. Grossman and K. Moldave, eds.), v. 21, pp. 413-428. Academic Press, New York.
26. Jeppesen, P.G.N. (1974). Analyt. Biochem. 58, 195.
27. Bonner, W.M., and R.A. Laskey. (1974). Eur. J. Biochem. 46, 83.
28. Laskey, R.A., and A.D. Mills. (1975). Eur. J. Biochem. 56, 335.
29. De La Maza, L.M., and Carter, B.J. (1978). In "Replication of Mammalian Parvoviruses" (D.C. Ward and P. Tattersall, eds.), pp. 193-204. Cold Spring Harbor Laboratory.
30. Müller, H.P., Gautschi, M., and Siegl, G. (1978). In "Replication of Mammalian Parvoviruses" (D.C. Ward and P. Tattersall, eds.), pp. 231-240.
31. Rhode, III, S.L. (1978). J. Virol. 27, 347.
32. Robinson, A.J., and Bellet, H.J.D. (1974). Cold Spring Harbor Symp. Quant. Biol. 39, 523.
33. Davoli, D., Ganem, E.D., Nussbaum, A.L., and Fareed, G.C. (1977). Virology 77, 110.
34. Daniell, E. (1976). J. Virol. 19, 685.
35. Tattersall, P., and Ward, D.C. (1976). Nature 263, 106.

COMPLEX STRUCTURES AND NEW SURPRISES IN SV40 mRNA[1]

Michael Piatak, Prabat K. Ghosh, V. Bhaskara Reddy,
Paul Lebowitz, Sherman M. Weissman

Departments of Human Genetics, Internal
Medicine,
Yale University School of Medicine,
New Haven, CT. 06510

ABSTRACT We have determined the nucleotide sequences of many of the SV40 mRNAs produced in virus transformed cells, and early and late in lytically infected cells.Late lytic mRNAS exhibit a multiplicity of 5' ends. 16S mRNAs all contain contiguous stretches of RNA coding for VP1 spliced onto three classes of 5' terminal leaders. One class of leaders contains an intraleader splice while another contains a duplication of 93 nucleotides at the 3' terminus of the leader. 19S mRNAs contain coding information for VP2 and VP3 and fall into four classes, one of which is comprised of unspliced species and three of which contain different 5' terminal splices. 16 and 19S mRNAs share only one major leader. The late mRNA leaders are transcribed from a region of approximately 10% of the viral genome. SV40 large and small T antigen early mRNAs isolated from infected and transformed cells exhibit single well-defined overlapping splices and a limited degree of 5' terminal heterogeneity.

INTRODUCTION

Simian Virus 40 (SV40) is a small DNA virus that replicates in the nucleus of certain primate cells,transforms cells from a variety of species *in vitro* and induces tumors in certain rodents.

[1]This work was supported by a grant from the American Cancer Society and by grant #CA-16038 from the National Cancer Institute, DHEW.

ISBN 0-12-198780-9

This virus has been intensely studied for more than a decade and is in many respects the best understood of the animal viruses. The full nucleotide sequence of the SV40 DNA has been known for some time (1,2) and the exact boundaries of the genes coding for five recognized proteins are well defined. However, there is little specific understanding of the mechanisms of regulation of SV40 gene expression. Recent studies, some of which are described below, have indicated that the virus produces a complex array of mRNAs. Much of the complexity of the mRNAs is presumably a reflection, not of complexity of the translation products, but of the manner in which the mRNAs are produced.

These observations suggest speculations concerning regulation of mRNA production and gene expression in SV40. A major unanswered question is whether the complexity of the SV40 mRNAs is entirely a reflection of the specialization of this virus to optimize the use of DNA whose total informational content is limited by requirements for effective packaging, or whether mechanisms operative in the biogenesis of SV40 mRNA may also be relevant for host gene expression.

The genome of SV40 virus (Fig. 1) may be divided into early and late gene regions, consisting of contiguous opposite halves of the circular viral genome. An origin of DNA replication overlaps the boundary between early and late gene regions and is composed of sequences which also code for 5' terminal sequences of early and late mRNAs, respectively. A series of tandem repeated nucleotide sequences lie adjacent to the origin of DNA replication.

In permissive cells, the early gene region of SV40 DNA is transcribed into cytoplasmic mRNA from shortly after infection through the remainder of the lytic cycle. It is also expressed in infected non-permissive cells and in cells transformed by SV40. Presumably any viral protein contributing to maintenance of the transformed state is encoded in this region. Under physiologic circumstances, the late region is expressed in lytically infected cells only after the initiation of DNA replication. However, continuing DNA replication may not be necessary for continued production of late message. Late mRNA is not synthesized in infected non-permissive cells or transformed cells.

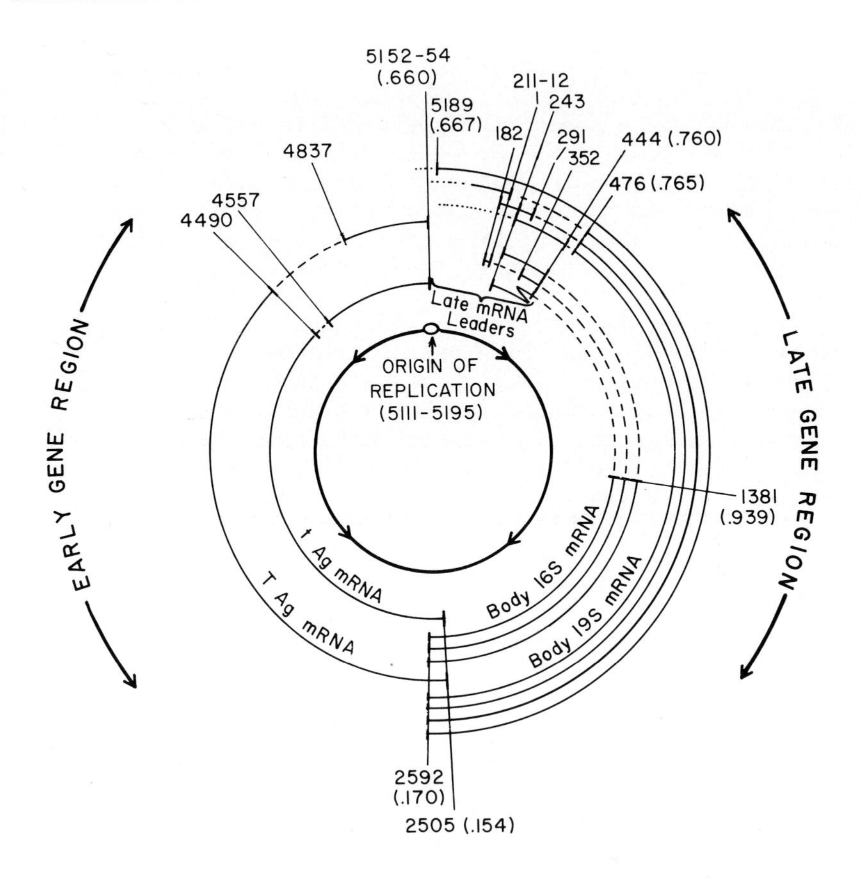

FIGURE 1. Schematic representation of the SV40 genome (central dark circle) and its early and late mRNA products (lighter concentric rings). The DNA is oriented with the origin of replication at the top of the figure and the early and late gene regions occupying the left and right halves of the genome. The direction of transcript of the early and late gene regions is indicated by arrows on the inner circle. Splices in mRNAs are indicated by dashed lines. Extension of late mRNA leaders for undetermined distances in a 5' direction is indicated by dotted lines. Whole numbers refer to nucleotide numbers on the sequence of the SV40 genome. These indicate sites of 3' and 5' termini and fused nucleotides of splices in the mRNAs.

Two early proteins of SV40 have been recognized and mapped on the genome. These are the large T antigen, whose codons are contained in two non-contiguous segments of the early region separated by a small intervening sequence that includes termination codons in all three phases; and a small t antigen encoded by a contiguous segment of DNA located in the early region near the origin of replication (Fig. 1). A number of immunoprecipitable peptides smaller than large T antigen appear to be produced by *in vitro* proteolysis during T antigen extraction, but there is a suggestion that a slightly larger form of T antigen may be produced in certain transformed cells as a primary translation product(3). In the related virus, polyoma, a T antigen of intermediate size, middle T, has been clearly identified by immuno-precipitation and suggestions have been made for the coding regions for this antigen (4,5). In SV40 it is uncertain if there is any middle T antigen of a size comparable to that of polyoma virus.

There are three late proteins of SV40 that contribute to the structure of the viral capsid. The codons for the major structural proteins of VP1 have been identified and are contained in a contiguous segment of DNA from .947-.155 map units on the circular genomic map (Fig. 1). VP2 and VP3 are both minor capsid proteins. A number of lines of evidence suggest that the carboxy terminal 70% of VP2 and the entirety of VP3 contain an identical sequence of amino acids and are encoded in an identical set of codons extending on the SV40 map from approximately .83-.97 map units. Codons for the amino terminal 30% of VP2 appear to lie in the genomic region from approximately .76-.83 map units. VP3 is not derived from VP2 by post-translational processing; rather VP3 is translated from an initiation codon within the coding region for VP2, perhaps at the first internal methionine AUG triplet within the VP2 coding region. In polyoma, the analogous VP2 and VP3 proteins appear to be encoded by partially separable mRNA species (6, 7). However, it has been very difficult to separate mRNAs for VP2 and VP3 in SV40 infected cells.

The present paper reviews our recent results concerning the structure of the multiple forms of SV40 early and late mRNAs and considers the significance of these mRNAs in expression of viral genetic information. These topics are also discussed in a recent review of the organization and transcription of the SV40 genome (8).

MATERIALS AND METHODS

These have all been described in detail previously (9). The SV40 used in these experiments was derived by low multiplicity passage of a small plaque strain 776 virus originally isolated by Professor Daniel Nathans and his colleagues. The genome of this virus has been sequenced and contains 5226 base pairs (1). For the production of early and late lytic mRNAs, the VERO continuous line of African Green Monkey kidney cells was grown in monolayers in roller bottles and infected with SV40 at a multiplicity of infection of 10 to 20 plaque forming units per cell. For late mRNA preparations, cells were generally harvested 40 hr after infection. For early mRNA, cytosine arabinoside, 20 ug per ml, was added to cultures two hours after infection and cells were harvested 18 hr after infection. Transformed cell mRNA was obtained from cultures of the human transformed line, SV80, and mouse transformed 3T3 line, SVT2, kindly provided by Drs. David Livingston and George Todaro.

To prepare RNA cells were disrupted and nuclear and cytoplasmic extracts prepared by slight modification (9) of the procedure of Penman (10). Tween-40-deoxycholate washes of nuclei were added to cytoplasmic fractions. Nuclei were extracted with DNAse prior to RNA extraction. RNA was extracted from nuclei and cytoplasm by treatment with Proteinase K and .05% sodium dodecyl sulfate, followed by phenol-chloroform extraction. Polyadenylated mRNA was isolated on oligo-(dT)-cellulose columns.

SV40 cRNA was prepared by *E. coli* RNA polymerase-catalyzed transcription of SV40 Form I DNA. This procedure yields predominantly early strand RNA. The resulting cRNA was extracted with .2% SDS-phenol, alcohol precipitated and fractionated on a sucrose gradient. RNA with a sedimentation coefficient greater than 16S was recovered, phenol extracted and alcohol precipitated.

Analyses of the sequences in polyadenylated RNAs were performed with the use of SV40 DNA restriction fragments labelled at their 5' termini with ^{32}P. These were annealed to RNAs in 80% formamide under conditions favoring RNA-DNA hybrid formation (9). In each hybridization reaction, a molar excess of the end-labelled restriction fragment over mRNA was used. After hybridization, RNA molecules containing annealed DNA fragments were again selected on oligo-(dT)-cellulose. The DNA primers annealed to RNA were then elongated towards the 5' end of the RNA with a relatively RNAse-free preparation of reverse transcriptase. Following degradation of template RNA with NaOH, the products of primer elongation (cDNAs) were separated by acrylamide-7M urea gel electrophoresis. The sequence of each cDNA was then analyzed by the method of Maxam and Gilbert (11). cDNA syntheses using SV40 cRNA as a template were performed by the identical method with the omission of the isolation of DNA-RNA hybrids on oligo-(dT)-cellulose.

RESULTS AND DISCUSSION

A. Late mRNA. Early studies reviewed in reference (8) demonstrated that there are two principal classes of SV40 late mRNA: 19S and 16S mRNAs. 16S mRNA is translated *in vitro* and almost certainly *in vivo* to form VP1, while 19S mRNA is translated *in vitro* principally to form VP2 and, in more recent studies, VP3 as well (12, 13). The 3' ends of 19 and 16S mRNA appeared to be coterminous by initial hybridization studies and we have obtained direct confirmation for this by sequence analysis (14). The early sequence studies (14) also demonstrated that a minor fraction of late mRNA is transcribed from templates extending up to and probably across the origin of SV40 DNA replication. The major part of late mRNA, however, is transcribed entirely from templates at least 200 nucleotides removed from the origin of replication. These studies also demonstrated the existence of a 5' terminal sequence of approximately 200 nucleotides in late mRNA which was termed a leader sequence (14) on the basis of several formal analogies to leader sequences described for the tryptophan operon of *E. coli* (15) and presumptive leader sequences for lambda bacteriophage late mRNAs (16). The SV40 leader sequence is of similar chain length to the leaders found on these prokaryotic mRNAs.As with the prokaryotic sequences, the SV40 leader sequence contains an initiation codon followed by a number of sense codons and so could potentially encode a peptide. There are also possibilities for multiple base-paired conformations within the SV40 leader and uridylic acid rich sequences are present near its 3' end. These early studies also showed that beyond the 3' end of the SV40 leader sequence there is a stretch of about 30 nucleotides present in DNA whose transcript is essentially entirely missing, or spliced out, from the cytoplasmic mRNA. Beyond this point SV40 late mRNA was found to contain an initiation codon for translation of the VP2 capsid protein.

A number of lines of evidence suggested initially that the principal late leader sequence of SV40 is attached predominantly to the 5' end of 16S mRNA and, to a lesser extent, to the 5' end of 19S mRNA. More recent studies, however, have shown that there are a multiplicity of leader sequences for both 16S and 19S mRNAs (1, 17, 20). All forms of 16S mRNA are now known to contain a splice which joins residues transcribed from nucleotides 444 and 1381 of SV40 DNA (18). Furthermore, the 5' terminal leaders of 16S mRNA are of three types. The principal leader in 16S mRNA is a 202-long nucleotide sequence transcribed colinearly from nucleotides 243-444 on the SV40 genome (18, 21). Secondly, there is a family of leaders whose 5' ends are encoded in DNA nearer the origin of replication than the 5' end of the principal leader. All of these leaders have a second splice within the leader sequence which removes nucleotides from positions 212 to 351. Indeed, all 16S leaders whose template lies upstream from nucleotide 243 contain this intra-leader splice (18). Finally, there is a minor form of SV40 late 16S leader which contains a tandem duplication of the 3' terminal 93 nucleotides of the leader sequence from residues 351-443 (18). It has been suggested that this duplication may be derived by splicing out large tracts of sequences from transcripts containing tandem repeats of the SV40 sequence. These transcripts could be formed either by RNA polymerase traversing the SV40 circle more than one time or by transcription of DNA templates containing tandem copies of SV40 DNA. Transcripts with duplicated segments (not yet shown to be tandem duplications) are present in much larger relative amounts in polyoma infected cells (22).

19S mRNAs fall into four classes on the basis of their 5' terminal structures. One class of 19S species contains 5' terminal sequences which are colinear with SV40 DNA.The remaining three classes are spliced and contain leaders with 3' termini at positions 211, 291 and 444 joined to the body of 19S mRNA with a 5' terminus at residue 476. From this data it is apparent that most 19S mRNA leaders are different from those of 16S mRNA. First, none of the 19S mRNA leader sequences contains the splice within the leader that was described for the minor leaders of 16S mRNA. Secondly, leaders terminating at residues 211 and 291 are not found in 16S mRNA species. Thirdly, the most abundant single family of 19S leaders (leaders with 3' terminus at residue 291) lacks the 3' terminal portion of the major 16S leader However, one species of 19S mRNA does contain the same leader as the principal species of 16S mRNA.

Our primer elongation experiments have suggested the possible presence of very minor components of late lytic mRNA that contain the initiation codon for VP3 but do not contain the initiation codon for VP2 (19). It appears unlikely that the amounts of these RNA species are adequate to account for the several-fold greater synthesis of VP3 than of VP2 observed in infected cells or for the similar amounts of VP2 and VP3 synthesized in cell-free systems with purified mRNA. Examination of leader sequences between residues 291-444 has revealed several possibilities for base-pairing with sequences about the presumptive site for initiation of VP2 translation at residues 480-482. Since such base-pairing could block the accessibility of the VP2 initiation site to ribosomes, we have speculated that the class of 19S mRNAs with leaders containing sequences from residues 291-444 code for VP3 while 19S splices with leaders lacking these sequences code for VP2 (19). However, at this time, direct experimental evidence in support of this hypothesis is lacking.

It is noteworthy that the nuclear polyadenylated RNAs that contain the VP2 initiation codon all represent contiguous transcripts of SV40 DNA (19). Of additional note, both nuclear and cytoplasmic late RNAs contain a number of 5' ends (Fig. 1). Furthermore, with the exception of species with 5' termini at residue 243, the distribution of 5' terminal sequences estimated by reverse transcriptase elongation is the same for the 16S mRNA species, for the 19S spliced cytoplasmic mRNA species and for the nuclear RNA species. The fact that cytoplasmic 16 and 19S species and nuclear late RNAs appear to contain the same 5' termini suggests that the various cytoplasmic species may be derived by splicing events from nuclear unspliced RNAs with the corresponding 5' termini. Whether the 5' termini of nuclear RNAs represent transcription initiation sites, or whether caps are added or new 5' ends created by post-transcriptional cleavage and processing is at present totally unknown.

DNA encoding the 5' ends of the leaders of the principal 16 and 19S mRNAs can be deleted from the SV40 genome without significantly altering the viability of the mutant viruses (23, 24, 25). We have investigated mRNA production in two such deletion mutant (810 and 1470). This investigation was performed in collaboration with Drs. K. N. Subramanian, Tom Shenk and Janet Mertz. Our salient observation was that 16S late mRNA from cells infected with these mutants always contained a leader with its 3' terminal residue transcribed from nucleotide 444 joined to the main body of mRNA transcribed from nucleotide 1381 exactly as in wild type 16S mRNA. However, the 5' ends of the leader sequences specified by the mutants were found to lie nearer the origin of DNA replication than the 3' terminal boundaries of the deleted segments of the DNA. The 5' ends of several species of RNA in 810 infected cells were encoded by DNA 50 nucleotides upstream of the deletion terminus (about residue 210) and were present in at most a very small fraction of late mRNA of cells infected with wild type virus. Furthermore, mutant 1470 with a deletion extending towards the origin of DNA replication produced a family of late mRNAs with relatively increased amounts of 5' ends copied from DNA sequences in the region upstream from some of the most abundant 5' ends of the 1470 mRNA. It thus appears that the virus can compensate for the deletion of templates for the 5' ends of late mRNA by using other DNA templates closer to the origin of DNA replication for synthesis of 5' termini of late mRNA. One speculation is that there is in some sense an entry point for the production of late mRNA, either at the RNA polymerase level or at the level of RNA cutting and recapping, and that after the entry process has initiated the active complex can migrate down the DNA or RNA chain generating heterogeneity of 5' ends. In this regard, it is interesting that a principal 5' end of the longer leaders seems to lie within the origin of DNA replication;thus it is possible that late RNA production may really be initiated primarily at this position.

B. Early mRNA. Our earliest structural studies of early mRNA were performed by fingerprinting ^{32}P labelled RNA isolated from lytically infected cells grown in the presence of ^{32}P phosphate. These studies demonstrated the approximate location (1, 26) of the 5' end of early mRNA and also an overlap of 70 nucleotides between the 3' ends of early and late mRNAs. We have now extended these studies by sequencing cDNAs transcribed on early SV40 mRNA isolated from infected cells early and late in the lytic cycle and from transformed cells. These studies have established the exact sites (1, 27) of two internal splices (1, 28) in early mRNA isolated from these three sources: the first splice fusing nucleotides transcribed from DNA residues 4490 and 4837 and the second fusing nucleotides copied from DNA residues 4490 and 4557. The former splice, present in one mRNA species, serves to join codons for the amino-terminal portion of large T antigen in phase with the codons for the remainder of this protein. The latter splice, present in a second mRNA, occurs immediately beyond the termination codon for translation of small t antigen.

Early lytic mRNA whether isolated late or early in the infectious cycle, shows the same pattern of 5' ends (Fig. 2).

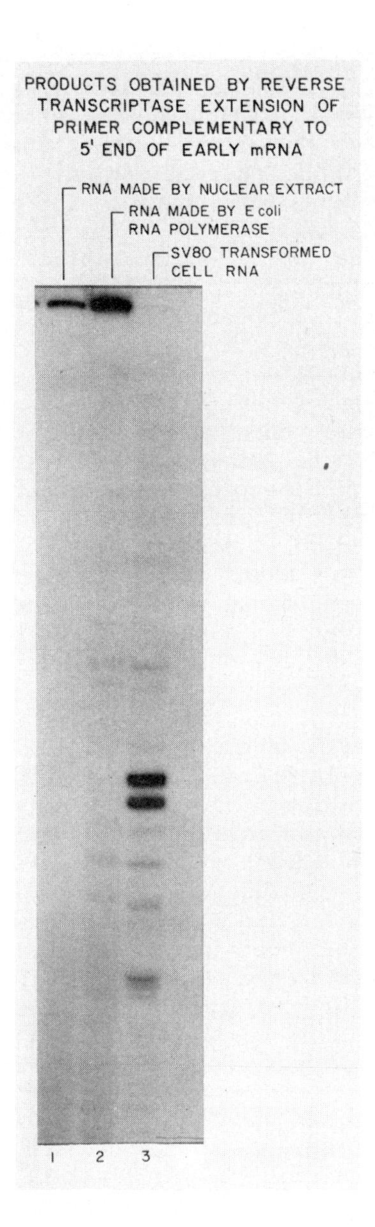

FIGURE 2. Autoradiogram of acrylamide gel electrophoretic separation of cDNA products transcribed from the 5' ends of early SV40 mRNA. The DNA primer used had its 3' end at residue 5092 and its 5' end at residue 5054. Electrophoresis was from above downward. Channels 1, 2, and 3 represent

There appear to be two principal ends located 5-7 nucleotides apart whose template (residues 5147-48 and 5152-54) is part of the DNA encoding the long, inverted repeat that constitutes part of the origin of SV40 DNA replication. These ends both fall within the location of the 5' ends of early mRNA estimated by the oligonucleotide mapping procedures (28). A third, minor 5' end of SV40 RNA also appears to be present downstream from these two ends but still upstream from the initiation codon used for large T and small t antigen production (Fig. 2). In control experiments, cDNAs synthesized on SV40 cRNA have not terminated at any of the three sites at which cDNAs transcribed on early lytic mRNAs terminate (Fig. 2). It thus appears likely that these three termini represent true 5' termini of separate species of early mRNA. Since lytically infected cells contain threefold more large T than small t mRNA, it seemed likely that both of the principal 5' ends of early mRNA were derived at least in part from mRNA for the large T antigen. To investigate whether the small t antigen mRNA also had the same 5' ends as large T mRNA, we used as primer the Mbo I restriction fragment extending from nucleotides 4629 to 4689 on the SV40 DNA as a primer for cDNA synthesis. This primer is complementary to sequences retained in small t antigen but spliced out of large T antigen mRNA. The extension products still contained the same termini as total early mRNA suggesting that both large and small T antigen mRNAs contain the same principal 5' termini.

We then investigated the 5' ends of mRNA from the human SV40-transformed cell line, SV80, and observed that some of the early mRNA had the same 5' ends present in early mRNA from infected

---

the products obtained when the template was (1) RNA produced when SV40 DNA was incubated with a crude nuclear extract (29), (2) RNA transcribed from SV40 DNA by *E. coli* RNA polymerase, and (3) Poly(A) terminal cytoplasmic RNA from the SV40 transformed human cell line SV80. In each case approximately equal amounts (1 ug) of SV40 specific RNA were used. The two dark bands in lane 3 correspond to cDNAs ending at about positions 5047 and 5052 on the SV40 genome.

cells. However, a portion of the early mRNA appeared to have additional 5' ends located further downstream from the origin of replication, including some falling within the region encoding the amino-terminal portion of small and large T antigens. When we examined the SVT2 transformed line of mouse cells, these new downstream 5' ends were the most abundant 5' termini. Therefore, the possibility exists that a portion of large T and perhaps even small t antigen molecules from transformed cells is actually somewhat shorter than the predominant proteins in infected cells. This could occur as a result of initiation at an AUG codon that is normally translated to produce an internal methionine of the large T and small t antigens present in lytically infected cells. Certain transformed cell lines produce a large T antigen of higher molecular weight than that made in lytically infected cells, but as noted the presence of lower molecular weight forms of T antigen has been largely attributed to proteolysis.

We have explored the structure of early lytic and transformed cell mRNAs further by the use of a large number of DNA primers complementary to different regions of early mRNA. These primers have been annealed either to early mRNA or cDNA and extended with reverse transcriptase. The resulting products have been separated by acrylamide gel electrophoresis, either directly or after cutting with *Haemophilus aegyptius* III restriction endonuclease. A particularly complex pattern was noted when we used the DNA fragment *Hinf*-J (nucleotides 2743 to 2769) as primer(Fig. 3). This fragment is complementary to a region of early RNA approximately 200 nucleotides from the 3' end of the mRNA. A multiplicity of extension products of *Hinf*-J, between approximately 90 and 500 nucleotides long, was noted. Extension products corresponding to full length small t and large T mRNAs were also present. Qualitatively, the position of extension products in the acrylamide gels was consistent from preparation to preparation of the same type; however, the relative abundance of the products varied somewhat from preparation to preparation. We do not know the significance of this variation.

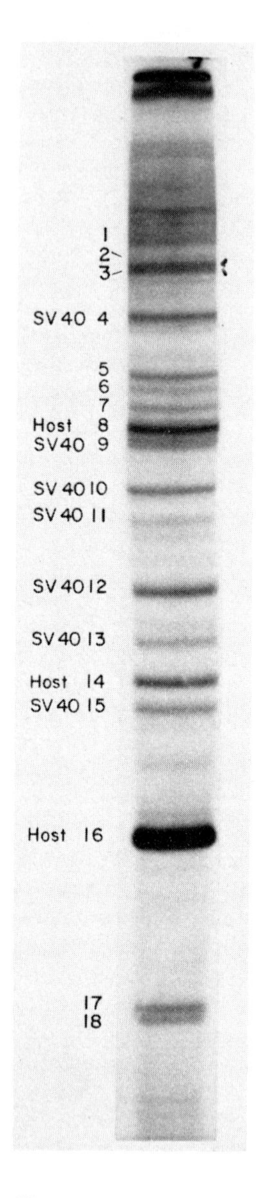

FIGURE 3. Autoradiogram of acrylamide gel electrophoretic separation of products of extension of the Hinf-J restriction fragment (see text). The RNA used was total polyadenylated cytoplasmic RNA from transformed cells. Numbers refer to products of increasing chain length. Electrophoresis was from above downward "SV40" and "Host" in

On sequence analysis, most of the *Hinf*-J extension products contained SV40 sequences extending varying distances in a 5' direction with respect to the mRNA template. This suggests the possible existence of mRNAs consisting of transcripts of the 3' end of the SV40 early gene region, but with heterogeneous 5' ends falling within the coding region for large T antigen. In this regard, both SV40 and BK virus (R. Dhar, personal communication) contain a second open reading frame at the 3' end of the early region in addition to that coding for large T antigen. In the case of SV40 this reading frame begins with four consecutive AUG triplets in the mRNA and contains 91 sense triplets in sequence. Most of the short reverse transcriptase extension products of *Hinf*-J stopped upstream to these four AUGs. Thus the possibility may be raised that if these cDNA products correspond to short RNAs which function as mRNAs *in vivo* they might be translated in an alternative reading frame to that encoding T antigen and produce a short protein that would have no amino acid sequences of, and hence no immune cross-reactivity with large T or small t antigens. Such a peptide has not been detected so far in cells infected with or transformed by SV40 or in studies of protein synthesis directed by SV40 mRNA in cell-free extracts.

Some of the more prominent products obtained by extension of *Hinf*-J on early mRNA from lytically infected and transformed cells exhibited an unusual feature upon further analysis. In reading the sequence beyond the end of the primer, such products shortly transferred from the SV40 sequence into a nucleotide sequence that did not correspond to any portion of SV40 DNA (Fig 3). Similar products have been obtained from RNA of uninfected monkey cells and non-transformed mouse cells but these have not yet been sequenced. This suggests that host cell mRNA may include sequences with partial homology to the *Hinf*-J segment of SV40 DNA.

---

the left channel designate cDNA products whose sequence was examined and found to correspond to either a contiguous stretch of SV40 DNA, or to a sequence differing at a number of positions from any sequence present in SV40 DNA.

As judged by the primer extension procedure, this RNA appears to be relatively prominent in the SVT2 transformed line as compared with non-transformed 3T3 cells and may be variably increased in infected as compared with uninfected cells. Experiments are currently underway to establish conditions under which this effect is regularly seen and to test whether it is dependent on viral protein synthesis.

REFERENCES

1. Reddy, V. B., Thimmappaya, B., Dhar, R., Subramanian, K. N., Zain, B. S., Pan, J., Ghosh, P. K., Celma, M. L., Weissman, S. M.(1978) Science 200, 494.
2. Haegeman, G. and Fiers, W. (1978) Nature 273, 70.
3. Prives, C., Gluzman, Y. and Winocour, E. (1978) J. Virol. 25, 587.
4. Hutchinson, M. A., Hunter, T., Eckhart, W. (1978) Cell 15, 67.
5. Kamen, R., Ito, Y.,personal communication.
6. Hunter, T., Gibson, W. (1978) J. Virol. 28, 240.
7. Sidell, S. G., Smith, A. E. (1978) J. Virol. 27, 427.
8. Lebowitz, P., Weissman, S. M. In press, Current Topics in Microbiology and Immunology.
9. Ghosh, P.K., Reddy, V. B., Lebowitz, P. Piatak, M. and Weissman, S. M. In press, Methods in Enzymology.
10. Penman, S. (1966) J. Mol. Biol. 17, 117.
11. Maxam, A. and Gilbert, W. (1977) Proc. Natl. Acad. Sci. USA 74, 560.
12. Prives, C. (1975) INSERM 47, 305.
13. Smith, A., personal communication, Prives, C., personal communication, Hunter, T., personal communication.
14. Dhar, R., Subramanian, K.N., Pan, J., and Weissman, S. M. (1977) Proc. Natl. Acad. Sci. USA 74, 827.
15. Bertrand, K., Squires, C. and Yanofsky, C. (1976) J. Mol. Biol. 103, 319.
16. Lebowitz, P., Weissman, S. M. and Radding, C. M.(1971) J. Biol. Chem. 246, 5120.
17. Lai, C. T., Dhar, R., Khoury, G. (1978) Cell

14, 971.
18. Reddy, V. B., Ghosh, P. K., Lebowitz, P., Weissman, S. M. (1978) Nucleic Acids Res. 5, 4195.
19. Ghosh, P. K., Reddy, V. B., Swinscoe, J., Lebowitz, P. and Weissman, S. M. (1978) J. Mol. Biol. 126, 813.
20. Bratosin, S., Horowitz, M., Laub, O., Aloni, Y. (1978) Cell 13, 783.
21. Ghosh, P. K., Reddy, V. B., Swinscoe, J., Lebowitz, P. and Weissman, S. M. (1978) J. Biol. Chem. 253, 3643.
22. Kamen, R., personal communication.
23. Mertz, J. E., Berg, P. (1974) Proc. Natl. Acad. Sci. USA 71, 4879.
24. Shenk, T., Carbon, J., Berg, P. (1976) J. Virol. 18, 664.
25. Subramanian, K. N., personal communication.
26. Reddy, V. B., Ghosh, P. K., Lebowitz, P., Piatak, M., Weissman, S. M. In press, J. Virol.
27. Berk, A., Sharp, P. (1978) Proc. Natl. Acad. Sci. USA 74, 1274.
28. Dhar, R., Subramanian, K. N., Pan, J., and Weissman, S. M. (1977) J. Biol. Chem. 252, 368.
29. Wu, G.-J., Luciw, P., Mitra, S., Zubay, G., and Ginsberg, H. S. Barzelon Symp. on Nucleic Acid Protein Interaction. In"Nucleic Acid-Protein Recognition" (Vogel, H., ed.), (1977), pp. 171-186. Academic Press,New York.

STRUCTURE AND ORGANIZATION OF ADENOVIRUS 5 TRANSFORMATION GENES

A.J. van der Eb, H. van Ormondt, P.I. Schrier, H. Jochemsen, J.H. Lupker, P.J. van den Elsen, J. Maat, C.P. van Beveren and A. de Waard

Sylvius Laboratories, University of Leiden, Wassenaarseweg 72, 2333 AL Leiden, The Netherlands.

ABSTRACT The results presented in this paper show that the transforming region of adenovirus 5 (Ad5) DNA can be divided into at least two parts, which each appear to have a distinct role in transformation. To obtain an understanding of the nature and organization of the transforming region of this virus, we have determined the primary structure of the transforming area and identified the proteins and RNAs encoded by it. It will be shown that the left-most 4.5% of the genome codes for a series of partially overlapping proteins, which have the ability to convert a diploid cell with a limited life span into a permanent cell line. The adjacent segment (4.5 - 11%) codes for at least two proteins (the major 19K and 65K antigens) which appear to be responsible for the induction of a number of properties that are characteristic of transformed cells.

INTRODUCTION

Transformation by the small and medium sized DNA tumor viruses (SV40 and adenoviruses) is a process in which several viral proteins are involved. These proteins appear to be encoded by partially overlapping genes, each of which consists of (at least) 2 DNA segments that are non-contiguous (1-4). The present paper summarizes the results of our recent studies on the organization of the transforming genes of human adenovirus 5 (Ad5), a non-oncogenic subgroup C adenovirus, and on the nature and possible functions of their gene products. Results will be presented showing that the region of Ad5 DNA involved in transformation consists of 2 segments. Each of these segments codes for several overlapping proteins and there is evidence that each segment has a distinct role in transformation.

ISBN 0-12-198780-9

RESULTS

Previous studies in our laboratory by Graham *et al.* (5,6) have indicated that fragments by Ad5 DNA produced by shear or by cleavage with restriction endonucleases contained transforming activity, provided the segments did not become smaller than approximately 1 x 10^6 daltons. These studies also demonstrated that the transforming activity of this virus is localized at the left end of the viral genome, in a segment mapping between 1% and about 8% from the left terminus. Very similar results were obtained also with the closely related Ad2 and with the oncogenic adenoviruses types 3, 7 and 12 (6-11). Figure 1 shows a summary of the cleavage maps of Ad5 DNA with a number of restriction endonucleases, in which the fragments with transforming activity have been indicated. It can be seen that all transforming fragments are localized at the left-hand end of the genome. Some of the fragments are large and contain the entire early region no.1 (E1), which maps between 0 and 11% of the viral genome, but four fragments are smaller than 10% and hence comprise only part of early

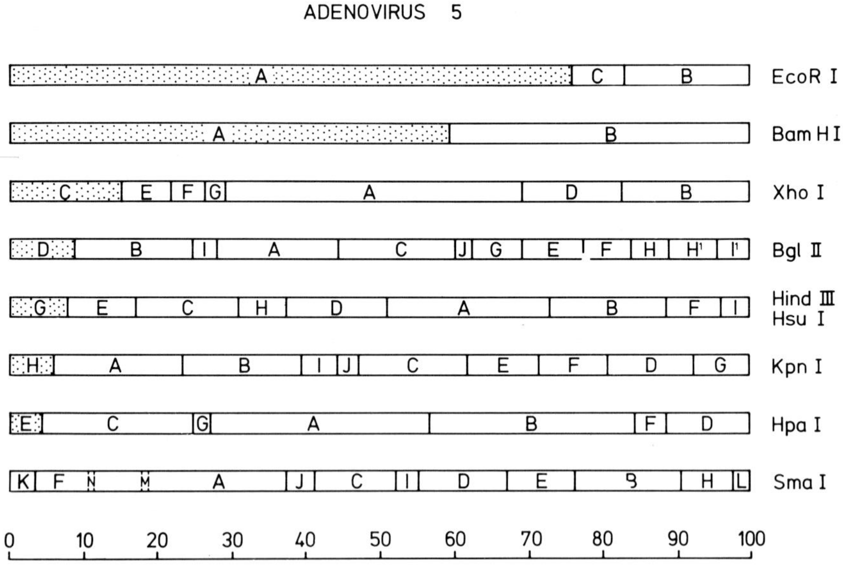

ADENOVIRUS 5

FIGURE 1. Maps of Ad5 DNA showing the cleavage sites of a number of restriction endonucleases. The dotted fragments have been shown to contain transforming activity.

region E1. Hybridization studies have indicated that these transformed cells contained DNA sequences homologous to the fragment that was used to transform the cells (cf 8). This shows that the transformation was not caused by contaminating intact DNA molecules or large DNA fragments.

Properties of the cell lines transformed by Ad5 DNA fragments of various sizes. Most transformation experiments that will be discussed in this paper were carried out using the calcium phosphate technique (12,13) and primary cultures of baby rat kidney cells. In the course of these studies, we noticed that the cells transformed by DNA fragments of various sizes were not identical in their phenotypical properties, but that certain differences occurred which were correlated with the size of the DNA fragment used to transform the cells. On the basis of these differences, we were able to distinquish 3 categories of transformed cells: (1) cells transformed by the larger DNA fragments that contain the entire left hand early region E1 (1-11%) or more. (2) Cells transformed by the 8% *Hin*dIII G fragment (and probably also the 9% *Bgl*II D fragment), which contain 70-80% of early region 1. (3) Cells transformed by the 4.5% *Hpa*I E fragment (and possibly the 6% *Kpn*I H fragment), which contain less than 50% of early region E1.

The first class of transformed cells is similar in properties to cells transformed by intact DNA or whole virus, and hence will be designated here as the standard type of transformed cell. The cells have a typical epithelial morphology, are capable of growing to high cell densities, and show the normal T antigen immunofluorescence staining pattern. The second class of cells transformed by the 8% *Hin*dIII G fragment usually are very similar to the standard type of transformed cells in morphology and growth properties, but they differ in the distribution of their T antigen in that it is localized predominantly in the cytoplasm rather than mainly in and around the nucleus (7). In addition, the concentration of T antigen in *Hin*dIII G fragment-transformed cells is clearly lower than in the standard transformed cells, as judged from the intensity of the immunofluorescence. The cells of the third category, transformed by the 4.5% *Hpa*I E fragment, are characterized also by the atypical cytoplasmic ditribution of the fluorescence, but the T antigen concentration appears to be even lower than in cells of the second category. Interesting, however, is that *Hpa*I E transformed cells differ from the other two categories in that they have a more or less pronounced fibroblastic appearance rather than the epithelial morphology characteristic of other Ad5 transformed cells (fig.2). In addition, *Hpa*I E fragment-transformed cells grow more slowly and are unable to reach high cell densities. These properties,

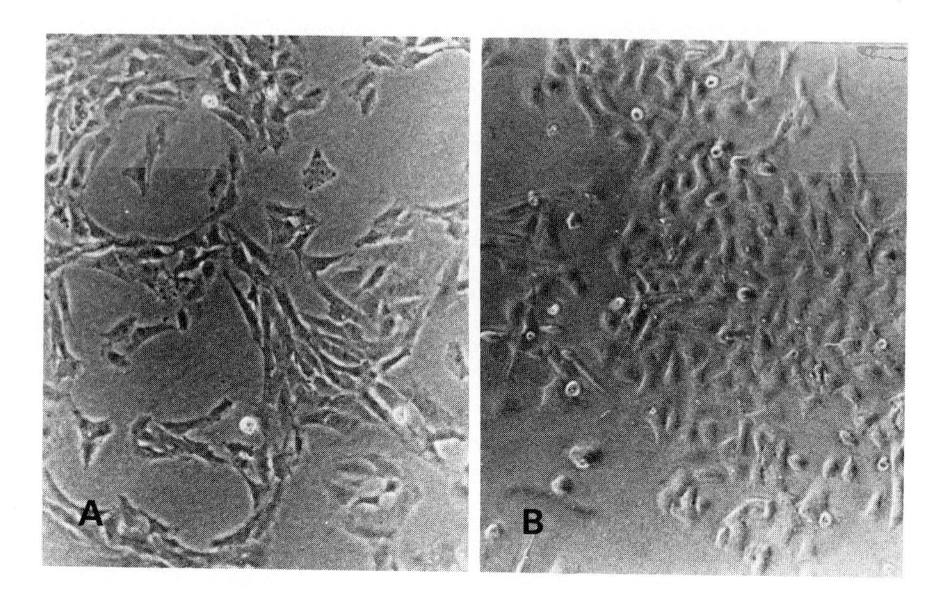

FIGURE 2. Cultures of baby rat kidney cells transformed by the 8% *Hin*dIII G fragment (B) and the 4.5% *Hpa*I E fragment (A). Note that the *Hpa*I E transformed cells have a somewhat fibroblastic appearance, and tend to grow in organized patterns at low cell densities.

however, often tend to disappear upon prolonged passage *in vitro*. Appart from the differences between the three categories of transformed cells mentioned in this paragraph, they also have properties in common: all transformed lines are aneuploid (P. Pearson, personal communication), and they all grow as immortal cell lines.

These results suggested that transformation by Ad5 is a process in which more than one viral gene function is involved. Cells transformed by DNA fragments that comprise only part of early region E1, (the segment that is expressed in cells transformed by whole virus (14)),may lack part of the viral information that is normally active in this process and this may result in an abnormal transformed phenotype. It appeared of interest,therefore, to obtain a detailed understanding of the nature of the transforming proteins and of the structural organization of the transforming segment of Ad5. In particular, we wanted to identify the viral gene products that are expressed in the cells transformed by DNA fragments of different sizes. In this way we hoped to obtain some insight in the role these viral proteins may play in transformation.

Identification of the proteins encoded by the transforming segment of Ad5 DNA. Two approaches were used to identify the proteins specified by the transforming region of Ad5: (1) *in vitro* protein synthesis using viral mRNA extracted from infected or transformed cells and (2), immunoprecipitation of T antigens from transformed cells.

In the first approach, cytoplasmic RNA was isolated from spinner KB cells 6-8 hrs after infection with Ad5. Viral RNA specific for the transforming region was isolated by hybridization to appropriate restriction endonuclease DNA fragments, and the selected RNAs were translated in cell-free systems for protein synthesis derived from wheat germ or rabbit reticulocytes. The 35S methionine labeled products were either analyzed directly by SDS polyacrylamide gel electrophoresis or were first immunoprecipitated and then electrophoresed. The results indicated that viral RNA transcribed from the transforming region in lytically infected cells is translated into 6 major proteins (15K, 19K, 34K, 36K, 40K, 42K). All proteins could be immunoprecipitated, although 15K appeared to be only weakly immunogenic. To map these proteins more precisely, translation experiments were carried out with viral RNAs that were selected to a number of small left hand DNA fragments. The results indicated that the proteins with molecular weights of 15K, and 34K, 36K, 40K and 42K were mapping between 0 and 4.5% (fragment *Hpa*I E) and that the 19K protein mapped between 4.5 and 11% (15). Similar results were reported for adenovirus 2 (16); the 15K protein described by these authors probably corresponds to our 19K protein). These results were surprising since the sum of the molecular weights of the left terminal proteins exceeds by far the maximum coding capacity of the 4.5% *Hpa*I E fragment (mol. weight 1×10^6). This discrepancy was solved, however, by the demonstration that the 4 largest proteins (34-42K) are related as judged from partial proteolysis experiments and hence are encoded to a large extent by a common DNA sequence (17; J.H.Lupker, unpublished results). In addition to these 6 proteins, several minor polypeptides were detected with molecular weights of 16K, 17K, 18K and 25K.Since these proteins were not detected reproducibly in each experiment and were present at low concentrations, it was not possible to map them unequivocally.

To determine which of these proteins are expressed in the cells transformed by the various DNA fragments, virus-specific RNA was isolated from representative lines of each of the 3 categories of transformed cells, and the RNAs were translated in cell-free systems. Figure 3 shows that RNA from cells containing the entire left hand early region (in this example, the Ad5 transformed HEK line 293 (18)) is translated into essentially the same proteins as is RNA from lytically infected

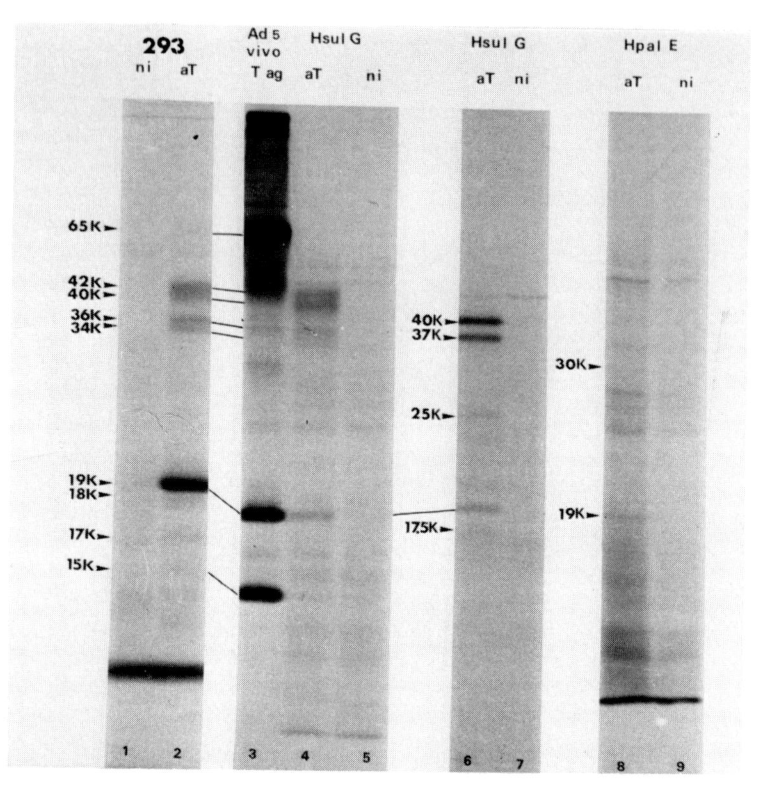

FIGURE 3. Autoradiographs of ^{35}S labeled proteins synthesized *in vitro* using virus-specific RNA from cells transformed by Ad5 DNA fragments. RNAs were translated in cell-free systems from wheat germ or rabbit reticulocytes, and the viral proteins were immunoprecipitated with sera from Ad5 tumor bearing hamsters (aT) or with non-immune sera (ni). The proteins were electrophoresed in 13.4% SDS polyacrylamide gels. Virus specific RNA was obtained from: Ad5 HEK 293, lanes 1 and 2; Ad5 *Hsu*I G cVI AcI, lanes 4, 5, 6, 7; Ad5 *Hpa*I E cXI, lanes 8 and 9; lane 3 shows the proteins immunoprecipitated from Ad5 transformed rat cell 5RK20.

cells: 15K, 19K, 34-42K as well as 17 and 18K. RNA from cells transformed by the 8% *Hin*dIII G fragment was translated into a similar set of proteins, although 15K was not found in this case. In some experiments only two proteins of 37K and 40K were present rather than the 4 viral proteins of 34-42K. If it is assumed that the 19K protein found with RNA from *Hin*d III G transformed cells is the same polypeptide as the 19K protein from lytically infected cells (which was mapped between 4.5 and 11.5%), then we can conclude that its genetic information must be localized in a segment between 4.5 and 8%.

Our initial attempts to detect specific polypeptides among the products synthesized with RNA selected from several *Hpa*I E transformed-cells were unsuccessful. This suggested either that the concentration of the virus-specific RNA was very low or that the viral DNA was not expressed in the cells we had used. When RNA was extracted from very large numbers of E fragment-transformed cells (corresponding to about 30 mg of total cytoplasmic RNA), we were able to detect two immunoprecipitable polypeptides of 19K and 30K (fig.3). These proteins do not correspond to the products that are known to be encoded by the *Hpa*I E fragment. This can be explained in two ways: (1) Berk and Sharp (3) found that the 3' ends of the mRNAs specified by the left most 4.5% of Ad5 DNA extend slightly beyond the *Hpa*I cleavage site at 4.5%, so that the stop signal

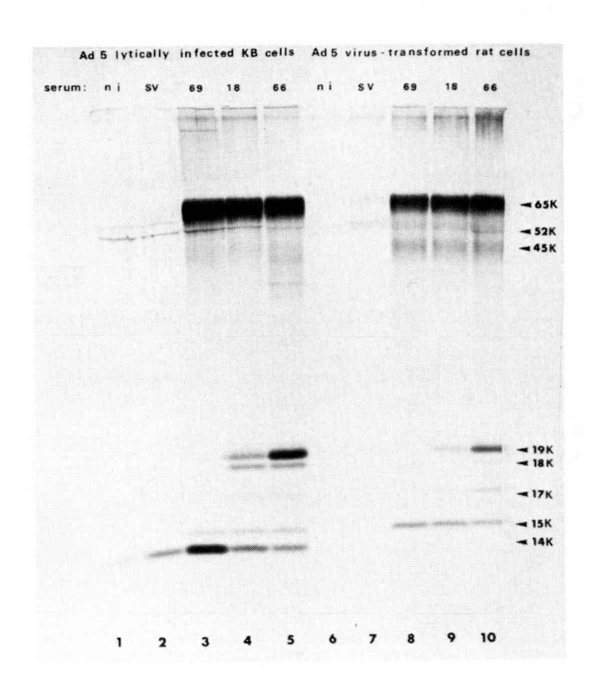

FIGURE 4. Proteins precipitated from extracts of spinner KB cells infected with Ad5 (labeled from 6-8 hr p.i.) (lanes 1-5), or from rat cells transformed by UV-irradiated Ad5 (lanes 6-10), using 3 different sera obtained from individual Ad5 tumor-bearing hamsters, (lanes 3-5 and 8-10). The ^{35}S-labeled proteins are fractionated by electrophoresis in 13% SDS polyacrylamide gels. Lanes 1,6 and 2,7 represent control precipitations with sera of a normal hamster and a hamster bearing a SV40 tumor, respectively.

for RNA transcription would be missing in the integrated DNA. (2) Integration of viral DNA fragments into cellular DNA might be accompanied by a loss of terminal DNA sequences. In the case of the *Hpa*I E fragment, this would result in the loss of aminoacids of the carboxy terminal end of the proteins, leading to the synthesis of shortened products (see Fig.7).

To obtain more information on the viral proteins expressed in the different classes of transformed cells, viral T antigens were also analyzed directly using the immunoprecipitation technique and sera from Ad5 tumor-bearing hamsters. Figure 4 shows the T antigen species precipitated with 3 different Ad5 anti T sera from lytically infected KB cells and from an Ad5 virus transformed cell line. It can be seen that almost identical sets of proteins are precipitated from those two types of cells: two major T antigen species of 65K and 19K and a number of minor species of 52K, 45K, 18K, 17K, 15K (and 14K in lytically infected cells). In addition, a number of proteins in the range of 34-42K, and a 25K protein can be detected at low concentration (cf Fig.3). Figure 4 also illustrates that the specificity of idividual sera is quite variable, particularly with respect to the T antigens in the lower molecular weight range. The 52K and 45K proteins could represent degradation products of the 65K protein, since preliminary results have shown that these proteins share common peptides. Figure 5 shows a comparison of the T antigens precipitated from representative cell lines of the 3 categories of transformed cells. The most important difference between standard transformed cells and cells transformed by the 8% *Hin*dIII G fragment is the absence of the 65K major T antigen. The minor band at 65K, which is visible in lanes 6 and 9 is different from the 65K in normal transformed cells, since it does not precipitate with serum 63 (lanes 5 and 8) as the normal 65K does (lane 2). Similarly, the 52K protein present in G and E fragment-transformed cells must be different from the 52K found in standard transformed cells. The 18K protein is present in small amounts in G fragment transformed cells but 17K is usually not detected. Cells transformed by the 4.5% *Hpa*I E fragment differ from cells transformed by the *Hin*dIII G fragment in that they lack the major 19K T antigen, as well as the 18K protein. In some E fragment-transformed lines, a number of proteins in the range of 34-42K are seen, which could correspond to the 34-42K proteins specific for the *Hpa*I E fragment. In other lines such proteins could not be detected. All *Hpa*I E fragment-transformed cells contain a 52K antigen.

Table I summarizes the T antigen species that have been identified in cells transformed by DNA fragments of different sizes. The results show that the number of the T antigen species becomes smaller as the size of the transforming fragment

TABLE I

SUMMARY OF Ad5 T ANTIGENS PRECIPITATED
FROM LYTICALLY INFECTED AND TRANSFORMED CELLS

Lytically infected cells	Virus-transformed cells	Cells transformed by DNA fragments				
		e.g. ≥15% *Xho*I-C	9% *Bgl*II-D	8% *Hsu*I-G	6% *Kpn*I-H	4.5% *Hpa*I-E
65K	**65K**	**65K**				
52K[1]	52K[1]	52K[1]	**52K**	**52K**	**52K**	**52K**
			47K	47K	47K	47K
45K[1]	45K[1]	45K[1]				
34-42K	(34-42K)[2]	34-42K	34-42K	34-42K	(34-42K)[2]	(34-42K)[2]
25K	25K	25K	25K	25K	25K	25K
19K	**19K**	**19K**	**19K**	**19K**		
18K	18K	18K	18K	18K		
17K	17K	**17K**	**17.5K**	**(17.5K)**[2]		
		16K				
15K	15K	(15K)[2]				
14K						

[1]Probably degradation products of 65K.
[2]Results variable; in some cell lines these proteins are not detected.

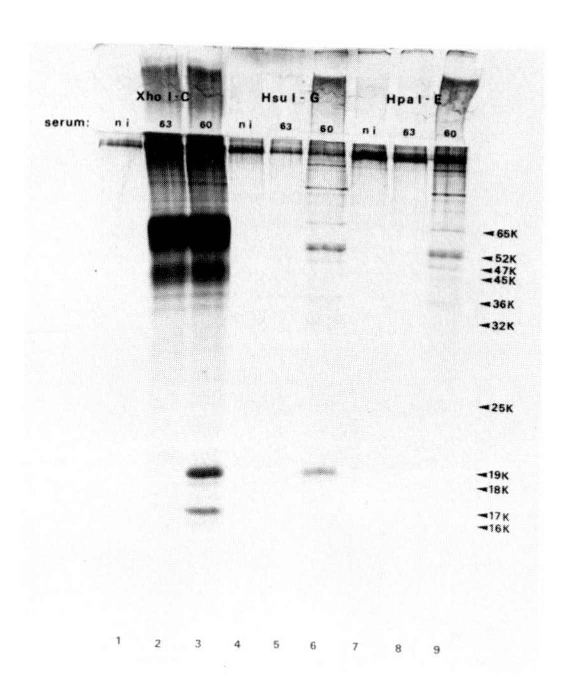

FIGURE 5. Proteins immunoprecipitated from rat cells transformed by Ad5 DNA fragments of different sizes, using serum from a normal hamster (lanes 1,4,7) or from 2 different Ad5 tumor bearing hamsters (lanes 2,5,8 and 3,6,9). Rat cells were transformed by Ad5 *Xho*I C fragment: (lanes 1,2,3), *Hsu*I G fragment (lanes 4,5,6) and *Hpa*I E fragment (lanes 7,8,9). The ^{35}S-labeled proteins were analyzed by electrophoresis in 13% SDS polyacrylamide gels.

decreases. The atypical T antigen pattern in cells transformed by the 8% *Hin*dIII G fragment is correlated with the absence of the 65K major T antigen. Furthermore the abnormal transformed phenotype in cells transformed by the 4.5% *Hpa*I E fragment may be due to the absence of the 19K major T antigen and/or the 18K T antigen.

A comparison between the results obtained by immunoprecipitation and *in vitro* translation shows that both methods generally identify the same set of proteins: 15K, (16K), 17K, 18K, 19K, (25K) and 34-42K. A major difference was noticed for the 65K and the 52K T antigens which were detected only using the immunoprecipitation technique. A close inspection of the electropherographs of *in vitro* synthesized products, however, occasionally revealed the presence of a weak 65K protein, both with RNA from lytically infected cells and from transformed cells (cf. Fig.3). Apparently, this protein is inefficiently synthesized *in vitro*, or the RNA coding for 65K is present at a low concentration.

Detection of viral RNA in HpaI E transformed cells. Since many of the *Hpa*I E fragment-transformed cells contained only very low concentrations of viral T antigen, some of which could in fact be of cellular origin (e.g. the 47K and 52K proteins), we decided to test some of the transformed lines for the presence of viral RNA. Poly-A containing cytoplasmic RNA was isolated from two *Hpa*I E-transformed cell lines and

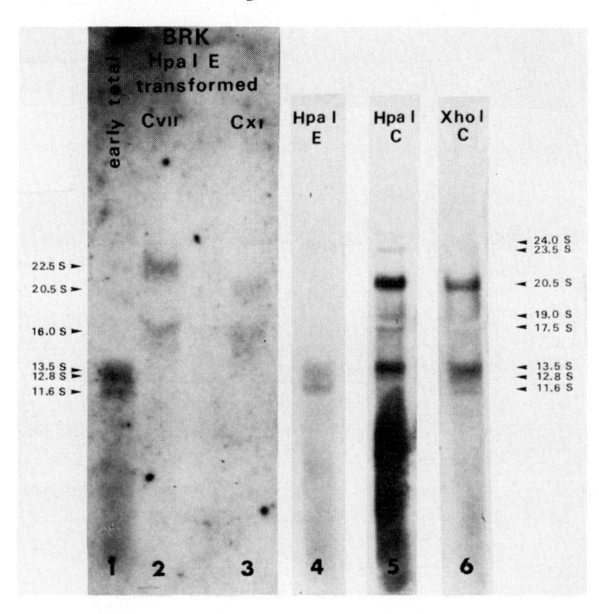

FIGURE 6. Identification of Ad5-specific RNAs in rat cells transformed by the 4.5% *Hpa*I E fragment (lanes 2 and 3) and of RNAs corresponding to early region 1 of Ad5 DNA in lytically infected KB cells (6 hr p.i.;lanes 1,4,5 and 6). Poly A-containing cytoplasmic RNA was isolated from the transformed cells and the RNA was fractionated by electrophoresis in a 1.2% agarose slabgel containing methyl mercury hydroxide as denaturing agent (19). After electrophoresis the RNAs were transfered to DBM paper (20) and the viral RNAs were identified by hybridization with ^{32}P-labeled fragment *Hpa*I E (lanes 2 and 3). Early cytoplasmic RNA from lytically infected cells was treated in a similar way and viral RNAs were identified by hybridization with ^{32}P-labeled fragments *Hpa*I E (0-4.5%, lanes 1 and 4) *Hpa*I C (4.5-25%, lane 5) or *Xho*I C (0-15.5% lane 6).

the RNA was electrophoresed and blotted onto DBM paper, as described in the legend to Fig.6. Viral RNA was detected by hybridization with ^{32}P-labeled Ad5 DNA fragments. Fig.6 shows that ^{32}P-labeled fragment *Hpa*I E detects 3 major RNA species in extracts from lytically infected cells of 13.5 S, 12.8 S, and 11.6 S and one or two minor species and that ^{32}P-labeled fragment *HpaI* C (the adjacent fragment) detects two major RNAs (20.5 S and 13.5 S) and 4 minor RNAs. RNA isolated from two *Hpa*I E-transformed lines contained two RNA species each, which are larger than the major RNAs normally transcribed from the E fragment. These results show that the *Hpa*I E fragment indeed is transcribed in the transformed cells. The aberrant sizes of the transcripts may be explained by assuming that the RNAs represent cotranscripts of cellular and viral DNA sequences, due to the fact that the stopsignal for RNA transcription may be absent from the *Hpa*I E fragment. Alternatively, the larger RNAs could represent transcripts of tandemly integrated viral DNA segments.

Nucleotide sequence analysis of Ad5 transforming region. In yet another approach, we decided to determine the nucleotide sequence of the leftmost early block E1 of Ad5 DNA (23,24). This sequence is now completed from the left terminus of the genome up to about the endo R·*Bgl*II site at map position 9.5. It was determined mainly by the procedure of Maxam and Gilbert (21) and in the latter stages also by a variant of the chain terminator method of Sanger *et al.*(22,23). In this region of the genome the r-strand serves as a template for mRNA synthesis, and the l-strand sequence corresponds to that of the mRNAs (26,27). By arranging the termination codons (TAA, TGA, TAG) found in the l-strand of the determined sequence, according to the reading frames in which they would block protein synthesis, we found three tracts devoid of stop codons (see Fig.7). These extend in reading frame 1 from position 1141 to 1543, and from 1594 to 3322, and in reading frame 2 from 518 to 1118.

From the work of Berk and Sharp (3), Kitchingman *et al.* (2), and Chow *et al.* (4) we know that the mRNAs originating from the corresponding region of the closely related Adenovirus 2 consist of segments transcribed from non-contiguous areas of the DNA; some of these mRNA mapping data are also incorporated in Fig.7. As a result of this splicing-together of non-adjacent parts of the primary transcripts, mRNA I (Fig.7) has lost a small segment containing three termination codons, and can be translated from the potential initiation triplet ATG at 560 to the termination codon TAA at 1543, yielding a 33K polypeptide. In the absence of splicing the primary transcript would allow the synthesis of at most a

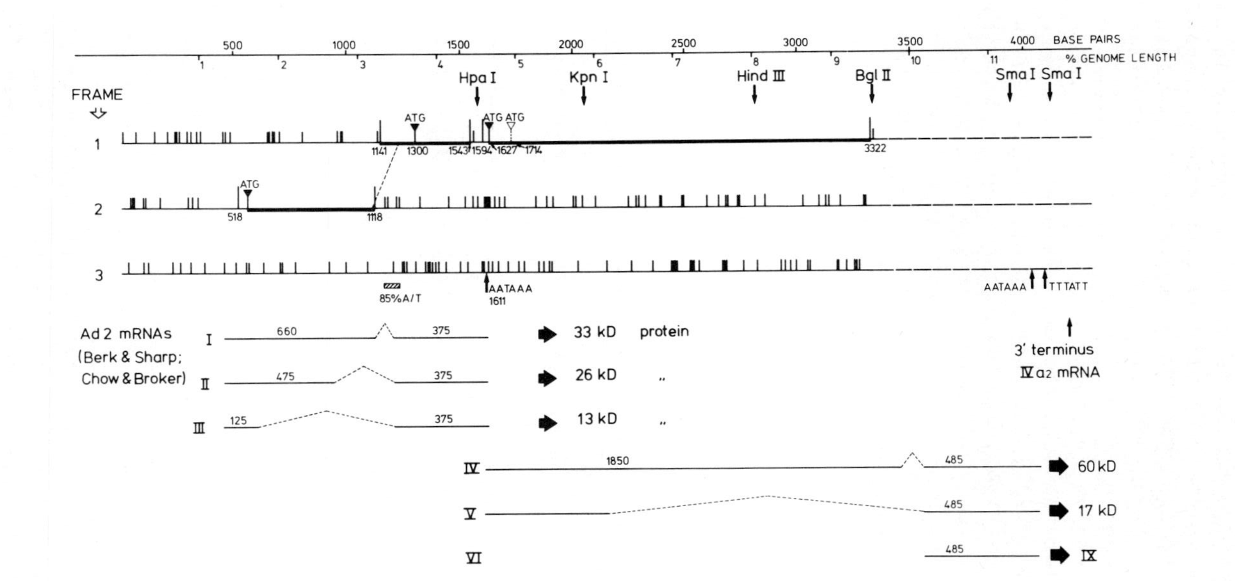

FIGURE 7. The stopcodons (TAA,TAG and TGA) in the l-strand of the *Bgl*II D fragment of Ad5 DNA (0-9.5%) arranged according to their reading frames. In the longer stopcodon-free stretches the ATG triplets are indicated which would initiate polypeptides of maximum length. In the bottom of the figure we summarize the results of (3) and (4) on the mapping of early mRNAs; the continuous lines represent the areas found to anneal to mRNA, the interrupted lines are the intervening sequences.

20.5K protein (560-1118), and possibly a 9K protein (1300-1543). Similarly mRNA II could contain the information for a 26K protein, and mRNA III for a 13K protein. Possibly, the 26K and 33K proteins correspond to the 34-42K proteins synthesized *in vitro* from Ad5 early mRNAs. The discrepancy in molecular weights between these *in vitro* synthesized proteins, and the proteins deduced from the DNA sequence may be explained by their high proline content predicted by the nucleotide sequence; this would decrease the electrophoretic mobilities of the proteins in SDS-polyacrylamide gels and hence increase their apparent molecular weights. mRNA IV contains a long open reading frame (1594-3322); the leftmost ATG triplet in this open frame is at position 1627, but, on comparison with the corresponding sequence of Ad7 DNA which was also determined in our laboratory (R. Dijkema, unpublished results) we feel that a following ATG triplet at position 1714 is a better candidate as initiator for protein synthesis. Starting then at his ATG triplet, mRNA IV would specify a protein of ca. 60K daltons. mRNA V has space for a 16-19K protein depending on the amount of information deleted from this messenger. Finally mRNA VI contains the information for viral protein IX. Evidently knowledge of the DNA sequence alone is not sufficient to predict the exact molecular weights of the proteins specified by this part of the Ad5 genome. It will also be necessary to determine the sequences of the various mRNAs around the splice points; such experiments are currently being performed in our laboratory. The above-mentioned RNA mapping studies place the site for the 3' ends of the leftmost family of messengers just beyond the *Hpa*I site at map position 4.5, and that of the second family round map position 11.5. This is in agreement with our DNA sequence data, since at position 1611 and also in the small *Sma*I N fragment located between positions 11.2 and 11.8, we found the sequence A-A-T-A-A-A; in all characterized eucaryotic mRNAs the sequence A-A-U-A-A-A has been found 10-20 nucleotides before the start of the 3' poly-A tail.

DISCUSSION

The results reported in this paper demonstrate that rat cells transformed by Ad5 DNA fragments containing different parts of the leftmost early region of the adenovirus genome (E1) exhibit different transformed phenotypes. This observation suggested that transformation by Ad5 is a process in which several viral genes are involved and that the abnormal transformed phenotypes are the result of the deficiency (or absence) of one or more viral gene products. To obtain a better understanding of the nature and functions of the adenovirus transformation genes, we have identified the proteins and RNAs encoded by this region of the genome and determined its nucleotide sequence.

The results of these investigations have shown that the transforming region of Ad5 DNA (region E1) which maps between 1.5 and 11.5%, can be divided into two parts (corresponding to regions E1a and E1b), each of which encodes for a set of proteins. Figure 8 summarizes the approximate coding regions that we were able to assign to the various proteins. It can be seen that the left part of early region 1 (region E1a) which extends from 1.5-4.5%, codes for at least five overlapping proteins. The four largest of these proteins (34K, 36K, 40K, 42K), and possibly all five proteins, are closely related and hence they should be encoded to a large extent by common DNA sequences. These polypeptides are specified probably by the three mRNA species of 11-13S that have been identified for this region (Fig. 6 and 7).

The adjacent segment which extends from 4.5 to 11.5%, contains the information for at least two early proteins (19K and 65K), as well as for one late protein (protein IX). The nucleotide sequence results and the RNA mapping data (ref.2. 3,4) suggest that 19K and 65K are derived from the two (overlapping) major RNAs that have been identified for this region (Fig.6 and 7), and that these proteins may possibly overlap.

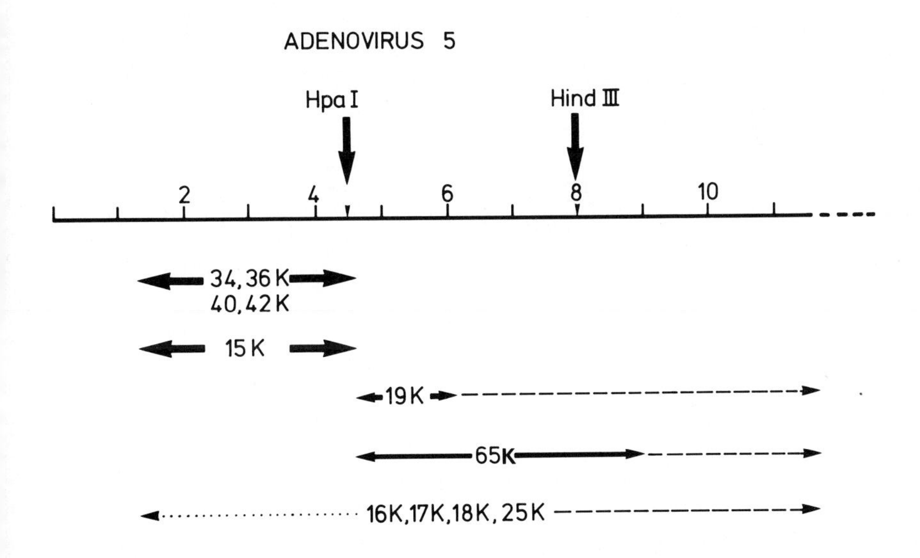

FIGURE 8. Approximate coding regions of proteins encoded by the transforming segment of Ad5 DNA (early region 1, extending from 1.5-11.5%).

Preliminary results from our laboratory, however, have now indicated that 19K and 65K are not structurally related, which indicates that their map positions as proposed in Fig.8 may not be correct. Definitive conclusions concerning the coding regions for these polypeptides will have to await results of amino acid sequencing, which are in progress.

We were surprised by the finding that the 65K T antigen, which is abundantly present in transformed and lytically infected cells, is either not synthesized at all in cell-free protein synthesizing systems or is made in only very limited amounts. In our opinion, the best candidate for the 65K mRNA would be the 20.5 S RNA which is encoded by the region between 4.5 and 11.5%, also because only this stretch can accommodate a protein of that size (Fig. 6 and 7). RNA preparations that were known to contain high concentrations of both 13.5 S and 20.5 S RNA (the main species encoded by the 4.5-11.5% region), however, when translated, yielded very little 65K protein, but gave rise to large quantities of 19K. We have no explanation for this phenomenon.

Our results indicate that the two regions E1a and E1b have different biological functions. The smallest DNA fragment still containing transforming activity is the *Hpa*I E fragment, which corresponds to region E1a. Rat kidney cells transformed by fragment *Hpa*I E are distinguished from other Ad5 transformed cells by their relatively "normal" appearance and their inability to grow to high cell densities in monolayer culture. Transformed cells that in addition contain the adjoining 3.5% or more, generally exhibit the typical transformed phenotype. The observations suggest that the proteins encoded by the *Hpa*I E fragment (or a part of these proteins) are capable of converting diploid cells with a limited lifespan into aneuploid cells with an unlimited lifespan (but still lacking the characteristic transformed phenotype). The proteins encoded by the adjacent segment (4.5-11.5%) in addition, will induce a number of properties that are typical for transformed cells. The 19K protein may play an important role in this respect. Since cells transformed by the *Hind*III G fragment (0-8%) are rather similar in appearance to the standard transformed cells, the region between 8 and 11.5% does not seem to be very essential for transformation. As the absence of this segment of the genome is correlated with the absence of the 65K T antigen, we conclude that this protein should not have an important role in transformation, although it may contribute to the transformed phenotype

Further studies are in progress to localize precisely the coding regions of the various adenovirus T antigens by amino acid sequence analysis and by determining the exact splice points in the mRNAs, and to assess their role in transformation.

ACKNOWLEDGEMENTS

We thank Ada Houweling, Jacqueline Hertoghs, Mieke Lupker, Arja Davis and Ben Dekker for expert technical assistance and dr. B.N. Bachra for critically reading the manuscript. This work was supported in part by funds from the Foundation for Medical Reasearch (FUNGO), the Foundation for Chemical Research (SON) and the Koningin Wilhelmina Fonds.

REFERENCES

1. Berk, A.J.,and Sharp, P.A. (1978). Proc.Natl.Acad.Sci. USA. 75, 1274.
2. Kitchingman, G.R., Lai, S.-P., and Westphal, H. (1977). Proc.Natl.Acad.Sci.USA 74, 4392.
3. Berk, A.J., and Sharp, P.A. (1978). Cell 14, 695.
4. Chow, L.T., Lewis, J.B., and Broker, T. Unpublished results.
5. Graham, F.L., Van der Eb, A.J., and Heijneker, H.L.(1974). Nature 251, 691.
6. Graham, F.L.,Abrahams, P.J., Mulder, C., Heijneker, H.L., Warnaar, S.O., de Vries, F.A.J., Fiers, W., and Van der Eb, A.J. (1974).Cold Spring Harbor Symp.Quant.Biol. 39, 650.
7. Van der Eb, A.J., Mulder, C., Graham, F.L., and Houweling, A. (1977).Gene 2, 132.
8. Van der Eb, A.J., and Houweling, A. (1977) Gene 2, 146.
9. Sekigawa, K., Shiroki, K., Shimojo, H., Ojima, S., and Fujinaga, K. (1978). Virology 82, 214.
10. Yano, S., Ojima, S., Fujinaga, K., Shiroki, K., and Shimojo, H. (1977). Virology 82, 214.
11. Shiroki, K., Handa, H., Shimojo, H., Yano, H., Ojima, S., (1977). Virology 82, 462.
12. Graham, F.L.,and Van der Eb, A.J. (1973). Virology 52, 467.
13. Graham, F.L.,and Van der Eb, A.J. (1973). Virology 54, 539.
14. Flint, S.J., Gallimore, P.H., and Sharp, P.A. (1975). J.Mol.Biol. 96, 68.
15. Lupker, J.H. Davis, A., Jochemsen, H., and Van der Eb, A.J. (1977) Colloques INSERM 69, 232.
16. Lewis, J.B., Atkins, J.F., Baum, P.R., Solem, R., Gesteland, R.F., and Anderson, C.W. (1976). Cell 7, 141.
17. Harter, M.L., and Lewis, J.M. (1978). J.Virol. 26, 736.

18. Graham, F.L., Smiley, J., Russell, W.C. and Nairn, R. (1977).J.Gen.Virol. 36, 59.
19. Bailey, J.M., and Davidson, N. (1976).Anal.Bioch. 70, 75.
20. Alwine, J.C., Kemp, D.J., and Stark, G.R. (1977). Proc. Natl.Acad.Sci.USA. 74, 5350.
21. Maxam, A.M., and Gilbert, W. (1977). Proc.Natl.Acad.Sci. USA. 74, 560.
22. Sanger, F., Nicklen, S., and Coulson, A.R. (1977). Proc. Natl.Acad.Sci.USA 74, 5463.
23. Maat, J., and Smith, A.J.H. (1978). Nucleic Acids Res. 5, 4537.
24. Van Ormondt, H., Maat, J., De Waard, A., and Van der Eb, A.J. (1978).Gene 4, 309.
25. Maat, J., and Van Ormondt, H. (1979).Gene, in the press.
26. Philipson, L., Pettersson, U., Lindberg, U., Tibbetts, C., Vennström, B., and Persson, T. (1974). Cold Spring Harbor Symp.Quant.Biol. 39, 447.
27. Sharp, P.A., Gallimore, P.H.,and Flint,J.S. (1974). Cold Spring Harbor Symp.Quant.Biol. 39, 457.

THE EPSTEIN-BARR VIRUS PLASMID

Joseph S. Pagano, M.D.

Cancer Research Center, The University of North Carolina, Chapel Hill, North Carolina 27514

ABSTRACT The Epstein-Barr Virus plasmid, the molecular weight of which is approximately 100 x 10^6 daltons, persists in non-virus-producing cells in a stable regulated state. In the Burkitt-Lymphoma- cell line, Raji, there are 50-60 copies of EBV genomes per cell. The EBV DNA is situated in the nucleus and it is chromosomally associated, but not covalently integrated into host-cell DNA. At least two-thirds of the viral DNA can be recovered in a supercoiled form with contour lengths approximating 100 x 10^6 daltons. Such molecular forms exist in Burkitt's lymphoma tissue and nasopharyngeal carcinomas *in vivo*. Supercoiled forms of EBV DNA have also been recovered from the EBV-producing Burkitt's lymphoma cell line, P3HR-1, by suppression of virus production with the antiviral drug, acycloguanosine, and from the EBV-producing marmoset lymphocyte line, B-95-8, with the use of phosphonoacetic acid. The contour length of P3HR-1 supercoiled genomes appears to be less than the length of Raji EBV genomes. Restriction-endonuclease digestion analyses of these forms of the EBV genome are now becoming available. This form of EBV DNA, which has been called the EBV plasmid or episome, appears to be replicated by host DNA polymerases rather than by virus-induced polymerases, can be induced to a limited extent with IUDR, but cannot be cured by any known treatment. The function of the stable EBV plasmid, whether it is expressed, particularly in relation to lymphocyte proliferation and transformation, and its relation to possible integrated viral DNA sequences are probably central issues in the cellular biology and pathobiology of EBV-associated diseases.

THE EBV PLASMID

The Epstein-Barr Virus and its Genome. The genome of the Epstein-Barr Virus (EBV) exists in at least two states: within the virions as a linear double-stranded molecule

ISBN 0-12-198780-9

approximately 100 x 10^6 daltons and intracellularly as a supertwisted circular form.[1] In addition there is indirect evidence that EBV genes may be integrated into cellular DNA by covalent bonding. Circular replicative intermediate forms of the EBV genome including concatameric forms may also exist (2). The EBV plasmid[2], intriguing because of its uniqueness in eucaryotic cells, poses novel questions about its structure, its maintenance in the infected cell and its function. These molecular issues must relate to central biologic features of the herpes-group viruses. The hallmarks of these viruses are their ability to enter a true latent state and later to undergo reactivation spontaneously or in response to inducing agents or immunologic circumstances. The Epstein-Barr Virus is also a likely candidate for a human tumor virus, and it is tempting to think that the EBV plasmid plays a role in the molecular pathogenesis of malignant transformation on the cellular level.

The first linear maps of EBV genomes have been constructed by Kieff _et al_. (4). Since temperature-sensitive mutants of EBV are not available the relation of these crude divisions of the genomes to function is only beginning to be surmised. For example, Kieff has proposed fragments that may contain the EBV transforming gene or genesbased on lack of such sequences in the genome of a non-Burkitt lymphoma strain of the virus, B95-8, of infectious mononucleosis (IM) origin and their retention in a BL-derived line (5).

It is not clear whether EBV has a structural organization of its genome resembling the unique arrangement found in herpes simplex virus and other herpes-group viruses including human cytomegalovirus. The preliminary data of G. Hayward and E. Kieff do suggest that there is terminal complementarity revealed by exonuclease treatment that would permit the linear form of the genome to circularize.

Discovery of the EBV Plasmid. The initial work with EBV was carried out by cultivation of Burkitt's lymphoma cells in a line called Jijoya; P3HR-1 is a non-transforming substrain of this cell line. Such cells, grown in suspension cultures, continually shed virus into the medium. Another Burkitt's lymphoma cell line called Raji is not virus-producing, but was shown to contain EBV DNA by DNA-DNA hybridization (6) and to contain approximately sixty genome equivalents per cell by

[1]For general reviews see (1, 2, 3).

[2]I follow the definition of plasmid as applied to EBV by A. Adams (2) in which plasmid DNA may exist independently or insert into cellular DNA non-reversibly whereas episomal DNA and integrated sequences are reversibly exchangeable.

by cRNA-DNA hybridization (7). Since the number of copies of EBV genomes in the cell line was regulated and constant it was assumed that the viral DNA was probably integrated into the cellular DNA, which proved not to be the case. Although the viral DNA could be quantitatively recovered in partially purified Raji chromosomes, it was not covalently bonded to the cellular DNA as shown by its separation from high molecular weight Raji-cell DNA in alkaline glycerol velocity sedimentation gradients (8). At this point the idea that EBV DNA might exist in such cells in the form of a plasmid or episome was first proposed. This notion was attractive because it seemed to be in keeping with the biologic properties of this, a herpes-group virus, namely latency and reactivation after a period of dormancy -- behavior that was reminiscent of plasmids or episomes in procaryotic cells.

General Characteristics. The plasmid form of EBV, first isolated from Raji cells by Adams and Lindahl, exists in a supercoiled configuration; approximately two-thirds of the EBV DNA in the Raji cells could be recovered in this form (9). Most of the rest of the DNA was presumed to exist also as plasmids, but because of shearing during extraction this DNA could not be separated from linear cellular DNA in the neutral glycerol gradients first used to identify the plasmids. The contour length of EBV DNA from the Raji cells was consistent with a molecular weight of 100 x 10^6 daltons, about equal to virion DNA. Since then EBV plasmids have been recovered from a variety of cell lines and tissue sources as indicated later.

Replication. The EBV plasmid replicates in synchrony with cellular DNA replication. The time when the plasmid DNA begins to replicate has been determined by Hampar *et al.* (10) to be early in the S-1 phase of the cell cycle. The EBV plasmids are almost certainly replicated by host DNA polymerases; this inference comes from the capacity of the EBV episomes to be maintained in the same copy number in cells treated with inhibitors of EBV DNA replication such as acycloguanosine (11); (Colby, Shaw, Elion and Pagano, unpublished data) and inhibitors of virus-specific DNA polymerase such as phosphonoacetic acid (12).

a) Induction. EBV episomes are induced into replication and an increase in copy number only with difficulty. Bromodeoxyuridine and iododeoxyuridine do induce the appearance of antigens that generally accompany free viral DNA replication, but the effect is quite variable, and there is at best only a doubling of the genome number (13). The phorbol ester, TPA (12-O-tetradecanoyl-phorbol-13-acetate), a chemical carcinogen which has a pronounced inducing effect in a virus-producing

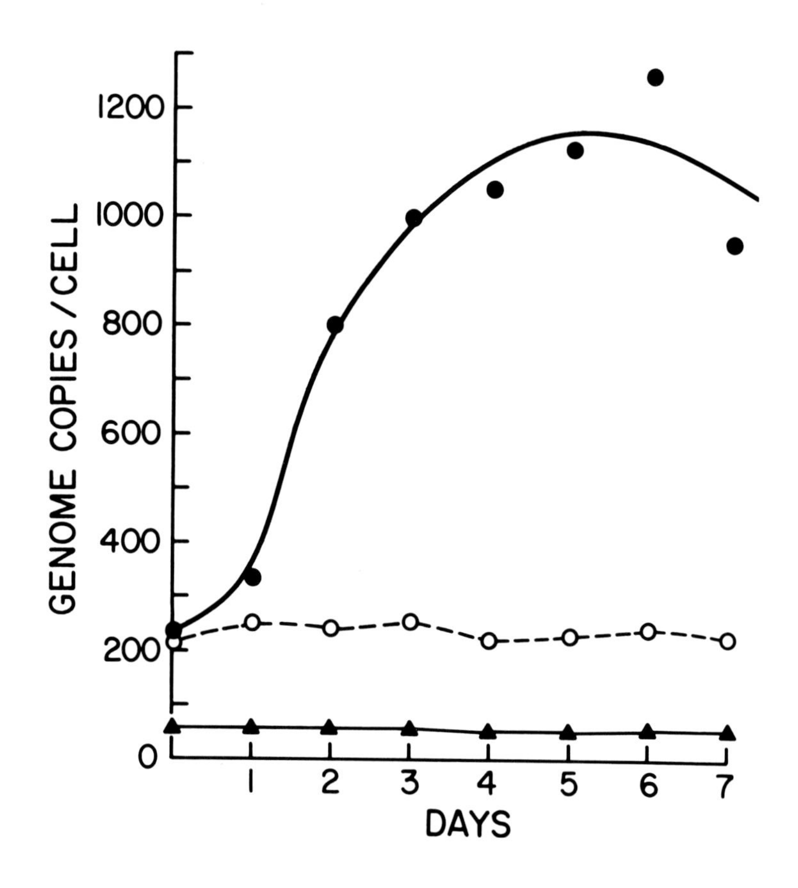

Figure 1. Induction of EBV DNA Replication by (12-O-tetradecanoyl phorbol-13-acetate (TPA). The number of EBV genome equivalents per cell was determined by cRNA-DNA hybridization with EBV-specific cRNA (7). Data of Lin _et al_. (15).

cell line (14) has little effect on the Raji cells beyond induction of a small percentage of cells to display EBV early antigen (EA) without however an accompanying increase in the number of EBV genome copies (15) (Figure 1).

b) Chromosomal association. EBV plasmid DNA is associated with chromosomes (7). From early in situ cytohybridization studies it was thought that EBV DNA might be associated with specific chromosomes in Raji cells (16). Moreover there is a recurrent defect in chromosome No. 14 in Burkitt's lymphomas (17). However somatic cell hybridization work to date has failed to disclose any consistent chromosomal association of EBV DNA (18, 19). One problem with this work is that much of it has been done with virus-producing rather than non-virus-producing cells. Also, the marker for the presence of viral genome has been the viral antigen, EBNA, rather than direct tests for viral DNA.

c) Introduction of plasmids into cells _in vitro_. Lymphocytes can be infected _in vitro_ with transforming strains of EBV with the result that plasmid forms of the genome become emplaced, usually in a fixed copy number, in the new cell line that is established in the process. The transforming viruses are of two general types: fresh virus isolated from the throats of patients with acute infectious mononucleosis and IM virus that has been passaged through marmoset lymphocytes.

The target cells, which are human B-lymphocytes, are also of two general classes: permissive and non-permissive. Lymphocytes taken from the umbilical cords of newborn infants are generally not permissive, and transformation or immortalization of these cells by EBV does not lead to a virus-producing line. In such cells although the Epstein-Barr Virus Nuclear Antigen (EBNA) appears, early antigen (EA) and viral capsid antigen (VCA) are not detectable. The number of EBV genome copies in such lines is generally low, 10-16 copies or less (20). Mature lymphocytes taken from EBV antibody-negative persons can also be transformed with EBV into continuous lines carrying the EBV genome, but these lines are often virus-producing, as are those cell lines that can be established spontaneously from the peripheral blood of EBV antibody-positive persons (1).

The establishment of the non-permissive cell lines thus poses a paradox inasmuch as the limited amount of viral DNA retained in these cells speaks against the induction and functioning of virus-induced DNA polymerase. I have made the suggestion before that, inasmuch as the plasmid form of the genome appears to be maintained in the cell lines by host DNA polymerase, perhaps the few molecules of EBV DNA found in the non-permissive cells are also replicated initially by host enzymes. This situation would be quite different in permissive cells in which virus-induced polymerase is probably responsible for replication of the large number of copies of EBV genomes in a subset of the cells. There is already some evidence to suggest that cell transformation and lymphocyte transformation can occur in the presence of viral DNA inhibitors (21, 22, 22a).

Plasmids in Virus-Producing Cells.

a) Existence. Virus-producing cell lines such as P3HR-1 contain two populations: a minor population (5-15%) which is replicating viral DNA and produces virus and major population (85-95%) which discloses EBNA but does not display other viral antigens or give evidence of abundant viral DNA replication as shown by in situ cytohybridization (1). However all the cells in both populations contain the viral genome.

Recently we have succeeded in recovering supercoiled forms of EBV DNA from the P3HR-1 cell line by cultivating in the presence of an antiviral drug, acycloguanosine (11) (Colby, Shaw, Elion and Pagano, unpublished data). Supercoiled and open circular forms of EBV DNA from P3HR-1 lines are shown in Figures 2 and 3. There are approximately 12 copies per cell (Colby and Shaw). Andersson et al. (2) have used a similar approach with phosphonacetic acid (PAA). Yajima et al. (23) found 10 copies per cell in the presence of phosphonoacetic acid. Thus the difference between virus-producing and non-producing lines is basically that the plasmid forms in the productive line continually tilt into free replication spontaneously, whereas the plasmids in the non-permissive line never spontaneously move into a free replicative phase. The two cell lines thus correspond to the biological situations of latency on the one hand and reactivation on the other hand.

The mechanism whereby the plasmid is regulated in the one cell type and escapes regulation in the other circumstance is not understood, but we surmise that the EBV nuclear protein, EBNA, (24) may play a role in either the repression or derepression process. The first approach to understanding the molecular bases of the differences between these two classes of cells is to examine whether there are structural or gross sequence differences in their plasmids, and this work is underway (Shaw et al., unpublished data).

b) Induction. Efforts to induce viral DNA synthesis to a higher level in the virus-producing cells have followed many paths, all of them empirical. A shift to low temperature (32°C), arginine deprivation, dexamethasone treatment and therapeutic intercalating dyes such as chloroquine all have some effect (see ref. 15). The most striking effect however is produced by TPA (12-0-tetradecanoyl-phorbol-13-acetate) which gives at least a ten-fold increase in the number of

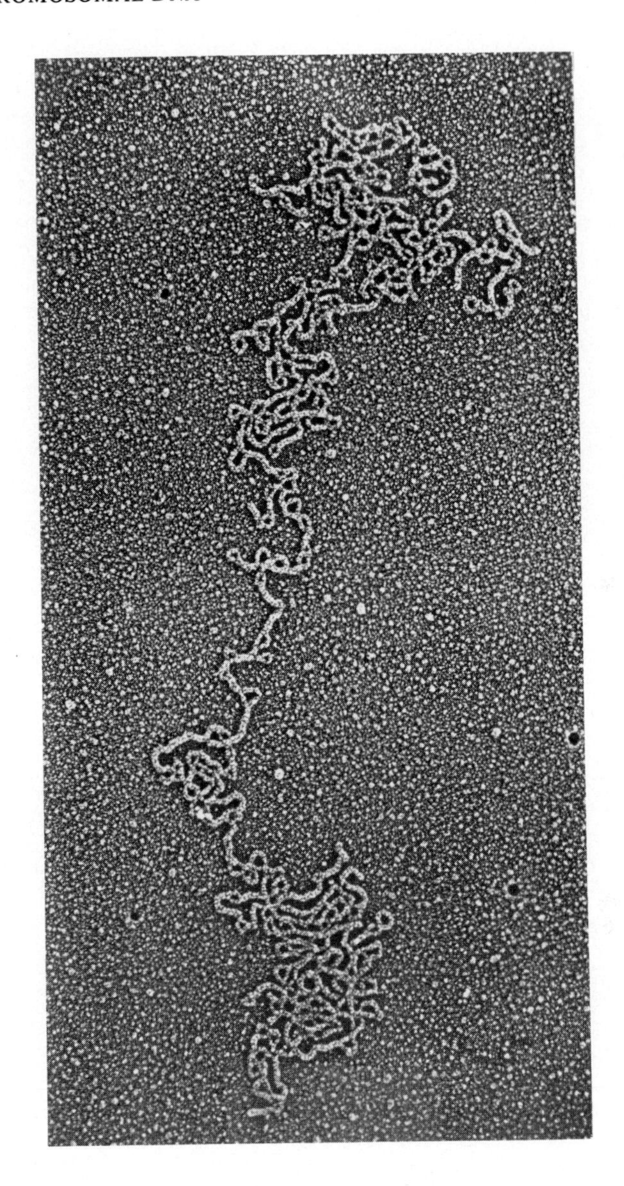

Figure 2. Supercoiled EBV DNA recovered from virus-producing P3HR-1 cells treated with acycloguanosine (9-[2-hydroxethoxymethyl]guanine). Cells were treated with 100 μM of the drug for 7 days. Colby _et al._, unpublished data; photomicrograph courtesy of C. Moore and J. Griffith.

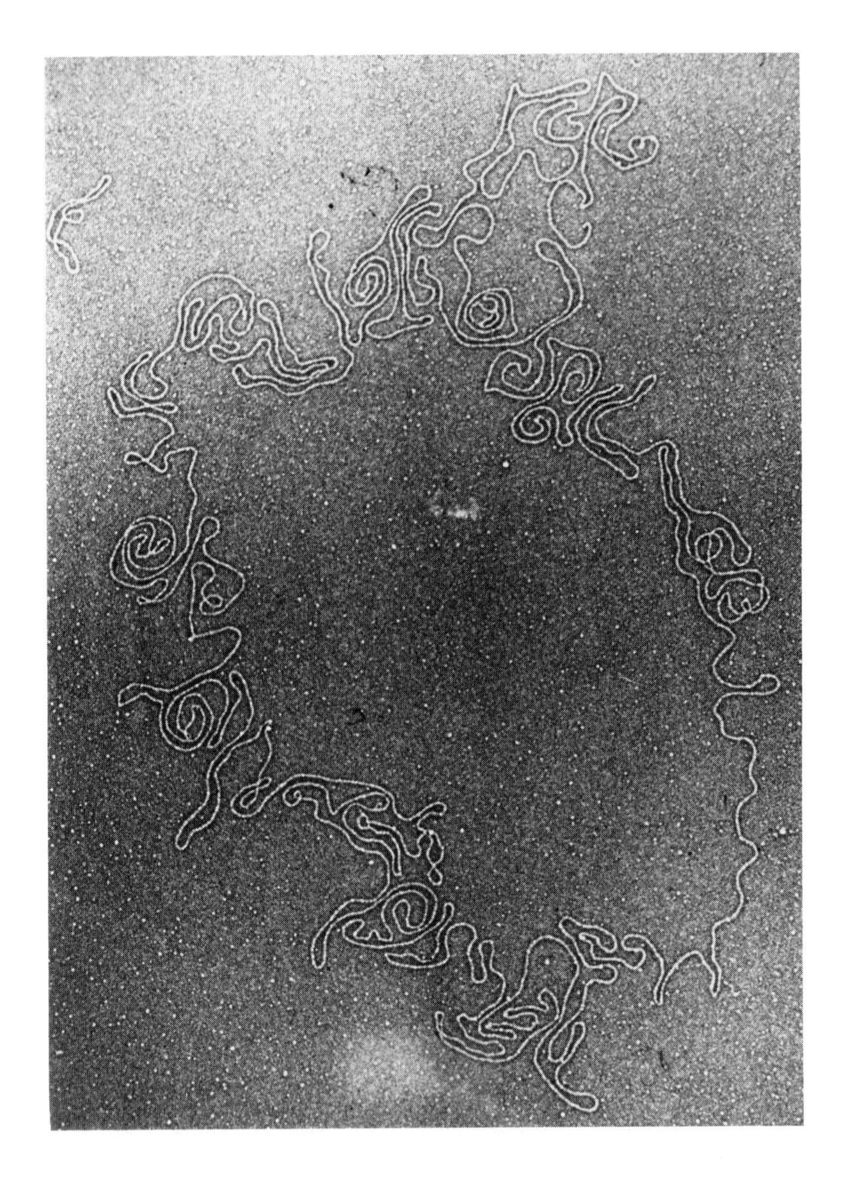

Figure 3. Open circular form of EBV genome recovered from P3HR-1 cells treated with acycloguanosine. Colby et al., unpublished data; photomicrograph courtesy of C. Moore and J. Griffith.

genomes obtainable from P3HR-1 cell (See Figure 1). In situ cytohybridization and counting of cells killed by viral replication indicate that essentially every cell in the major population containing latent genome is induced into replication and consequently is killed. This is an unprecedented result which has practical utility. Recently we have been able to recover 160 μg of purified EBV DNA from less than 4 liters ofP3HR-1 culture fluid, about thirty-fold better than before.

c) Cure of episomes. The plasmid forms are quite resistant to the action of specific antiviral inhibitors such as phosphonacetic acid and acycloguanosine. This stability in the face of specific drug treatment is true both of virus-producing and non-producing cells. With the latter drug maintenance of Raji cells and P3HR-1 cells for as long as six months has been without effect on the number of copies of plasmid DNA contained in either type of cell line. Raji cells kept in the presence of cycloheximide reduce the number of genome copies to approximately 20, but this level is restored to normal once the drug has been removed (25). More recently Tanaka _et al._ (26) have produced a P3HR-1 cell line that carries only 2 copies of EBV genomes by cultivation in the presence of cycloheximide.

Existence in Nature. Fresh tissues taken from Burkitt's lymphomas and nasopharyngeal carcinomas contain the plasmid form of the EBV genome (27). This molecular form is not an artifact of culture _in vitro_; it must have a natural role. The EBV plasmid is also found _in vitro_ in lymphocytic lines established from the peripheral blood of patients with infectious mononucleosis; the plasmid exists in the lymphocytes _in vivo_ as well (28). We do not know whether viral replication is aborted and the plasmid form ever arises in normal epithelial cells in the oropharynx where EBV is believed to replicate initially (29, 30).

The character of the plasmids found in different diseases is a matter of great interest. There is some evidence from DNA-DNA renaturation kinetics analyses, which reflect mainly the viral sequences contained in the plasmid, of differences in the percentage of the EBV genome found in both NPC (31) and infectious mononucleosis (32). Some such samples appeared to lack up to 30% of the sequences found in the P3HR-1 probe, whereas other specimens of NPC and IM origin contained at least 90% of the P3HR-1 sequences (31, 33, 34). None of these results can be considered definitive.

Adams _et al._ have made contour-length measurements of the various "species" of plasmids; the molecules from three IM lines were close to the unit genome length of P3HR-1 virion DNA. The contour-lengths of most BL-derived plasmids were strikingly uniform. The plasmids from a single IM line and

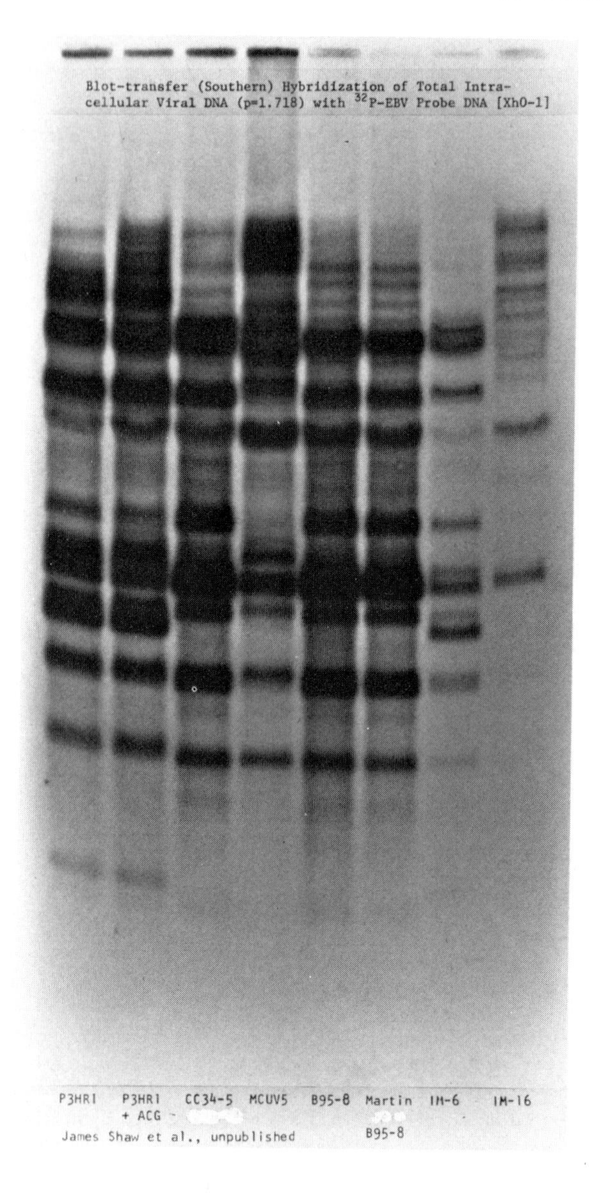

Figure 4. Blot-transfer hybridization of intracellular viral DNA with EBV DNA. Total cellular DNA after digestion with restriction-endonuclease XhO-1. From left to right: P3HR-1 cells treated with acycloguanosine; the CC34-5 cell line derived from Nyevu virus (courtesy of George Miller); the MCUV5 cell line; B95-8 cell line; Martin cells + B95-8; IM 6 cell line from human peripheral blood (29; 32); IM 16 cell line derived from human peripheral blood (29; 32). Shaw *et al.*

the B95-8 line are slightly smaller. In preliminary information obtained by Southern transfer technique these investigators have not so far discerned consistent patterns of restriction-enzyme cleavage in intracellular viral DNA from the three diseases (See ref. 2).

Recent work from our laboratory shows that the restriction patterns of the intracellular DNA in several types of cell lines differ significantly as shown in Figure 4 (Shaw and Pagano, unpublished data). The cell lines include virus-producing and non-producing Burkitt's lymphoma lines, two marmoset lymphocyte lines in which human EBV deriving from infectious mononucleosis has been passaged (gift of George Miller), and two infectious mononucleosis cell lines established from the peripheral blood. The interpretation of these results requires further work, but the data suggest the existence of EBV strains.

Function. The function of the EBV plasmids is in fact a mystery; their general significance can only be surmised. The only other system in which similar forms have been detected is in herpesvirus samiri-infected lines (35). This is a simian herpesvirus that produces malignant lymphomas in alternate simian hosts. Presumably the plasmid is the molecular basis for herpesvirus genetic information as it is retained in cells in the latent form or in latent infection. By extension we deduce that plasmid forms of other herpes-group viruses such as herpes simplex and cytomegalovirus might also exist in nature. The fact that such forms of other herpesvirus genomes have not been discovered may be due to the lack of suitable cell lines or fresh tissues in which to look directly. Alternatively it may be a peculiarity of malignant or proliferating tissue to retain plasmid forms that do not reactivate spontaneously, at least *in vivo*. In fact there is indirect evidence from serologic data (high EA titers and high VCA titers) that virus does reactivate in Burkitt's lymphoma and nasopharyngeal carcinoma. In any case the restringent state of the plasmid is a necessary condition for survival of the transformed cell inasmuch as free replication of the EBV genome kills the cell.

We infer that transpositions might take place between the plasmid or portions of it and the cellular genome. Adams *et al*. have proposed that there are, in addition to the plasmid sequences, integrated viral sequences in Raji and other non-producing cell lines (36). The identity of the putative integrated viral sequences has not been ascertained. Such transpositions would certainly be facilitated by the proximity of the viral plasmid in the chromosome and the use by the plasmid of host enzymes with which to replicate. Thus the plasmid form might provide a crucial link in the molecular pathogenesis of malignant transformation of cells (37).

Infectious mononucleosis cell lines also contain plasmid forms of the EBV genome, but such cell lines are not considered to be malignant, either in terms of the disease with which they are associated, which is benign, or the polyclonal nature of such cell lines. This situation thus raises the question of whether the plasmid is not merely latent but is in fact expressed. Although Traub and Kieff have gone a considerable way towards mapping the polyadenylated RNA found in Raji cells and also in BL tumor tissue -- with very interesting results-- (4) there has been no way of determining the origin of these transcripts inasmuch as they might arise from integrated viral sequences. However the other possibility is appealing, namely, that the plasmid is not entirely dormant, but it is transcribed in part. We do know of two and possibly as many as three viral functions expressed in cells that harbor the plasmids, namely EBNA, LYDMA (lymphocyte-determined membrane antigen; (38)) and cell proliferation or immortalization with whatever cell surface changes are implied by this behavior. We have suggested before that the proliferating non-malignant IM cells might be under the control of the plasmid. Malignant transformation might be tied in with a shift of control of cell proliferation from plasmid genes to specific integrated transforming genes. These are of course speculations, but they should eventually prove accessible to fruitful experimentation.

A final thought on the importance of the episome comes from the strongly based deduction that the molecule is replicated by host DNA polymerase. We have to assume that either the host polymerase "jumps track," somehow moving to the separate nearby episomal molecule to replicate it once during every cell cycle, or that there is a site in the episome where host DNA polymerase may initiate replication. The conclusion from the second possibility is that EBV plasmids contain a host-cell DNA replication initiation site. Shaw with very precise measurements demonstrated that purified Raji plasmids are slightly lighter in buoyant density than virion DNA (3). If this intriguing possibility is true the plasmids could provide a convenient molecule with which to study cellular DNA replication, and my guess is that we are only beginning to scratch the surface of the contribution of Epstein-Barr virology to eucaryotic cell biology.

REFERENCES

1. Pagano, J.S. (1974). In "Viruses, Evolution and Cancer" (E. Kurstak and K. Maramorosch, eds.), pp. 79-116. Academic Press, New York.
2. Adams, A. (1979). In: The Epstein Barr Virus. (in press) (M.A. Epstein and B.G. Achong, eds.), Springer-Verlag KG, Heidelberg.

3. Pagano, J.S., and Shaw, J.E. In: "The Epstein Barr Virus." (in press). (M.A. Epstein and B.G. Achong, eds.), Springer-Verlag KG, Heidelberg, West Germany.
4. Kieff, E., Given, D., Powell, A.L.T., King, W., Dambaugh, T., Rabb-Traub, N. (1978). Biochem. Biophys. Acta, Reviews on Cancer (in press).
5. Kieff, E., Rabb-Traub, N., Given, D., Pritchett, R., Powell, A., King, W., Dambaugh, T. (1979). In: "Oncogenesis and herpesviruses III. (F. Rapp and G. de Thé, eds.), Lyons, IARC (in press).
6. zur Hausen, H., Shulte-Holthausen, H. (1970). Nature (London) 227, 245.
7. Nonoyama, M., Pagano, J. (1971). Nature New Biol. 223, 103.
8. Nonoyama, M., Pagano, J.S. (1972b). Nature New Biol. 238, 169.
9. Adams, A., Lindahl, T. (1975). Proc. Natl. Acad. Sci. USA 72, 1477.
10. Hampar, B., Lenoir, G., Nonoyama, M., Derge, J.G., Chang, S.-Y. (1976). Virol. 69, 660.
11. Elion, G., Furman, P., Fyfe, J.S., deMiranda, P., Beauchamp, l., Schaeffer, J.J. (1977). Proc. Natl. Acad. Sci. USA 74, 5616.
12. Summers, W.C., Klein, G. (1976). J. Virol. 18, 151.
13. Hampar, B., Tanaka, A., Nonoyama, M. Derge, J. (1974). Proc. Natl. Acad. Sci. USA 71, 631.
14. zur Hausen, H., O'Neill, F.J., and Freeze, U.K. (1978). Nature 272, 373.
15. Lin, J.C., Shaw, J.E., Smith, C. and Pagano, J.S. (submitted to Virology).
16. zur Hausen, H., Schulte-Holthausen, H. (1972). In: Oncogenesis and Herpesviruses I. (P.M. Biggs, G. de Thé and L.N. Payne eds.), pp. 321-325. IARC Scientific Publication No. 2. Lyons.
17. Jarvin, J.E., Ball, G., Rickinson, A.B., et al. (1974). Int. J. Cancer 14, 716.
18. Glaser, R., Nonoyama, M., Shows, T.B., Henle, G., Henle, W. (1975). In: "Oncogenesis and Herpesviruses II." (G. de Thé, M.A. Epstein, and H. zur Hausen eds.), pp. 457-466. IARC Scientific Publication No. 11. Lyons.
19. Spira, J., Povey, S., Wiener, F., Klein, G., Andersson-Anvret, M. (in press). Int. J. Cancer.
20. Miller, G., Coope, D., Niederman, J., Pagano, J. (1976). J. Virol. 18, 1071.
21. Lemon, S., Hutt, L., Pagano, J.S. (1978). J. Virol. 25, 138.
22. Thorley-Lawson, D.A., Strominger, J.L. (1978). Virol. 86, 423.

22a. Rickinson, A.B. and Epstein, M.A. (1978). J. Gen. Virol. 40, 409.
23. Yajima, Y., Tanaka, A., Nonoyama, M. (1976). Virol. 71, 352.
24. Klein, G., Giovanella, B.C., Lindahl, T., Failkow, P.J., Singh, S., Stehlin, J.S. (1974a). Proc. Natl. Acad. Sci. USA 71, 4737.
Klein, G., Lindahl, T., Jondal, M., Leibold, W., Menzés, J., Nilsson, K., Sunstrom, C. (1974b). Proc. Natl. Acad. Sci. USA 71, 3283.
25. Pagano, J.S., Nonoyama, M., Huang, C.H. (1973). In: "Possible Episomes in Eukaryotes." (1973). pp. 218-228. Amsterdam, North-Holland.
26. Tanaka, A., Nonoyama, M., Hampar, B. (1976). Virol. 70, 164.
27. Kaschka-Dierich, C., Adams, A., Lindahl, T., Bornkamm, G. W., Bjursell, G., Klein, G., Giovanella, B.C., Singh, S. (1976). Nature 260, 302.
28. Adams, A., Bjursell, G., Kaschka-Dierich, C., Lindahl, T. (1977). J. Virol. 22, 373.
29. Lemon, S., Hutt, L., Shaw, J., Li, J.-L.H., Pagano, J.S. (1977). Nature 268, 268.
30. Pagano, J.S. (1978). ICN-UCLA Symposia on Molecular and Cellular Biology 1978. In: "Persistent Viruses." (J.G. Stevens, G.J. Todaro, and C.F. Fox eds.), Academic Press, San Francisco.
31. Pagano, J., Huang, C.-H., Klein, G., de Thé, G., Shanmugaratnum, K., Yang, C.-S. (1975). In: "Oncogenesis and Herpesviruses II. G. de Thé, M.A. Epstein, H. zur Hausen eds.), pp. 179-190. IARC, Lyon.
32. Pagano, J., Huang, C.-H., Huang, Y.-T. (1976). Nature 263, 787.
33. Kawai, Y., Nonoyama, M., Pagano, J.S. (1973). J. Virol. 12, 1006.
34. Kieff, E., Levine, J. (1974). Proc. Natl. Acad. Sci. USA 71, 355.
35. Werner, F.J., Bornkamm, G.W., Fleckenstine, B. (1977). Amer. Soc. for Microbiology 22, 794.
36. Adams, A., Lindahl, T., Klein, G. (1973). Proc. Natl. Acad. Sci. USA 7-, 2888.
37. Pagano, J.S., Okasinski, G.F. (1979). In: Proceedings of the 3rd International Symposium on Oncogenesis and Herpesviruses. Boston, Mass. (in press).
38. Ernberg, I., Masucci, A., Klein, G. (1976). Int. J. Cancer 17, 197.

THE INFLUENCE OF THE NUCLEAR AND MITOCHONDRIAL GENOMES ON THE EXPRESSION OF MITOCHONDRIAL DNA IN *SACCHAROMYCES CEREVISIAE*

Anthony W. Linnane, Anne Astin, Sangkot Marzuki, Mark Murphy and Stuart C. Smith

Department of Biochemistry, Monash University, Clayton, Victoria, 3168, Australia

ABSTRACT Two aspects which may be important in the regulation of mitochondrial gene expression will be discussed. Data will be presented which indicates that phenotypic interactions can occur between enzyme complexes of the inner mitochondrial membrane such that the formation of one enzyme complex can be influenced by a defect in another complex encoded in remote regions of the mitochondrial DNA. Results will also be described showing how different nuclear genomes can modulate mitochondrial gene expression.

INTRODUCTION

The mitochondrial DNA (mtDNA) of the yeast *Saccharomyces cerevisiae* has been shown to code for approximately 26 tRNAs, two rRNAs (21S and 15S), and at least six protein components of the inner mitochondrial membrane; these are subunits I, II and III of cytochrome oxidase, the apoprotein of cytochrome *b*, subunit 9 of the mitochondrial oligomycin-sensitive ATPase (mtATPase, OS-ATPase), and a variant protein of 43-47 kd (*var1* protein) [for most recent reviews, see 1, 2]. Other proteins might also be encoded in the mt-genome since more than six and perhaps up to 20 polypeptides are synthesized on the mt-ribosomes [3]. The positions of genetic loci and rRNA coding genes on the circular mitochondrial genome, as well as the gene products of some regions of the mtDNA are shown on the most recent version of the physical map from our laboratory, in Fig. 1 [see refs.1 and 4 for the construction of the physical map].

Very little is understood concerning the control of the expression of the mt-genome. The controlling mechanisms must be highly sophisticated since all the characterized products of mitochondrial protein synthesis are part of multi-subunit inner membrane enzyme complexes, the assembly of which into functional units is dependent on concomitant synthesis of several other subunits in the extra-mitochondrial cytoplasm under the influence of the nucleus [see 5,6 for review]. Furthermore, since the co-operative activities of

ISBN 0-12-198780-9

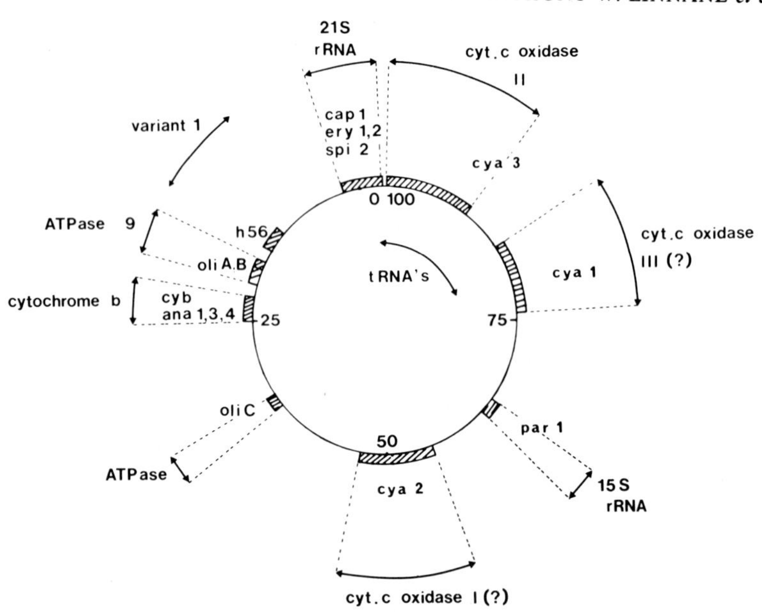

FIGURE 1. Physical Map of the Mitochondrial DNA of *Saccharomyces cerevisiae*. Genetic loci represent (a) *ery1,2*, *cap1*, *spi2*, *par1* and *ana1,3,4*: mutations conferring resistance to erythromycin, chloramphenicol, spiramycin, paromomycin and antimycin A, respectively. (b) *oliA*, *oliB* and *oliC*: three classes of mutations conferring resistance to oligomycin. (c) *cyb*: mit^- mutations with primary lesion in the cytochrome bc_1 complex. (d) *cya1,2* and *3*: mit^- mutations affecting cytochrome oxidase. (e) h56: a mutation conferring temperature sensitivity for growth on non-fermentable substrates. One map unit represents 750 base pairs or 250 amino acids.

these enzyme complexes are required for electron transport and oxidative phosphorylation, the synthesis of their protein components together with their transport and subsequent integration must be co-ordinated in a precise manner in order to maintain the intricate membrane structure and function.

In this communication, we discuss two aspects which may be important in the regulation of mitochondrial gene expression. Firstly, we will present some data which indicates that interactions occur between different inner mitochondrial membrane enzyme complexes so that the formation of one enzyme complex can be influenced by a defect in another complex. Secondly, preliminary results on how nuclear genomes modulate mitochondrial gene expression will be described.

PHENOTYPIC INTERACTIONS IN MITOCHONDRIAL GENE EXPRESSION

Interactions between Mutations in the Mitochondrial Genome. Two mitochondrial mutations affecting the same gene product can interact to alter the phenotypic expression of either mutation. Table I summarizes some interactions which have been observed in this laboratory. A cross between two strains carrying different mutations conferring resistance to the Complex III inhibitor, antimycin A (*ana1* and *ana3*, mapping in *cyb* region of mtDNA), results in the appearance of some unusual diploids as well as the usual sensitive recombinants [7]. These diploids which presumably carry both the *ana1* and *ana3* mutations, not only show resistance to a high level of antimycin A but also exhibit a partial dependence on this inhibitor for growth on ethanol. Another

TABLE 1

PHENOTYPIC INTERACTIONS BETWEEN MUTATIONS IN mtDNA*

Mutations Interacting	Functions Affected	Result of Interaction
ana3 - *ana1*	cytochrome bc_1 complex	Antimycin dependence for growth
ana4 - unidentified	cytochrome bc_1 complex	Antimycin resistance enhanced
ery2 - *spi2*	mt-ribosomes	Increased resistance to carbomycin. Decreased resistance to oleandomycin
oli1 - *ery1* / *oli1* - *ery2*	mtATPase/mt-ribosomes	*oli1* reduces resistance to protein synthesis inhibitors

* Summarized from ref.[7]

antimycin A mutation (*ana4*) interacts with as yet an unidentified mitochondrial genetic marker such that in the presence of this latter marker the resistance to antimycin A of the *ana4* strain is enhanced.

These types of phenotypic interactions are not limited to only the *ana-cyb* region of mtDNA. For example, interactions can be detected between two mutations conferring resistance to the mitochondrial protein synthesis inhibitors, erythromycin (*ery2*) and spiramycin (*spi2*) [7]. These mutations are located in the 0 - 5 map unit segment of mtDNA which contains the 21S rRNA gene [see Fig. 1]. *In vitro* assays of mitochondrial protein synthesis show that the *ery2* mutant is weakly resistant to carbomycin, erythromycin, lincomycin and spiramycin, but is resistant to high concentrations of oleandomycin. The *spi2* mitochondria show resistance to spiramycin and carbomycin only. Recombinant strains carrying both mutations have resistance levels to erythromycin, lincomycin and spiramycin which are comparable to those of the *spi2* strain. The resistance to carbomycin, however, is significantly increased (4-fold increase above the *spi2* mutant) and there is a marked increase in the sensitivity to oleandomycin (25-50 times more sensitive than *spi2* or *ery2* strains).

A mutation in one structural gene can also affect the expression of another mutation in a remote region of the mtDNA. Thus, a mutation in the structural gene of subunit 9 of the mtATPase (*oli1*) [8], conferring resistance to oligomycin, influences the expression of erythromycin resistant mutations (*ery1* and *ery2*) [7]. In the presence of the *oli1* mutation the *in vivo* erythromycin resistance of strains carrying either the *ery1* or *ery2* mutation is substantially reduced.

The indirect effect of the mutation affecting the mtATPase is probably mediated through the membrane. The mtATPase complex and mt-ribosomes are functionally associated with the inner mitochondrial membrane [9]. It is therefore possible that an alteration in the conformation of the mtATPase as the result of a mutation, causes a change in the membrane which in turn affects the function of the mt-ribosomes. This type of interaction might also exist between other enzyme complexes of the inner mitochondrial membrane.

Single Mutations in mtDNA affect the Formation of Enzyme Complexes encoded in other Regions of the Genome.

(a) Mutations in *cyb-box* region of mtDNA: The *cyb-box* region of mtDNA [see Fig. 1] contains the structural gene for the cytochrome *b* apoprotein [10,11]. Some mutations in this region (*cyb* and *box* mutations) result in a defect in a defect in both the cytochrome bc_1 complex and cytochrome oxidase. In some mutants, the inability to synthesize the latter complex is due to an increased sensitivity to catabolite repression as the result of the *cyb* or *box* mutation [12, 13]. Thus, no spectral cytochrome aa_3 or cytochrome oxidase activity is observed in these strains when grown in batch cultures under conditions which derepress a wild-type strain as well as some other *cyb* mutants. However, when these mutants are grown under strictly derepressing conditions in glucose-limited chemostat cultures, cytochrome oxidase is present [12].

Some *cyb* mutants however do not synthesize cytochrome oxidase even when grown under the above derepressing conditions. Analysis of the mitochondrial translation products on SDS-polyacrylamide gels shows that, in the majority of these strains, both the cytochrome *b* apoprotein (32 kd) and subunit I of cytochrome oxidase (42 kd) are absent [see Fig. 2].

The loss of cytochrome oxidase subunit I, however, is not conditional on the absence of the apo-cytochrome *b*. In one mutant a reduced amount of cytochrome oxidase subunit I is observed even though the apoprotein of cytochrome *b* is apparently present, perhaps in an altered form [Fig. 2]. In contrast the loss of the cytochrome *b* apoprotein does not always result in the disappearance of the subunit I of cytochrome oxidase.

It has been suggested from both physical studies of the mtDNA as well as genetic and biochemical analysis of *mit*$^-$ mutants affecting the synthesis of cytochrome *b* [14, 15] that the structural gene of the apo-cytochrome *b* is split into several distinct segments ("exons") separated by as yet functionally undefined intervening sequences ("introns") [14]. The majority of Slonimski's *box* mutants which are defective in both the cytochrome bc_1 and cytochrome oxidase complexes have been shown to map between the "exons", within the intervening sequences [14, see also paper by Mahler and Perlman in this volume]. This results in either an absence of subunit I of cytochrome oxidase and the cytochrome *b* apoprotein, or to the appearance of new polypeptides concomitant with the loss of cytochrome *b* together with or without the disappearance of cytochrome oxidase subunit I [16].

Models to explain how mutations in intervening sequences result in an absence of both cytochrome *b* and cytochrome oxidase subunit I have been proposed for the *cyb-box* region of mtDNA [14, 2]. Briefly, these are: (i) mutations in intervening sequences are in tRNA genes used selectively or frequently in the synthesis of apo-cytochrome *b* and cytochrome oxidase subunit I; (ii) an intervening sequence could be an

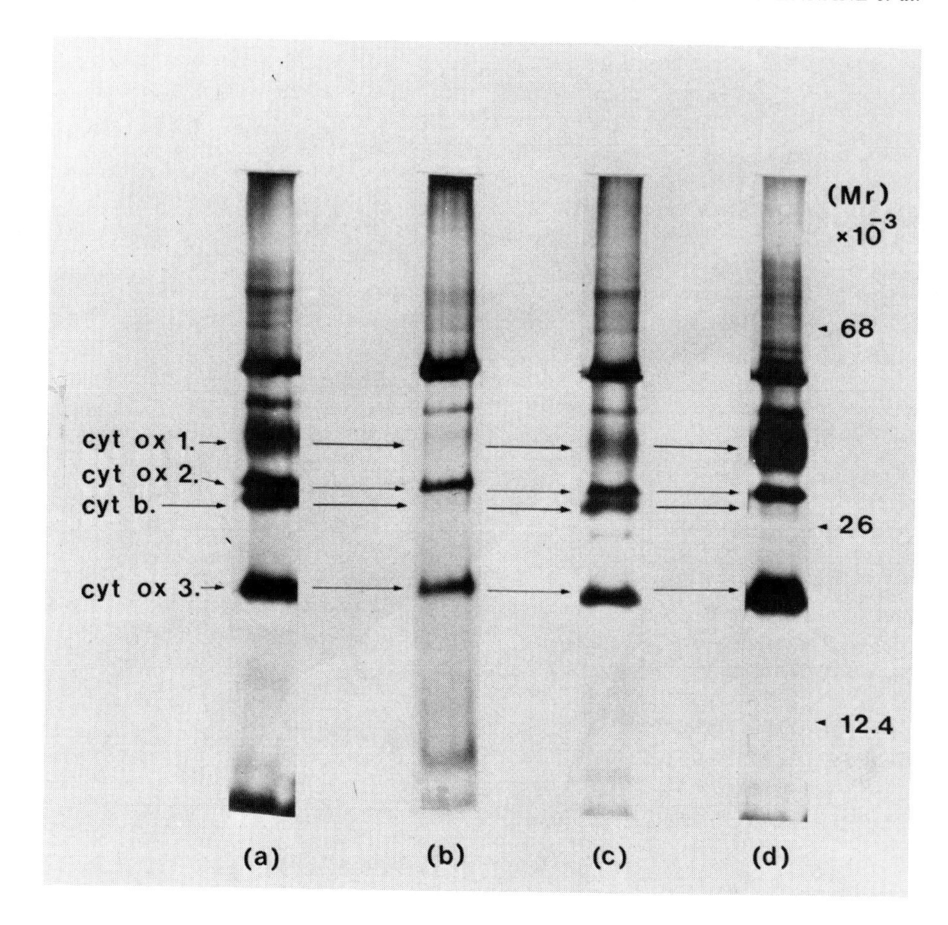

FIGURE 2. SDS-polyacrylamide Gel Analysis of Mitochondrial Translation Products of *cyb* Mutants. All strains were grown in glucose-limited chemostat cultures under derepressing conditions [9]. Products of mitochondrial protein synthesis were labelled with [^{35}S]-sulphate [8], mitochondria were isolated [12] and analysed on SDS-polyacrylamide gels [3]. Mitochondrial translation products were detected by fluorography [13].
(a) wild type strain, J69-1B
(b) *cyb* strains, 1508, 1510, 1515, 2111, 2108, 2230, 2229
(c) *cyb* strain, 1208
(d) *cyb* strains, 41-2-4, 1319, 2223, 2426, 2425, 17-62-2

"exon" coding for a part of the cytochrome oxidase subunit I structural gene; (iii) the intervening sequences could code for a protease involved in post-translational processing of polypeptide precursors, or for a "switch-on-and-off" protein involved in transcription of mtDNA, or for adaptor molecules essential for correct folding and/or splicing of pre-mRNAs of cytochrome *b* and subunit I of cytochrome oxidase; and (iv) a common mechanism is perhaps responsible for the processing of pre-mRNAs of cytochrome *b* and cytochrome oxidase subunit I, and a mutation in the intervening sequences affects the processing of the cytochrome *b* pre-mRNA. Accumulation of this non-processable RNA would then prevent the processing of pre-mRNA of cytochrome oxidase subunit I.

All of these models have been proposed to specifically explain the concomitant loss of the cytochrome *b* apoprotein and cytochrome oxidase subunit I as the result of single mutations in the *cyb-box* region of mtDNA. Results in the following sections, however, will demonstrate that the pleiotropic effects of a single mutation are not unique to the *cyb-box* region and can occur in mtDNA regions coding for protein subunits of other enzyme complexes.

(b) Mutations in the *oliC* region of mtDNA: The formation of the cytochrome oxidase complex is also affected by some mutations in the *oliC* region of mtDNA. This region is one of two located on the mt-genome in which mutations conferring oligomycin resistance to the mtATPase have been mapped (*oliAB* and *oliC* regions, see Fig. 1)[19]. While the *oliAB* region has been shown to contain the structural gene of the mtATPase subunit 9 [8, 20, 21] the gene product of the *oliC* region is still unidentified.

Some mutations in the *oliC* region confer not only oligomycin resistance to the mtATPase, but also cold-sensitivity for growth at 18° on non-fermentable substrates. Furthermore, two such cold-sensitive mutants in our laboratory grow poorly on ethanol even at the permissive growth temperature of 28°. At this temperature these two strains have respiratory and mtATPase activities comparable to wild-type strains. Mitochondria isolated from these mutants, however, have no capacity for oxidative phosphorylation. Thus, the lesions in these *oliC* strains primarily affect the mtATPase resulting in the uncoupling of oxidative phosphorylation. At the restrictive growth temperature (18°), however, a secondary effect on cytochrome oxidase is observed. No spectral cytochrome aa_3 and cytochrome oxidase activity could be detected; subunit I of this complex apparently was not present in this strain at 18° [see Fig.3].

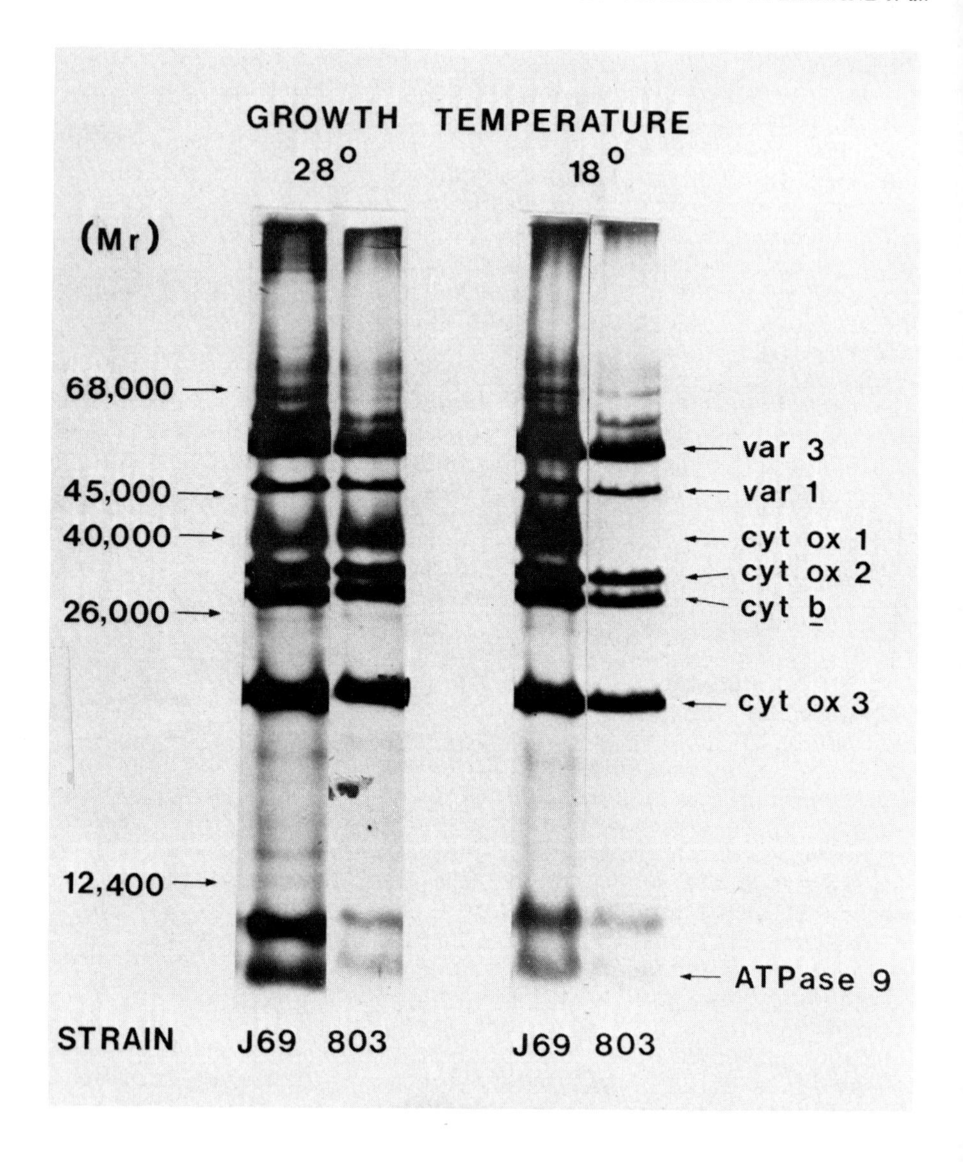

FIGURE 3. Mitochondrial Translation Products of Strains J69-1B (wild type) and 803 (*oliC* mutant). Cells were grown at 28° and 18°, and mitochondrial translation products analysed as described in Fig. 2.

(c) Mutations in the *cya3* region of mtDNA: Cytochrome oxidase is not the only complex which can be affected by a mutation in other remote regions of the mtDNA. For example, a *cya3* mutation (89 - 100 map units) [22], presumably in the structural gene of cytochrome oxidase subunit II [23] has been shown to affect the formation of the cytochrome bc_1 complex. One particular strain carrying this mutation lacks spectral cytochrome aa_3, has no cytochrome oxidase activity and subunit II of this complex (34 kd) cannot be detected on SDS-polyacrylamide gels [Table 2]. In addition, spectral cytochrome *b* levels are reduced and although a low level of NADH-cytochrome *c* reductase activity can still be detected (10% of wild-type) this activity is completely antimycin A insensitive. However, SDS-polyacrylamide gel analysis of mitochondrial translation products indicates that the cytochrome *b* apoprotein is apparently still present in this mutant. Thus, the loss of subunit II of cytochrome oxidase in this strain results in the inability to assemble the cytochrome bc_1 complex into an active form.

(d) Mutations in the *var1* region of mtDNA: A mutation in the mtDNA can indirectly affect the formation of more than one enzyme complex encoded in other regions of the mt-genome. This phenomenon can be observed in a temperature-sensitive mutant (strain h56) recently isolated in our laboratory. The mutation in this strain has been mapped physically very close to the *oli1* locus (an *oliA* mutation) [19], between *oli1* and the 21S rRNA gene within a segment of mtDNA of about 3 kbp (K.B. Choo, personal communication). This mtDNA segment overlaps a 2.8 kbp restriction fragment within which the *var1* gene lies [24]. This gene codes for a polymorphic protein (var1 protein) which can exist in variant forms of slightly differing molecular weights [3], ranging from 40,000 to 47,000 daltons. This variant protein has been shown to co-purify with the small subunit of the mt-ribosomes [25].

The mutation in strain h56 confers conditional respiratory deficiency on cells grown at high temperature (36°). Thus, while at the permissive growth temperature (28°), the mutant strain has normal respiratory enzyme activities, at the restrictive temperature both the cytochrome oxidase and mtATPase activities in this strain are very low. Analysis of the mitochondrial translation products by SDS-polyacrylamide gel electrophoresis shows that at the permissive (28°) and restrictive (36°) growth temperatures, mitochondria isolated from the mutant contain only trace amounts of the var1 protein (45 kd) [see Fig. 4]. Instead, a new mitochondrial product of 25 kd, which corresponds to another variant protein (var2 protein) is observed [3]. The *var2* gene is

TABLE 2

PHENOTYPIC INTERACTIONS IN *CYA3* REGION OF mtDNA

Strain	Enzyme Activity (μmoles/min/mg protein)		Mitochondrial Spectra (μmoles haem/mg protein)			Mitochondrial Translation Products	
	NADH-cyt.*c* reductase	Cytochrome oxidase	aa_3	*b*	cc_1	Cytochrome oxidase Subunit II	Apo-cytochrome *b*
J69-1B	0.46 (88%)	0.42	0.23	0.27	0.23	+	+
1203	0.06 (3%)	trace	trace	0.11	0.31	-	+

The wild-type strain (J69-1B) and the *cya3* mutant (strain 1203) were grown in glucose-limited chemostat cultures under derepressing conditions [9]. Mitochondria were isolated [12], the activities of NADH-cytochrome *c* reductase [26] and cytochrome oxidase [27] were determined, and mitochondrial spectra were quantitated [28]. Figures in brackets indicate the antimycin A sensitivity of NADH-cytochrome *c* reductase activity.

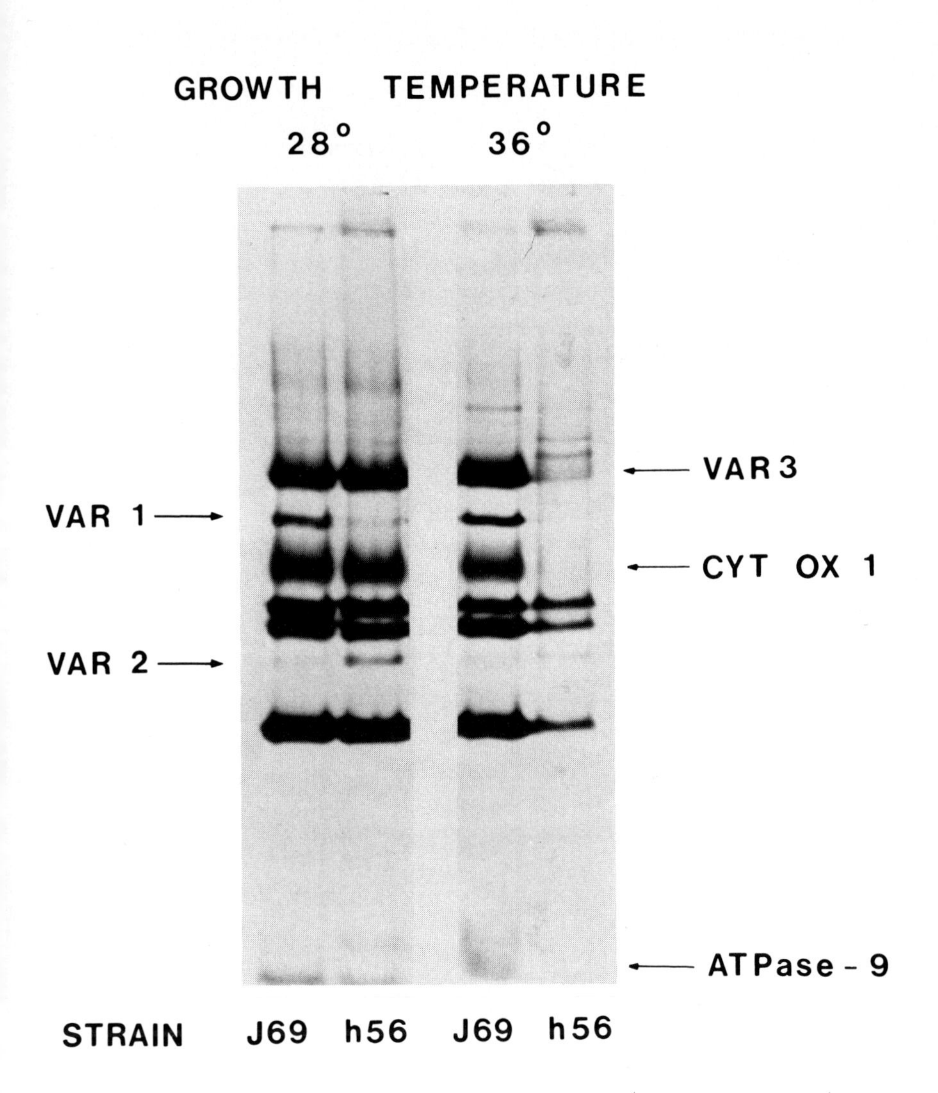

FIGURE 4. Mitochondrial Translation Products of Strains J69-1B (wild type) and h56 (*var1* mutant). Cells were grown at 28° and 36° and mitochondrial translation products analysed as described in Fig. 2.

allelic to that of *var1* [29]. This observation, in agreement with the physical mapping data, suggests that the mutation in strain h56 is probably in the structural gene of the var1 protein.

Secondary effects of the mutation in strain h56 can be observed at 36°. At this temperature, the production of subunit I of cytochrome oxidase and subunit 9 of the mtATPase are also affected [Fig. 4].

The Nature of Phenotypic Interactions in Mitochondrial Gene Expression. As we have illustrated in this communication the phenomenon originally observed in the *cyb-box* region of mtDNA is a general one, and can occur between any number of gene products of the mtDNA. Thus, the synthesis of cytochrome oxidase subunit I is not only influenced by a mutation in the *cyb-box* region, but also by *oliC* mutations which primarily affect the mtATPase. Furthermore, the formation of a functional cytochrome bc_1 complex can be impaired by a mutation in the structural gene of subunit II of the cytochrome oxidase (*cya3* mutation). Two other mutants mapping in the *cap-oxi1* region of mtDNA have also been described which show pleiotropic effects on the three subunits of cytochrome oxidase and apo-cytochrome *b* [15].

We have previously observed that there is a functional interdependence between components of the inner mitochondrial membrane. As described earlier in this paper, a mutation in the structural gene of the mtATPase subunit 9 can alter the sensitivity of the mt-ribosomes to protein synthesis inhibitors, presumably due to interaction between the two complexes in the inner mitochondrial membrane. The same type of inter-complex interaction might also be responsible for at least some of the pleiotropic effects of the mutations described above. For example, an *oliC* mutation which primarily affects the mtATPase might simply, through a complex series of interactions in the membrane, affect the integration of subunit I of cytochrome oxidase, which subsequently inhibits the synthesis of this subunit. Since interactions can occur between any complexes in the mitochondrial membrane, then potentially a mutation in any region of the mtDNA could indirectly affect the formation of other complexes encoded in remote regions of the mt-genome.

These inter-complex interactions might also account for the lack of the cytochrome bc_1 complex activity in the *cya3* mutant as well as the absence of cytochrome oxidase subunit I when the cytochrome *b* apoprotein is altered as a result of mutations in its structural gene.

However, the above mechanism might not be completely responsible for all the phenotypic interactions described above. For instance, the absence of apo-cytochrome *b* in some *cyb* and *box* mutants is not always associated with the loss of subunit I of cytochrome oxidase. Therefore, the absence of this latter subunit cannot simply be due to a lack of integration as the result of abnormal interaction between a "cytochrome *b*-less respiratory complex III" and cytochrome oxidase. In this case, one of the previously proposed models (see section on *cyb-box* mutations in this paper) might explain the phenomena observed. These models could also be extended to explain the pleiotropic effects of single mutations in other regions of mtDNA provided that the structural genes of the mitochondrial translation products affected are organized into distinct coding segments and the lesions are in intervening sequences.

Furthermore, while it is possible that the mutation in strain h56, which primarily affects the var1 protein, might through interaction of the defective protein with mtATPase and cytochrome oxidase in the membrane interfere with the integration of mtATPase subunit 9 and subunit I of cytochrome oxidase, it should be kept in mind that the var1 protein is associated with the mt-ribosomes. This protein therefore might be directly involved in the regulation of mitochondrial protein synthesis in such a way that the mutation in strain h56 results in the preferential loss of the mtATPase subunit 9 and cytochrome oxidase subunit I. If this is the case, then this mutant will be important for the study of the function of the var1 protein which at present remains unknown.

It is very likely that there is no common mechanism to explain all of the phenotypic interactions observed by ourselves and others. Two distinct mutations might give rise to similar phenotypes through completely different mechanisms. On the other hand, the phenotypic expression of a single mutation could arise as a result of interaction of any two or more of the mechanisms discussed.

MODULATION OF MITOCHONDRIAL GENE EXPRESSION BY NUCLEAR DNA

From both genetic and physical studies of yeast mtDNA it has become apparent that there are significant differences (insertions, deletions) in the mitochondrial genomes of different strains of *Saccharomyces cerevisiae*. Sequence differences have been shown by Butow to be associated with the formation of variant polypeptides [27]. It may be further asked whether sequence differences in the mtDNA in particular grande strains can lead to different gene products being produced, and concomitantly, does the appearance of potentially new mitochondrial gene products depend on particular nuclear genomic influences on the mtDNA?

We have addressed ourselves to these questions by constructing sets of new yeast strains containing specified mtDNAs and specified nuclear genomes. This has been achieved by constructing sets of isonuclear strains containing different mitochondrial genomes, making use of the nuclear mutation *kar1-1* [30].

Construction of Isonuclear and Isomitochondrial Strains. The general procedure for the construction of haploid strains carrying the nucleus of one strain and the mitochondrial genome of another strain has been described in detail by Nagley and Linnane [31]. Briefly, the technique involves crosses between one parent which is a petite of the *rho*^{o} type (complete absence of mtDNA) and a respiratory competent grande strain (*rho*$^+$) [see Fig. 5]. If the *rho*^{o} parent carries the nuclear mutation *kar1-1* and the *rho*$^+$ parent is selected from a variety of laboratory strains then it is possible to construct a set of haploid strains which have different mitochondrial genomes in a single common nuclear background. Conversely, if the *rho*$^+$ parent carries the *kar1-1* mutation a single mitochondrial genome can be transferred into a variety of nuclear backgrounds making it possible to systematically study the influence of any nucleus on the expression of any mitochondrial genome.

The Expression of Different Mitochondrial Genomes in a Single Nuclear Background. We have studied the expression of several mitochondrial genomes in a single nuclear genomic background by analysing the products of mitochondrial protein synthesis of constructed isonuclear haploid strains on SDS-polyacrylamide gels. In the majority of strains examined the gel patterns were essentially identical: bands corresponding to the var3 oligomer of the mtATPase subunit 9 (50 kd), the var1 protein (45-47 kd), subunits I, II and III of cytochrome oxidase (42, 34, 22 kd respectively), the cytochrome *b* apoprotein (32 kd) and subunits 8 and 9 of the mtATPase (9 and 7.5 kd respectively) could be easily detected. However, in three of the isonuclear strains additional new bands were observed. The exact number of new bands was difficult to determine because some of them had mobilities very close to var3. To overcome this problem we increased the concentration of β-mercaptoethanol in the mitochondrial membrane solubilization procedure from 1% (v/v) to 10% (v/v) final concentration. This modification resulted in the almost complete converstion of the oligomer to the mtATPase subunit 9 monomer. Under these conditions, three new bands (59, 55 and 29 kd) were clearly visible in all three strains [see Fig. 6]. Since

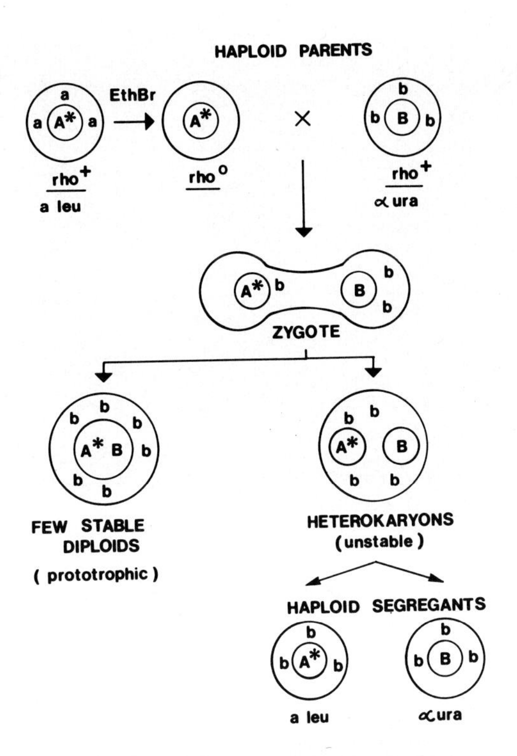

FIGURE 5. Construction of Strains containing Specified Nuclear and Mitochondrial Genomes. The figure illustrates how a particular mtDNA (b) can be transferred into a different nuclear background (from nucleus B to nucleus A*), using the *kar1-1* mutation [30]. The mtDNA (a) of a haploid strain (nuclear DNA, A* *kar1-1*) was eliminated using ethidium bromide (EthBr) [31]. The mtDNA-less strain (*rho°*) was crossed with a haploid strain carrying mtDNA (b). The zygotes formed give rise to a small number of stable diploids and unstable heterokaryons, which subsequently segregate haploid cells, all carrying mtDNA (b) and either nuclear DNA, A* or B. The above crosses can be performed whether either or both nuclei contain the *kar1-1* mutation.

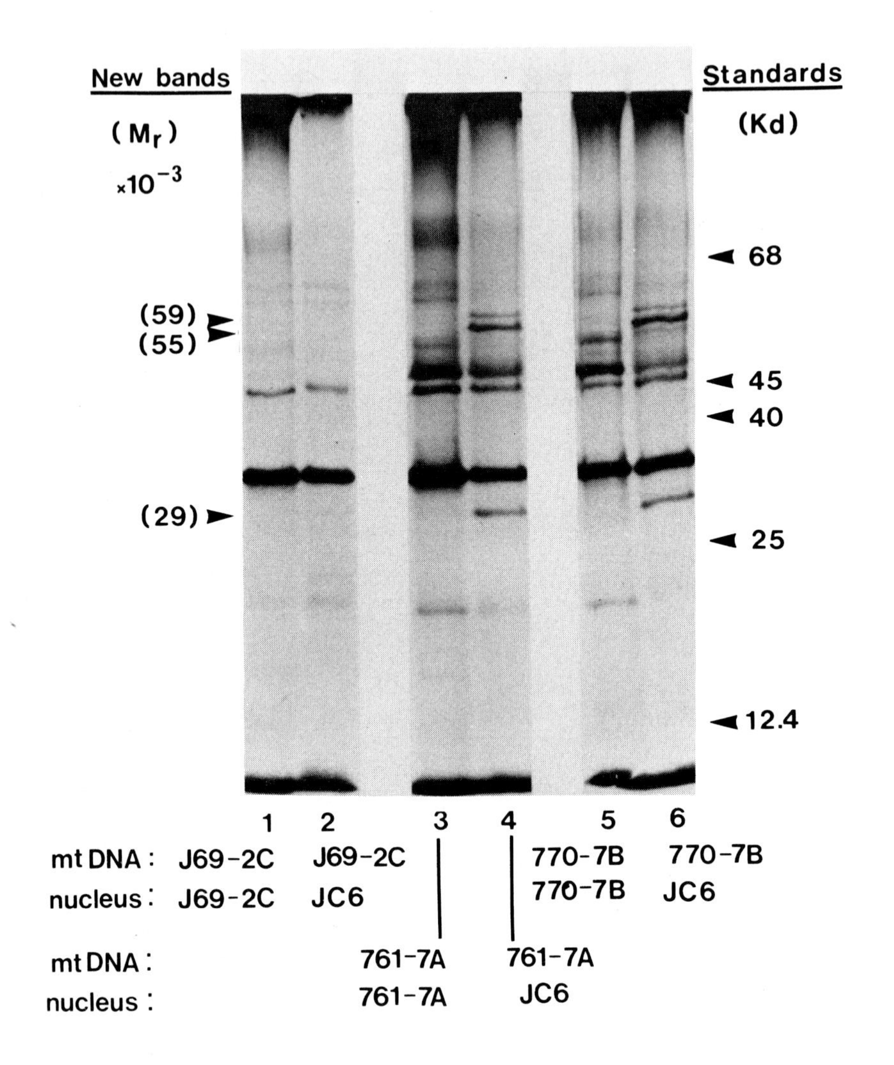

FIGURE 6. The Influence of Nuclear Genomic Backgrounds on the Expression of mtDNA. Isomitochondrial strains were constructed as described in Fig. 5. Cells were grown on glucose (2% w/v) to 4 mg cell dry weight/ml. Mitochondrial translation products were analysed as described in Fig.2 except that 10% (w/v)β-mercaptoethanol was used in the solubilization of the mitochondrial membranes (see text) together with sodium dodecylsulphate (2% w/v). The nuclear and mitochondrial genomes of the strains examined are indicated under each gel display.

these bands were not observed in their original nuclear backgrounds, the appearance of these novel mitochondrial products which are apparently coded for by mtDNA, is under nuclear influence. Furthermore, the appearance of these bands is not due to alterations in the mtDNA as a result of the genetic manipulations since reconstruction of strains containing the original nuclear and mitochondrial genomes produces a gel pattern identical with that of the parental strains themselves (G. Stephenson, manuscript in preparation). The synthesis of these novel products is therefore due to the influence of particular nuclear backgrounds on mitochondrial gene expression.

Future investigations will be the further identification and characterization of these new bands. One might postulate that these proteins are indeed novel products of mitochondrial protein synthesis. All three strains examined so far, in which these bands have been observed, grow very slowly when ethanol is used as an energy source. These polypeptides therefore, might be responsible for, or appear as the result of this slow growth. However, they may also be precursors of other mitochondrial translation products in which case the question may be posed as to why there is an accumulation of these products when a mitochondrial genome is influenced by a particular nuclear background. At present, studies of the immunological properties of these novel polypeptides as well as the construction and characterization of more isonuclear and isomitochondrial strains are in progress in order to resolve the above questions.

REFERENCES

1. Linnane, A.W., and Nagley, P. (1978). Plasmid. 1, 324.
2. Borst, P., and Grivell, L.A. (1978). Cell. 15, 705.
3. Douglas, M.G., and Butow, R.A. (1976). Proc. Nat. Acad. Sci. USA. 73, 1083.
4. Linnane, A.W., Lukins, H.B., Molloy, P.L., Nagley, P., Rytka, J., Sriprakash, K.S., and Trembath, M.K. (1976). Proc. Nat. Acad. Sci. USA. 73, 2082.
5. Schatz, G., and Mason, T.L. (1974). Ann. Rev. Biochem. 43, 51.
6. Tzagoloff, A., Rubin, M.S., and Sierra, M.F. (1973). Biochim. Biophys. Acta. 301, 71.
7. Linnane, A.W., Nagley, P., Hall, R.M., Marzuki, S., and Trembath, M.K. (1978). In "Biochemistry and Genetics of Yeasts" (M. Bacila, B.L. Horecker and A.O.M. Stoppani, eds.), pp. 489-512. Academic Press, New York.

8. Murphy, M., Gutowski, S.J., Marzuki, S., Lukins, H.B., and Linnane, A.W. (1978). Biochem. Biophys. Res. Commun. 85, 1283.
9. Marzuki, S., Cobon, G.S.,Crowfoot, P.D., and Linnane, A.W. (1975). Arch. Biochem. Biophys. 169, 591.
10. Claisse, M.L., Spyridakis, A., and Slonimski, P.P. (1977). In "Mitochondria 1977" (W. Bandlow, R.J. Schweyen, K. Wolf, and F. Kaudewitz, eds.), pp. 337-344. de Gruyter, Berlin.
11. Mahler, H.R., Hanson, D., Miller, D., Bilinski, R., Ellis, D.M., Alexander, N.J., and Perlman, P.S. (1977). In "Mitochondria 1977" (W. Bandlow, R.J. Schweyen, K. Wolf, and F. Kaudewitz, eds.), pp. 345-370. de Gruyter, Berlin.
12. Cobon, G.S., Groot Obbink, D.J., Hall, R.M., Maxwell, R., Murphy, M., Rytka, J., and Linnane, A.W. (1976). In "Genetics and Biogenesis of Chloroplasts and Mitochondria" (Th. Bucher, W. Neupert, W. Sebald, and S. Werner, eds.), pp. 453-460. North-Holland Publishing Co., Amsterdam.
13. Pajot, P., Wambier-Kluppel, M.L., Kotylak, Z., and Slonimski, P.P. (1976). In "Genetics and Biogenesis of Chloroplasts and Mitochondria" (Th. Bucher, W. Neupert, W. Sebald, and S. Werner, eds.), pp. 443-451. North-Holland Publishing Co., Amsterdam.
14. Slonimski, P.P., Claisse, M.L., Foucher, M., Jacq, C., Kochko, A., Lamourous, A., Pajot, P., Perrodin, G., Spyridakis, A., and Wambier-Kluppel, M.L. (1978). In "Biochemistry and Genetics of Yeasts" (M. Bacila, B.L. Horecker, and A.O.M. Stoppani, eds.), pp. 391-401. Academic Press, New York.
15. Mahler, H.R., Hanson, D., Miller, D., Lin, C.C., Alexander, N.J., Vincent, R.D., and Perlman, P.S. (1978). In "Biochemistry and Genetics of Yeasts" (M. Bacila, B.L. Horecker, and A.O.M. Stoppani, eds.), pp. 513-547. Academic Press, New York.
16. Claisse, M.L., Spyridakis, A., Wambier-Kluppel, M.L., Pajot, P., and Slonimski, P.P. (1978). In "Biochemistry and Genetics of Yeasts" (M. Bacila, B.L. Horecker, and A.O.M. Stoppani, eds.), pp. 369-391. Academic Press, New York.
17. Roberts, H., Choo, W.M., Smith, S.C., Marzuki, S., Linnane, A.W., Porter, T.H., and Folkers, K. (1978). Arch. Biochem. Biophys. 191, 306.
18. Bonner, W.M., and Laskey, R.A. (1974). Eur. J. Biochem. 46, 83.

19. Trembath, M.K., Molloy, P.L., Sriprakash, K.S., Cutting, G.J., Linnane, A.W., and Lukins, H.B. (1976). Molec. gen. Genet. 145, 43.
20. Groot Obbink, D.J., Hall, R.M., Linnane, A.W., Lukins, H.B., Monk, B.C., Spithill, T.W., and Trembath, M.K. (1976). In "The Genetic Function of Mitochondrial DNA" (C. Saccone, and A.M. Kroon, eds.), pp. 163-173. North-Holland Publishing Co., Amsterdam.
21. Wachter, E., Sebald, W., and Tzagoloff, A. (1977). In "Mitochondria 1977" (W. Bandlow, R.J. Schweyen, K. Wolf, and F. Kaudewitz, eds.), pp. 441-449. de Gruyter, Berlin.
22. Rytka, J., English, K.J., Hall, R.M., Linnane, A.W., and Lukins, H.B. (1976). In "Genetics and Biogenesis of Chloroplasts and Mitochondria" (Th. Bucher, W. Neupert, and S. Werner, eds.), pp. 427-434. North-Holland Publishing Co., Amsterdam.
23. Cabral, F., Solioz, M., Rudin, Y., Schatz, G., Clavilier, L., and Slonimski, P.P. (1978). J. Biol. Chem. 253, 297.
24. Hatefi, Y., and Rieske, J.S. (1967). Methods Enzymol. 10, 225.
25. Wharton, D.C., and Tzagoloff, A. (1967). Methods Enzymol. 10, 245.
26. Chance, B. (1957). Methods Enzymol. 4, 273.
27. Butow, R.A., Vincent, R.D., Strausberg, R.L., Zanders, E., and Perlman, P.S. (1977). In "Mitochondria 1977" (W. Bandlow, R.J. Schweyen, K. Wolf, and F. Kaudewitz, eds.), pp. 317-335. de Gruyter, Berlin.
28. Groot, G.S.P., Grivell, L.A., Van Harten-Loosbroek, N., Krieka, J., Moorman, A.F.M., and Van Ommen, G.J.B. (1977). In "Structure and Function of Energy Transducing Membranes" (K. Van Dam, and B.F. Van Gelder, eds.), pp. 177-186. Elsevier/North Holland Biomedical Press, Amsterdam.
29. Perlman, P.S., Douglas, M.G., Strausberg, R.L., and Butow, R.A. (1977). J. Mol. Biol. 115, 675.
30. Conde, J., and Fink, G.R. (1976). Proc. Nat. Acad. Sci. USA. 73, 3651.
31. Nagley, P., and Linnane, A.W. (1978). Biochem. Biophys. Res. Commun. 85, 585.

GENETIC AND BIOCHEMICAL ANALYSIS OF VAR1[1]

Ronald A. Butow, Peter Terpstra, and
Robert L. Strausberg

Department of Biochemistry, The University of Texas Health
Science Center at Dallas, Dallas, Texas 75235

ABSTRACT Var1 polypeptide is an unusual product of the yeast mitochondrial genome displaying a strain-dependent size polymorphism in the range of 40,000 to 44,000 daltons. Different molecular weight forms of var1 can be generated in crosses by asymmetric gene conversion formally involving two DNA segments which account for roughly 10% of the total protein size. These segments appear capable of inserting independently and at different frequencies from one allele to another. In some crosses we observe the insertion of only a portion of one of the segments; thus, a number of different molecular weight forms by var1 can be generated by various combinations of insertions. Based in part on our results with var1, a model is presented to account for asymmetric versus symmetric gene conversions determined by insertions at different sites along the mitochondrial genome. Biochemical studies show that var1 polypeptide is an integral protein of the small 38S mitochondrial ribosomal subunit. From inhibitor studies and *in vivo* pulse-chase experiments, var1 polypeptide is found early in the maturation of the small ribosomal subunit and appears to be required for its assembly.

INTRODUCTION

One of the major objectives in the analysis of genetic elements is to understand the relationship between the organization of DNA segments (genes, regulatory sequence etc.) and function. By function we include not only the molecular mechanisms by which gene expression occurs and is regulated to meet specific cellular demands, but also all of those processes involved in genetic transmission such as recombination, replication and segregation.

From considerations at the population level to molecular details, the yeast mitochondrial genome presents some inter-

[1]This work was supported by Grants GM 22525 and GM 26546 from the U. S. Public Health Service and Grant I-642 from the Robert A. Welch Foundation.

ISBN 0-12-198780-9

esting challenges to a number of important questions concerning genome organization, function and activity:

1) Wild-type yeast cells contain many molecules of DNA which are extremely active in genetic recombination (1,2); hence, allelic differences must be capable of spreading efficiently through genome populations.

2) There are phenomenologically distinguishable types of recombination where in some regions of the genome, the production of recombinant classes can be highly asymmetric, while in others, apparent reciprocal recombination events predominate (2).

3) Although gene order seems to be invariant, there is a striking strain dependent variation in the absolute size of the mitochondrial genome ranging between 67,000 and 78,000 base pairs (3,4). From restriction endonuclease analysis, it is evident that numerous insertions and deletions occur quite frequently in genetic crosses such that recombinant species are apparent in the profile of restriction endonuclease fragments (5,6).

4) The mitochondrial genome is organized quite "heterogenously" with dAT segments interspersed with gene regions and regions of high dGC (7).

5) Some mitochondrial genes appear to contain non-coding, intervening sequences (8). In one case, the 21S rRNA cistron, the intervening sequence (ω^+) behaves as an insertion element which can be transferred to other 21S rRNA alleles (ω^-) by gene conversion (9-11).

Clearly, it is of considerable interest to know how these insertion/deletion variants of the mitochondrial genome arise and what role variable DNA segments might play in mitochondrial genome activity.

We have considered these questions in some detail in the analysis of the mitochondrial gene var1 and its gene product, var1 polypeptide. The protein is unusual in that it is polymorphic with respect to size, showing a strain-dependent range of molecular weights on SDS-polyacrylamide gels between 40,000 and 44,000. We have taken advantage of this size polymorphism to map genetically the allelic differences between the various molecular weight forms of the protein to the ery oli1 segment of the genome (12), and physically, to a 2.06 kb DNA segment of Hinc II-band 10 (5,13).

Our major conclusion concerning the generation of different forms of var1 polypeptide is that they arise in crosses by asymmetric gene conversion similar to the conversion of ω^- to ω^+ alleles (14). Further, from genetic considerations, we suggest that there are at least two variable DNA segments which appear capable of inserting from one allelic form of var1 into another, either wholly or in part, by asymmetric gene conversion (15). A detailed genetic and biochemical analysis of this phenomenon has led us to propose a model describing how segments of the mitochondrial genome could undergo either symmetric asymmetric gene conversion depending upon the insertion of DNA segments at specific sites along the DNA.

Finally, we will briefly review data which confirm and extend the results of Groot (16,17) that var1 polypeptide is a protein associated with the 38S mitochondrial ribosomal subunit.

RESULTS AND DISCUSSION

Analysis of Mitochondrial Translation Products in Crosses. We have developed methods for the rapid analysis of mitochondrial translation products in random as well as segregant diploid populations. These procedures have been applied in the case of *var1* to evaluate the transmission of different forms of var1 polypeptide in crosses. For example, we may wish to know if a series of petite strains derived from a wild-type strain containing a particular molecular weight form of var1 polypeptide contain within their repeat unit of mtDNA, determinants for that *var1* allele. This procedure of petite deletion mapping requires that the petites be crossed to a wild-type tester strain with a different molecular weight form of var1 polypeptide and the progeny analyzed for transmission. A positive result would be evident by the presence of *two* different molecular weight forms of var1 polypeptide. We may also ask whether non-parental forms of var1 can be generated in crosses. Depending upon the frequency of such an event, a random diploid sample might be sufficient for analysis or many individual diploid progeny might have to be examined.

Fig. 1 shows three general protocols we use for scoring different molecular weight forms of var1 polypeptide. The analysis of individual segregants from a cross (following ≥ 20 generations of diploid clones) provides the most accurate but most time consuming measure of the relative distribution of different molecular weight species of var1 polypeptide; each individual clone must be labeled *in vivo*, mitochondria isolated and analyzed on a separate slot on an SDS-polyacrylamide slab gel. A reasonably accurate estimation of the distribution of the major molecular weight forms of var1 polypeptide can be determined by analyzing an aliquot from the random diploid population of a given cross. From reconstruction experiments (14) and comparison with the results of individual segregant analysis, we have verified that the relative amounts of the major var1 species can be accurately determined by this procedure. Although this approach is rapid in that only one sample needs to be analyzed, it is limited to the detection and quantitation of those species which are present at ≥ 2-5% of the total population. In order to pick up and quantitate minor species, homoplasmic diploid segregants can be pooled in groups of 10 or 20 and subsequently analyzed on SDS-polyacrylamide slab gels after in vivo labeling and isolation of the mitochondria in the pooled samples. All of the minor recombinant forms of var1 polypeptide described here were detected by this procedure.

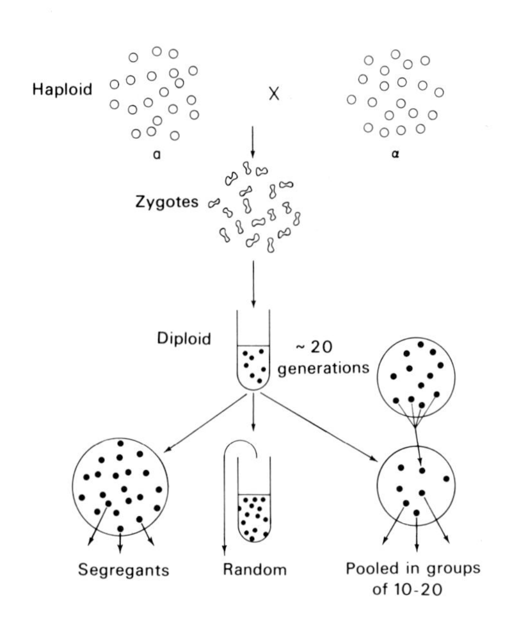

FIGURE 1. Analysis of Mitochondrial Gene Segregation.

The Generation of Different Molecular Weight Forms of Var1 Polypeptide. Fig. 2A, shows an SDS-polyacrylamide pattern of different molecular weight forms of var1 polypeptide present in parental stocks and in various progeny obtained from crosses. In all of the yeast strains we have examined thus far, whether obtained from independent sources or generated in crosses, the different var1 species all fall within the molecular weight range of 40,000 to 44,000. A detailed genetic and biochemical analysis has been carried out describing the mode of origin for these diverse var1 species (14,15). Here we will briefly summarize those conclusions and present

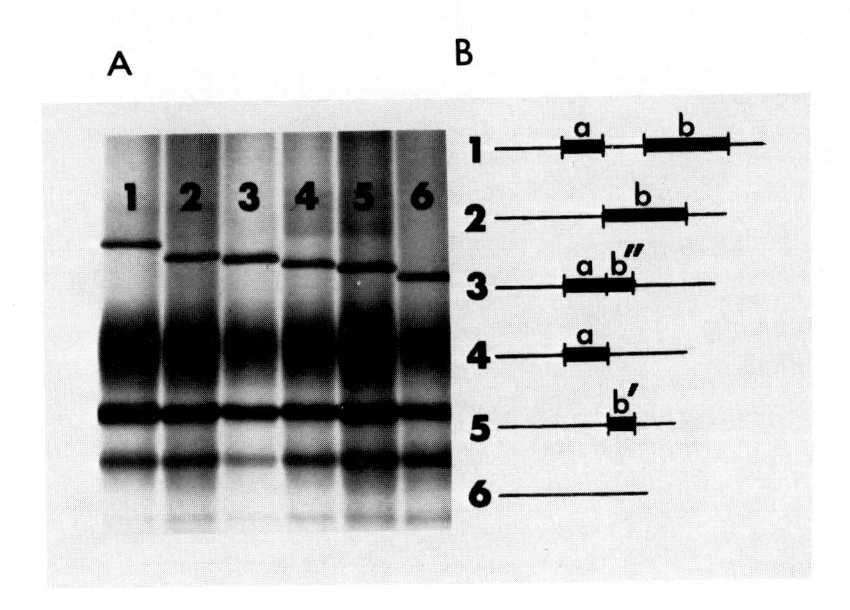

FIGURE 2. A. Mitochondrial translation products of wild-type yeast strains labeled in vivo [^{35}S]H_2SO_4 in the presence of cycloheximide. Each strain (1-6) has a unique allelic form of var1. B. Diagramatic representation of inserts in var1 for the six strains shown in the autoradiogram. Note that there are two inserts, termed a and b, and that b can insert as a partial in which case it is termed b' (or b'' when it inserts together with a). The positions of the inserts are arbitrary.

new data which shed additional light on the unusual properties of the var1 gene.

1. We can account for all of the molecular weight forms of var1 polypeptide observed to date, whether in strains obtained from independent sources or derived from genetic crosses, by two DNA segments which behave operationally as transposable elements capable of inserting from one genome to another. Based on polypeptide molecular weights, we estimate these DNA segments to be about 35-40 and 55-60 bp in length, designated a and b, respectively; they are completely absent from some strains (var1 molecular weight = 40,000) or exist in others in various combinations to give higher molecular weight forms up to 44,000. Each of the samples in Fig. 2A has been selected to illustrate a representative form of var1 polypeptide where the presence or absence of the a and b insert has been deduced from crosses.

Fig. 2B, shows diagramatically the arrangement of the inserts corresponding to each representative form of var 1 polypeptide. This picture has been deduced from genetic data alone; the actual location and order of the inserts is not known. To arrive at these conclusions we have established certain criteria to assist in the interpretation of the data obtained from crosses designed to determine the identity of the different forms of var1 polypeptide. These are summarized as follows and diagramatically illustrated in Fig. 3:

i) When two strains with the identical size var1 polypeptide are crossed, the var1 alleles are considered to be identical if no new molecular weight forms of the protein are generated (Fig. 3A).

ii) When two strains with different molecular weight forms of var1 polypeptide are crossed, the variable segments are considered to be overlapping with one end in common if no molecular weight form appears which is larger than the larger parental species; however, forms of var1 polypeptide with molecular weights intermediate in size between the parentals can be generated by partial insertions within the regions of difference (Fig. 3C).

iii) When two strains are crossed which have different molecular weight forms of var1 polypeptide, the variable segments are non-overlapping if new forms of the protein are generated which are larger in size than the larger of the two parental species (Fig. 3B). This result could obtain as well if the inserts were of equal size but non-overlapping, or overlapping with non-identical ends (Fig. 3D).

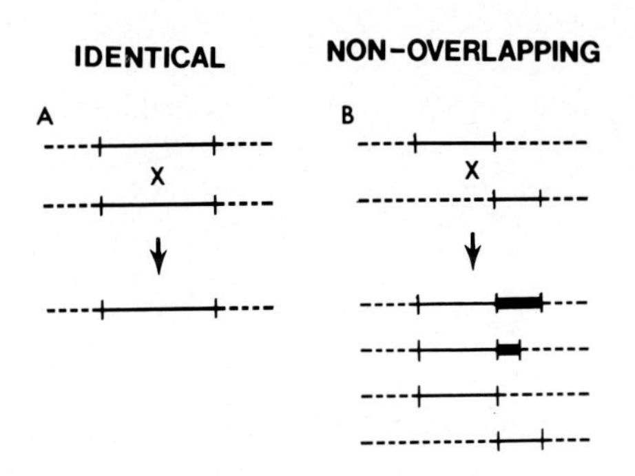

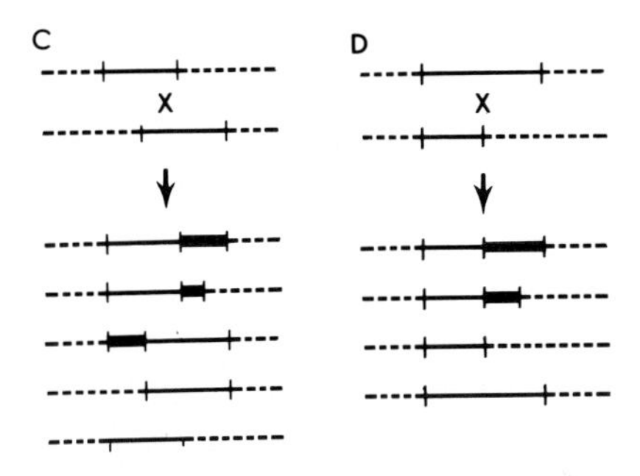

FIGURE 3. Diagramatic representation of the origin of non-parental forms in crosses. The Figure shows four configurations of DNA insertions in parental alleles. When the alleles contain insertions which are overlapping and identical in size (A), no new forms arise. With non-overlapping (B) or overlapping (C,D) segments, new forms arise by the insertion of either an entire allelic segment or a portion of a segment (both indicated by the heavy lines) to give non-parental species. If the parental insertions are non-overlapping (B) progeny larger than either parental allel can be generated. If the parental alleles are overlapping (C,D), no larger forms of varl are generated when the parental insertion have one end in common (D); however, larger varl species may be generated if neither insertion end is in common (C).

By carrying out the appropriate backcrosses, we have been able to confirm all of the examples presented in Fig. 2 and predicted according to the diagramatic representation shown in Fig. 3.

2. The segments a and b can insert independent of one another. For example in a cross between a strain lacking inserts (designated $a^- b^-$) and one containing both ($a^+ b^+$), some progeny are found which are $a^+ b^-$ or $a^- b^+$. Furthermore, the frequency of insertion is much greater for a than for b. In a standard cross, about 25% of the total diploid progeny are $a^+ b^-$ whereas only 2-4% are $a^- b^+$. These observations hold true whether the cross has a cis (a^+b^+ x a^-b^-) or trans (a^+b^- x a^-b^+) configuration.

3. The a segment always appears to insert as a whole while in some cases, parts of the b segment can insert at low frequency (∿0.15-1.0% of the total diploids) to give shorter forms designated $b^{+'}$ (Figs. 2 and 3). Not all of the $b^{+'}$ insertions are of the same size.

4. In certain configurations of $a^+ b^{+'}$, the a segment and a portion of the $b^{+'}$ segment are inserted together to give a new form designated $a^+b^{+''}$ (Fig. 2). Since this form may have a molecular weight indistinguishable from some strains which are $b^{+'}$, conformation that the configuration is indeed $a^+b^{+''}$ can be established by backcrosses to a strain which is a^-b^-; in such cases a and b'' can be recovered independently in the progeny of the cross (Fig. 2).

5. Based on quantitative marker transmission analysis, all insertions occur by asymmetric gene conversion similar to the conversion of ω^- to ω^+ (14).

6. In the conversion of a shorter form of var1 polypeptide to a longer form, the restriction fragment carrying determinants for the allelic differences (Hinc II - band 10) increases in size (14).

The Recipient Can Modulate the Output of Recombinant Forms of Var1. In screening for var1 transmission in a series of petites all derived from an a^-b^- wild-type strain, one transmitting petite was found which behaved quite differently from the others and from the parental wild-type. As shown in Fig. 4, the wild-type parent and a petite representative of most of the var1 transmitting petites show a similar behavior with respect to the transmission of the a and b segments to the diploid progeny: the insertion of a occurs at a much higher frequency than that of b. However, in the case of petite A14-41, the frequency of b insertion is now about the same as that of the a segment; together, a^+ and b^+ progeny may comprise as much as 40% of the total diploid population. The same situation holds for crosses with strains which are $a^+b^{+'}$ and $a^+b^{+''}$.

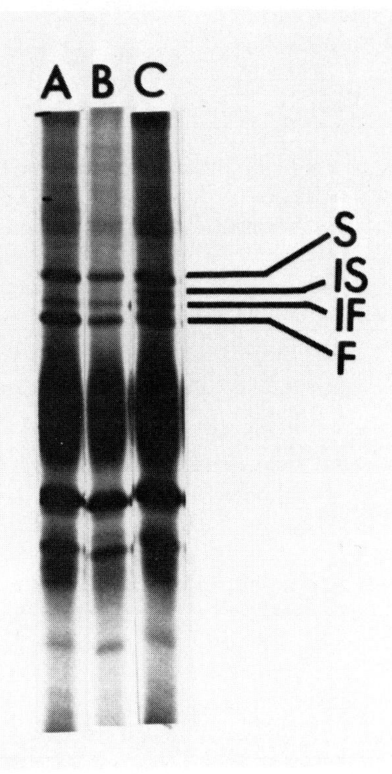

FIGURE 4. The recipient can influence the extent of gene conversion. Petites A14-41 and A12-6 were isolated from wild-type strain A10, which has a *var1* a^-b^- allele. The two petites and A10 were each crossed to a a^+b^+ *var1* tester and random diploids were isolated. The autoradiograph shows mitochondrial translation products from random diploids of the three crosses. *Var1* a^-b^+ is present in all crosses, but is not present at a high enough level to be seen in lanes A and B. In the figure, the allelic forms are indicated as: $var1^S$, a^+b^+; $var1^{IS}$, a^-b^+; *var1* I^F a^+b^-; and $var1^F$, a^-b^-. A. A10 x *var1* a^+b^+; B. A12-6 x *var1* a^+b^+; C. A14-41 x *var1* a^+b^+

We have not been able to test whether the frequency of transmission of the *b* segment alone is altered in crosses with A14-41 since this petite does not have any outsidc markers; consequently, quantitative transmission data cannot be obtained. Nevertheless, the results we do have on A14-41 clearly demonstrate that the nature of the recepient partner can dramatically affect the relative frequency of *a* and *b* insertions. Thus, we conclude that the recipient allele must determine with some specificity, the gene conversion event. We have some indication, however, that this feature of petite A14-41 may be complex since we have not yet been able to observe the A14-41 phenotype in a ρ^+ configuration. In other words, when A14-41 is crossed into a wild-type strain and ρ^+ segregants with the A14-41 form of var1 polypeptide are selected (a^-b^-), sporulated, and crossed into strains which are a^+b^+, we see only the "normal" relative frequency of a and b insertions. Although the molecular mechanism accounting for the behavior of A14-41 is unclear, further analysis of the relevant DNA segments in A14-41 as well as wild-type strains should be most informative in resolving the mechanisms underlying these conversions events.

A Role for Insertions in Gene Conversion. Our results on the recombination involving different var1 alleles are similar in many respects to the events occurring at the ω locus; in both cases asymmetric gene conversion seems to be the predominant recombination process. Based on estimates of the number of mating rounds and the frequency of conversion of ω^- to ω^+, Dujon et al. (2), suggested that all recombinations of the yeast mitochondrial genome may well be gene conversions. At some alleles, however, the conversions are symmetric since equal numbers of reciprocal recombinants are obtained at the population level. In others, the conversion is unidirectional, resulting in the preferential loss of one allele in a mating round. Here, we propose a model which can account for both asymmetric and symmetric gene conversion (Fig. 5). The model bears some similarity to the earlier suggestions of Perlman and Birky (18) and more recently to the proposal of Heyting and Menke (19) to account for gene conversion at ω. Two important considerations in our model attempt to accomodate the experimental observations that 1) DNA insertions/deletions are intimately associated with asymmetric gene conversion and, 2) modifications in the "recipient" strand can greatly affect the recombination process. Specific examples of the latter include results presented here with petite A14-41 and the ω^n variant which appears to be a mutation in an ω^- allele which eliminates conversion of ω^- to ω^+ (20).

As illustrated in Fig. 5, we propose that conversion events are initiated at specific sites along the mitochondrial genome. These sites would have two important properties: 1) they are recognized and cut by a specific endonuclease, and 2) they are not arranged per se as an inverted repeat (although they could be part of a larger palindromic structure). Consequently, only a single strand cut is made by the putative endonuclease recognizing that sequence. In the case of symmetric

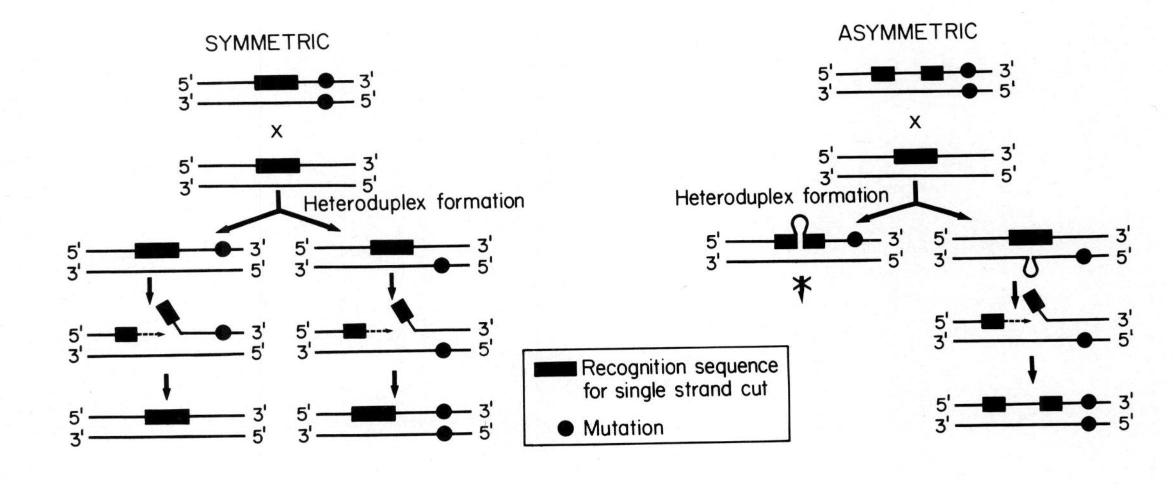

FIGURE 5. A model for symmetric vs. asymmetric gene conversion.

gene conversion, each of the reciprocal heteroduplexes of the recombination intermediate would be cut at the recognition sequence. Reciprocal gene conversion events would then follow by nick translation replication.

The essential feature of our model is that in some cases, the recognition sequence is split by a DNA insertion so that the sequence no longer is a substrate for the site specific endonuclease. This feature determines whether conversions are symmetric or asymmetric. In recombination, one of the heteroduplexes would be an abortive intermediate since the recognition sequence is split by the insertion and not cut by the endonuclease; in the reciprocal heteroduplex, cutting of the recognition sequence followed by nick translation results in the unidirectional conversion of some proximal allele as well as the insertion into the shorter strand, of the DNA segment splitting the recognition site. The controlling element in the conversion process is then the presence or absence of an insertion within the recognization sequence. The frequency of gene conversion would be controlled by the specific recognition sequence, i.e. the probability of cutting at that sequence over the average life-time of the recombination intermediate (heteroduplexes). Base changes within the sequence (or in nearby bases) could markedly affect the cutting frequency resulting in an increase or decrease in the total output of converted alleles. The phenotype of petite A14-41 described here and ω^n strains could readily be explained by such a mechanism. In addition, the frequent appearance of non-parental DNA bands in the restriction endonuclease pattern of mtDNA of the progency of crosses could be explained by the type of insertion mechanism proposed here.

Obviously, the model predicts the existence in yeast of an enzyme with site-specific single strand endonuclease activity. It is of considerable interest that such an enzyme has been reported to be present in *Chlamydomonas* by Burton *et al.* (21) with the implication that it might function in chloroplast recombination.

A Function for Var1 Polypeptide. An earlier report by Groot (16) showed an association of an unidentified product of mitochondrial protein synthesis with the small mitochondrial ribosomal subunit in yeast. Later, Groot *et al.* (17) showed labeled polypeptide was associated with the small mitochondrial ribosomal subunit which had the same mobility as var1 on SDS-polyacrylamide gels. Recent work from Lambowitz and his colleagues (22,23) showed that in *Neurospora crassa*, a mitochondrial ribosomal protein is made on mitochondrial ribosomes. We have confirmed and extended the results of Groot (24,25) to show that var1 polypeptide is tightly associated with the 38S yeast mitochondrial ribosomal subunit.

As shown in Fig. 6, high salt washed mitochondrial 38S ribosomal subunits contain a Coomassie blue stainable band with the same electrophoretic mobility as var1. Isonuclear strains with three different molecular weight forms of var1 show an exact correlation between the electrophoretic mobility of this stainable protein and var1 (24). On the basis of Coomassie Blue stain intensity, var1 appears to be roughly stoichiometric with the other ribosomal proteins. Moreover, by two-dimensional electrophoretic analysis, var1 is a basic protein like most other ribosomal proteins (24). These observations lead us to the conclusion that var1 is an intrinsic protein of the small mitochondrial ribosomal subunit.

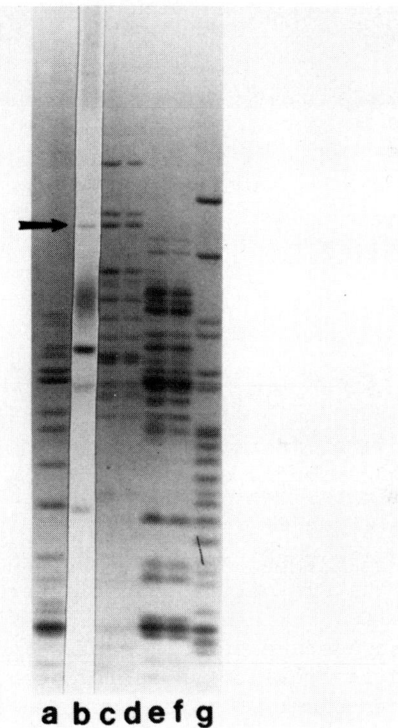

FIGURE 6. Var1 polypeptide is a ribosomal protein. Mitochondrial ribosomal subunits from yeast strain D^{IS} were prepared according to Faye and Sor (26) and purified by sedimentation through a high salt, 1.8 M sucrose cushion. The subunits were electrophoresed on an 11% SDS-polyacrylamide gel and stained with Coomassie Blue. a) cytoplasmic 40S subunits; b) autoradiograph of $^{35}SO_4^{=}$-labeled products of mitochondrial protein synthesis. The arrow points to var1 polypeptide; c) and d), mitochondrial 38S subunit; e) and f) mitochondrial 54S subunit; g) cytoplasmic 60S subunit.

If var1 is an intrinsic ribosomal protein, then one would expect it to have a function in ribosome assembly or in protein synthesis. That var1 has a role in assembly is illustrated by the following series of experiments: when cells are labeled in vivo in the presence of cycloheximide (CHX), the bulk of var1 appears to sediment at 15-20S (Fig. 7 panel A). After a cold chase in the absence of CHX, var1 now sediments at 38S (Fig. 7, panel B) indicating that in the absence of a pool of cytoplasmically made mitochondrial ribosomal proteins, var1 appears in a particle which behaves as a precursor to the mature 38S subunit. This conclusion is supported by the observation that the 15-20S particle is completely sensitive to RNAse (25). Conversely, one would predict that inhibition of mitochondrial protein synthesis should lead to a reduction in the amount of mature small subunit. To test this, cells were grown for about 9 generations in the presence of 2 mg/ml erythromycin. Mitochondrial protein synthesis was inhibited over 99% and there was no detectable cytochrome oxidase activity. When the cells were harvested, ≥92% of the cells could be scored as ρ^{+}. Analysis of the mitochondrial ribosomes shows that the cells grown in the presence of erythromycin contain a normal amount of the large mitochondrial subunit, but a reduced amount of the small subunit (Fig. 8). The peak observed in the 38S region is composed primarily of contaminating cytoplasmic small subunit; the large subunit peak, however, contains nearly pure mitochondrial large subunits with a protein profile indistinguishable from the control (25). Clearly, the assembly of small subunits is inhibited by inhibition of mitochondrial protein synthesis.

The maturation of the small mitochondrial ribosomal subunit appears to be strongly dependent on the extent of overall inhibition of mitochondrial protein synthesis. Chloramphenicol for example, inhibits mitochondrial protein synthesis an average of 90% compared to 99% for erythromycin. Cells grown in presence of chloramphenicol appear to have a normal complement of ribosomal proteins, including var1. An explanation for this result probably lies in the fact that the synthesis var1 polypeptide is considerably less sensitive to inhibition than the other products of mitochondrial protein synthesis; at an average inhibition of mitochondrial protein synthesis 90%, var1 synthesis is inhibited by only 40% (25).

Together, our observations lead to the conclusion that var1 polypeptide is needed for the assembly and maturation of the small mitochondrial ribosomal subunit; hence, var1

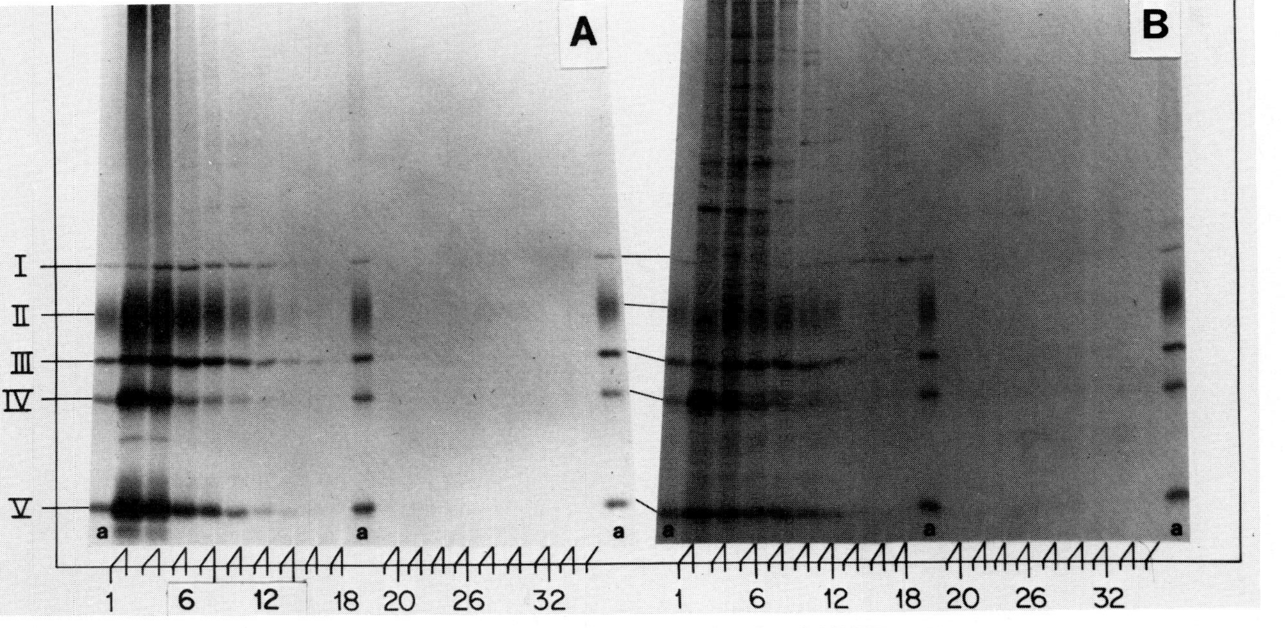

FIGURE 7. Sucrose gradient centrifugation of products of mitochondrial protein synthesis. A: Cells were labeled *in vivo* with $^{35}SO_4^=$ in the presence of cycloheximide. Isolated mitochondria were lysed with Triton X-100 and sedimented over a high salt isokinetic sucrose gradient. Fractions were combined and electrophoresed on an 11% SDS-polyacrylamide gel a) input to gradient. B: Same as in A except labeled cells were washed and regrown for one generation in rich medium without cycloheximide and label. The arrows point to the 40S position.

probably has a similar role as the mitochondrially synthesized ribosomal protein in *Neurospora crassa* (22,23). Whether var1 functions in protein synthesis is not clear.

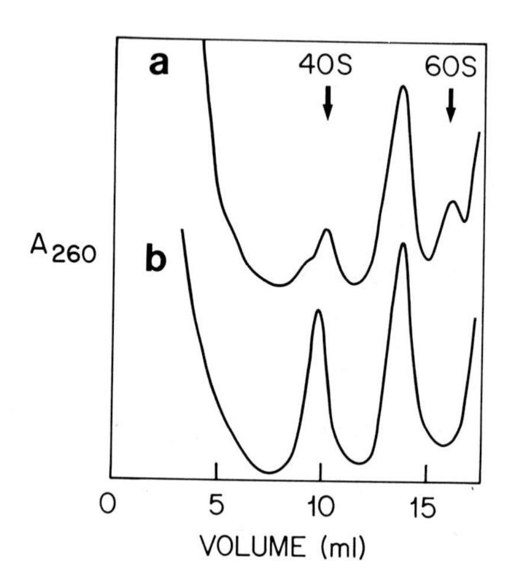

FIGURE 8. Sucrose gradient analysis of mitochondrial ribosomal subunits. Mitochondrial ribosomes from cells grown in the presence (a) and absence (b) of erythromycin (2 mg/ml) with 2% galactose as carbon source. Mitochondrial lysates were sedimented over isokinetic sucrose gradients containing 500 mM NH_4Cl; 10 mM Tris-Cl, pH 7.4; 10 mM $MgCl_2$ and 6 mM β-mercaptoethanol. The arrows point to the position of cytoplasmic ribosomal subunits run in a parallel gradient.

ACKNOWLEDGMENTS

We thank Philip Perlman, Nancy Alexander and Robert Vincent for making strains available to us and for very helpful discussion.

REFERENCES

1. Williamson, D. H., and Fennell, D. J. (1974). Molec. Gen. Genet. 131, 193-207.
2. Dujon, B., Slonimski, P. P., and Weill, L. (1974). Genetics 78, 415-437.
3. Borst, P., Bos, J. L., Grivell, L. A., Groot, G. S. P., Heyting, C., Moorman, A. F. M., Sanders, J. P. M., Talen, J. L., Van Kreijl, C. F., and Van Ommen, G. J. B. (1977). In "Mitochondria 1977, Genetics and Biogensis of Mitochondria" (Bandlow, W., Schweyen, R. J., Wolf, K. and Kaudewitz, F., eds.), pp. 213-254, de Gruyter, Berlin.
4. Morimoto, R., Merten, S., Lewin, A., Martin, N. C., and Rabinowitz, M. (1978). Molec. Gen. Genet. 163, 241-255.
5. Butow, R. A., Vincent, R. D., Strausberg, R. L., Zanders, E., and Perlman, P. S. (1977). In "Mitochondria 1977. Genetics and Biogenesis of Mitochondria" (Bandlow, W., Schweyen, R. J., Wolf, K. and Kaudewitz, F., eds.), pp. 317-335. de Gruyter, Berlin.
6. Fonty, G., Goursot, R., Wilkie, D., and Bernardi, G. (1978). J. Mol. Biol. 119, 213-235.
7. Prunell, A., and Bernardi, G. (1977). J. Mol. Biol. 110, 53-74.
8. Borst, P., and Grivell, L. A. (1978). Cell 15, 705-723.
9. Jacq, C., Kujawa, C., Grandchamp, C., and Netter, P. (1977). In "Mitochondria 1977, Genetics and Biogenesis of Mitochondria" (Bandlow, W., Schweyen, R. J., Wolf, K., and Kaudewitz, F., eds.), pp. 255-270, de Gruyter, Berlin.
10. Faye, G., Dennebouy, N., Kujawa, C., and Jacq, C. (1979). Molec. Gen. Genet. 168, 101-109.
11. Heyting, C., Meijlink, F. C. P. W., Verbeet, M. P., Sanders, J. P. M., Bos, J. L., and Borst, P. (1979). Molec. Gen. Genet. 168, 231-250.
12. Perlman, P. S., Douglas, M. G., Strausberg, R. L., and Butow, R. A. (1977). J. Mol. Biol. 115, 675-694.
13. Vincent, R. D., Perlman, P. S., Strausberg, R. L., and Butow, R. A. (1979). Manuscript in preparation.
14. Strausberg, R. L., Vincent, R. D., Perlman, P. S., and Butow, R. A. (1978). Nature 276, 577-583.
15. Strausberg, R. L. and Butow, R. A. (1979). Manuscript in preparation.

16. Groot, G. S. P. (1974). In "The Biogenesis of Mitochondria" (Kroan, A. M. and Saccone, S., eds.), pp. 443-452. Academic Press, New York.
17. Groot, G. S. P., Grivell, L. A., Van Harten-Loosbroek, N., Kreike, J., Moorman, A. F. M., and Van Ommen, G. J. B. (1977). In "Structure and Function of Energy Transducing Membranes" (Van Dam, K. and Van Gelder, B. F., eds.), pp. 117-186. Elsevier Scientific, Amsterdam.
18. Perlman, P. S., and Birky, C. W., Jr. (1974). Proc. Nat. Acad. Sci. USA 71, 4612-4616.
19. Heyting, C., and Menke, H. H. (1979). Molec. Gen. Genet. 168, 279-291.
20. Dujon, B., Bolotin-Fukuhara, M., Coen, D., Deutsch, J., Netter, P., Slonimski, P. P., and Weill, L. (1976). Molec. Gen. Genet. 143, 131-165.
21. Burton, W. G., Roberts, R. J., Myers, P. A., and Sager, R. (1977). Proc. Nat. Acad. Sci. USA 74, 2687-2691.
22. Lambowitz, A. M., Chua, N. H., and Luck, D. J. L. (1976). J. Mol. Biol. 107, 223-253.
23. LaPolla, R. J., and Lambowitz, A. M. (1977). J. Mol. Biol. 116, 189-205.
24. Terpstra, P., Zanders, E., and Butow, R. A. (1979). Manuscript in preparation.
25. Terpstra, P., and Butow, R. A. (1979). Manuscript in preparation.
26. Faye, G., and Sor, F. (1977). Molec. Gen. Genet. 155, 27-34.

TRANSCRIPTION OF YEAST MITOCHONDRIAL DNA[1]

David Levens, John Edwards, Joseph Locker, Arthur Lustig, Sylvie Merten, Richard Morimoto, Richard Synenki, and Murray Rabinowitz

Departments of Medicine, Biochemistry, Pathology, and Biology and The Franklin McLean Memorial Research Institute,[2] The University of Chicago, Chicago, Illinois 60637

ABSTRACT Yeast mtRNA was analyzed by means of 6M urea-agarose, 6M urea-acrylamide, and 10 mM methyl mercuric hydroxide-agarose gels, and RNA was transferred from gels to diazebenzyloxymethyl paper for subsequent hybridization to labeled DNA. In grande yeast, more than forty different species of RNA, ranging from 200 to 9500 nucleotides, were identified, including 11 species larger than 21S rRNA. Petites contain many of these RNA species, making deletion mapping possible. Processed tRNA, however, is identified only in petites that contain a region of the genome near P. Petites were thus used not only for deletion mapping of mitochondrial RNA species, but also for identification of different types of mtRNA processing. Hybridization studies on RNA from petite F11 appear to identify precursors to 21S rRNA. Overall, the number and size of these mitochondrial transcripts imply the existence of large precursor molecules which are subsequently processed by cleavage to a mature form.

The promoter function of yeast mtDNA has been examined by analysis of a mitochondrial transcription complex, and by the use of purified yeast mitochondrial RNA polymerase. The transcription complex primarily synthesizes RNA from the same regions and from the same strand that encodes the *in vivo* rRNA's. Since the 21S and 14S rRNA genes are widely separated, transcription must occur from at least two promoters. When template-dependent RNA polymerase was isolated either from the

[1]This work was supported in part by USPHS Grants HL0442 and HL09172 and Grant NP-281 from the American Cancer Society.

[2]The Franklin McLean Memorial Research Institute is operated by The University of Chicago for the United States Department of Energy under Contract EY-76-C-02-0069.

ISBN 0-12-198780-9

transcription complex or from soluble protein fractions, we found that the two enzyme fractions expressed similar properties. The soluble polymerase has been extensively purified by a variety of chromatographic techniques; this purified enzyme is associated with a 45,000 D peptide. Prior to a final glycerol gradient centrifugation, a 65,000 D peptide is associated with the enzyme activity in a molar ratio. The enzyme from the glycerol gradient was found to be extremely labile. In further experiments, antibodies to the 45,000 D band precipitated that peptide and inhibited RNA polymerase activity. Promoter function and mapping have been studied by means of transcription *in vitro* of mtDNA as well as binding of the purified RNA polymerase to restriction fragments. The results show the purified RNA polymerase binds to and transcribes mitochondrial DNA in a non-random fashion.

INTRODUCTION

The physical and genetic organization of the yeast mitochondrial genome has been studied extensively in recent years (for recent reviews, see 1, 2). Only a limited number of gene products are known to be specified by the 70-75 kb yeast mitochondrial genome. These account for at most 30 to 40% of the information available in a single-strand equivalent of yeast mtDNA. Conserved high-AT sequences which are apparently transcribed make up most of the remainder of the genome. Recently, laboratories have demonstrated the presence of intervening sequences in the 21S ribosomal RNA (3, 4) and in the cytochrome b (COB) region of mtDNA (2, 5, 6). Thus, in many ways, yeast mtDNA appears to resemble eukaryotes with respect to the characteristics of transcription.

Our laboratory has been studying the transcription and organization of the yeast mitochondrial genome. Here we will describe two current aspects of our work; first, studies related to the properties and processing of mitochondrial transcripts; second, the isolation and characterization of yeast mitochondrial RNA polymerase and its potential use in the identification and mapping of mitochondrial promoters.

RESULTS AND DISCUSSION

A restriction endonuclease fragment map of yeast mtDNA containing the localization of genetic markers has been derived in our laboratories (7, 8). This map is similar to the genetic and physical maps derived by several other laboratories. The loci were mapped by restriction analysis of genetically characterized petite deletion mutants, and by hybridization of specific transcripts and cRNA transcribed from simple petites to restriction fragments. It is particularly important for further discussion that the 21S and 14S rRNA subunits are widely separated, and that the other genetic loci are distributed throughout the 75 kb circular genome.

We have previously shown by hybridization of highly labeled nick-translated mtDNA with excess mitochondrial RNA that at least 70% of a single-strand equivalent of mtDNA is transcribed (9). Since only 30-40% of the genome is required to code for the established gene products of mtDNA, considerable processing of the mitochondrial transcripts must occur. The mitochondrial genome is relatively simple, and most of the genes probably have been identified; it should therefore be possible to analyze transcripts, at least in part, by means of highly discriminating gel-electrophoretic systems.

Many mitochondrial transcripts can be visualized (10) on agarose-urea, acrylamide-urea, and methyl mercury gels (11). Figure 1 shows agarose-urea gels (12) which demonstrate the presence of multiple RNA species. Several high-molecular-weight transcripts are present which are larger than 21S rRNA. These high-molecular-weight transcripts persist in completely denaturing methyl mercury agarose gels and after heating in 75% formamide, they may be high-molecular-weight precursors of functional transcripts. Lower-molecular-weight transcripts are best shown by higher-% agarose-urea and acrylamide-urea gel electrophoresis. Altogether, more than 40 non-tRNA mitochondrial transcripts can be discerned with these procedures; it is very likely that many more are present.

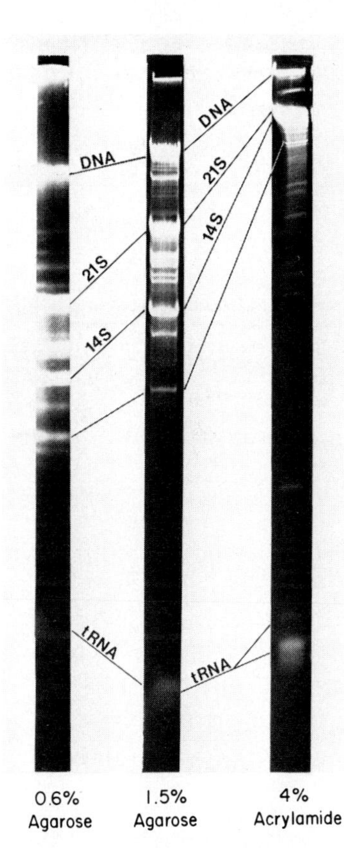

Figure 1. Electrophoresis of yeast mtRNA in gels containing 6M urea.

We have shown that many of these RNA species are transcribed from mtDNA by application of the RNA gel-blotting method recently described by Alwine et al. (13). In this method, RNA is electrophoretically separated on agarose gels containing 6M urea and transferred to diazotized paper by blotting. A radioactive DNA probe, labeled by nick translation, is hybridized to the RNA bands bound to the paper. Labeled grande mtDNA was hybridized with grande RNA transferred to the diazotized paper. A large number of transcripts hybridize and are thus specified by mtDNA; these transcripts are not present in the cytoplasmic RNA. Finally, many of these transcripts specifically appear in groups of petites containing common genetic markers.

We have been able to obtain information about the processing and mapping of some of these transcripts by hybridization of labeled DNA to RNA immobilized on diazotized paper, and by analysis of the RNA of several deletional mit$^-$ and cytoplasmic petite strains.

Cloned Eco R1 fragments 6 and 7 (14) were labeled to high specific activity by nick translation; they were then hybridized to grande mtRNA and transferred to diazotized paper. Eco R1 fragment 6 is associated with the COB locus; Eco R1 fragment 7 is associated with the O_{II} locus and perhaps with a small segment of OXI-3 (15, 16). Figure 2 shows that a series of transcripts hybridize with these two DNAs, with

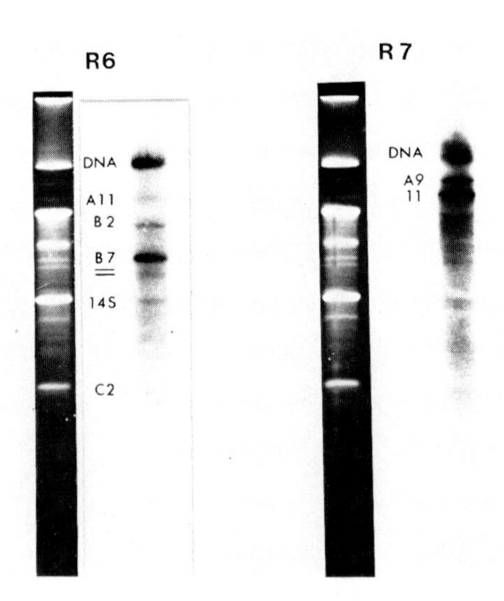

Figure 2. Transfer-hybridization of mtRNA to cloned DNA fragments. Grande mtRNA was transferred to diazobenzyloxymethyl paper (11) and hybridized to cloned Eco R1 DNA fragments. Fragment R6 is associated with the COB locus region, and R7 with the O_{II} region and perhaps a small segment of OXI-3.

sizes ranging from 900 to 6500 nucleotides. Thus, the larger transcripts may represent processing intermediates transcribed from these genes.

We have examined the transcripts associated with the OXI-3 gene locus by analyzing the RNA from mit⁻ deletion mutants which have defects in the biosynthesis of peptide 1 of cytochrome oxidase (17). Comparison of the mitochondrial RNA from these mutants with grande shows that the OXI-3 mutants M10-150 and M5-16, which contain large 5- to 7-kb deletions of mtDNA, lack three RNA bands of 3100, 2600, and 840 nucleotides. Mitochondrial RNA from the OXI-3 mutant M12-120, which has a very small deletion not detected by restriction analysis of the DNA, has the 3100-nucleotide transcript, but still lacks the 2600- and 840-nucleotide species. These data suggest that the 2600- and 840-nucleotide species are related to the OXI-3 gene products. Mitochondrial RNA from the point mutant M11-244 contain all of these RNA species. The 3100- and 2600-nucleotide RNA species missing in OXI-3 deletion mutants correspond to the transcripts present in petite deletion mutants which have retained the OXI-3 locus. The interrelationship of these transcripts is under study.

It is also clear that cytoplasmic petites, each of which retains only a fraction of the total mitochondrial genome, process their RNA effectively (18-20). Methyl mercuric hydroxide-agarose gel electrophoresis of RNA from a series of petites is illustrated in Figure 3. Petites contain some transcripts which are identical in size to those present in the grande and thus can be mapped. They also contain "group-specific" transcripts which are not apparent in the grande, but which are associated with specific regions of the genome in different petites. These transcripts may represent accumulation of intermediates that are present only in low concentration in grande strains. "Strain-specific" transcripts were also present in individual petites; they are probably related to the site of the deletion and are not useful for mapping purposes.

Figure 4a shows that the RNA of petite F11 contains 21S RNA and 2 high-molecular-weight bands above it. These bands appear to be 21S rRNA precursors, as judged by hybridization of labeled E41 DNA to petite F11 transcripts transferred to diazotized paper. The 1500 bp E41 DNA sequence is almost completely within the 21S RNA gene and hybridizes strongly to the two high-molecular-weight bands of F11 RNA as well as to the 21S RNA. It thus appears that the two larger transcripts represent precursors of the 21S RNA.

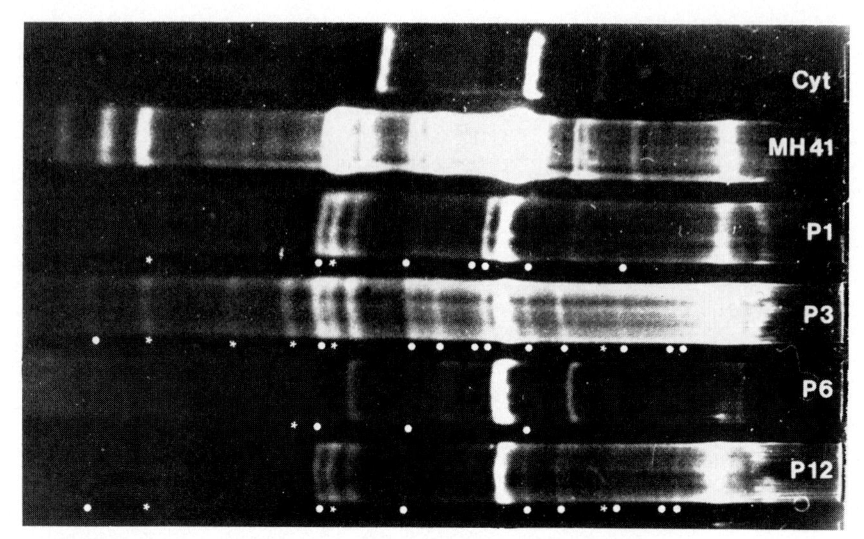

Figure 3. 10 mM methyl mercuric hydroxide-agarose gel electrophoresis of mtRNA from a series of petites containing the P region. *Dots* indicate bands that comigrate with grande, and *asterisks* indicate "group-specific" bands.

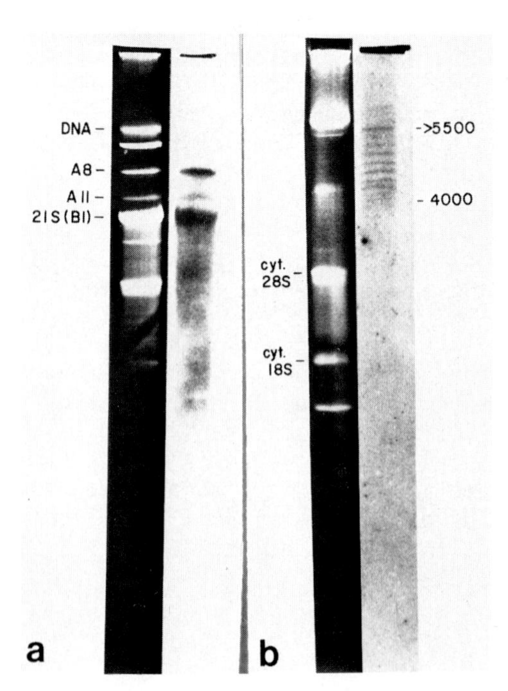

Figure 4. Transfer-hybridization of mtDNA from petites. *a*. RNA from petite F11, which contains the 21S rRNA gene, was transferred to diazobenzyloxymethyl paper and hybridized to ^{32}P DNA from petite strain E41, which is almost entirely within the 21S gene. *b*. RNA from petite O_I-3 (6600 base pairs) was transferred and hybridized to ^{32}P DNA from petite O_I-2, which contains a much larger DNA segment from the same region of the genome.

We observed that the processing of tRNA in petites may be related to a region near the P locus. Only petites which contain this P region contain processed mitochondrial tRNA. Mitochondrial RNA from genetically characterized petite strains was analyzed by 5% acrylamide-6M urea gel electrophoresis. Mitochondrial and cytoplasmic tRNA can be distinguished in this system, since the cytoplasmic species migrate more rapidly. Analysis of mtDNA from 21 petite strains revealed that only petites that retain a part of the tRNA II region which is near P locus transcribe and process their tRNAs (Fig. 5). Small amounts of tRNA, however, may be undetected in petities retaining other regions. We have previously shown that the high-molecular-weight RNA of petite O_I-3 contains $tRNA^{glu}$ sequences which are not processed into functional tRNA (21).

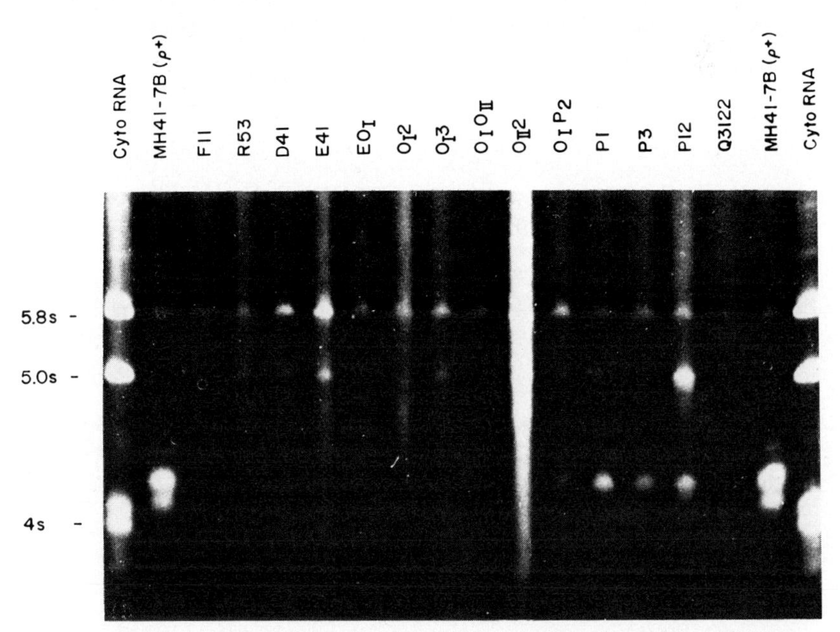

Figure 5. Low molecular weight mtRNA from 14 petites. On this 4% acrylamide-6M urea gel, mitochondrial tRNA can be resolved from cytoplasmic tRNA. A mitochondrial tRNA band is detectable only in RNA from strains O_IP-2, P1, P3, and P12, although most of the other illustrated strains contain tRNA genes.

The enzymes involved in the transcription of mtDNA and processing of RNA are not mitochondrial gene products, since petites process their RNA normally. It is not clear why tRNA transcription is associated with the retention of specific regions of the petite genome. Processing of tRNA would appear to depend upon the synthesis of specific RNA transcripts. From the analysis of 18 petites, a deletion of transcripts was derived (Fig. 6); the map also shows the tRNA-processing

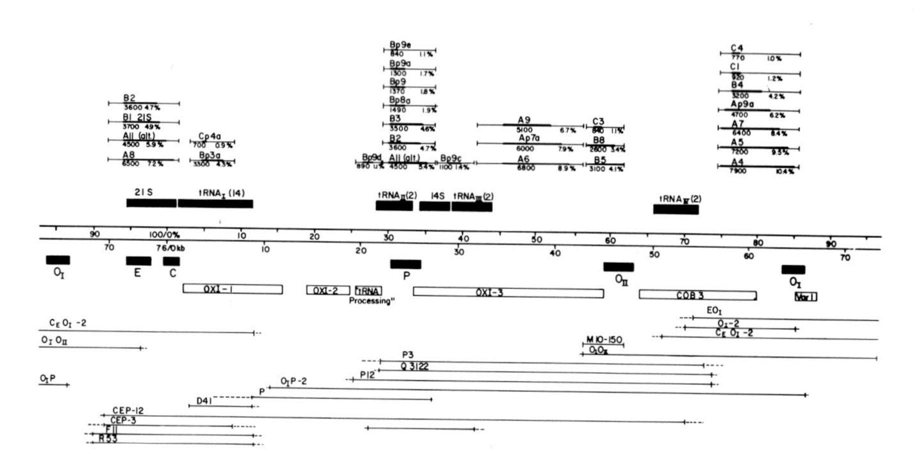

Figure 6. Deletion map of mitochondrial transcripts. The upper region of the figure shows the size and approximate location of individual transcripts. Transcript size is indicated by the heavy line, and the deletion interval to which it has been mapped, by a lighter line. It should be noted that part of the transcript may extend out of this interval. The middle region of the figure shows a current map of mitochondrial genes, and the bottom part of the figure shows the extent of the petite genomes used to construct the deletion intervals.

locus. For several regions in the genome, for example, in the COB region, there are many transcripts which exceed the coding capacity of the region. Probably these represent a series of partially processed intermediates. The transcript map obtained is similar in many ways to that observed by Van Omen et al. (22), who used a different procedure.

Processing of RNA can also be seen in a transfer hybridization of O_I-2 DNA to the transcripts of strain O_I-3 (Fig. 4b). A series of transcripts differing among each other in molecular size by 100-200 bases is observed. The largest transcript is approximately the size of the entire genome. Thus, processing appears to occur which removes small, discrete segments of the transcript.

We are further investigating processing of RNA in specific genetic loci by using the RNA transfer technique, R-loops, and fine mapping of transcripts.

A second aspect of our research involves another approach to the regulation of mitochondrial transcription. We have purified yeast mitochondrial RNA polymerase and are studying its interactions with mtDNA in order to obtain information about the characteristics and location of mitochondrial promoters.

Initially, we tried to purify the mitochondrial RNA polymerase from yeast mitochondria by isolating a specific complex between mtDNA and the RNA polymerase that retained transcriptional activity. The complex gave us important information about mitochondrial transcription, namely, that transcription is asymmetric and originates from at least two promoters.

When lysates of yeast mitochondria from grande or petite strains are chromatographed over Sepharose 4B columns, two peaks of RNA polymerase activity are observed (Fig. 7). The first peak, which elutes in the void volume, contains a

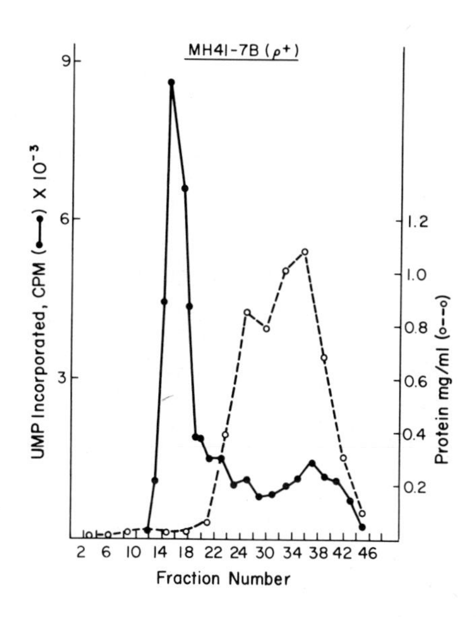

Figure 7. Sepharose 4-B chromatography of mitochondrial lysates of grande strain MH41-7B. Mitochondria were lysed with 0.5% NP-40 and 0.25M KCl, centrifuged for 30 min. at 27,000 rpm. The supernatant was chromatographed on Sepharose-4B. Fractions were assayed for protein and RNA polymerase activity.

template-independent RNA polymerase. This transcription complex is separated from more than 99% of the mitochondrial protein. Cells harvested in the exponential phase of growth contain more template-bound polymerase than those harvested in the stationary phase. The soluble RNA polymerase in the second peak is template-dependent.

When fractions from the excluded volume of the Sepharose 4B column are incubated with all four ribonucleotides and magnesium, an RNA is synthesized that is similar in many ways to the RNA that is synthesized *in vivo* (Fig. 8). RNA transcribed by the complex hybridizes principally to restriction

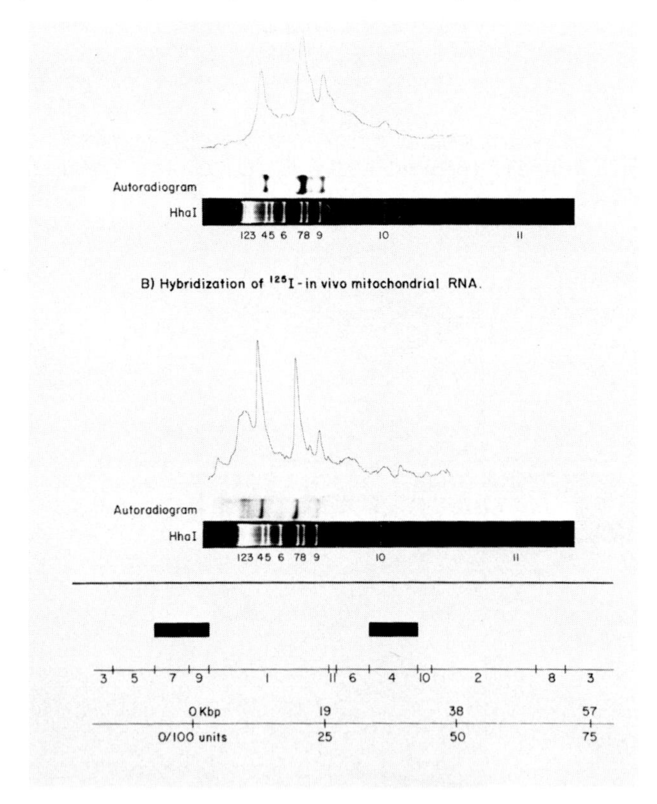

Figure 8. Hybridization of RNA synthesized by the transcription complex compared to hybridization of *in vivo* mitochondrial RNA. The top figure shows the pattern of hybridization of RNA synthesized by the transcription complex to an Hha I digest of MH41-7B. The middle shows the hybridization of ^{125}I-labeled total mitochondrial RNA hybridized to the same digest. The bottom shows an Hha I restriction map of MH41-7B. The solid bars show the location of the 21S and 14S rRNA's.

fragments containing the 14S and 21S rRNA genes, and not to the sequences that widely separate the two genes. *In vivo* RNA exhibits a similar pattern of hybridization. The same result is obtained with RNA synthesized for one or twenty minutes by the complex. RNA synthesized *in vitro* continues to accumulate for at least fifteen minutes. Furthermore, the RNA synthesized *in vitro* does not anneal with excess mitochondrial RNA. Since the complex does not initiate *in vitro*, the synthesized RNA must reflect *in vivo* initiation events. We concluded that mitochondrial transcription is asymmetric and originates from at least two promoters. Yeast mitochondrial transcription thus differs from that seen in HeLa mitochondria, in which complete symmetric transcription is observed (23, 24).

The second peak of RNA polymerase activity from the Sepharose 4B column contains a completely template-dependent RNA polymerase (Table 1). This activity was purified extensively by heparin Sepharose 4B, phosphocellulose, and DEAE-Sephadex chromatrography. The activity from the transcription complex can also be purified by means of heparin Sepharose-4B and DEAE-cellulose chromatography, following an autolysis step that renders the transcriptional activity template-dependent. The two enzymes have similar properties with respect to ionic strength, divalent cations, inhibitors, and template preferences; the enzymes can be distinguished from yeast nuclear RNA polymerase by these properties and by their chromatographic characteristics (25-29). The properties of this enzyme differ

Table 1

PURIFICATION OF YEAST MITOCHONDRIAL RNA POLYMERASE

STEPS	PROTEIN (MG)	SPECIFIC ACTIVITY* (UNITS)	YIELD (%)
MITOCHONDRIA	630	0.24	100
SUPERNATANT	400	0.06	10
SEPHAROSE 4-B	220	0.13	112
HEPARIN SEPHAROSE 4-B	16	18	150
PHOSPHOCELLULOSE	2.8	115	160
DEAE-SEPHADEX	0.4	300	65

*1 unit = $\frac{\text{1 nmole UMP incorporated}}{\text{1 mg protein — 10 minutes}}$ with poly d(AT)

considerably from those of the other preparations of mitochondrial polymerase previously described (30-33). The mitochondrial RNA polymerase from Xenopus ovaries possesses characteristics similar to those of the yeast mitochondrial enzyme (34).

The enzyme from the second peak of activity from the Sepharose 4B column, after purification on the final DEAE Sephadex column, contains two bands of 65,000 D and 45,000 D on SDS polyacrylamide electrophoresis (Fig. 9). Examination of the active fractions from the various purification steps

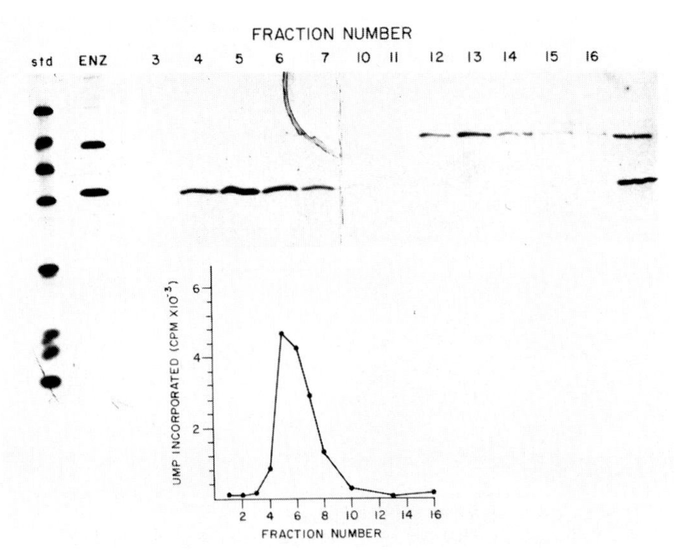

Figure 9. Final purification steps for mitochondrial RNA polymerase. Top left: 15% polyacrylamide-SDS gel electrophoresis of DEAE-Sephadex purified mitochondrial RNA polymerase and adjacent, protein standards. Top right: Electrophoretic profile of fractions from a 25-50% glycerol gradient spun for 19 hrs. at 59,000 rpm. Bottom: Activity profile from the gradient.

suggests that the RNA polymerase activity is associated with the 45,000 D band, but is either stabilized or further activated by the 65,000 D band. This is further indicated by glycerol gradient centrifugation of the purified RNA polymerase, in which the activity co-sediments with the 45,000 D band at about 6S as shown by SDS polyacrylamide electrophoresis of the fractions. The active fractions contain only trace amounts of the 65,000 D band, which self-aggregates and sediments at about 20S. It appears likely that the separation of the two bands contributes to the extreme lability of the activity recovered from the glycerol gradients in contrast to the stability of the material loaded onto the gradients. The small molecular size of the active enzyme and its electrophoretic pattern conclusively distinguish this activity from that of the nuclear polymerases (25, 26, 28, 35, 36).

Antibodies to the 45,000 D band were obtained by injection, into rabbits, of homogenized polyacrylamide gel strips containing this band (Fig. 10). The antibody raised in this

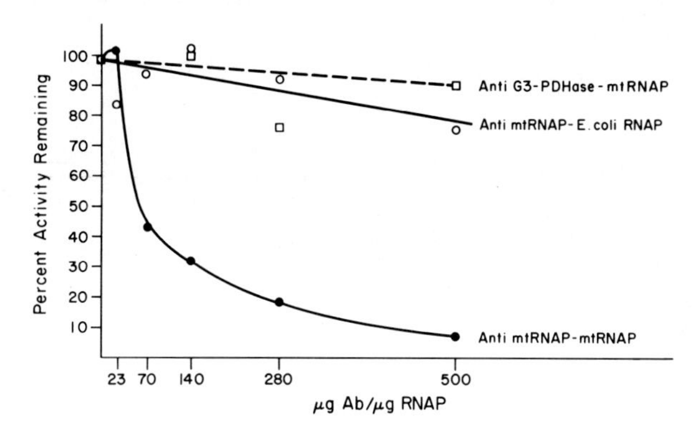

Figure 10. Immunological inhibition of mitochondrial RNA polymerase by antibody to the 45,000 D peptide.

manner precipitates the 45,000 D band in an indirect immunoassay and inhibits the activity of the purified enzyme. The antibody also inhibits the activity of the enzyme purified from the transcription complex. *E. coli* RNA polymerase is inhibited only at very high ratios of antibody to antigen. In contrast, control serum does not inhibit the RNA polymerase activity of either *E. coli* or mitochondrial RNA polymerase. The fact that the antibody to the 45,000 D band inhibits the activity of both the complex and the free peaks shows that both peaks of activity are related, if not identical.

The availability of a highly purified mitochondrial RNA polymerase allowed us to commence promoter mapping experiments in which we used the pure enzyme as a probe for promoters. We first sought to determine whether the purified enzyme possessed the capacity to interact specifically with certain DNA sequences, as is the case for prokaryotic RNA polymerases (37). The absence of specific RNA-polymerase-DNA interactions might implicate the involvement of additional factors that confer transcriptional specificity (38, 39). To determine whether the purified enzyme binds and transcribes mitochondrial DNA specifically, we used a template that allowed a relatively simple analysis of transcriptional selectivity.

We chose the mtDNA of the petite strain F11 as template. F11 retains a 12,000 base-pair region of the grande genome which includes the complete 21S rRNA and six tRNA genes. This ω^+ strain contains the 1100 bp intervening sequence in the 21S RNA genes and transcribes and processes its large rRNA normally.

An Hha I-Hinc II double digest of F11 mtDNA (Fig. 11) produces a pattern with five restriction bands visible on

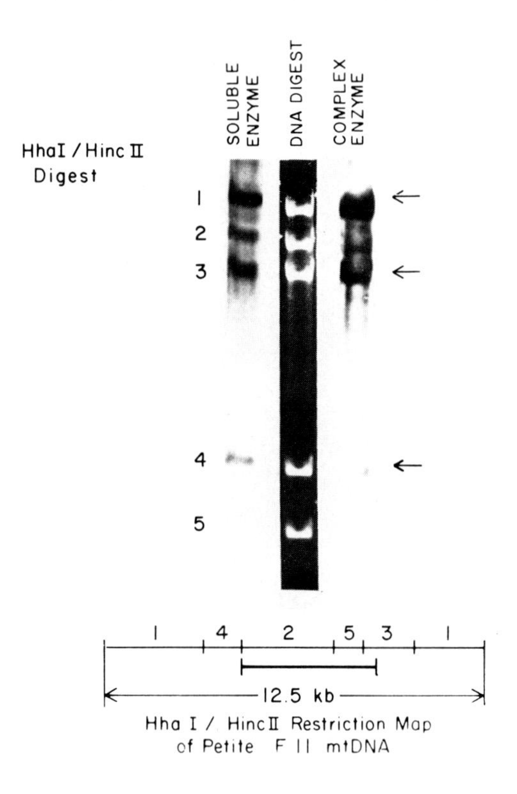

Figure 11. Comparison of RNA synthesized by soluble and complexed RNA polymerase. Left: hybridization of RNA synthesized by RNA polymerase purified from the second Sepharose 4B peak using F11 mtDNA as template; middle: 2% agarose gel of Hha I-Hinc II digest of F11 mtDNA: right: hybridization of RNA synthesized by the RNA polymerase purified from the complex.

2% agarose gels. F11 mtDNA was used as the template for transcription *in vitro* by the purified mitochondrial RNA polymerase. The labeled RNA hybridized to an Hha I-Hinc II restriction digest when the technique of Southern was used (40). In this manner, the most heavily transcribed regions were localized. The pattern of hybridization observed was distinctly non-random; bands one and three were heavily transcribed, whereas bands two and five were clearly underrepresented. Furthermore, the pattern of hybridization was the same whether the RNA polymerase was purified from the complex or from the second Sepharose-4B peak. Therefore, the two enzymes appear to have the same transcriptional specificity. The Hha I-Hinc II fragments which are poorly transcribed appear to lie within the rRNA gene. Since the 21S rRNA is derived from a large precursor, and since the *in vitro* transcripts are short, those fragments internal to the gene and close to its 3' end should be underrepresented in the hybridization. Fragment five, which is composed of sequences found primarily within the intervening sequence of the large rRNA gene, is highly underrepresented. These results indicate that the mitochondrial RNA polymerase may be a suitable probe for promoter localization.

The term "promoter" is a functional definition that involves the binding of RNA polymerase to template and the initiation of transcription (41). A search for promoters must use the RNA polymerase as either a physical or a genetic probe. Regulatory sequences cannot be rigorously defined without confirmation by either physical or genetic means. Therefore we studied the interaction of the mitochondrial RNA polymerase with F11 mtDNA (Fig. 12). F11 mtDNA was cut with

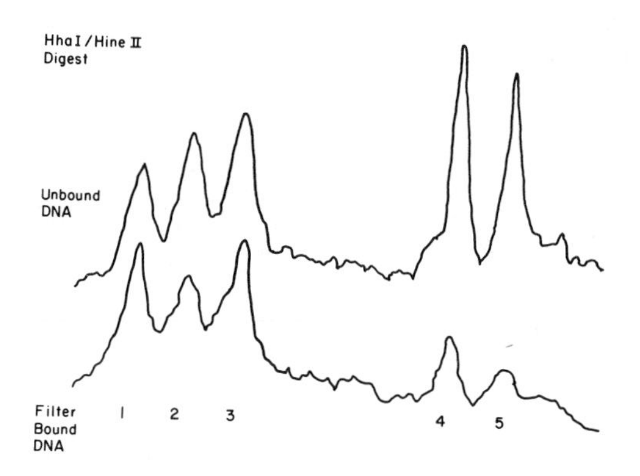

Figure 12. Retention of Hha I-Hinc II DNA fragments to filters by mitochondrial RNA polymerase. ^{32}P-DNA from petite F11 was used.

Hha I and Hinc II endonuclease and then 5' end-labeled with ^{32}P by means of polynucleotide kinase. The labeled DNA was used for nitrocellulose filter binding experiments. In these experiments, complexes are formed between DNA and protein; upon filtration through nitrocellulose, protein DNA complexes are retained, whereas free DNA passes through the filter. In the scans shown in Figure 13, there appears to be a specific interaction of the RNA polymerase with the same DNA restriction fragments that are heavily transcribed *in vitro*. However, the specificity is not as marked as in the transcription experiment, probably due to non-specific binding of the RNA polymerase to the restriction fragments. To enhance our ability to see the specifically bound fragments, we used Hpa II, which makes many cuts to distribute the non-specifically bound enzyme over many fragments. As shown in Figure 13, specific binding to at least three fragments is clearly seen.

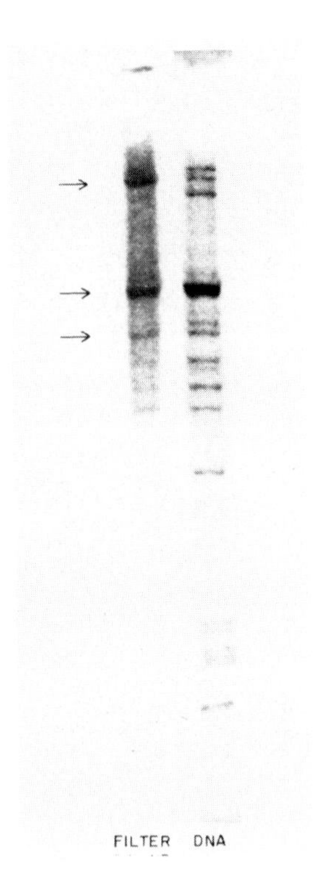

Figure 13. Retention of Hpa II DNA fragments to filters by mitochondrial RNA polymerase. ^{32}P-DNA from petite F11 was used.

The existence of specific binding sites does not prove that these sites correspond to promoters; however, they are likely candidates. Their positive identification requires the localization of the 5' ends of the primary transcripts on mitochondrial DNA.

REFERENCES

1. Locker, J., and Rabinowitz, M. (1979) Methods in Enzymology 56, 1-19.
2. Borst, P., and Grivell, L.A. (1978) Cell 15, 715-723.
3. Bos, J.L., Heyting, C., Borst, P., Arnberg, A.C., and Van Bruggen, E.F.J. (1978) Nature 275, 336-338.
4. Faye, et al. (1979) Molec. Gen. Genet. 168, 101-109.
5. Slonimski, P.P., et al. (1978) In "Biochemistry and Genetics of Yeast" (M. Bacilla et al., eds.), in press.
6. Mahler, H.R., et al. (1978) In "Biochemistry and Genetics of Yeast" (M. Bacilla, et al., eds.), in press.
7. Morimoto, R., Lewin, A., and Rabinowitz, M. (1977) Nucl. Acids Res. 4, 2331-2351.
8. Morimoto, R., Merten, S., Lewin, A., Martin, N.C., and Rabinowitz, M. (1978) Molec. Gen. Genet. 163, 241-255.
9. Jakovcic, S., Hendler, F., Halbreich, A., and Rabinowitz, M., submitted for publication.
10. Locker, J., Morimoto, R., Synenki, R., and Rabinowitz, M., submitted for publication.
11. Bailey, J.M., and Davidson, N. (1977) Anal. Biochem. 70, 75-85.
12. Locker, J., submitted for publication.
13. Alwine, J.C., Kemp, D.J., and Stark, G.R. (1977) Proc. Nat. Acad. Sci. USA 74, 5350-5354.
14. Berg, P.E., Lewin, A., Christiansen, T., and Rabinowitz, M., submitted for publication.
15. Lewin, A., Morimoto, R., Rabinowitz, M., and Fukuhara, H. (1978) Molec. Gen. Genet. 163, 257-275.
16. Borst, et al. (1977) In "Mitochondria 1977" (W. Bandlow, et al., eds.), pp. 213-224. DeGruyter, Berlin.
17. Morimoto, R., Lewin, A., and Rabinowitz, M. (1979) Molec. Gen. Genet., in press.
18. Faye, G., Kujawa, C., and Fukuhara, H. (1974) J. Mol. Biol. 88, 185-203.
19. Morimoto, R., Locker, J., Synenki, R., and Rabinowitz, M., submitted for publication.
20. Morimoto, R. (1978) Ph.D. Thesis, University of Chicago.
21. Martin, N.C., Rabinowitz, M., and Fukuhara, H. (1976) J. Mol. Biol. 101, 285-296.
22. Van Ommen, G.J.B., and Groot, G.S.P. (1977) In "Mitochondria 1977" (W. Bandlow, et al., eds.), pp. 415-424. DeGruyter, Berlin.
23. Aloni, Y., and Attardi, G. (1971) J. Mol. Biol. 55, 251.
24. Aloni, Y., and Attardi, G. (1971) Proc. Nat. Acad. Sci. USA 68, 1757.
25. Ponta, H., Ponta, U., and Wintersberger, E. (1972) Eur. J. Biochem. 29, 110.

26. Hager, G., Holland, M.J., and Rutter, W.J. (1977) Biochemistry 16, 1.
27. Chambon, P. (1974) In "The Enzymes" (P.D. Boyer, ed.), 10, 261.
28. Valenzuela, P., et al. (1976) Proc. Nat. Acad. Sci. USA 73, 1024.
29. Valenzuela, P., et al. (1976) J. Biol. Chem. 251, 1464.
30. Rogall, G., and Wintersberger, E. (1974) FEBS Letters 46, 333.
31. Eccleshall, T.R., and Criddle, R.S. (1974) In "Biogenesis of Mitochondria" (A.M. Kroon and E. Saccone, eds.), p. 31.
32. Tsai, M., and Michaelis, G. (1971) Proc. Nat. Acad. Sci. USA 68, 473.
33. Scragg, A.H. (1976) Biochem. Biophys. Acta 442, 331.
34. Wu, G., and Dawid, I.B. (1972) Biochemistry 11, 3589.
35. Valenzuela, P., et al. (1978) In "Methods in Cell Biology" (G. Stein and J. Stein, eds.), 19, 1.
36. Sentenac, A., et al. (1976) In "RNA Polymerases" (R. Losick and M. Chamberlin, eds.), 763.
37. Chamberlin, M. (1976) In "RNA Polymerase" (R. Losick and M. Chamberlin, eds.), 159.
38. Teakamp, P.A., et al. (1979) J. Biol. Chem. 254, 955.
39. Ng, S.Y., Parker, E.S., and Roeder, R.G. (1979) Proc. Nat. Acad. Sci, USA 76, 136.
40. Southern, E.M. (1975) J. Mol. Biol. 98, 503.
41. Gilbert, W. (1976) In "RNA Polymerase" (R. Losick and M. Chamberlin, eds.), 193.

TRANSCRIPTS OF YEAST MITOCHONDRIAL DNA AND THEIR PROCESSING[1]

L.A. Grivell, A.C. Arnberg[2], P.H. Boer, P. Borst,
J.L. Bos, E.F.J. van Bruggen[2]. G.S.P. Groot, N.B. Hecht,
L.A.M. Hensgens, G.J.B. van Ommen and H.F. Tabak

Section for Medical Enzymology and Molecular Biology,
Laboratory of Biochemistry, University of Amsterdam,
Jan Swammerdam Institute, P.O. Box 60.000,
1005 GA AMSTERDAM (The Netherlands)

ABSTRACT Recent advances in our knowledge of gene organization and transcripts of yeast mtDNA are reviewed. Evidence is presented for various post-transcriptional modifications of mtRNAs, including splicing and circularisation. The importance of processing in the regulation of mitochondrial gene expression is discussed.

INTRODUCTION

In the three years subsequent to the construction of the first restriction maps of yeast mtDNA (1), emphasis of research has gradually shifted from the phenomenology of mitochondrial biogenesis and the listing of mitochondrial gene products to a detailed examination of mitochondrial gene organization and the mechanism whereby gene expression is regulated. As a first step in this direction, we set out to identify and characterize the transcripts of mtDNA. This article summarizes our findings to date and shows that the yeast mitochondrial genome conceals an unexpectedly high complexity of organization which is not only interesting in itself, but may also serve as a useful model in the study of corresponding phenomena in the eukaryotic nuclear genome.

[1]This work was supported in part by a grant to P.B. and G.S.P.G. from The Netherlands Foundation for Chemical Research (S.O.N.) with financial aid from the Netherlands Organization for the advancement of Pure Research (Z.W.O.).

[2]Biochemical Laboratory, University of Groningen, Zernikelaan, Groningen (The Netherlands)

ISBN 0-12-198780-9

RESULTS AND DISCUSSION

1. THE TRANSCRIPTION MAPS OF mtDNA

Our present knowledge of transcription of mtDNA results largely from the hybridization mapping studies of Van Ommen et al. (2-4) and is summarized in the maps shown in Figs 1a and 1b. The maps are for the mtDNAs of *S. carlsbergensis* and *S. cerevisiae* KL 14-4A and show all transcripts longer than about 350 nucleotides present in concentrations exceeding roughly 0.01% of total mtRNA and tRNAs. They were constructed by hybridization of fractionated RNAs with restriction digests of mtDNA transferred to nitrocellulose strips by standard blotting procedures. The two DNAs differ from each other by a number of insertions and deletions varying in length from 500 to 3500 bp at sites previously identified by Sanders et al. (1). Although only the hybridization of each mtDNA with its homologous RNA is presented, much useful information was obtained from the analysis of the hybridization of heterologous combinations. Major points of interest can be summarized as follows:

a. There are transcripts which correspond in position to all genetic markers. It seems reasonable to assume that these represent mRNAs of the structural genes in which the markers are located. This assumption has been directly borne out by in vitro translation of RNAs from the *oli-1*, *oxi-1* and *oxi-3* loci (6) and for the *cob* region (see below). Whether these RNAs are the biologically active species in the intact mitochondrion still has to be verified.

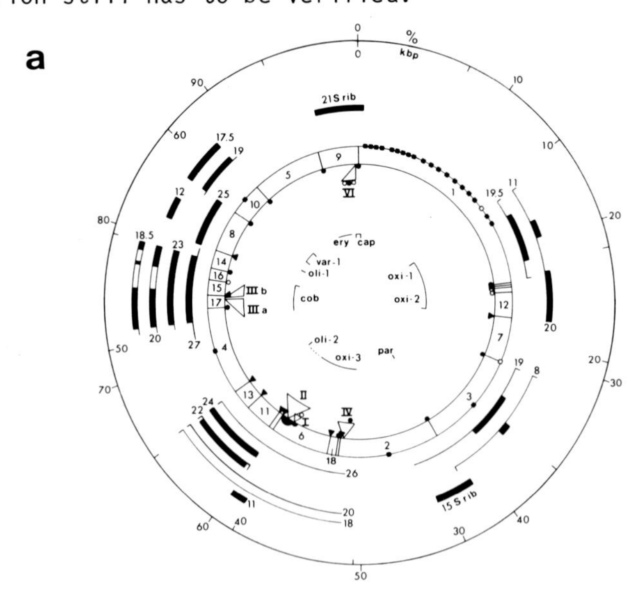

fig. 1

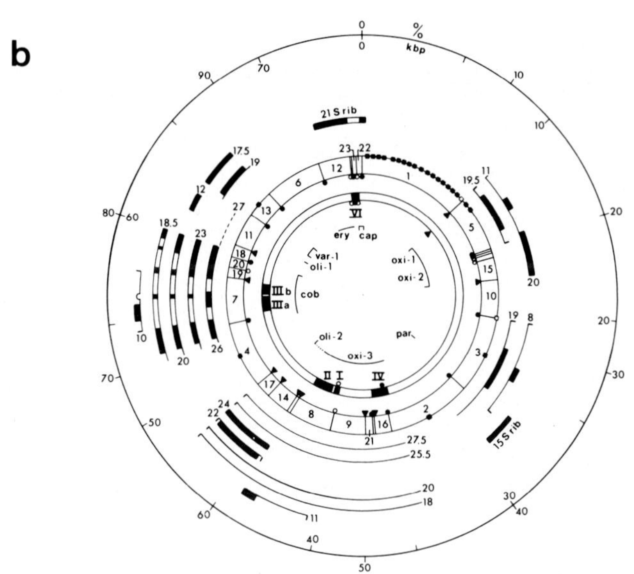

FIGURE 1. Transcription maps of the mtDNA of *S. carlsbergensis*, NCYC 74(A) and *S. cerevisiae* KL 14-4A (B). The inner ring gives the position of genetic markers, determined by co-retention analysis in petite mutants (5) for KL 14-4A and extrapolated to *S. carlsbergensis*. The physical maps were constructed by Sanders et al. (1) and show recognition sites for HindII (●), HindIII (O) and EcoR1 (▽). I, II, III etc denote inserted sequences present in KL 14-4A, but absent from *S. carlsbergensis*. On the outer ring, (■) and (□) give the approximate positions of 4S RNA and tRNA met genes respectively (2). The positions of other transcripts are shown outside the outer ring (3,4), the thin lines indicating uncertainty in the exact location. Open blocks represent sequences removed in a cut-and-splice process. Lengths of transcripts are indicated by an approximate electrophoretic S value, calculated in relation to 21 and 15S rRNAs.

b. Transcripts are generally much longer than minimally necessary to specify the known protein product of the locus concerned, e.g. 20S (approx. 2600 nucl) for subunit III of cytochrome *c* oxidase (mol. wt 20 000); 18.5S (2200 nucl) for apo-cytochrome *b* (mol. wt 30 000); 12S (850 nucl) for subunit 9 of the ATPase complex (mol. wt 8000). We know for the latter two cases, that most of the extra sequences are present in long, possibly untranslated leaders (see below).

c. Several loci, which in KL 14-4A mtDNA contain, or are close to major insertions, are transcribed into long RNA species (>25S). These RNAs contain transcripts of the insertions. The same loci also give rise to smaller transcripts

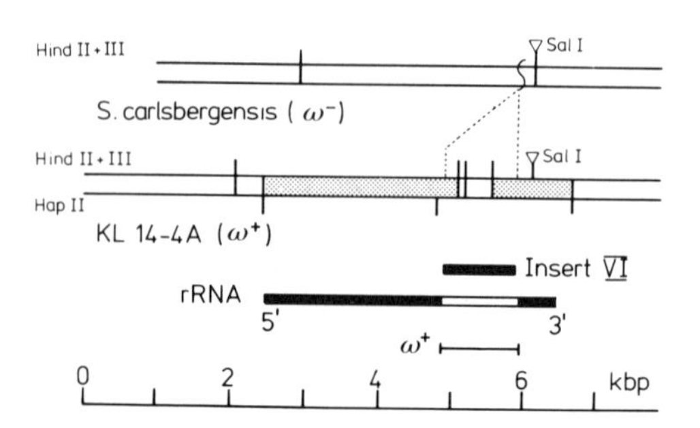

FIGURE 2. Map of the 21S rRNA gene in ω^+ and ω^- mtDNAs. Shaded fragments are those which hybridize to 21S rRNA (7). The orientation of the 21S rRNA gene was determined by hybridization to the separated strands of the 290 bp Hap X HindIII fragment to the left of the intervening sequence (see text).

(generally <21S), which are of corresponding size in both *S. carlsbergensis* and KL 14-4A and which lack the inserted sequences. All transcripts of KL 14-4A have their counterparts in *S. carlsbergensis*, with one notable exception: a 10S RNA is mapped in insert III; this RNA and the sequences coding for it are absent from *S. carlsbergensis*. Its function is unknown.

d. Multiple transcripts, showing overlapping hybridization, are found at several other loci. Their significance will be discussed later, in relation to the organization of their respective genes.

2. SPLIT GENES AND SPLICED TRANSCRIPTS

At least two genes on yeast mtDNA are split in a similar fashion to various genes in eukaryotic nuclear DNA. The first gene for which unambiguous evidence for this type of sequence organization was obtained is that for the 21S rRNA of the large subunit of the mitochondrial ribosome (7). This RNA hybridizes to non-contiguous fragments on the restriction map of ω^+ mtDNAs (Fig. 2) and the presence of an intervening sequence was confirmed directly by electron microscopy of the hybrids between 21S RNA and appropriate restriction fragments. The intervening sequence corresponds exactly in position to insert VI of Sanders et al. (1) and is 1100 bp long. This insert is closely linked (or identical) to the ω^+ allele of the mitochondrial polarity locus and is about 500 bp from one end of the 21S rRNA gene. From experiments in which the direction of transcription of this gene has been found to be clockwise on the physical maps shown in Figs 1a and 1b (Osinga, K. and Bos, J.L. unpublished), we identify this as the 3'-end. The insertions in other split rRNA genes, including those of chloroplasts, are also located near to the 3'-end (8-10) and this might imply that the intervening sequences of these diverse rRNA genes are at functionally analogous sites. DNA sequence analysis of the intervening sequence is largely complete (J.L. Bos, unpublished). Contrary to the predictions made by Sanders et al. (11), the region is relatively GC-rich (27 mole present GC). Work is now in progress to identify boundaries between the rRNA gene and intervening sequence.

An intriguing question concerns the organization of the 21S rRNA gene in ω^- strains, which lack insert VI. Electron microscopy of 21S rRNA, hybridized to ω^- mtDNA, shows hybrids base paired over their whole length (12) and the simplest explanation would be that the organization of the gene is identical in ω^+ and ω^- strains, but that in the latter case, the intervening sequence is too small to be detectable by electron microscopy. This explanation is, however, ruled out by the finding that S1 nuclease treatment (13) of hybrids between ω^- mtDNA and its homologous 21S rRNA produces only fragments with a length corresponding to that of intact 21S rRNA (J.L. Bos and N.B. Hecht, unpublished). We conclude that the 21S gene in these strains is continuous, with the consequence that the process of rRNA maturation must differ in ω^+ and ω^- strains. A situation analogous to this has been reported in *Tetrahymena* (10), in which the gene for 25S cell sap rRNA is split in some strains, but not in others. Both forms of the gene are transcribed. In *Drosophila*, 28S rRNA genes of the same organism differ by the presence or absence of an intervening sequence (8), but in this case, it is not known which is transcriptionally active.

S1 nuclease mapping has also shown that the gene for 15S rRNA is continuous (N.B. Hecht and H.F. Tabak, unpublished). This observation, together with the difference between the 21S rRNA gene of ω^+ and ω^- strains, makes it difficult to see what advantage (if any) can be derived from the possession of an intervening sequence.

By analogy with the transcription of split genes in eukaryotic nuclear DNA, it might be expected that the 21S rRNA of ω^+ strains is synthesized in the form of a precursor containing a transcript of the intervening sequence. The existence of such a precursor is suggested by study of the RNA species synthesized by isolated KL 14-4A mitochondria (G.S.P. Groot, G.J.B. van Ommen and N. van Harten-Loosbroek, unpublished). Isolated mitochondria synthesize a range of RNAs, which resemble RNAs synthesized in vivo in both their migration and hybridization behaviour. The putative precursor can be de-

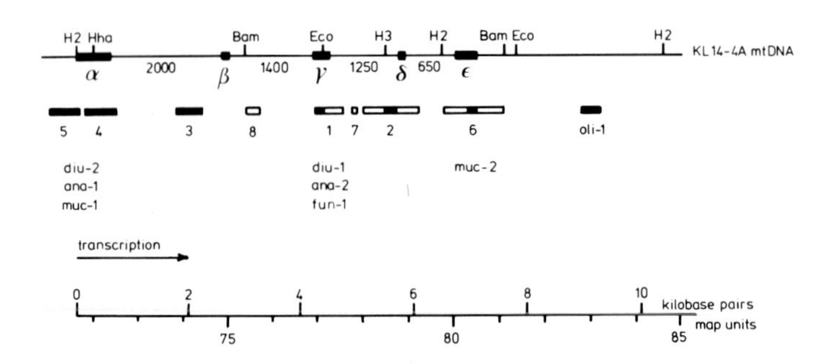

FIGURE 3. The structure of the cytochrome *b* gene. The positions of the loci box 1-6 are taken from Slonimski et al. (14). The positions of box 7 and 8 are inferred from genetic mapping (ref. 15 and R.J. Schweyen, personal communication). Heavy bars indicate the approximate positions of sequences specifying 18.5S RNA, deduced from electron microscopy of hybrids between *S. carlsbergensis* 18.5S RNA and KL 14-4A mtDNA (see also Plate I).

tected by its hybridization to DNA fragments derived from the intervening sequence. It has a length of about 5000 nucleotides - long enough to accommodate a complete transcript of the intervening sequence. Possibly a shortage of cytoplasmic components required for ribosome biosynthesis is the cause of its accumulation in this experimental system.

The Gene for Apo-Cytochrome *b*: The coding sequence for this protein shows a remarkably complex organization in being interrupted by four intervening sequences between 650 and 2000 bp long. This conclusion, together with the detailed picture of the gene and its transcripts shown in Fig. 3 is based on the following evidence:

a. Van Ommen et al. (3,4) have identified an 18.5S RNA as the major transcript of the cytochrome *b* region. This RNA is present in both *S. carlsbergensis* and KL 14-4A, hybridizes to non-contiguous fragments and extends over a large span of mtDNA. In KL 14-4A mtDNA, insert III is flanked by fragments which hybridize (see Fig. 1b).

b. In vitro translation experiments confirm that the 18.5 S RNA codes for cytochrome *b*. First, as shown by Fig. 4A, prehybridization of total mtRNA with mtDNA from the petite A15-1, which retains all genetic markers in the *cob* region, effectively suppresses the synthesis of antigenic determinants of apo-cytochrome *b*, while synthesis of other mitochondrial products is unaffected. Second, as shown in Fig. 4B and C the

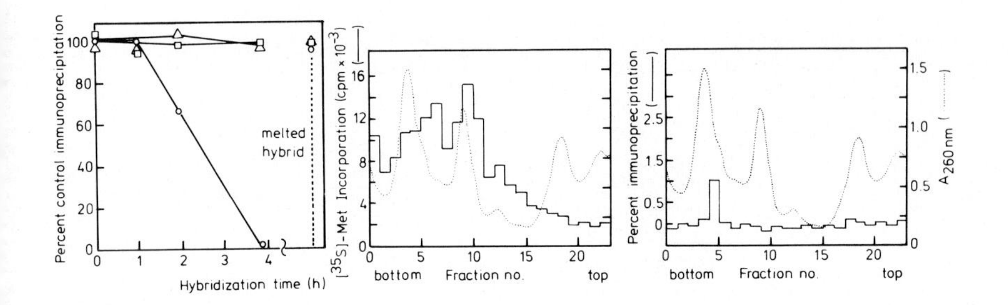

FIGURE 4. Identification of the mRNA for cytochrome *b*. (A) *Hybrid-arrested translation (16)*. Samples of 80 μg total mtRNA were translated in a cell-free system from *E. coli* (6), after hybridization at 43°C for the times shown with 4 μg mtDNA from the petite A15-1, which retains all genetic markers in the *cob* region. In vitro products were then submitted to immunoprecipitation. Control values in immunoprecipitation were 0.8%, 8.0% and 0.7% for anti-cytochrome *b* (O), anti-cytochrome *c* oxidase (□) and anti-ATPase (△). All values were corrected for precipitation with pre-immune serum (1.2%). (B and C). *Sucrose gradient analysis*. 180 μg mtRNA were fractionated by sedimentation through isokinetic sucrose gradients and RNA in each fraction was translated in a cell-free system from *E. coli* (6). (B) shows the A260 profile and total incorporation of 35S-methionine. (C) Products immunoprecipitated with anti-cytochrome *b*.

sole RNA capable of directing the synthesis in vitro of antigenic determinants of *b* sediments at approximately 18-20S.

c. Electron microscopy of hybrids between selected restriction fragments from the *cob* region with 18.5S RNA confirms the presence of intervening sequences (Plate I) and permits location of these on the restriction map of the *cob* region (Fig. 3). Five coding regions (α-ε), with lengths of 600, 110, 200, 80 and 400 bp are interspersed with intervening sequences of 2000, 1400, 1250 and 650 bp, respectively. Together, these coding regions account for approx. 1400 nucleotides of the 2200 nucleotide long 18.5S RNA. Preliminary data, based on a limited number of hybrid molecules, indicates that the remainder of the RNA, containing the 5'-terminus (see below), is specified by sequences extending to the left of Fig. 3 and that these are interrupted by two further intervening sequences (A.C. Arnberg, unpublished). A striking feature of this map is that it shows that insert III is in fact two inserts 2000 and 1400 bp long, separated by the coding segment of 110 bp. The hybrids shown in Plate I are

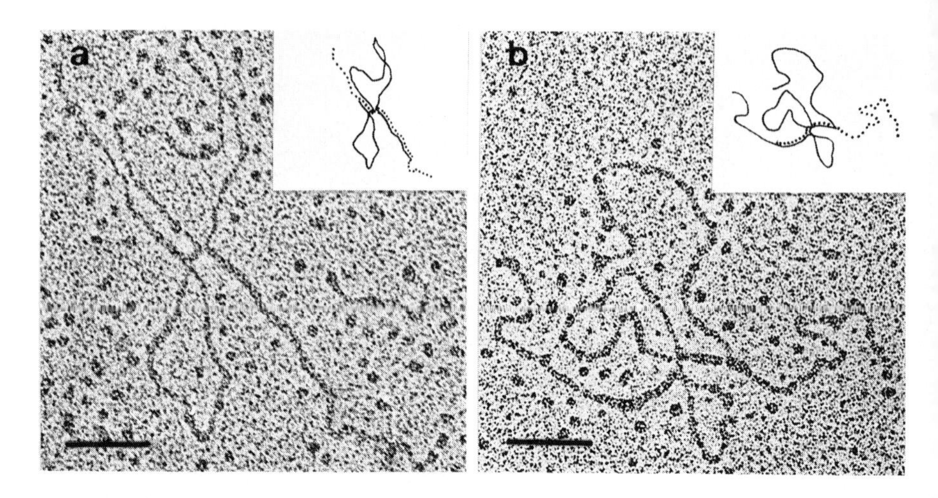

PLATE I. Electron micrographs of DNA-RNA hybrids between *S. carlsbergensis* 18.5S RNA and either (A) KL 14-4A Eco x HindII + III fragment 7 (4400 bp), or (B) petite A15-1 Bam fragment 1 (4600 bp; identical with KL 14-4A BB5, see ref. 1 and fig. 3). Hybidization was for 2 h at 40°C under conditions described previously (7). In the inserts, RNA is shown as a stippled line; DNA is continuous. The bar is 0.1 μm.

those formed between 18.5S RNA from *S. carlsbergensis* and mtDNA from Kl 14-4A, or KL 14-4A-derived petite strains. We therefore do not know at this stage whether mtDNAs which lack insert III also lack intervening sequences at this position, or whether such interruptions are merely much shorter.

A mosaic organization of the cytochrome *b* gene was predicted from the genetic properties of cytochrome *b* deficient and inhibitor resistant mutants (14,15). The electron microscopic data confirm this model. Alignment of EM and genetic maps show that the DNA sequences (α-ε) which specify the 18.5S RNA correspond in position to box loci thought to be situated in exons of the cytochrome *b* gene (14,15). α, γ, δ and ε correspond to box 4 + 5, 1, 2-1 and 6, respectively, while β would correspond to the recently discovered box 8 locus (R.J. Schweyen, personal communication). The intervening sequences of 2000 and 1250 bp extend across and could be identical with the box loci 3 and 7. No mutants have been found in the 1400 and 650 bp intervening sequences and these probably correspond to the silent interruptions postulated by Haid et al. (15) on the basis of the high intralocus recombination frequency between groups of mutants within their clusters C and E.

Mutations in the loci 4 + 5, 8, 1, 2-1 and 6 lead to the accumulation of novel polypeptides of characteristic molecular weight. These polypeptides are often shorter than the wild type apo-cytochrome *b* and decrease in length in the order box 6 > box 2-1 > box 1 > box 8 > box 4 + 5. The suggestion has been made (14,15) that these novel products represent premature termination fragments of apo-cytochrome *b* and thus that the direction of transcription-translation must be from box 4 + 5 towards box 6 (i.e. α to ε). This hypothesis is confirmed by the recent finding that these polypeptides cross-react with antisera mono-specific for cytochrome *b* (H. Bechmann, P.H. Boer, F.J. van Hemert and J. Kreike, unpublished). Assuming that in the majority of cases the lesions are nonsense mutations, or small deletions leading to early termination within ± 60 bp of the mutation as a result of frameshift (the expectation for a random occurrence of 3 termination triplets in every 64), it is clear that the order α, β, γ, δ, ε also represents the direction of transcription and translation. It is also possible to calculate that these segments should have lengths of about 450 bp (α), 30 bp (β), 120 bp (γ), 30 bp (δ) and 250 bp (ε). With the exception of α and ε, values predicted fit reasonably well with lengths determined by electron microscopy. Both α and ε are longer than predicted. It is therefore possible that ε contains an untranslated 3'-section about 150 bp long and that α contains not only sequences specifying the amino terminus of apo-cytochrome *b*, but also part of the 5'-leader sequence.

Consistent with a cut-and-splice model for mRNA maturation in the *cob* region, Van Ommen et al. (4) have been able to identify a number of RNAs, longer than 18.5S and present in low concentrations, which also hybridize to restriction fragments in the *cob* region (Figs 1a and 1b). All these species hybridize to a greater extent with one or more of the intervening sequences. The size of these RNAs and their hybridization behaviour are consistent with a major processing pathway in which the various inserts are successively removed, beginning at the 5'-end. However, due to the complexity of the hybridization patterns in the high molecular weight region of the gel and the difficulty of accurate molecular weight determination in this range, other pathways are not excluded.

The simplest explanation for the mutations in box 3 and 7, which lie in the intervening sequences of 2000 and 1250 bp respectively would be that these are defective in splicing. If so, it is somewhat surprising that mutants in the remaining intervening sequences have not been found. It may indicate that the splicing process in the former two regions is more critical, or proceeds by a different mechanism. The finding that the mutations in box 3 and 7 fall into separate complementation groups (14,15) points in the latter direction. An attractive hypothesis which accounts for the complementation

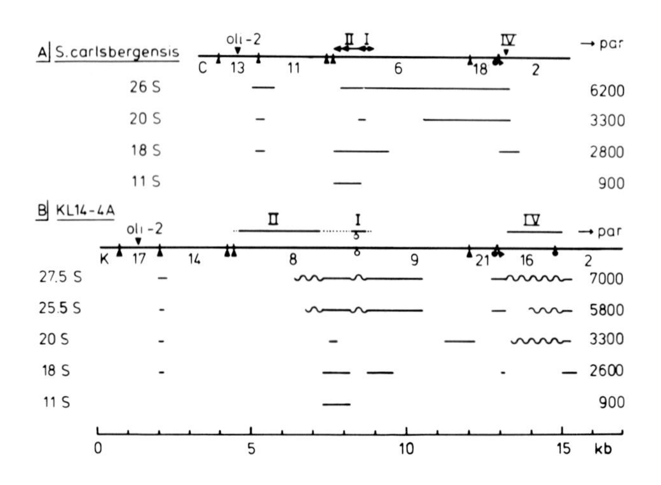

FIGURE 5. Transcripts of the oxi-3 region. Adapted from ref. 4. Identification of fragments is as in Fig. 1. (——) Hybridization to sequences present in both *S. carlsbergensis* and KL 14-4A. (∿∿∿) Hybridization to inserted sequences. See text for details.

data, is that splicing at these sites is bimolecular and makes use of an adaptor RNA (see ref. 17) for discussion.

Oxi-3: The *oxi-3* region covers an inordinately long stretch of mtDNA and is thought to contain the structural gene for subunit I of cytochrome *c* oxidase. KL 14-4A mtDNA contains three major insertions in this region, which are absent from *S. carlsbergensis* mtDNA. We have previously suggested (17,18) that this gene may be split in an analogous fashion to the cytochrome *b* and 21S rRNA genes and this is supported by detailed examination of the complex hybridization behaviour of the many stable transcripts of this region. Fig. 5 makes attempts to rationalize our findings. It is based on an assumption of a minumum number of transcripts, arranged to account most simply for the results of hybridization in both homologous and heterologous combinations of mtDNA and RNA, with a minimum number of splicing events. The result is remarkably informative in three respects. It shows that the longer RNAs in KL 14-4A contain transcripts of the inserted sequences. It suggests that the shorter, overlapping RNAs could arise by differential splicing events. It makes clear predictions as to the location of RNA-coding sequences. These predictions are open to direct experimental verification. Two puzzling aspects remain: first, there is no obvious explanation for the apparent linkage of transcription of this region

```
AATTATATATAATATATTATATATAATTATATATATATATATAAATAATAATAAATATATATATAATATAAAAATAAGAATAGATTAAATATTTAA
     -100        -90        -80        -70        -60        -50        -40        -30        -20

              f-met gln leu val leu ala ala lys tyr ile gly ala gly ile ser thr ile gly leu leu gly
TAAATAAATATT [ATG] CAA TTA GTA TTA GCA GCT AAA TAT ATT GGA GCA GGT ATC TCA ACA ATT GGT TTA TTA GGA
-10           0          10          20          30          40          50          60

ala gly ile gly ile ala ile val phe ala ala leu ile asn gly val ser arg asn pro ser ile lys asp
GCA GGT ATT GGT ATT GCT ATC GTA TTC GCA GCT TTA ATT AAT GGT GTA TCA AGA AAC CCA TCA ATT AAA GAC
         70          80          90         100         110         120         130

leu val phe pro met ala ile phe gly phe ala leu ser glu ala thr gly leu phe cys leu met val ser
CTA GTA TTC CCT ATG GCT ATT TTT GGT TTC GCC TTA TCA GAA GCT ACA GGT TTA TTC TGT TTA ATG GTT TCA
    140         150         160         170         180         190         200

                                  stop        stop
phe leu leu leu phe gly val
TTC TTA TTA TTA TTC GGT GTA [TAA] TAT ATA [TAA] TATATTATAAATAAATAAAAAATAATGAAATTAATAAAAAAAAAAATAAAAT
 210         220         230         240        250        260        270        280        290

AAAACCAGTT
       300
```

FIGURE 6. Nucleotide sequence of one strand of the gene for ATPase subunit 9 and the derived amino acid sequence of this protein. Singly underlined sequences: Recognition sites for Alu I. There is no doubly underlined residue. See ref.21.

with that of the *oli-2* locus. Second, if the various transcripts of intermediate length represent intermediates in a splicing pathway, it is not clear why these are so stable and why these should be present in concentrations comparable to the 'mature' 18S and 11S species. Any model which ascribes functional differences to the various, differentially spliced transcripts requires that more genes be present in *oxi-3* than are currently known to exist (see ref. 17 for discussion).

3. SEQUENCE ANALYSIS OF mtDNA

a) <u>The Anatomy of the Structural Gene for ATPase Subunit 9</u>. The structural gene for ATPase subunit 9 contains the genetic locus *oli-1* and possibly also *oli-3*, both conferring resistance of the mitochondrial ATPase to oligomycin. Wachter et al. (19,20) have determined the complete amino acid sequence of this subunit, a protein of molecular weight 8000, and more recently, Hensgens et al. (21) have determined the nucleotide sequence of the structural gene and its surrounding regions (Fig. 6). This information, coupled with the knowledge that the major transcript of the gene is an RNA sedimenting at 12S in sucrose gradients, makes this gene an ideal object of analysis in the study of the organization and expression of a mitochondrial gene. Unusual features of organization revealed by DNA sequence analysis and fine structure mapping of the 12S transcript include:

1. Both the sequences in the structural gene and those specifying the 12S transcript are continuous, since not only do DNA and protein sequences correspond exactly, but S1 nuclease treatment (13) of hybrids between 12S RNA and mtDNA yields protected fragments with a length corresponding to the full length of the RNA (A. Sonnenberg, unpublished).

2. The gene is surrounded by quasi-repetitive, highly AT-rich sequences. On both sides of the gene, the nucleotide sequence contains less than 4% G + C for at least 250 bp. All of the AT-rich sequence preceding the gene is present in 12S RNA (see below).

3. Codon usage within the gene is highly non-random. Only 27 of the 61 amino acid specifying codons are used, so that it is still possible that the mitochondrion economizes on tRNA genes by not using all codons. When a codon choice is possible, an AT-rich codon is preferred and the result of this bias is reflected in the overall G + C content of the gene, which at 33% is close to the theoretical minimum of 29% for this amino acid sequence. Without selection for AT-rich codons, the mole percent GC would be 48%. Clearly, the low GC content of yeast mtDNA is not restricted to non-coding 'spacer' regions (cf. Ref. 22) and the low G + C content of rRNAs and tRNAs (23,24), but is also a feature of protein-coding sequences, perhaps as a result of selective pressure.

4. Hybridization of restriction fragments containing 5'- and 3'-portions of the structural gene has permitted its orientation on the wild type genome to be determined. Like all genes for which data are so far available, the direction of transcription and translation is clockwise on the maps in Figs 1a and 1b. This information permits interpretation of electron micrographs of hybrids between 12S RNA and mtDNA from the petite RP6, a strain which retains the complete structural gene for subunit 9, and which was used for DNA sequence analysis (Plate II and Fig. 7). From these can be seen that the 3'-end of the RNA is located about 70 bp from the end of the structural gene and that the gene is approximately 500 nucleotides from the 5'-end of the RNA. A long (untranslated) 5'-leader sequence as also observed for the 18.5S cytochrome *b* mRNA may thus turn out to be general feature of mitochondrial mRNAs. Why such a long leader should be necessary is of course unknown. However, in the case of 12S

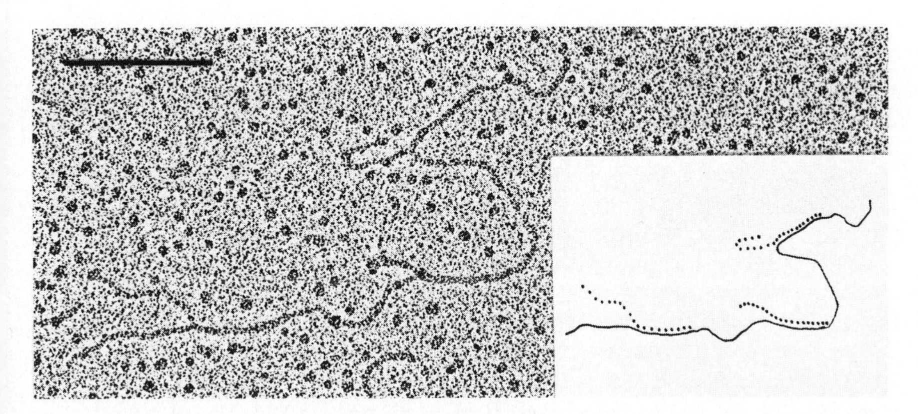

PLATE II. Electron micrograph of a DNA-RNA hybrid between *S. carlsbergensis* 12S RNA and a 3075 bp fragment of RP6 mtDNA produced by partial digestion with HapII (see Fig. 7). Hybridization conditions were as described in the legend to plate I. The bar is 0.2 μm.

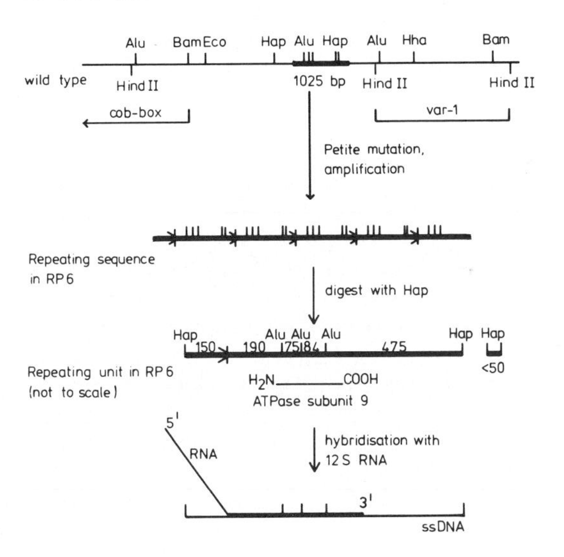

FIGURE 7. Structure and transcription of petite RP6 mtDNA. RP6 is an *oli-1* petite mutant containing a single 1025 bp fragment of wild type mtDNA. The figure shows the position of the RP6 sequence on mtDNA (thickened bar) and the physical map of the fragment retained after petite mutation and amplification. For the sake of clarity, only a limited number of the restriction sites in wild type mtDNA are shown. The position of the ATPase subunit 9 gene is deduced from DNA sequence analysis. The lower part of the figure shows a hybrid between 12S RNA and RP6 mtDNA digested with HapII. See also plate II. From ref. 21 and unpublished.

RNA, not all the leader sequence is essential for function, since the homologous RNAs from *S. carlsbergensis* and KL 14-4A differ in size by about 50 nucleotides due to the presence of an insertion in *S. carlsbergensis*, situated 120-190 bp in front of the gene (A. Sonnenberg, unpublished).

5. We have scanned the sequences preceding the structural gene for possible control elements. Apart from a 14 nucleotide palindrome, following two lonely G's at positions -31 and -26, there is nothing immediately recognizable as a ribosome binding site and there are no binding sites for bacterial ribosomes within 100 nucleotides upstream of the gene. At this stage, we cannot exclude the possibility that mitochondrial ribosomes initiate in a fashion similar to eukaryotic ribosomes (25) by binding close to the 5'-end and sliding down the intervening 500 nucleotides to the structural gene.

6. An unresolved point is the question of whether the 12S RNA is in fact the primary transcript of the subunit 9 structural gene. Larger transcripts of the region are present in low concentrations and can be detected both by hybridization of labelled RP6 mtDNA to mtRNA fractionated and transferred to DBM-paper (26) (data not shown) and by the standard transcript mapping technique. These larger transcripts must extend in the direction of the cytochrome *b* gene, since no hybridization with the *var-1* gene fragment is detectable. It remains to be seen whether such transcripts are precursors, or whether they result from readthrough, or the action of a weak promotor in *cob*.

b) Two Genes for Mitochondrial tRNAs: The mitochondrial tRNAs form a group of transcripts with a number of interesting facets. The mitochondrial translational system may economize on tRNA genes by not using all codons, and it is also possible that the system uses fewer tRNAs to recognize all codons than would be predicted from normal wobble-pairing rules (see ref. 17 for discussion). Further, mtRNAs contain a low mole percent G + C and fewer than average modified bases (24), so that it is of interest to ask how this is compatible with normal function. Finally, genes for mitochondrial tRNAs are scattered over the mitochondrial genome (2). The mechanism of their transcription and its regulation, if any, are as yet unanswered questions.

We hope that DNA sequence analysis of mitochondrial tRNA genes will provide information on some of these points and have begun a study of those genes forming a small cluster between 12 'o' clock and 1 'o' clock' on the map. Physical mapping (2) has shown that 3-4 genes are located here and hybridization studies with petites have identified some of these as genes for $tRNA_{thr}$, $tRNA_{cys}$ and $tRNA_{his}$ (27). A search for

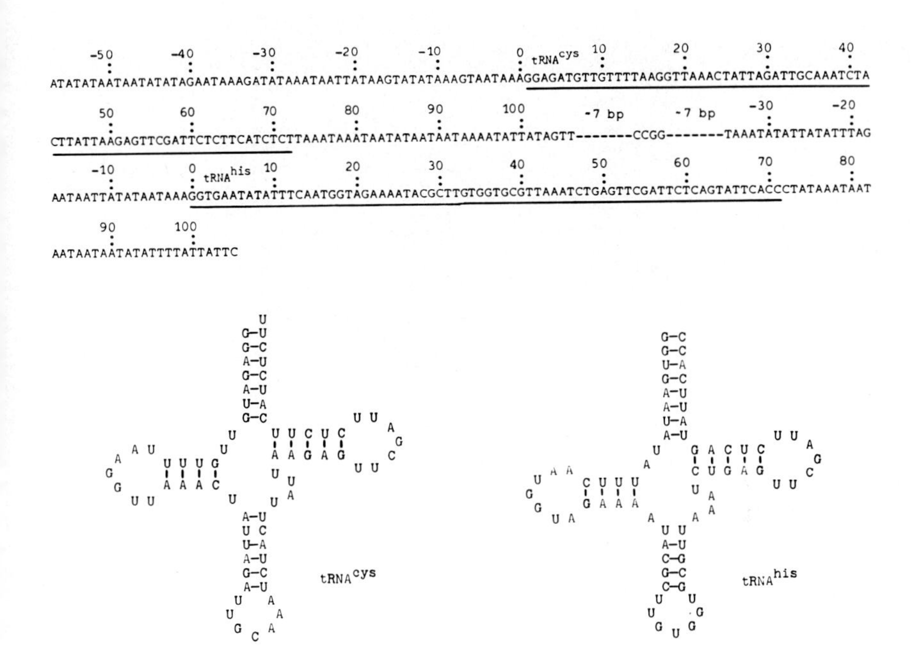

FIGURE 8. DNA sequence of the genes for $tRNA_{cys}$ and $tRNA_{his}$ and deduced structures. See text for explanation.

these genes on restriction fragments of this region has resulted in the sequences of $tRNA_{cys}$ and $tRNA_{his}$ shown in Fig. 8 (ref. 28). The genes are about 80 bp apart, are in the order predicted by petite mapping and are set in highly AT-rich surroundings. Fig. 8 also shows the clover-leaf structures possible when maximum base-pairing is assumed. From these, it would seem that $tRNA_{his}$ has a fairly orthodox structure, in possessing a number of the features of other tRNAs both pro- and eukaryotic (29), although a slightly unusual feature is the presence of a U-U pair at the top of the anticodon stem.

The deduced structure of $tRNA_{cys}$ displays a number of interesting features. Unusual, but not unique, is the replacement by a U residue of the normally invariant A_{21} (also seen in *E. coli* $tRNA_{cys}$; nomenclature according to ref. 30) and the absence of a normal Watson-Crick base-pair at the top of the aminoacyl stem. Unique is the replacement by a U of the invariant A_{14}, which in other tRNAs pairs with U8. The most unusual feature is, however, the length of the anticodon stem, which if maximum base-pairing is realised will consist of 6 bp, instead of the normal 5 bp. This extra base-pair alters the normal configuration of the anticodon arm and a tRNA possess-

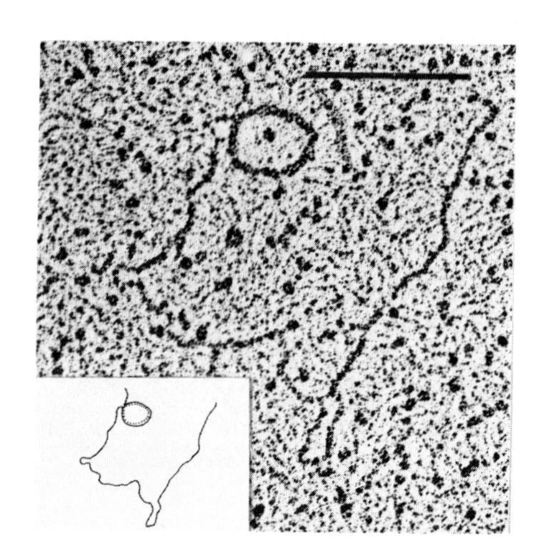

PLATE III. Electron micrograph of a DNA-RNA hybrid between circular *S. carlsbergensis* 11S RNA and KL 14-4A Eco x HindII+III fragment 8 (4000 bp; see fig. 1b). See legend to plate I for details of hybridization. The bar is 0.2 μm.

ing it is unlikely to be active in protein synthesis. It is of course possible that the structure proposed is incorrect: rearrangements within the D-arm to give a less base-paired structure could eliminate the superfluous base-pair. Alternatively, this feature could be compensated for by other unusual features of the molecule, or by the mitochondrial ribosome. Finally, we cannot exclude that this $tRNA_{cys}$ is involved in a function other than protein synthesis: hybridization data show that there are two cistrons for distinct cysteinyl tRNAs on yeast mtDNA and that both are transcribed (31).

The direction of transcription of both genes is clockwise on the physical map. Striking features of the sequences preceding each are that the sequence -28 to -18 and -6 to 0 for $tRNA_{cys}$ is identical to the sequence from -18 to -1 for $tRNA_{his}$ and that each contains a six base pair true palindrome. Clearly it will be of the greatest interest to verify whether these tRNAs are synthesized as precursors containing these sequences and to pinpoint possible promotor sites for these genes.

4. POST-TRANSCRIPTIONAL MODIFICATION OF mtRNA; ARE SOME MITOCHONDRIAL mRNAs CIRCULAR?

Our knowledge of post-transcriptional modifications undergone by mitochondrial RNA is sketchy and has been summarized in a recent review (17). However, evidence for what may be one of the most bizarre types of modification was found during

EM analysis of mtDNA-RNA hybrids (A.C. Arnberg, unpublished). Some mtRNA preparations contain a high number of circular molecules, with a contour length roughly equal to that of the linear RNA molecules. The circles are single-stranded, devoid of secondary structure and seldom possess anything resembling a panhandle structure containing the annealed ends of the RNA. They are present in RNA fractions spread directly after extraction from gels, even when conditions which might permit annealing of the ends are avoided. They are not eliminated by DNase or pronase treatment of the RNA. The circles are so stable that they can participate as such in DNA-RNA hybrids, or form R-loops (A.C. Arnberg, J.E. Bollen-de Boer and H.W.A. J. Garretsen, unpublished). In Plate III, an example of such a hybrid is shown, in which the RNA - an 11S transcript of the *oxi-3* region - hybridizes over its whole length with the DNA (fragment 8 in Fig. 1b) and gives rise to a circular DNA-RNA hybrid. Evidently the association between the RNA ends is so strong that this configuration is preferred over a linear hybrid. This strongly suggested to us that the ends of the RNA might be covalently joined and this belief was strengthened by the fact that the circles are not eliminated by a number of denaturing conditions including boiling, glyoxal treatment and spreading in the presence of 2.8 M urea plus 70% formamide.

These facts are most easily explained by the assumption that some mtRNAs are circular as a result of a post-transcriptional covalent linkage of the ends. The RNAs showing this phenomenon and studied in most detail are the 11S and 18S RNAs from the *oxi-3* region and the 18.5 S of the *cob* region. 12S RNA preparations show a lower proportion of circles which most probably represent contaminating 11S molecules. Circles have not been seen in preparations of 21S rRNA. At this stage, we cannot rule out the possibility that the circular RNAs are not active as mRNAs, but represent storage forms, or arise as intermediates (or errors) in a cut-and-splice process. Even so, they constitute a biological curiosity, since circular RNAs have only been found previously as genomes of viroids (32).

5. HOW IS GENE EXPRESSION REGULATED IN YEAST MITOCHONDRIA?

In HeLa cells, both strands of mtDNA are transcribed completely and the resulting transcripts are rapidly processed to yield rRNAs, tRNAs and mRNAs, whose relative concentrations must be controlled by processing (33). In yeast, there is little evidence that symmetric transcription occurs, although it cannot be formally ruled out (17). On the other hand, it is becoming increasingly clear that RNA processing plays a central role in the regulation of expression of mitochondrial genes.

a. At least two and probably more mitochondrial genes contain intervening sequences. In the case of the genes for cytochrome *b* and 21S rRNA, there is reasonable evidence that biologically active transcripts of these genes are created by a cut-and-splice mechanism. In the *oxi-3* region, alternative splicing may regulate more functions than simply the synthesis of cytochrome oxidase subunit I. These findings also imply that mitochondria must contain a complete set of the necessary processing enzymes, which are likely to be distinct from those employed in the nucleus.

b. Long transcripts originate from many regions of the transcription map, either overlapping shorter transcripts in apparent simple precursors-product relationship (e.g. *var-1*, 15S rRNA), or more obscurely, even extending over different loci. This is most clearly seen in Figs 1a and 1b for the *oli-2* and *oxi-3* loci, but in fact, there are few real gaps in the transcription map and high molecular weight transcripts hybridize weakly to all DNA fragments between loci. Studies with isolated mitochondria show that the transcripts of these 'poorly transcribed' regions have a high turnover. A plausible explanation would be that control of the start and stop of transcription is sloppy and that a backstop, or proofreading function is provided by a processing system which produces active, gene-specific transcripts. It is worth noting that the orientation of all genes studied to date - 21S rRNA, $tRNA_{cys}$, $tRNA_{his}$, subunit II of cytochrome *c* oxidase (B. Weiss-Brummer, personal communication) cytochrome *b*, ATPase subunit 9 - is the same. If this holds for the other genes, a single round of transcription would suffice for coordinate expression. Promotor/attenuator control and processing could then regulate the steady state levels of the various transcripts.

c. Finally, the finding that some of the putative mRNAs of yeast mitochondria are circular adds yet another step to the various post-transcriptional modifications undergone by mtRNA and raises questions as to the nature and function of this circularity. Clearly, the study of mtDNA is in an extremely exciting phase and may be expected to remain so for some time to come.

ACKNOWLEDGEMENTS

Our thanks are due to P.M.M. Enthoven, K.A. Osinga, E. Roosendaal and A. Sonnenberg for allowing us to quote unpublished data, to M. de Haan, N. van Harten-Loosbroek, G. van der Horst and H.H. Menke for technical assistance and to Ms. Ans Brouwer-Honselaar for typing the manuscript.

REFERENCES

1. Sanders, J.P.M., Heyting, C. Verbeet, M.Ph., Meijlink, F.C.P.W., and Borst, P. (1977). Mol. Gen. Genet. 157, 239-261.
2. Van Ommen, G.J.B., Groot, G.S.P., and Borst, P. (1977). Mol. Gen. Genet. 154, 255-262.
3. Van Ommen, G.J.B., and Groot, G.S.P. (1977). In 'Mitochondria 1977', Genetics and Biogenesis of Mitochondria (W. Bandlow, R.J. Schweyen, K. Wolf and F. Kaudewitz, eds.), pp. 415-424. De Gruyter, Berlin.
4. Van Ommen, G.J.B., Groot, G.S.P., and Grivell, L.A. (1979). Cell, submitted.
5. Schweyen, R.J., Weiss-Brummer, B., Backhaus, B., and Kaudewitz, F. (1978). Mol. Gen. Genet. 159, 151-160.
6. Moorman, A.F.M., Van Ommen, G.J.B., and Grivell, L.A. (1978). Mol. Gen. Genet. 160, 13-24.
7. Bos, J.L., Heyting, C., Borst, P., Arnberg, A.C., and Van Bruggen, E.F.J. (1978). Nature 275, 336-338.
8. Glover, D.M., and Hogness, D.S. (1977). Cell 10, 167-176.
9. Rochaix, J.D., and Malnoe, P. (1978). Cell 16, 661-670.
10. Wild, M.A., and Gall, J.G. (1979) Cell, in press.
11. Sanders, J.P.M., and Borst, P. (1977) Mol. Gen. Genet. 157, 263-269.
12. Faye, G., Dennebouy, N., Kujawa, C., and Jacq, C. (1979) Mol. Gen. Genet. 168, 101-109.
13. Berk, A.J., and Sharp, P.A. (1977) Cell 12, 721-732.
14. Slonimski, P.P., Claisse, M.L., Foucher, M., Jacq, C., Kochko, A., Lamouroux, A., Pajot, P., Perrodin, G., Spyridakis, A., and Wambier-Kluppel, M.L. (1978). In 'Biochemistry and Genetics of Yeast' (M. Bacilla, B.L. Horecker and A.O.M. Stoppani, eds.), in press, Academic Press, New York.
15. Haid, A., Schweyen, R.J., Bechmann, H., Kaudewitz, F., Solioz, M., and Schatz, G. (1979). Eur. J. Biochem. in press.
16. Paterson, B.M., Roberts, B.E., and Kuff, E.L. (1977). Proc. Natl. Acad. Sci. U.S. 74, 4370-4374.
17. Borst, P., and Grivell, L.A. (1978) Cell 15, 705-723
18. Grivell, L.A., and Moorman, A.F.M. (1977). In 'Mitochondria 1977', Genetics and Biogenesis of Mitochondria' (W. Bandlow, R.J. Schweyen, K. Wolf and F. Kaudewitz, eds.), pp. 371-384, De Gruyter, Berlin.
19. Wachter, E., Sebald, W., and Tzagoloff, A. (1977). In 'Mitochondria 1977', Genetics and Biogenesis of Mitochondria' (W. Bandlow, R.J. Schweyen, K. Wolf and F. Kaudewitz, eds.), pp. 441-449, De Gruyter, Berlin.

20. Sebald, W., and Wachter, E. (1978). In 29th Mosbacher Colloquium on 'Energy Conservation in Biological Membranes (G. Schäfer and M. Klingenberg, eds.), in press, Springer-Verlag, Berlin.
21. Hensgens, L.A.M., Grivell, L.A., Borst, P., and Bos, J.L. (1979). Proc. Natl. Acad. Sci. U.S., in press.
22. Prunell, A., and Bernardi, G. (1977). J Mol. Biol. 110, 53-74.
23. Reijnders, L., Kleisen, C.M., Grivell, L.A., and Borst, P. (1972). Biochim. Biophys. Acta 272, 396-407.
24. Martin, R., Schneller, J.M., Stahl, A.J.C., and Dirheimer, G. (1976). Biochem. Biophys. Res. Commun. 70, 997-1002.
25. Kozak, M. (1978). Cell 15, 1109-1123.
26. Alwine, J.C., Kemp, D.J., and Stark, G.R. (1977). Proc. Natl. Acad. Sci. U.S. 74, 5350-5354.
27. Martin, N.C., Rabinowitz, M., and Fukuhara, H. (1977). Biochemistry 16, 4672-4677.
28. Bos, J.L., Osinga, K.A., and Van der Horst, G. (1979). Nucleic Acid Res., submitted.
29. Rich, A., and RajBhandary, U.L. (1976). Ann. Rev. Biochem. 45, 805-860.
30. Gauss, D.H., Grüter, F., and Sprinzl, M. (1979). Nucleic Acid Res. 6, r1-r19.
31. Rabinowitz, M., Jakovcic, S., Martin, N., Hendler, F., Halbreich, A., Lewin, A., and Morimoto, R. (1976). In 'The Genetic Function of Mitochondrial DNA' (C. Saccone and A.M. Kroon, eds.), pp. 219-230, North-Holland, Amsterdam
32. Sänger, H.L., Klotz, G., Riesner, D., Gross, H.J., and Kleinschmidt, A.K. (1976). Proc. Natl. Acad. Sci. U.S. 73, 3852-3856.
33. Amalric, F., Merkel, C., Gelfand, R., and Attardi, G. (1978). J. Mol. Biol. 118, 1-25.

THE EXPRESSION IN *SACCHAROMYCES CEREVISIAE* OF BACTERIAL ß-LACTAMASE AND OTHER ANTIBIOTIC RESISTANCE GENES INTEGRATED IN A 2-μm DNA VECTOR

Cornelis P. Hollenberg
Max-Planck-Institut für Biologie, Abt. Beermann
Spemannstr. 34, 7400 Tübingen, FRG

ABSTRACT We have studied the question whether bacterial genes can be functionally expressed in yeast in order to possibly use such genes as markers for 2-μm DNA in functional studies as well as in yeast transformation studies. Three genes that code for resistance to ampicillin, chloramphenicol or kanamycin and are present on a number of bacterial cloning vectors were tested. Our results show that the ampicillin resistance gene derived from pBR325 is expressed in S.cerevisiae. The product, a ß-lactamase activity, is able to degrade benzylpenicillin and nitrocefin. The presence of the enzyme can be demonstrated in cell-free extracts as well as in intact yeast cells. In non-denaturating polyacrylamide gels the ß-lactamase product that is synthesized in the yeast cell has a mobility slightly slower than the ß-lactamase active in E.coli. This suggests that the yeast cell is not able to process the primary gene product, a preprotein that in E.coli contains a signal peptide of 23 amino acids. Preliminary experiments suggest that bacterial genes coding for chloramphenicol or kanamycin resistance are expressed in yeast as well.

Yeast 2-μm DNA is a closed circular extra-chromosomal DNA element of which 50-100 copies are normally found in several strains of Saccharomyces cerevisiae (1-8) The molecule contains a non-tandem inverted duplication, approximately 600 base pairs long, that is involved in intramolecular recombination leading to the inversion of the enclosed unique DNA segments (2-8). No gene products of 2-μm DNA are known and no function has yet been established. When inserted in different bacterial plasmids and introduced into E.coli, the 2-μm DNA

ISBN 0-12-198780-9

promotes the synthesis of at least 4 discrete polypeptides (9,10). The DNA regions coding for these polypeptides have been mapped (10). In view of the functional expression of several yeast genes in E.coli (11-15), it is likely that some of these polypeptides represent products of genes located on the 2-μm DNA.

Recently, efficient transformation of yeast has been achieved (16-18) by using 2-μm DNA as a vector and the cloned yeast leu2 or his3 gene as a selective marker.

We have used this technique to analyse the expression of some bacterial antibiotic resistance genes in the yeast cell in order to study the following questions:

(1) Are prokaryotic genes functionally expressed in the yeast cell and can such genes be used as selective markers for yeast transformation?

(2) Can bacterial genes on the 2-μm DNA be exploited to study the normal function of 2-μm DNA in the yeast cell?

In this paper, I present data to show that a bacterial ampicillin resistance gene is expressed in S.cerevisiae and discuss evidence indicating that bacterial chloramphenicol and kanamycin resistance genes confer their resistance on a yeast cell.

The advantages of bacterial resistance genes as selective markers on a 2-μm DNA vector are twofold. They can be used for most strains since they are not dependent on the presence of auxotrophic markers. In addition, the absence of homology with nuclear DNA sequences precludes chromosomal integration, which has been observed with homologous markers (21, 17).

RESULTS

Construction of recombinant plasmids. Beggs (16) has developed an efficient transformation system for yeast by using composite plasmids that consist of pMB9, 2-μm DNA and a yeast fragment bearing the leu2 gene. In this study, pJDB219 (kindly donated by Dr. J. Beggs) was digested with HindIII and ligated to pBR325 (19) that had been linearized with HindIII. The ligation mixture was used to transform E.coli JA221 leuB6, in order to isolate clones that contain pBR325 carrying the yeast leu2 gene. Transformants were selected on medium minus leucine and subsequently replica plated on L-broth plates containing tetracycline, ampicillin or chloramphenicol. The plasmid DNA from clones that were

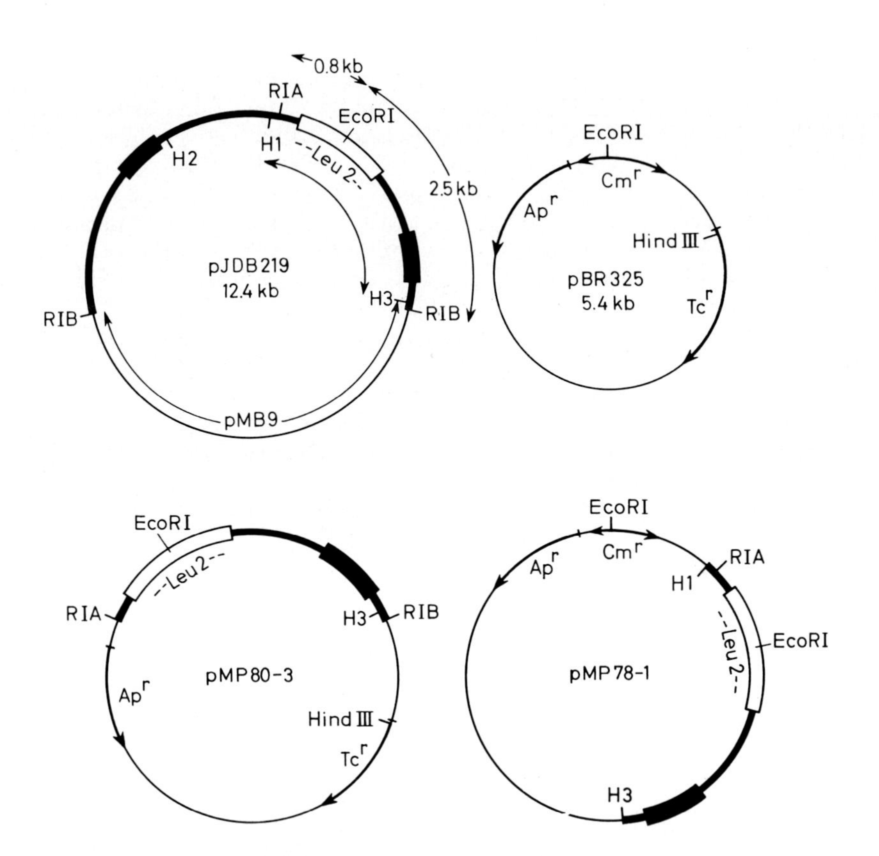

Figure 1. Schemes of recombinant plasmids. To construct pMP78-1, the HindIII fragment of 3.3 kb from pJDB219 (16) was inserted into the HindIII site of pBR325. The 3.3 kb HindIII fragment carries the yeast leu2 gene integrated at the PstI site of HindIII fragment 3 present on 2-μm DNA type 14. pMP80-3 consists of pBR325 that has integrated at its EcoRI site two EcoRI fragments of pJDB together covering about the same region as the 3.3 kb HindIII fragment. The 2-μm DNA part is drawn thick and the inverted duplication sequences extra thick. RIA and RIB are the EcoRI sites and H1, H2, H3, are the HindIII sites of 2-μm DNA. Restriction enzyme digestions, ligations and E.coli JA221 (C600 recA1 leuB6 trpΔE5 hsdR^- hsdM^+ lacY) transformation were described previously (10). The Ap^r (ampicillin resistance), Cm^r (chloramphenicol resistance) and Tc^r (tetracycline resistance) genes are indicated with arrows.

sensitive to tetracycline but resistant to chloramphenicol and ampicillin was isolated and analyzed.

Plasmid pMP78-1 (Fig. 1) from clone T78-1 consists of pBR325 and a HindIII fragment of about 3.3 kb also present in the HindIII digest of pJDB219. This fragment carries the yeast leu2 gene and a segment of 2-µm DNA as indicated in Fig. 1. pMP80-3 (Fig. 1) was constructed in a similar way and consists of pBR325 with two EcoRI fragments of pJDB219 integrated at the EcoRI site. The two EcoRI fragments cover about the same region of pJDB219 as the HindIII segment. The integration at the EcoRI site of pBR325 inactivates the chloramphenicol resistance gene. A third plasmid, pMP81, was constructed by the integration of the 3.3 kb HindIII fragment in the HindIII site of plasmid pCRI (20). pMP81 thus carries a kanamycin resistance gene.

The intact bacterial resistance genes present on the three constructed plasmids, pMP78-1, pMP80-3 and pMP81, are listed in Table 1.

TABLE 1

RESISTANCE GENES ON RECOMBINANT PLASMIDS

Recombinant Plasmid	Bacterial vector	Integration site	Intact bacterial resistance genes	Yeast transformants
pMP78-1	pBR325	HindIII	Amp^R Cam^R	YT6
pMP80-3	pBR325	EcoRI	Amp^R Tet^R	YT8
pMP81-3	pCRI	HindIII	Kan^R	YT10

Yeast transformants. The purified pMP plasmids were used to transform S.cerevisiae AH22, a double leu2 mutant, (21) following the procedure described by Beggs (16). Colonies prototrophic for leucine were obtained with frequencies of 1-2 per 1000 regenerating cells. The same transformation frequency was obtained with plasmid pJDB219 in agreement with the results of Beggs (16) This means that the 2-µm DNA sequences required for the stable replication of the recombinant plasmids in yeast must reside on the 2.12 kb HindIII fragment 3, present on 2-µm DNA type 14 (2).

The stability of the pMP plasmids in the yeast transformants was studied by plasmid DNA and marker analysis.

Supercoiled DNA was isolated from the transformants after 50-100 generations growth on selective medium and analyzed on agarose gels. The main supercoiled band of the transformants YT6 and YT8 migrated at the same rate as the plasmids pMP78-1 and pMP80-3 that had been used for their transformation. In both cases 2-μm DNA molecules comprised only 10-20% of the circular DNA fraction. A different situation was found for YT10-2. Here the pMP81-2 plasmid made up only 10% of the total supercoiled DNA fraction, the rest being 2-μm DNA molecules.

The arrangement of the 2-μm DNA sequences in the supercoiled and the linear DNA fractions from the transformed yeast strains was examined by restriction analysis and Southern hybridization with labelled BTYP-1 DNA (BTYP-1 consists of pBR322 and total 2-μm DNA linked over the PstI site, ref 10). After HindIII digestion, both DNA fractions from YT6 showed the five HindIII fragments of 2-μm DNA plus the two fragments of pMP78-2. A few bands of molecular weights higher than the largest HindIII fragments were observed both in the supercoiled and the linear DNA digest. These could result from incomplete digestion or from minor amounts of 3-μm DNA (22) or rare multimeric forms (6). Recombination between the endogeneous 2-μm DNA and the imported pMP plasmids does not seem to have occurred in the course of these experiments as certain additional fragments to be expected in such recombinant molecules were not detected.

A preliminary comparision of restriction fragments from the supercoiled and the linear DNA fractions gave no indication for the integration of the plasmids in the chromosomal DNA. On Southern blots, no difference could be observed between the two DNA fractions.

The stability of the markers in the transformed cells was determined on non-selective medium. YT6 was grown continuously on non-selective medium with 2% glucose and cells were tested on medium minus leucine every 10 generations. About 15% of the cells were leu^- and over a period of 80 generations no increase was observed.

<u>Expression of the ß-lactamase gene in yeast.</u> Transformants YT6 and YT8 contain a relatively stable population of the recombinant plasmids pMP78-1 and pMP80-3, which contain the bacterial ampicillin resistance gene from pBR325. Plasmid pBR325 (19, 23) is derived from pBR322 and the ampicillin resistance gene

originates from a R plasmid from Salmonella paratyphi B (24, 25). The gene codes for a β-lactamase, an enzyme which hydrolyses the β-lactam bond in penicillin antibiotics.

The expression of the β-lactamase gene in yeast could be shown in two ways. Cell-free extracts of YT6 transformants were tested for β-lactamase activity. Benzylpenicillin-sensitive E.coli cells were plated in soft agar containing 150 μg of benzylpenicillin per ml. 25 μl of extracts of YT6 and AH22 were spotted on these plates, which were subsequently incubated at 37° C.

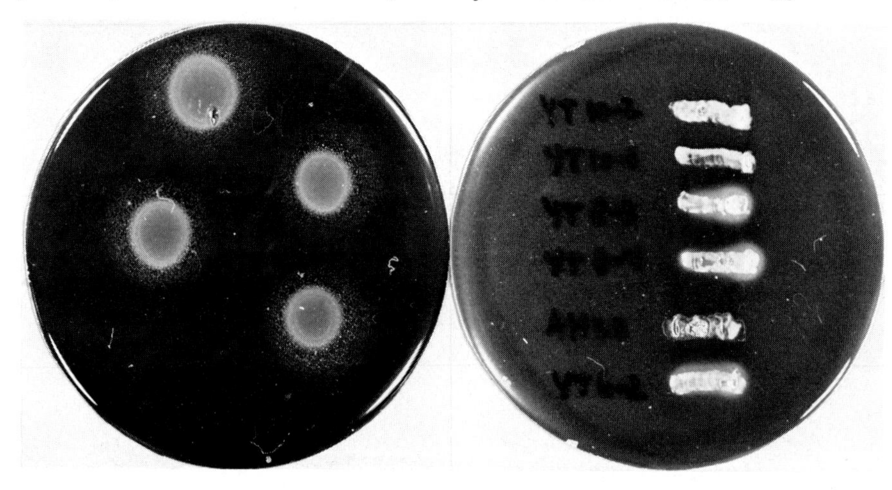

A B

FIGURE 2 A. Demonstration of β-lactamase activity in YT6 cell-free extracts. 1 g of YT6 cells was suspended in 0.05 M Tris-HCl (pH 7.4), 5 mM EDTA and disrupted by shaking with glass beads. The homogenate was centrifuged for 15 min at 16000 xg. The supernatant was spotted on soft agar plates containing benzylpenicillin (150 μg/ml) and sensitive E.coli cells. The control spot of the AH22 extract is invisible.

B. Demonstration of β-lactamase activity in intact YT6 and YT8 cells. The transformants and control AH22 were streaked on the soft agar plates described under A and grown at 37° C.

Fig. 2 A shows strong growth of E.coli cells at the spots of the YT6 extracts and absence of growth at the AH22 control spot. Clearly, the extracts of the transformed yeast cells, carrying the ampicillin resistance gene, contain a substance that degrades benzylpenicillin and thus allows growth of the sensitive E.coli cells. That the growing cells were still penicillin-sensitive and had not been transformed could be shown by replating on benzylpenicillin plates.

The synthesis of active ß-lactamase could also be demonstrated in intact YT6 and YT8 cells. The transformants were streaked on a plate of soft agar containing benzylpenicillin and sensitive E.coli cells. After 1-2 days incubation at 37° C or 30° C a clear growth of E. coli cells around the grown yeast streaks was visible (Fig. 2B). The control streaks of strain AH22 and YT10 did not induce growth of E.coli cells. On the contrary, a clear inhibitory effect on the background growth of E.coli cells after longer incubation times could be observed around the cell streak of AH22. This experiment shows that ß-lactamase activity is synthesized and can be demonstrated in intact yeast cells carrying the bacterial gene. Under certain conditions this assay can also be used to screen single yeast colonies for the presence of ß-lactamase.

O'Callaghan and Morris (26) developed a color reaction to assay ß-lactamase, in which a chromogenic derivative of cephalosporin (87/312), called nitrocefin, is used. ß-lactamase converts the yellow color of the intact molecule into the red cleavage product. When nitrocefin was applied to outgrown cell streaks, the color change could be observed after 15-60 min depending on the age of the cells. This reaction allows rapid detection of ß-lactamase-producing yeast cells.

To quantitate the relative amounts of ß-lactamase activity present in transformed yeast strains,1 g of stationary cells was disrupted and the homogenate was centrifuged for 15 min at 16000xg. The ß-lactamase activity of the supernatant was determined by measuring the cleavage of nitrocefin at 390 nm. Fig. 3 shows the activity curves of extracts from YT6-2 and YT8-3 and a shockate and sonicate of E.coli cells carrying pBR325. The E.coli (pBR325) sonicate contained about 10-20 times more ß-lactamase activity per g cells than the YT6 extract. How this value is influenced by extraction and assay conditions is not yet clear.Extracts of strain AH22 or YT10 did not show any activity in this assay.

In E.coli, the ß-lactamase of plasmid RI has been shown to be a periplasmatic enzyme of the type RTEM (27). The enzyme is synthesized as a preprotein with a 23 amino acid leader sequence (28, 29), which persumably serves as a signal to direct the transport of the protein

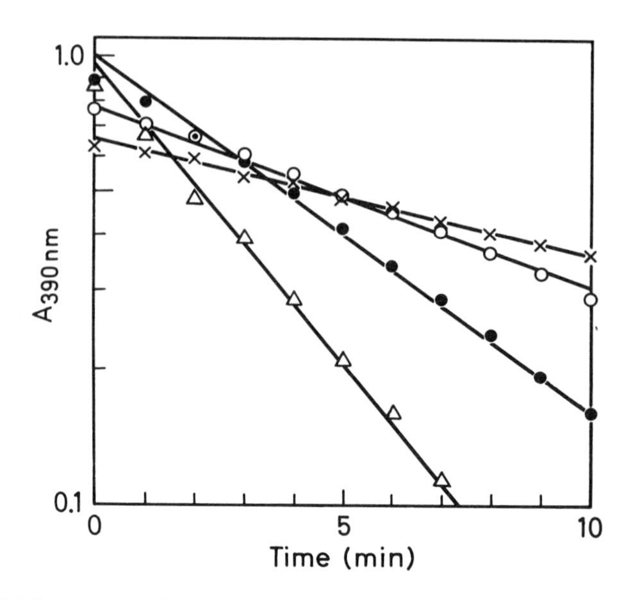

FIGURE 3. ß-lactamase activity curves of cell-free extracts from two yeast transformants, YT6-2 and YT8-3. As a control an E.coli strain carrying plasmid pBR325 was extracted by osmotic shock (28) or by sonication. ß-lactamase activity was determined in 1 ml at 37° C with cephalosporin 87/312 (nitrocefin) as substrate as described (26). 1 g of YT6-2 or YT8-3 was disrupted and suspended in a final volume of 5 ml of which 25 µl were used per assay. E.coli (pBR325) cells from 100 ml of an overnight culture with benzylpenicillin (100 µg/ml) were shocked (30) in 5 ml H_2O; 10 µl of the shockate were used per assay. 0.5 g E.coli (pBR325) was sonicated in 3 ml 50 mM Tris-HCl (pH 7.8), 50 µM 2-mercaptoethanol and 2 µl were used per assay. Δ-Δ, E.coli (pBR325) sonicate; •—•, E.coli (pBR325) shockate; o—o, Yt6-2; x—x, YT8-3.

through the cell membrane. The sizes of the ß-lactamase synthesized in E.coli and in the yeast were compared by electrophoresis of an E.coli shockate and yeast extracts in non-denaturing polyacrylamide gels. The ß-lactamase bands in the gel were detected with the nitrocefin reaction. Fig. 4 shows the gels after staining with Coomassie blue. The positions of the ß-lactamase bands are indicated with an arrow. The activity in the yeast extract migrates more slowly than the E.coli ß-lactamase. Although this type of gel gives only limited information on molecular weights, a larger size of the ß-lactamase

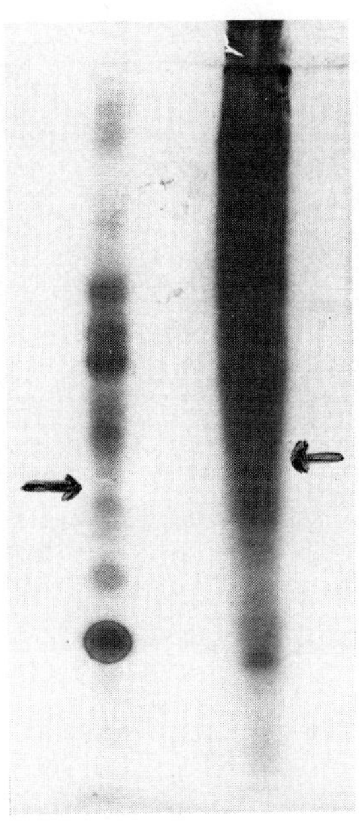

FIGURE 4. Detection of ß-lactamase activity after electrophoresis of an extract of YT6-2 in a non-denaturing polyacrylamide gel. Lane 1, 10 µl of E.coli (pBR325) extract. Lane 2, 40 µl of YT6 cell-free extract. See legend Fig. 2 A for extraction details.After electrophoresis the ß-lactamase bands were strained with nitrocefin (26). As such bands are difficult to photograph they were marked and indicated with arrows. Afterwards the gel was stained with Coomassie blue. Gelconcentration, 6% in 0.25 M Tris-HCl(pH 8.9), o.1 mM dithiotreitol; stacking gel 3% polyacrylamide.

synthesized in the yeast transformants seems to be indicated. An attempt to convert the larger yeast form with an E.coli extract was unsuccessful.

The culture medium of yeast transformants that produce ß-lactamase could not be shown to contain enzyme activity. This makes excretion of the enzyme unlikely. The large area of bacterial growth in Fig. 2 B could be explained by diffusion of the benzylpenicillin.

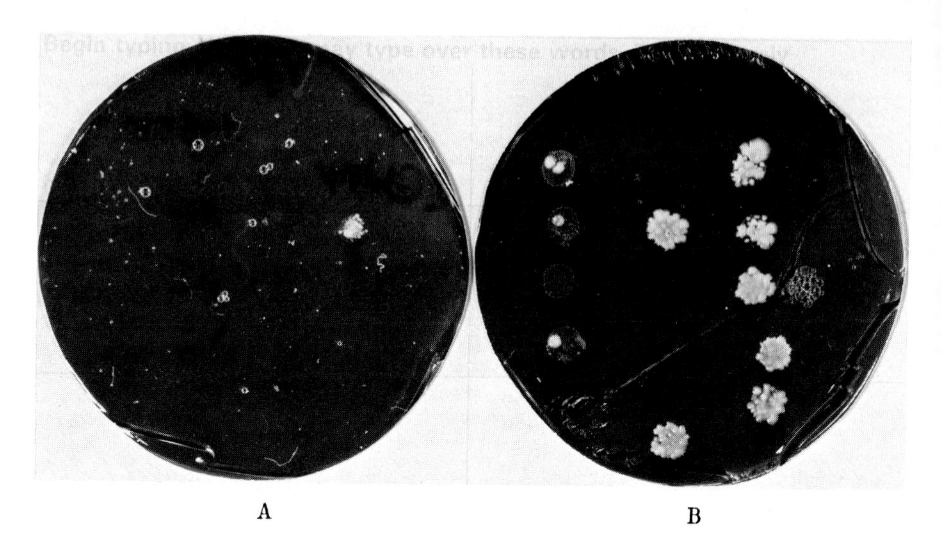

FIGURE 5 A. Yeast transformants, YT6-2 (right) YT8 (left), plated on a minimal medium containing histidine (10 μg/ml), 1% ethanol and chloramphenicol 1.8 mg/ml. Plates were grown at 30° C for 1-2 weeks.
B. Yeast transformants, YT8 (left) and YT10 (right), plated on minimal medium containing histidine (10 μg/ml), 0.2% glucose and neomycin (3 mg 90-95% neomycin B sulfate per ml). Plates were incubated for 1-2 weeks at 35° C.

Evidence for the expression of a chloramphenicol and a kanamycin resistance gene in yeast. YT6 and YT8 transformants were plated on minimal plates plus 1% ethanol and 1, 1.2, 1.4, 1.6, 1.8, or 2 mg chloramphenicol per ml. Both YT6 and YT8 grow on plates containing less than 1.8 mg chloramphenicol per ml. On the higher concentrations YT8 does not grow, but YT6 shows many growing colonies next to cells that are inhibited (Fig. 5A). The only difference between the plasmids of YT6 and YT8 (see Table 1) is the integration site in pBR325. YT8 contains plasmid pMP80-3 with an interrupted chloramphenicol resistance gene, whereas this gene is intact in YT6. The observed resistance of YT6, therefore, is provisionally interpreted as a result of the expression of the bacterial chloramphenicol resistance gene, which codes for a chloramphenicol transacetylase. Experiments to directly assay the enzyme activity in cell-free extracts are under way.

YT10, which contains pMP81-3 bearing a kanamycin resistance gene, shows a similar mode of resistance to neomycin. Kanamycin and neomycin are aminoglycoside antibiotics and can both be inactivated by a number of aminoglycoside modifying enzymes (3, 32). Fig. 5 B shows a plate with 3 mg of neomycin per ml grown at 35° C for 1 week. In the YT10 spots many growing colonies are present that on replating showed neomycin resistance. YT8 is largely sensitive; only a few colonies grew up.

DISCUSSION

The basic question whether prokaryotic structural genes can be expressed in an eukaryotic cell has been answered in the affirmative by an analysis of the expression of a bacterial ß-lactamase gene in S.cerevisiae. In addition, evidence has been presented for the functional expression in yeast of two other bacterial genes, coding for resistances to chloramphenicol and kanamycin.

We have chosen to study resistance genes because they are present on many bacterial cloning vectors and are provided with suitable restriction sites for integration. Functional expression of these genes in S.cerevisiae could enable them to assume a role in yeast transformation analogous to their function as selective markers in bacterial cloning systems.

Yeast is not sensitive to penicillin antibiotics and the resistance conferred by the ß-lactamase gene, therefore, cannot be shown directly. The presence of the ß-lactamase activity, however, can be readily demonstrated. The growth of indicator bacteria in the soft agar assay is easily detectable and with replica plating we were able to screen single colonies for segragation provided that the colony density was not too high. The nitrocefin assay is very rapid and works well for larger cell streaks, but we have not yet obtained satisfactory results on single colonies.

In E.coli, ß-lactamase is synthesized as a preprotein with a signal peptide of 23 amino acids (28, 29). The electrophoretic mobility of the enzyme synthesized in the yeast cell was analyzed in non-denaturing polyacrylamide gels to allow the detection of the enzyme activity in the gel with nitrocefin (26). In most cases, the ß-lactamase from the yeast extract migrated more slowly than the enzyme activity present in E.coli extracts. This suggests that the yeast cell does not produce the mature bacterial enzyme, but a slightly larger form that is active. Whether this form is identical with the bacterial preprotein remains to be seen. As the processing is a specific of the bacterial cell membrane, it seems unlikely that the yeast cell can perform this step.

Under the proper conditions, yeast is sensistive to chloramphenicol and neomycin and the presence of active bacterial genes in the cell coding for resistances to these compounds can be directly tested. Yeast transformant YT6-2 contains the intact chloramphenicol resistance gene as shown by bacterial transformation with cell-free extracts or with the supercoiled DNA fraction. Nutrient plates with 1% ethanol and high concentrations of chloramphenicol allow a high percentage of cells to grow. For YT8-3, which contains an interrupted chloramphenicol gene, very few growing cells were seen when plated under identical conditions. Some of these colonies were not resistant upon retesting. The same phenomen was observed on neomycin.

The presence of foreign genes on 2-μm DNA opens a new approach to the study of its function in the yeast cell. For this purpose the bacterial genes offer an alternative to the use of cloned yeast nuclear genes.

ACKNOWLEDGEMENTS

A gift of nitrocefin from Glaxo-Allenburys Research Ltd. by courtesy of Dr. C. H. O'Callaghan is gratefully acknowledged. I wish to thank Dr. J. Beggs for providing plasmid pJDB219, P. Hardy for critical reading of the manuscript and Professor W. Beermann for his interest and support.

REFERENCES

1. Hollenberg, C. P., Borst, P., and van Bruggen, E. F. J. (1970). Biochim. Biophys. Acta 209, 1.
2. Hollenberg, C. P., Degelmann, A., Kustermann-Kuhn, B., and Royer, H.-D. (1976). Proc. Natl. Acad. Sci. USA 73, 2072.
3. Guerineau, M., Grandchamp, C., and Slonimski, P. P. (1976) . Proc. Natl. Acad. Sci. USA 73, 3030.
4. Beggs, J. D., Guerineau, M., and Atkins, J. F. (1976) . Molec. Gen. Genet. 148, 287.
5. Livingston, D. M., and Klein, H. L. (1977). J. Bact. 129, 472.
6. Royer, H.-D.,and Hollenberg, C. P. (1977). Molec. Gen. Genet. 150, 271.
7. Cameron, J. R., Philippsen, P., and Davis, R. W., (1977)..Nucl. Acids. Res. 4, 1429.
8. Gubbins, E. J., Newlon, C. S., Kann, M. D., and Donelson, J. E. (1977). Gene 1, 185.
9. Hollenberg, C. P., Kustermann-Kuhn, B., and Royer, H.-D. (1976) . Gene 1, 33 .
10. Hollenberg, C. P. (1978). Molec. Gen. Genet. 162, 23.
11. Struhl, K., Cameron, J. R., and Davis, R. W. (1976). Proc. Natl. Acad. Sci. USA 73, 1471.
12. Ratzkin, B., and Carbon, J. (1977)..Proc. Natl. Acad. Sci. USA 74, 487.
13. Struhl, K., and Davis, R. W. (1977). Proc. Natl. Acad. Sci. USA 74, 5255.
14. Clarke, L., and Carbon, J. (1978). J. Molec. Biol. 120, 517.
15. Walz, A., Ratzkin, B., and Carbon, J. (1978). Proc. Natl. Acad. Sci. USA 75, 6172.
16. Beggs, J. D. (1978). Nature 275, 104.
17. Hicks, J. B., Hinnen, A., and Fink, G. R. (1978). Cold Spring Harbor Symp. Quant. Biol. in press.
18. Struhl, K., and Davis, R. W. (1978). in press.
19. Bolivar, F. (1978). Gene 4, 121.
20. Covey, C., Richardson, D., and Carbon, J. (1976). Molec. Gen. Genet. 145, 155.
21. Hinnen, A., Hicks, J. B., and Fink, G. R. (1978). Proc. Natl. Acad. Sci. USA 75, 1929.
22. Guerineau, M., Grandchamp, C., Paoletti, J., and Slonimski, P. (1971). Biochim. Biophys. Res. Commun. 42, 550.
23. Bolivar, F., Rodrîguez, R., Betlach, M., and Boyer, H. W. (1977). Gene 2, 75.
24. Meynell, E., and Datta, N. (1967). Nature 214, 885
25. So, M., Gill, R., and Falkow, S. (1975). Molec. Gen. Genet. 142, 239.

26. O'Callaghan, C. H., and Morris, A. (1972). Antimicrobiol. Agents and Chemotherapy 2, 442.
27. Matthew, M., and Hedges, R. W. (1976). J. Bacteriol. 125, 713.
28. Ambler, R. P., and Scott, G. K. (1978) . Proc. Natl. Acad. Sci. USA 75, 3732.
29. Suttcliffe, J. G. (1978). Proc. Natl. Acad. Sci. USA 75, 3737.
30. Neu, H. C., and Heppel, L. A. (1965). J. Biol. Chem. 240, 3685.
31. Benveniste, R., and Davies, J. (1973). Ann. Rev. Biochim. 42, 471.
32. Haas, M. J., and Dowding, J. E. (1975). Methods Enzymol. 43, 611.

ORGANIZATION OF MITOCHONDRIAL DNA IN YEAST[1]

Alexander Tzagoloff, Giuseppe Macino,
Marina P. Nobrega, and May Li

Department of Biological Sciences,
Columbia University,
New York, New York 10027

ABSTRACT The nucleotide sequence of a segment of mitochondrial DNA included between positions 79 and 87 of the wild type physical map has been determined. Even though the entire segment consists of over 5,000 base pairs, it contains only two discernible genes: the ATPase proteolipid (subunit 9) and a seryl-tRNA. The remainder of the DNA consists of long stretches rich in A+T (95%) plus numerous 30-70 nucleotide long sequences that have a very high content of G+C. The role of these distinct DNA regions and their significance in understanding the evolution of the mitochondrial genome are evaluated.

INTRODUCTION

Mitochondrial DNA (mtDNA) of *Saccharomyces cerevisiae* is known to code for ribosomal RNA's, transfer RNA's and a limited number of messenger RNA's (1). It is now reasonably well established that the messenger RNA's are translated on mitochondrial ribosomes and that the products are subunit polypeptides of cytochrome oxidase, coenzyme QH_2-cytochrome *c* reductase and the ATPase (1-4). In order to understand how mitochondrial genes of yeast are organized, we have undertaken to study the DNA sequence of several regions of the genome that carry genetic markers for different transfer RNA's and for the respiratory and ATPase complexes.

[1]This research was supported by Grants HL22174 and GM25250 from the National Institutes of Health, USPHS.

ISBN 0-12-198780-9

The most extensive sequence data obtained to date is in a region that contains the seryl-tRNA and ATPase proteolipid genes. The two genes have been identified in a segment of DNA that is approximately 5,000 base pairs long and does not appear to have any other genes. Based on the sequences adjoining the two genes, it is possible to propose some models for the transcription and organization of mitochondrial genes in general.

SEQUENCING STRATEGY

Spontaneous and chemically induced mutations in *Saccharomyces cerevisiae* can give rise to ρ^- mutants which are characterized by having long deletions in their mtDNA (5-7). The deletions may lead to the loss of as much as 99% or more of the wild type genome. An important attribute of ρ^- mutants is their ability to replicate the retained segment of DNA even if it is only several hundred base pairs long. In all known cases, the segment of DNA retained in a ρ^- mutant is reiterated to form a new genome whose length can equal that of wild type mtDNA (8). Although ρ^- genomes with palindromic (tail-to-tail) repeats have been reported to occur (9), in most ρ^- strains the retained segment is arranged in a tandem head-to-tail fashion. Since the total amount of mtDNA per cell remains constant, the amplification of the non-deleted segment is inversely proportional to its length. In practical terms this means that ρ^- mutants can be used to obtain sufficient amounts of mtDNA from any part of the genome for sequencing studies (10,11).

There are several considerations in deciding on the type of ρ^- mutant most suitable for sequence determinations. Mutants containing short segments of DNA (1,000 base pairs or less) are best suited for deducing the sequence of a single gene or a part thereof. Such mutants can be isolated by selecting strains that retain genetic markers in the gene of interest only. If, on the other hand, the purpose is to sequence an extended region of DNA, it is more practical to select mutants that already have fairly long segments of DNA since this reduces subsequent overlapping of segments. Retained segments that are 3,000-4,000 base pairs in length can in most cases be sequenced by current methods.

SELECTION OF ρ^- MUTANTS IN THE OLI 1/PHO 2 REGION OF THE MITOCHONDRIAL GENOME

Mutations in the oli 1 (12) and pho 2 (4) loci of the mitochondrial genome have recently been shown to cause amino acid substitutions in the proteolipid (subunit 9) component of the ATPase complex (13). These two genetically linked loci, therefore, provide convenient markers for selecting ρ^- mutants with segments of mtDNA containing the proteolipid gene. To enrich for this ATPase gene, a respiratory competent strain of *S. cerevisiae* (D273-10B/A21) carrying the oli 1 resistance marker was mutagenized with ethidium bromide to induce ρ^- mutations. A large number of independent ρ^- clones were collected and tested for the retention of the oli 1 resistance, pho 2, as well as other genetic markers in mitochondrial DNA. Based on the genetic tests, a small percentage of clones were found which retained the oli 1 and pho 2 loci but had lost all other markers. One clone (DS400/A4) was established to have retained only the oli 1 resistance marker.

The genotypes and complexity of the mtDNA in some representative clones are listed in Table I. Based on the DNA fragments generated by different restriction endonucleases it was estimated that the length of the repeat units of mtDNA in the clones studied ranged from 1,080-5,200 base pairs. The restriction patterns of the mtDNA from DS400/A4, DS400/A3 and DS401 are shown in Fig. 1. The mtDNA

TABLE I
COMPLEXITY OF mtDNA IN ρ^- CLONES CONTAINING THE OLI 1 MARKER

Clone	Markers		Size of mtDNA repeat unit in base pairs
	Oli 1	Pho 2	
DS400/A4	+	-	1,080
DS400/A3	+	+	1,800
DS401	+	+	5,200

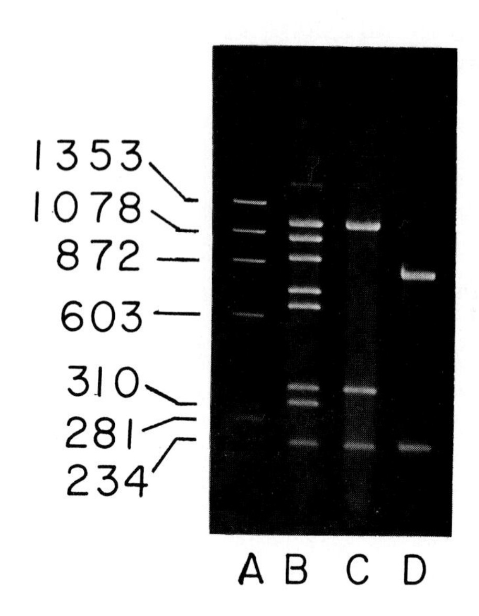

FIGURE 1. Agarose gel electrophoresis of HpaII digests of mtDNA from DS400/A3, DS400/A4 and DS401. The purified mtDNA's were digested with HpaII and separated on a 1.5% agarose gel. The sizes of the ethidium bromide stained bands were calculated from the known sizes of the HaeIII fragments of ϕX-174RF DNA. A: ϕX174-RF; B: DS401; C: DS400/A3; D: DS400/A4. (Taken from Macino and Tzagoloff, J. Biol. Chem. in press).

of DS400/A4 is cut by HpaII into two fragments whose total lengths correspond to 1,080 base pairs. This mutant contains the oli 1 but lacks the pho 2 marker. The DS400/A3 strain is of intermediate complexity with a repeat length of 1,800 base pairs. DS401 exhibits the most complex pattern. The eight HpaII bands of this mutant indicate that the repeat length of the mtDNA is approximately 5,200 base pairs. A number of other independently isolated ρ^- clones were found to have retained the same segment of mtDNA as DS401 and were not studied further.

RESTRICTION MAPS OF DS400/A4, DS400/A3 AND DS401

The restriction maps of the ρ^- mtDNA's were constructed by double and triple digests using combinations of various restriction endonucleases. In addition it was possible to purify from the wild type parental strain, two neighboring HincII fragments which included the entire sequences of DS400/A4, DS400/A3 and DS401. The results of the restriction analyses on the wild type HincII fragments and of the ρ^- mtDNA's were completely consistent indicating that the ρ^- strains retained continuous segments of DNA that did not have any internal secondary deletions. The restriction maps of the wild type and ρ^- DNA's showing the HpaII sites are presented in Fig. 2. The mtDNA of DS401

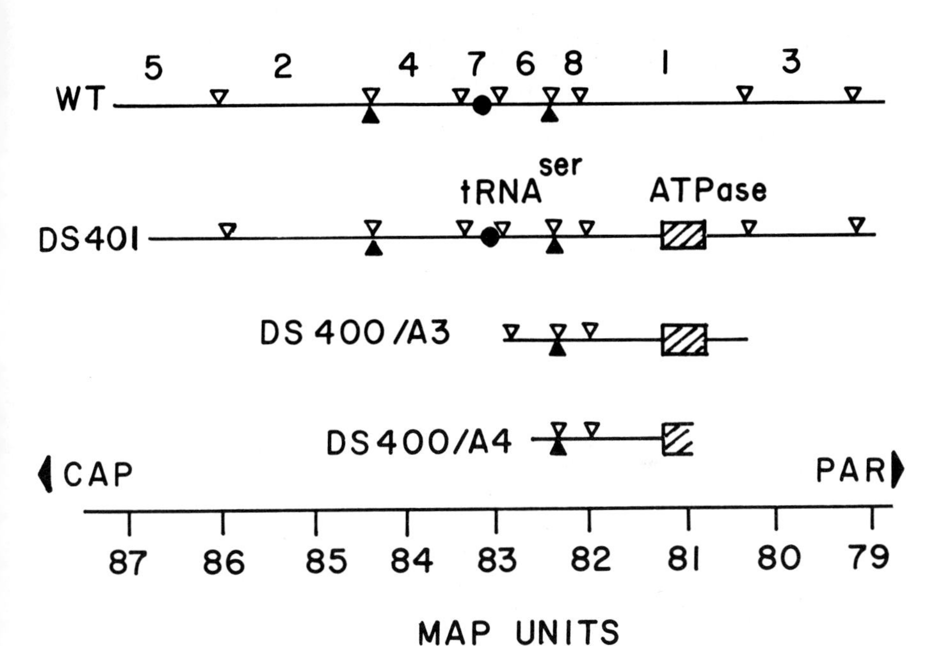

FIGURE 2. Restriction maps of DS401, DS400/A3 and DS400/A4. The restriciton maps of the ρ^- and wild type DNA's have been aligned to indicate correspondence of the restriction sites for HpaII (▽), HinCII (●) and HhaI (▲). The position of the HinCII site at 83 units is based on the physical maps of the D273-10B strain published by Morimoto and Rabinowitz (14). Each map unit is equivalent to 700 nucleotides. The HinCII site in HpaII fragment 7 is included in the seryl-tRNA gene sequence. The structural gene of the ATPase proteolipid in the HpaII fragment 1 is depicted by the hatched box. Other restriction sites for HinF, DpnII, HaeIII, AluI and MboII are not shown.

corresponds to a segment that starts at map position 79.1 and ends at position 86.9 of the wild type genome. As indicated in Fig. 2, the mtDNA's of DS400/A4 and DS400/A3 are completely included in the larger DS401 segment. This has been confirmed by the nucleotide sequences of the three mutant DNA's.

DNA SEQUENCES

The nucleotide sequences of the DS400/A4, DS400/A3 and DS401 segments were obtained by the Maxam and Gilbert method (15) of DNA sequencing. Restriction fragments were labeled at the 5'-ends and separated into single strands which were then used for sequence analysis. The sequences of the DNA's indicate that the ATPase proteolipid gene lies in the HpaII fragment 1 (1,200 base pairs) of the DS401 and DS400/A3 mutants. Since the deletion in DS400/A4 was initiated within the structural gene, the amino terminal end starting with the forty-fifth amino acid is missing. This agrees with the genetic data (Table I) which indicate that DS400/A4 lacks the pho 2 marker. In other studies, Sebald et al. (13) have shown that the pho 2 marker causes an amino acid substitution at amino acid residue 42.

The nucleotide sequence of the proteolipid gene with its adjoining A+T rich regions is shown in Fig. 3. It is evident that the sequence is in complete agreement with the previously determined primary structure of the protein (13). The one exception is the threonine at position 46 which according to the DNA sequence is encoded by a leucine codon. The explanation for this discrepancy will be discussed later in the paper. The complete nucleotide sequences of DS400/A4 and DS400A3 reveal that only the proteolipid gene is present in these segments. The larger mtDNA of DS401, however, has an additional short sequence which has been identified as a seryl-tRNA. This sequence (Fig. 4) occurs in HpaII fragment 7, some 1,250 nucleotides downstream from the end of the proteolipid gene. The seryl-tRNA gene straddles the only HincII site present in DS401. This gene has also been sequenced from a cloned fragment of mitochondrial DNA by Martin et al. (see this volume). It is of interest that the proteolipid and seryl-tRNA genes are transcribed from the same DNA strand. The

```
5'ATATATATATGAATT AATATTTAATAATAA ATAATAATATAATTA ATAATATTATTATTA TTATAATTTTTATTT ATAATATTATAAATA

  TTATTATATATATAT TATAATAATATTAAT AAGATATATAAATAA GTCCCTTTTTTTTTA TTTAAAATAAAGAAG ATAATTAATATATTT

  TAATAATTTAATTAA ATGTGTATTAAAAGA ATAATAAAAGATAA TATTAATATGTTAAT TATATATAATATATT ATATATAATTATATA

                                                                                  fMet-Gln-Leu-Val-
  TATATATATAAATAA TAATAAATATATATA TAATATAAAAATAAG AATAGATTAAATATT TAATAAATAAATATT ATG CAA TTA GTA

   5                   10                  15                  20                  25
  Leu-Ala-Ala-Lys-Tyr-Ile-Gly-Ala-Gly-Ile-Ser-Thr-Ile-Gly-Leu-Leu-Gly-Ala-Gly-Ile-Gly-Ile-Ala-Ile-
  TTA GCA GCT AAA TAT ATT GGA GCA GGT ATC TCA ACA ATT GGT TTA TTA GGA GCA GGT ATT GGT ATT GCT ATC

       30                  35                  40                  45                  50
  Val-Phe-Ala-Ala-Leu-Ile-Asn-Gly-Val-Ser-Arg-Asn-Pro-Ser-Ile-Lys-Asp-Thr-Val-Phe-Pro-Met-Ala-Ile-
  GTA TTC GCA GCT TTA ATT AAT GGT GTA TCA AGA AAC CCA TCA ATT AAA GAC CTA GTA TTC CCT ATG GCT ATT

           55                  60                  65                  70                  75
  Phe-Gly-Phe-Ala-Leu-Ser-Glu-Ala-Thr-Gly-Leu-Phe-Cys-Leu-Met-Val-Ser-Phe-Leu-Leu-Leu-Phe-Gly-Val-
  TTT GGT TTC GCC TTA TCA GAA GCT ACA GGT TTA TTC TGT TTA ATG GTT TCA TTC TTA TTA TTA TTC GGT GTA

  ochre      ochre
  TAA TATATA TAA   TATATTATAAATAA ATAAATAAATAAATA ATGAAATTAATAAAA AAATAAAATAAAATA AAATCTCATTTGATT

  AAATTAATAACATTC TTATAATTATATAAT TATTATAAATATATA AATATTATAATAATA ATAATATATAT...≈130 nucleotides

   to . . .  TTAT ATATATTATGATATT ATTATGTAACATTAT ATAATAATATAAATT ACCATAATGAAATAT ATTATTTATTAATAA

  TAAAATATTTATTAA TAATAGAATATATAT ATTATGATAATATTT ATTAATAAATAATAA ATTCTTTATATATAA ATATATTAAATATAT

  TTAATTGGACACAAT ATAATTTTTATTATA ATTATGATAATATTT ATTAATAAATAATAA ATTCTTTATATATAA ATATATTAAATATAT

  TTAATTGGACACAAT ATAATTTTTATTATA TTATTCATTTAATAA TATTAATATTAATAT TAATATTAATATAAT ATTGGTGAAACATCT

  CCTTTCGGGGTTCCG G
```

FIGURE 3. Nucleotide sequence of non-transcribed strand in HpaII fragment 1.

The nucleotide sequence of the proteolipid gene has also been determined in another yeast strain (see Grivell et al., this volume).

```
5' CCGGAACCCCG AAAGGAGTTTATTTA ATATTTATATTTATA
TTAATATTTATATTT ATATTTATATTCCTC TTAAGGATGGTTGAC
TGAGTGGTTTAAAGT GTGATATTTGAGCTA TCATTAGTCTTTATT
GGCTACGTAGGTTCA AATCCTACATCATCC GTAATAATACATATA
TATAATAATAATTTT AATATTATTCCTATA AAAATAAAATAAATA
AATAAATAATAATAA TTAATTAATTTTAAT AAATATAAAATATAT
AAAATAATAATAATA ATAATTATTATTTTA ATAATATTATTTATA
TAATAGTCCGG
```

FIGURE 4. Nucleotide sequence of HpaII fragment 7. The sequence of the seryl-tRNA gene is indicated by the italicized print. The sequence shown is that of the non-transcribed strand.

```
5' CCGGTTGTTCACCG GATTGGTCCCGCGGG GAATATTAATAATAA
   ATTACAACATTTAAA TAATATAAATAATTG AAATCTACAAATTTA
   TAATTATAATAAAAA TATAGAAATTATAAA TACTATAAATGATAA
   ATTAATTAATAAATT ATTATATAAAATAAT AACTTTAAAATTAAA
   TAATATAAATATTAA TAAAATTATTATAAG TAAACTTATTAATTA
   ACACAGTTTAAATAA ATTAAATATTAAATT TATTATTATAATAAT
   GATATTAATAATAAT AATAATAATAATAAT AATAATTATTATATA
   AATATAATAAATAAA TTAATAAATATTATA AATAATAATATAAAT
   AATAATTTATGTAAT ATTTTAAGTTATTAT TATAATAAAAAAGTA
   ACTATTGAACCTATT AAATTATCATATATT TATTTAAATAGTGAT
   ATTTTTAGTAAATAT ATTAGTTTAAATGAT ATAGATAAATATAAT
   AATGGTATCTTAACT AATTATCAACGTATA TTAAATAATATTATG
   CCTAAATTAAATGAT CAACCAACTAAATAT TTATATATTAAAATA
   TCTATAGGAATATTA TCAATAGTTATATTA TTATAAATATTATTA
   ATATTACCAATATAA TTATTATTATTATTA TTATAATTATTATTA
   TTATTAATATTATTA TTATTATTATTATTA TTATTATTATTTAAT
   AAATTAATTATATTA TTATATTTATTATTA TTAATATTATTAATA
   TTATTAATATAATTT ATAGAAATATTATGA TCTATTAAATTTAAA
   GGTAGTTTAAGTAAT AATAATGGTAGAACT AGTACACTTAATTTA
   TTAAATGGTACTTTT AATAATAAAAAATAT TTATGAAGTAATATT
   AATAATAATTATAAA TTAAATTATATCCCT TCTAATCATAATTTA
   TATAATAATTCTAAT ATTAATAAAAATGGT AAATATAATATTAAA
   GTTAAATTAAACTTT ATTTAATATATATAT TAATAGTCCGG
```

FIGURE 5. Nucleotide sequence of HpaII fragment 2. The sequence shown is for the same strand as in Figs. 3 and 4.

direction of their transcription is *par* → *cap* according to the convention of Fig. 2.

The segment of mtDNA in DS401 encompasses a region of the wild type genome that has been shown to contain the *var 1* gene product (16). Strausberg *et al*. (17) have localized the *var 1* gene by the zygotic gene rescue method in the vicinity of a HhaI site downstream from the seryl-tRNA gene. The nucleotide sequence of this region is rich in A+T (Fig. 5) and does not contain a reading frame that could code for a protein with a molecular weight of 40,000-43,000, the estimated size range of the *var 1* polypeptide (16). The explanation for the absence of the structural gene of *var 1* in the DS401 mtDNA is not clear at present (for further discussion see Butow *et al*., this volume).

AN UNUSUAL THREONYL tRNA IN YEAST MITOCHONDRIA

Yeast mitochondria contain two threonyl tRNA's ($tRNA_1^{Thr}$ and $tRNA_2^{Thr}$) that are encoded by separate mitochondrial genes (18,19). A mutant in $tRNA_1^{Thr}$ has been used to select ρ^- clones containing the threonyl tRNA gene. One such clone (DS243) has a retained segment of mtDNA only 468 base pairs in length. This segment of DNA has been found to contain only the $tRNA_1^{Thr}$ gene. The structure of the tRNA deduced from the DNA sequence has several unusual features (Fig. 6). 1) The anticodon for threonine is absent in the normal position of the anticodon loop. 2) Instead the tRNA has a GAU anticodon for leucine. 3) The anticodon loop is slightly asymmetrical due to the presence of an extra nucleotide on one side.

The leucine anticodon in the threonyl tRNA provides an explanation for the occurence of a threonine in a position (residue 46) of the proteolipid whose gene has a CTA codon for leucine in the corresponding position. Even though $tRNA_1^{Thr}$ charges with threonine it appears to recognize the CTA codon which normally encodes leucine. The presence of the leucine anticodon probably resulted from a mutation in the tRNA gene. The almost exclusive utilization of the UUA codon for leucine in yeast mitochondria allowed the mutant tRNA to be retained and to substitute threonine for leucine in those few instances where CTA is used.

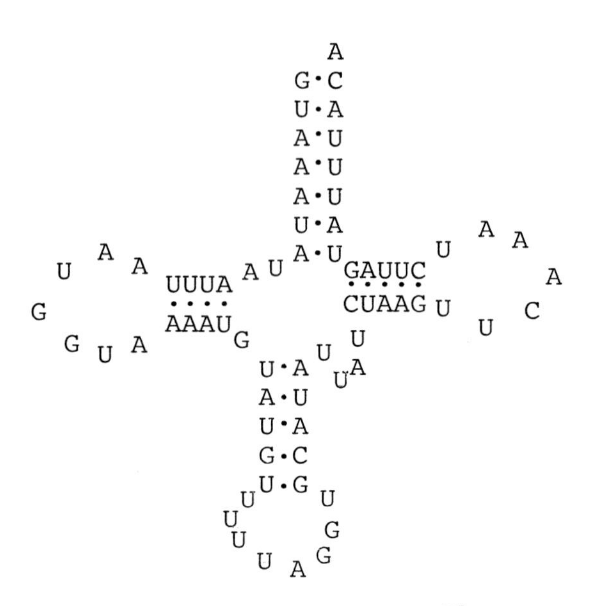

FIGURE 6. Structure of $tRNA_1^{Thr}$. Since the structure of the tRNA is based on the DNA sequence of the gene, the modified bases are not known.

MITOCHONDRIAL CODONS

Three mitochondrial genes have been sequenced to date; subunit 2 of cytochrome oxidase and two subunit polypeptides of the ATPase complex. In each case there is a high selectivity of codons that have an A or U in the third position of the letter code (Table II). The frequency of codons terminating in a G or C is only 10%. Furthermore, certain codons such as the CGN series for arginine do not appear to be used at all. These findings suggest that yeast mitochondria have a specialized code that may require relatively few tRNA's. In fact the 25-30 different mitochondrial tRNA species detected in yeast is somewhat less than the minimal requirement assuming all codons are used.

The absence of tRNA's capable of recognizing certain codons places severe constraints on the adaptability of the mitochondrial genetic system. For example, given an incomplete set of tRNA's, certain missense mutations which normally lead to amino acid substitutions would in effect cause premature chain termination. It is of interest

TABLE II
FREQUENCY OF MITOCHONDRIAL CODONS

Amino Acid	Codons	Observed # Codons	Amino Acid	Codons	Observed # Codons
Ala	GCA	13	Leu	UUA	83
	U	25		G	1
	C	1		CUA*	4
	G	0		U	2
Arg	AGA	10		C	0
	G	0		G	0
	CGA	0	Lys	AAA	14
	U	0		G	2
	C	0	Met	AUG	18
	G	0	Phe	UUU	24
Asn	AAU	23		C	16
	C	1	Pro	CCA	11
Asp	GAU	18		U	12
	C	1		C	0
Cys	UGU	6		G	0
	C	0	Ser	UCA	23
Gly	GGA	7		U	8
	U	29		C	0
	C	0		G	0
	G	2		AGU	6
Gln	CAA	14		C	0
	G	1	Thr	ACA	20
Glu	GAA	18		U	11
	G	2		C	0
Ile	AUU	61		G	0
	C	9	Val	GUA	18
	A	6		U	16
His	CAU	9		C	1
	C	0		G	0
			Tyr	UAU	24
				C	2

*CUA is listed under leucine but in yeast mitochondria acts as a codon for threonine.

that a large percentage of mitochondrial mutations, particularly in the structural gene of cytochrome b have been found to result in prematurely terminated polypeptides (20,21). As a consequence the ability of mitochondrial genes to undergo favorable mutations may be severely limited. It should be borne in mind, however, that mitochondria have a small number of genes which probably constitute a residue still in the process of being transferred to the nuclear genome. In view of the small number of genes involved, the evolutionary advantage of maintaining a complex translational machinery may have been sacrificed in favor of a simpler system that utilizes a restricted set of codons and a bare minimum of tRNA's.

FUNCTION OF A+T SPACERS

Based on the few regions of yeast mtDNA studied, it is apparent that only a small fraction of the total genome contains coding sequences. Most of the DNA consists of A+T rich sequences that border and separate a sparce population of genes. This is most dramatically illustrated by the DNA segment of the DS401 clone. The 5 kilobase segment of this mutant has only the ATPase proteolipid and seryl tRNA genes which are encoded in two short sequences totaling 317 nucleotides of 6% of the entire DNA.

The function of the A+T rich sequences or what may be considered to be gene spacers is not known. It is tempting to think that they may have some regulatory function but there is no evidence to support this at present. Thus, some of the strain specific deletions and insertions that have been noted in yeast mtDNA (22,23) are probably attributable to differences in the distribution and size of the A+T spacers. Such variations in the length and sequences of the A+T regions tends to argue that they may not have a specific function. The simplest assumption at present is that most of the yeast mitochondrial genome consists of vestigial genes that have degenerated to A+T rich sequences.

The retention of vestigial gene sequences has certain evolutionary implications. First, it suggests that the transfer of mitochondrial genes to chromosomal DNA may have involved an initial duplication of a gene in the nucleus followed by a silencing and eventual degeneration of the mitochondrial copy. The failure of mitochondria to rid themselves of the vestigial gene sequences is difficult to explain but could be related to the mechanism of mtDNA replication in yeast. It is known that the size of yeast mtDNA falls within rather strict limits in both wild type as well as in ρ^- mutants. Most ρ^- mutants have a genome that approximates the size of wild type mtDNA, even when the retained segment is only a few base pairs long. This suggests that there is a fairly stringent requirement of genome size irrespective of the DNA sequence itself. The pecularities of the DNA replication mechanism in yeast could account for the retention of sequences which have no present day function. Presumably further evolution did allow for the eventual reduction of the genome to a more economic size (viz. mammalian mtDNA).

POSSIBLE ROLE OF G+C CLUSTERS IN RNA PROCESSING

A puzzling feature of the mitochondrial genome is the frequent occurence of short sequences (30-70 nucleotides) with a G+C content of 70-90%. The existence of these clusters was predicted by Bernardi and collaborators (22,24)from their studies of wild type mtDNA. Most of the G+C clusters contain multiple HpaII and HaeIII sites, although there are some that have no known restriction sites. Some 15-20 clusters have been sequenced in our laboratory with the remarkable finding that many are homologous to each other (11). Furthermore, in some instances clusters that are separated by A+T regions or by genes, exhibit inverted homologies (25). In the DS401 ρ^- DNA for instance, there are 10 G+C clusters, 8 of which correspond to the HpaII sites shown in Fig. 2. The other two clusters have no restriction sites and occur within the HpaII fragments 3 and 6. Some examples of G+C sequences are shown in Fig. 7.

Prunell and Bernardi (22) have estimated that there are some hundred G+C clusters in the wild type genome of yeast. The large number of such sequences as well as their random distribution makes it unlikely that they act as promoters or transcriptional stops. A more tenable interpretation is that repetitive G+C sequences and in particular the inverted repeats are involved in the processing of primary RNA transcripts. A hypothetical scheme showing some possible processing steps is shown in Fig. 8. In this model, it is assumed that mitochondrial DNA is transcribed from a few promoter sites giving rise to high molecular weight RNA. Some of the G+C clusters present in

```
     T
  G     T
 T       C
   T·A
   G·C
   G·C
   C·G
   C·G
   T·A
   A·T
   A·T
   C·G
  GC·G  T
  C
   G·C
   G·C
   G·C
   C·G
   G·C
   C·G
   C·G
   C·G
   C·G
5' TATAAA   AATATTA 3'
```

```
3' CAAGG   TTGGG 5'
        G·C
        C·G
        C·G
        G·C
        G·C
           A
           C
        C·G
        C·G
        C·G
        T·A
        C·G
        G·C
        G·C
        C·G
        C·G
        T·A
        T·A
        G·C
        G·C
        G·C
        G·C
        C·G
        T·A
        T·A
        T·A
        C·G
        C·G
        T·A
        C·G
5' AACAT   TTTAT 3'
```

FIGURE 7. Nucleotide sequences of three different G+C clusters present in DS401. The sequence on the left is a palindrome that includes the HhaI site at 84.1 map units. The two sequences on the right are inverted repeats that occur at 82 and 82.9 units of the map (see Fig. 2).

in the primary transcripts will form double stranded RNA either through base pairing of inverted repeats or of palindromic G+C clusters (see two examples in Fig. 7). The double stranded RNA regions may provide cleavage sites for specific RNAses similar to RNase III that has been implicated in the processing of the 16S ribosomal RNA gene of *E. coli* (26). Further processing of the initial

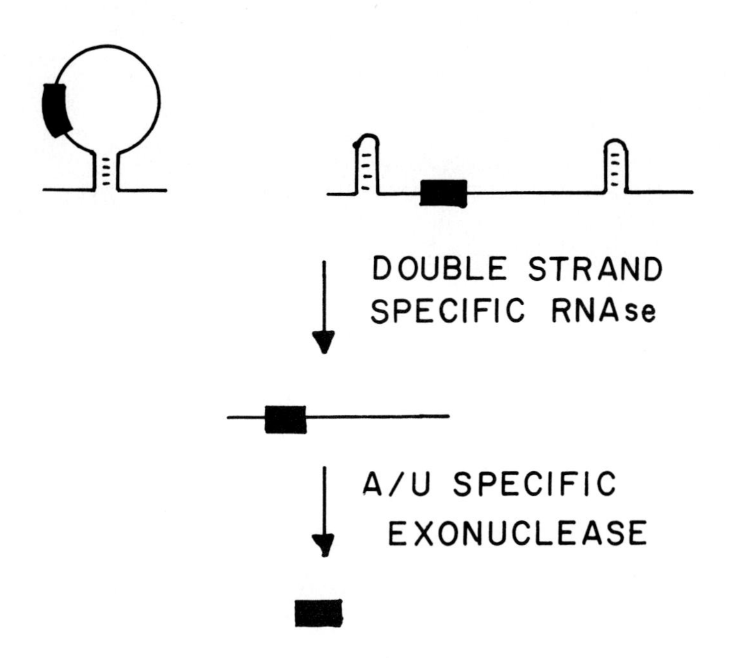

FIGURE 8. Hypothetical scheme for the processing of mitochondrial transcripts.

cleavage products may entail exonuclease trimming of A+U sequences bordering the structural gene.

REFERENCES

1. Borst, P., and Grivell, L.A. (1978). Cell 15, 705-723.
2. Tzagoloff, A., Akai, A., Needleman, R.B. and Zulch, G. (1975). J. Biol. Chem. 250, 8236-8242.
3. Slonimski, P.P., and Tzagoloff, A. (1976). Eur. J. Biochem. 61, 27-41.
4. Coruzzi, G., Trembath, M.K., and Tzagoloff, A. (1978). Eur. J. Biochem. 92, 279-287.
5. Borst, P. (1972). Ann. Rev. Biochem. 41, 333-376.
6. Goldring, E.S., Grossman, L.J., Krupnick, D., Cryer, D.R., and Marmur, J. (1970). J. Mol. Biol. 52, 323-335.
7. Faye, G., Fukuhara, H., Grandchamp, C., Lazowska, J., Michel, F., Casey, J., Getz, G.S., Locker, J., Rabinowitz, M., Bolotin-Fukuhara, M., Coen, D., Deutsch, B., Dujon, P., Netter, P., and Slonimski, P.P. (1973). Biochimie 55, 779-792.
8. Locker, J., Rabinowitz, M., and Getz, G.S. (1974). J. Mol. Biol. 88, 489-502.
9. Lazowska, J., Jacq, C., Cebrat, S., and Slonimski, P.P. (1976). In "The Genetic Function of Mitochondrial DNA" (C. Saccone and A.M. Kroon, eds.), pp. 325-335. North Holland Press, Amsterdam.
10. Macino, G., and Tzagoloff, A. (1979). Proc. Nat. Acad. Sci. USA 76, 131-135.
11. Cosson, J., and Tzagoloff, A. (1978). J. Biol. Chem. 254, 42-43.
12. Avner, P.R., Coen, D., Dujon, B., and Slonimski, P.P. (1973). Mol. Gen. Genet. 125, 9-52.
13. Sebald, W., Wachter, E., and Tzagoloff, A. (1979). Eur. J. Biochem. In Press.
14. Morimoto, R., and Rabinowitz, M. (1979). Mol. Gen. Genet. 170, 25-48.
15. Maxam, A., and Gilbert, (1977) Proc. Nat. Acad. Sci. USA 74, 560-564.
16. Douglas, M.G., and Butow, R.A. (1976). Proc. Nat. Acad. Sci. USA 73, 1083-1086.
17. Strausberg, R.L., Vincent, R.D., Perlman, P.S., and Butow, R.A. (1978). Nature 276, 577-583.
18. Macino, G., and Tzagoloff, A. (1979). Mol. Gen. Genet. 169, 183-188.
19. Martin, N.C., and Rabinowitz, M. (1978). Biochemistry 17, 1628-1634.

20. Claisse, M.L., Spyridakis, A., Wambier-Kluppel, M.L., Pajot, P., and Slonimski, P.P. (1978). In "Biochemistry and Genetics of Yeast" (M. Bacila, B.L. Horecker, and A.O.M. Stoppani, eds.), pp. 369-390. Academic Press, New York.
21. Alexander, N.J., Vincent, R.D., Perlman, P.S., Miller, D.H., Hanson, D.K., and Mahler, H. (1979). J. Biol. Chem. 254, 2471-2479.
22. Prunell, A., and Bernardi, G. (1977). J. Mol. Biol. 110, 53-74.
23. Sanders, J.P.M., Heyting, C., Verbeet, M.Ph., Meijlink, F.C.P.W., and Borst, P. (1977). Mol. Gen. Genet. 157, 239-261.
24. Fonty, G., Goursot, R., Wilkie, D., and Bernardi, G. (1978). J. Mol. Biol. 119, 213-235.
25. Macino, G., and Tzagoloff, A. (1979). J. Biol. Chem. In Press.
26. Young, R.A., and Steitz, J.A. (1978). Proc. Nat. Acad. Sci. USA 75, 3593-3597.

IDENTIFICATION AND SEQUENCING OF YEAST MITOCHONDRIAL tRNA GENES IN MITOCHONDRIAL DNA - pBR322 RECOMBINANTS[1]

Nancy C. Martin[+], Dennis L. Miller[*], John E. Donelson[*], Chris Sigurdson[+], James L. Hartley[*], Patrick S. Moynihan[*], and Hung Dinh Pham[+]

Department of Biochemistry, University of Minnesota[+], Minneapolis, MN and Department of Biochemistry, University of Iowa[*], Iowa City, Iowa

ABSTRACT We have cloned yeast mitochondrial DNA in the *E. coli* plasmid pBR322 by the poly(dA):poly(dT) tailing method and found that 62 of the 347 transformants obtained contain mitochondrial tRNA genes. In order to correlate the cloned mitochondrial tRNA genes with their gene products, we screened the recombinants with nick translated DNA isolated from petites known to carry choramphenicol (C), erythromycin (E), paramomycin (P),or oligomycin (O) markers as well as a limited subset of tRNA genes. A serine tRNA gene mapping in the O_I region of the mitochondrial DNA was identified by hybridizing DNA from O_I positive clones with [^{3}H]seryl tRNA. Other tRNA genes were identified by hybridizing [^{32}P]tRNA to Southern transfers of restriction enzyme digests of recombinant DNA molecules. Fragments selected for sequencing contained tRNA genes for two serine tRNAs, a phenylalanine tRNA, a glycine tRNA, a valine tRNA and an arginine tRNA. None of the tRNA genes appear to have intervening sequences, none have the CCA end encoded and all are surrounded by long stretches of very AT rich DNA.

INTRODUCTION

Mitochondrial DNA (mt DNA) codes for messenger RNAs for some respiratory components as well as the ribosomal RNAs and transfer RNAs (tRNAs) used in mitochondrial protein synthesis (1). Mitochondrial coded tRNAs corresponding to all of the 20 common amino acids have been identified by the hybridization of ^{3}H amino acyl tRNAs to mtDNA (2-4).

[1]This work was supported by grant number PCM77-17694 from the National Science Foundation to N.C.M. and grant number GM-21696 from the National Institutes of Health to J.E.D.

ISBN 0-12-198780-9

Isoaccepting mitochondrial tRNAs have been identified by column chromatography (5) and two dimensional gel electrophoresis (6) and it now appears that yeast mtDNA codes for a complete set of tRNAs for use in mitochondrial protein synthesis.

Mitochondrial tRNA genes have been located on the circular mitochondrial genome by hybridization of bulk labeled tRNA to restriction fragments of mtDNA according to the method of Southern (7,8) and by deletion mapping using a series of petite mutants (2-5). The data from these two techniques enabled the ordering of tRNA genes with respect to each other and indicated that about two thirds of the tRNA genes are located within 15% of the genome while the others are widely dispersed.

To provide a ready source of mtDNA for our studies on the structure and transcription of mitochondrial tRNA genes we have cloned random fragments of yeast mtDNA in *E. coli* and screened the transformants for tRNA genes with [^{32}P] labeled mitochondrial tRNA. Mitochondrial tRNA genes were present in 62 of the 347 transformants we obtained. We then screened the transformants with petite DNAs that had been characterized with respect to the tRNA genes they retained (Figure 1). This strategy enabled us to quickly identify clones in the bank carrying genes for a mitochondrial serine tRNA.

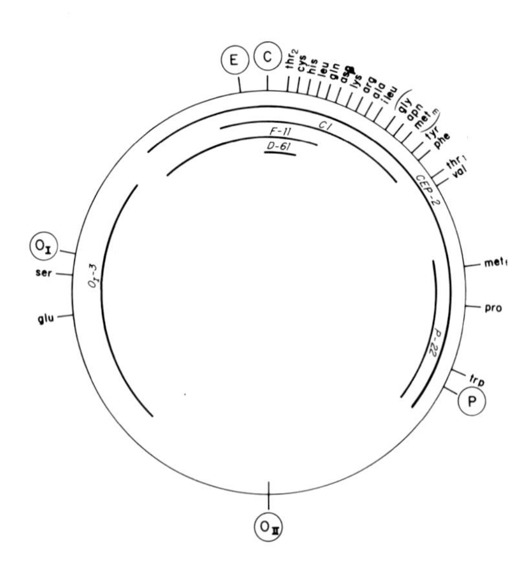

Figure 1. A genetic map of yeast mtDNA taken from the data of Fukuhara *et al.* (2), Martin *et al.* (3), and Wesolowski *et al.* (4).

DNA isolated from clones which contain the serine gene as well as from clones known to contain other tRNA genes was digested with various restriction enzymes, run on gels and transferred to nitrocellulose filters by the method of Southern (9). We have sequenced a 320 base pair HpaII fragment containing a serine tRNA gene from the O_I region of mtDNA. A 680 base pair HpaII fragment containing a phenylalanine tRNA gene and a 420 base pair HpaII fragment containing a second serine tRNA gene as well as an arginine tRNA gene from the CEP region have also been sequenced. Finally, we have sequenced portions of a 1029 base pair HpaII fragment and a 1800 base pair HaeIII fragment containing a glycine and valine tRNA gene respectively.

METHODS

Isolation and Radioactive Labeling of Mitochondrial Nucleic Acids. Mitochondrial DNA from the wild type strain D273-10B and from the petite strains shown in figure 1 was isolated either by the procedure of Hirt as modified for yeast by Livingston and Klein (10) or from isolated mitochondria as described previously (3). MtDNAs were labeled with [^{32}P] to a specific activity of 10^7 cpm/μg by nick translation as described by Schachat and Hogness (11). Mitochondrial tRNA was isolated from mitochondria (3) and labeled to a specific activity of 10^6 cpm/μg at the 3' end with [α^{32}P]ATP and *E. coli* nucleotidyl transferase.

Construction and Transformation of Recombinant DNAs. The bacterial plasmid pBR322 was isolated, cut with EcoR1 and tailed with approximately 300 dAMP residues using terminal deoxynucleotidyl transferase. Random fragments of yeast mtDNA from the wild type D273-10B were tailed with apprcximately 300 dTMP residues. The two DNAs were allowed to anneal overnight at 37°C and were then used to transform the *E. coli* strain HB101 (12). Transformants were isolated by selection for ampicillin resistance.

Resolution of DNA Fragments and Filter Hybridizations. Plasmid DNA and yeast mtDNA were digested with various restriction enzymes according to procedures suggested by the commercial sources of the enzymes. Analytical gel electrophoresis was done according to standard procedures and restriction fragments were transferred to nitrocellulose filters as described by Southern (9). Colony hybridizations were performed as described by Grunstein and Hogness (13). [^{32}P] tRNA was hybridized in 5 X SSC for 16 hours at 62°C and

[^{3}H] amino acyl tRNA in 33% formamide, 2 x SSC, pH 5.2 for 4 hours at 37°C.

Nucleic Acid Sequencing. Restriction fragments were labeled either by transfer of the γ phosphate from [γ^{32}P] ATP using polynucleotide kinase or by filling in staggered cuts with the appropriate [α^{32}P] deoxynucleotide triphosphate and DNA polymerase I. Sequencing was by the method of Maxam and Gilbert(14). The serine tRNA was isolated by RPC-5 chromatography (5) followed by gel electrophoresis and labeled at the 3' end with [α^{32}P] ATP. Partial nuclease digests and electrophoresis were as described by Simoncsits et al. (15)

Containment Conditions. In accordance with the N.I.H. Guidelines for Recombinant DNA Research the experiments involving recombinant plasmids between pBR322 and yeast mtDNA were conducted under P2 + EK1 containment conditions.

RESULTS

Characterization of the Recombinant Plasmids Containing Yeast mtDNA Inserts. Three hundred and forty-seven transformants were obtained and an analysis of about 50 different transformants indicated that the size of the inserts ranged from 0.5-5 kilobases with an average size of about 2.5 kilobases. Sixty-two transformants containing one or more yeast mitochondrial tRNAs genes were identified by the colony hybridization procedure of Grunstein and Hogness (13). An example of one such screening experiment is shown in Figure 2.

The transformants were also screened with nick translated petite DNAs. Figure 3 summarizes the results of this screening. As predicted, colonies containing DNA that hybridized to CEP-2 DNA also hybridized to D61 DNA. Conversely, colonies containing DNA that hybridized with O_I-3 DNA did not hybridize to CEP-2 DNA. The hybridizations of the different individual petite DNAs are, in general, internally consistent with their physical sizes as determined by Lewin (16). For example, CEP-2 DNA represents about 2.5 times more of the wild type genome than O_I-3 DNA (37 kb vs 14 kb) and about 2.2 times more colonies contained DNA that hybridized to CEP-2 DNA than to O_I-3 DNA (57 colonies vs 18 colonies). It should be noted, however, that the DNA retained in CEP-2 and O_I-3 petites corresponds to 70% of the wild type genome and only about 30% of the 347 transformants contained DNA that hybridized strongly to these petite DNAs.

Eight colonies contained DNA that hybridized to mitochondrial tRNA but not to any of the petite DNAs, thought to contain all of the mitochondrial tRNA genes.

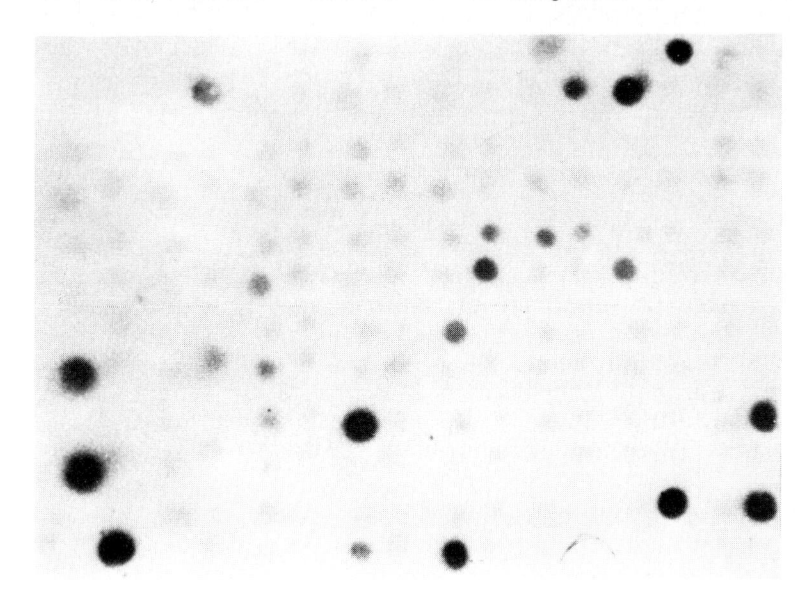

Figure 2. An autoradiogram of the hybridization of [^{32}P] tRNA to colonies fixed to nitrocellulose filters.

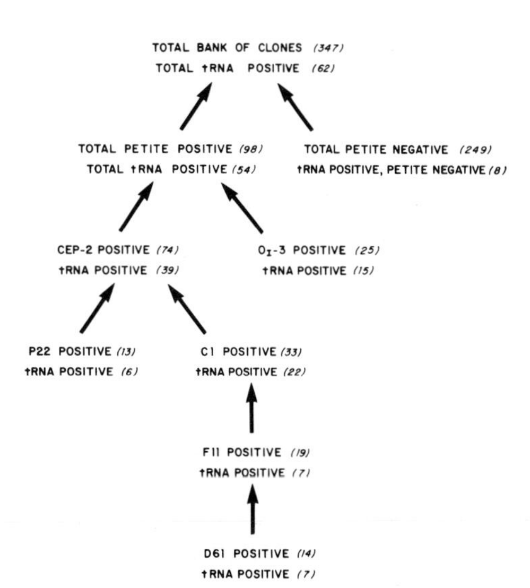

Figure 3. A summary of the tRNA and petite DNA hybridizations. The number of total petite positives (98) is one less than the sum of CEP-2 and O_I-3 positives (74 + 25). This is because both petite DNAs hybridized to a clone with a 3.5 kb insert. The closest end point of these petites may be less than 2 kb (15) so this cloned fragment might span the distance between these two end points.

Identification and Sequencing of Mitochondrial tRNA Genes. We have analysed a number of the recombinant plasmids that were positive for hybridization to mitochondrial tRNA by the Grunstein screening shown in figure 2. The results obtained for the recombinants which contain tRNA genes that we have sequenced are summarized here.

Serine tRNA Gene. Figure 4, panel A, lane a, displays the HpaII digest pattern obtained when pYm 214 DNA is digested and run on an agarose-acrylamide gel. Panel B, lane a, shows an autoradiogram of a Southern transfer (9) of this DNA followed by the hybridization of [^{32}P] mitochondrial tRNA. Lane b is the pattern obtained when the recombinant DNA digest is mixed with HpaII digested uncloned mtDNA and lane c is the digested uncloned mtDNA alone. A 320 base pair fragment that hybridizes tRNA is present in all three lanes. Since pYm 214 also hybridized with DNA from the O_I-3 petite we predicted that it contained either a tRNA^{ser} or tRNA^{glu} gene. The data in Table I shows that this recombinant contains a tRNA^{ser} gene. Eleven other transformants also contain this 320 base pair fragment.

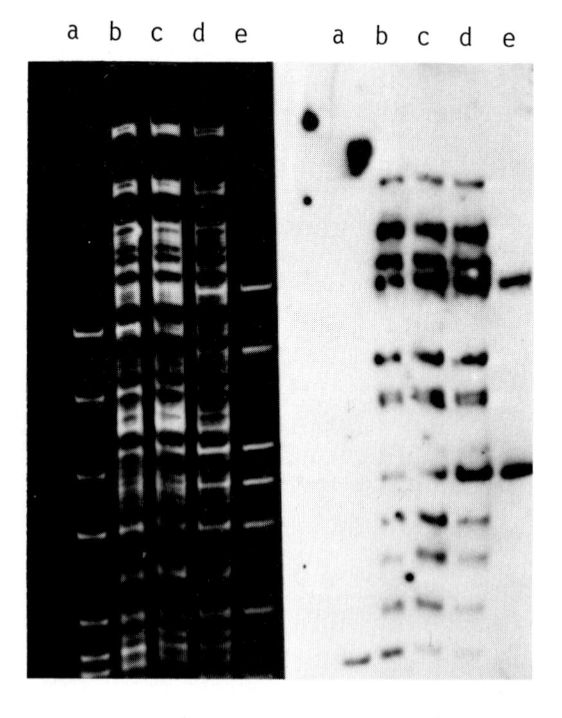

Figure 4. HpaII digestions (A) of plasmid and mtDNA and an autoradiogram (B) showing hybridization with [^{32}P] mitochondrial tRNA. See text for discussion.

TABLE I
HYBRIDIZATION OF [^{3}H] tRNAs TO mtDNAs

DNA	[^{3}H] tRNA^{glu}	[^{3}H] tRNA^{ser}
mtDNA (5 μg)	392	555
O_I-3DNA (10 μg)	702	4437
pYm 214 (1 μg)	0	412
pYm 215 (1 μg)	-	659

Blank values have been subtracted

The sequence of the Hpa 320 bp fragment was determined. A sequencing gel of a part of this fragment is shown in Figure 5 and its complete nucleotide sequence is given in Figure 6.

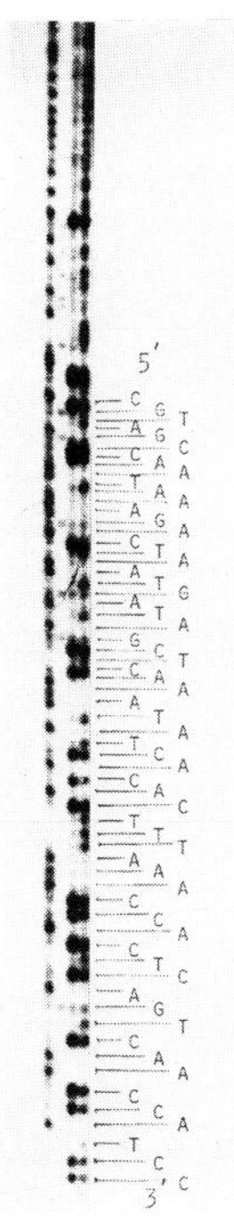

Figure 5. A polyacrylamide gel autoradiogram showing the pattern of cleavage products obtained by the technique of Maxam and Gilbert. This gel pattern is of one strand of the 3' labeled end of the Hpa 320 fragment and covers the sequence of the tRNA^{ser}.

```
HpaII
1                             30                            61
CCGGGCCCCGAAAGGTGTTTATTTAATATTTATATTTATATTAATATTTATATTTATATTT
→

                  HincII      90                  AluI      122
62
ATATTCCTCTTAAGGATGGTTGACTGAGTGGTTTAAAGTGTGATATTTGAGCTATCATTAG
                    pGGAUGGUUGACUGAGUGGUXUAAAGXGUGAUAXXUGAXXUAUCAUUAG
123                           150                       ←     183
TCTTTATTGGCTACGTAGGTTCAAATCCTACATCATCCGTAATAATACATATATATAATAA
UCUUUAUUGGXUACGUAGGXUCAAAUCCUACAUCAUCCG
184                           210                           244
TAATTTTAATATTATTCCTATAAAAATAAAATAAATAAATAAATAATAATAATTAATTAAT

245                           270                           305
TTTAATAAATATAAAATATATAAAATAATAATAATAATAATTATTATTTTAATAATATTAT

306
TTTATTATTATAGCCGG  3'
                ←
```

Figure 6. The sequence of the yeast mitochondrial tRNA^{ser} gene. In this and all subsequent sequences only the noncoding strand is presented. The arrows indicate the direction of sequencing away from the labeled end. Double lines indicate the areas where overlapping sequences were obtained. Italics represent tRNA^{ser} sequence.

In order to compare the DNA sequence with the tRNA sequence we isolated the mitochondrial tRNA^{ser} that hybridized to this fragment and determined its primary sequence by running gels of 3' end labeled tRNA^{ser} after partial nuclease digestions as described by Simoncists (15). Figure 7 shows one such gel. Data from several tRNA isolations and sequencing gels have been compiled to give the tRNA sequence shown below the gene sequence in figure 6. Positions with an X rather than a nucleotide are those that did not show a band in any of the nuclease digests even though a band was present in the formamide ladder at that position. A comparison of the DNA sequence and the tRNA sequence demonstrate that they are colinear and that the CCA end is not encoded. This tRNA^{ser} is 38% GC and the nucleotide sequence on either side is greater than 90% AT.

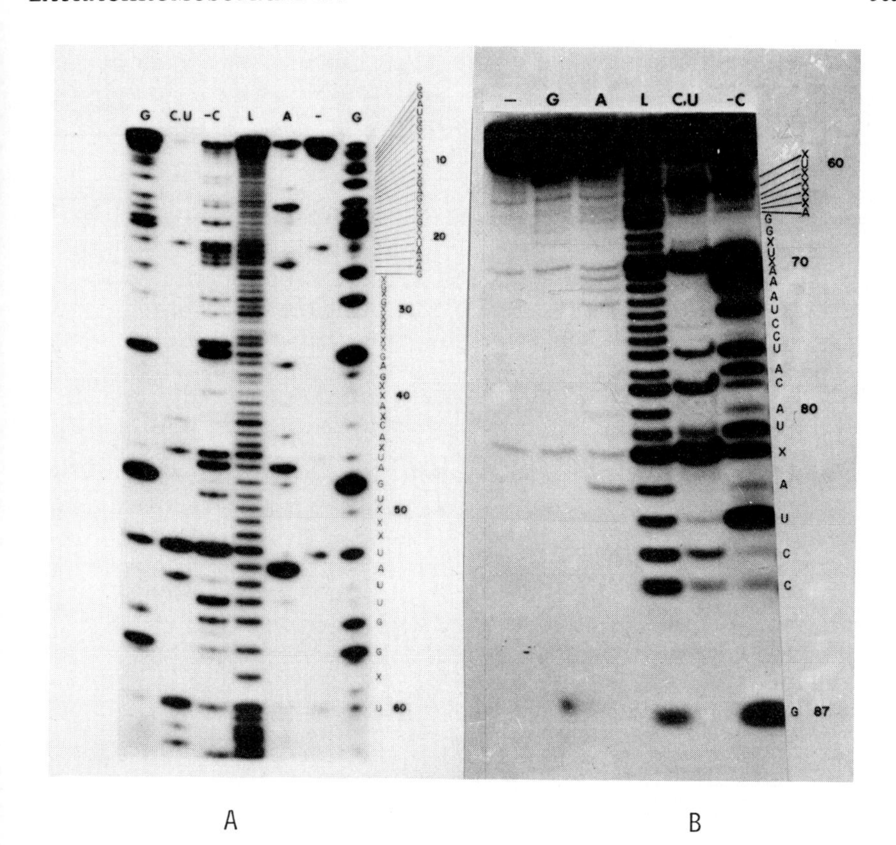

Figure 7. RNA sequencing gels of yeast mitochondrial $tRNA^{ser}$. (A) 10% gel; (B) 25% gel.

Phenylalanine tRNA Gene. Figure 4, panel A, lane e displays the HpaII digest pattern obtained when pYm 518 DNA is digested and run on an agarose-acrylamide gel. Panel B shows an autoradiogram of a Southern transfer (9) of this DNA followed by hybridization of [^{32}P] mitochondrial tRNA. Lane d is the pattern obtained when digests of pYm 518 and uncloned mtDNA are mixed and lane c is a digest of mtDNA alone. Two fragments, one 1450 and one 680 base pairs long hybridize mitochondrial tRNA. 5 transformants contained both of these fragments, one contained the 680 fragment alone and one contained the 1450 fragment alone. Sequence analyses of both fragments enabled us to demonstrate that the 680 base pair fragment contains the $tRNA^{phe}$ gene. Figure 8 gives the sequence of a portion of this fragment. A comparision of the the DNA sequence with the $tRNA^{phe}$ sequence determined by R.P. Martin (17) demonstrates that the DNA sequence and tRNA sequence are colinear and again that the CCA end is not encoded. The $tRNA^{phe}$ is 33% GC and the surrounding nucleotide sequence is again almost completely AT.

```
 HpaII
1                          30                            61
CCGG...about 300 bp to...TAATTTAATAATATAAATATATATATTATATATTAT

62                         90                            122
GTTTTAATTATATATATATATATATATTATGTATTATTATATAAATATATATATATATTATAT

123                        150            AluI           183
TATAAGTAATAATAAGTATTATATTATATATAGCTTTTATAGCTTAGTGGTAAAGCGATAA

                     TaqI   HinfI
184                         210                          244
ATTGAAGATTTATTTACATGTAGTTCGATTCTCATTAAGGGCAATAATAATAATATATTAA

245                        270                           305
TTAATAATTAATTTATAATAAATATATTATAATAATTAATATATATATATAATATATTTAA

306                        330                           366
TACAAAAGAAAATATATATTATATCTCTTATTTATTTATTTATTAATATTTTAATAAATAT

367
AATATTATAAAAAAAAAAGTTT...CCGG 3'
```

Figure 8. The nucleotide sequence of the mitochondrial $tRNA^{phe}$ gene. The dots signify the tRNA structural gene.

Valine tRNA Gene. The 1450 base pair fragment seen in Figure 4 was partially sequenced and found to contain a tRNA gene that specifies a tRNA with a valine anticodon. The DNA sequence containing this tRNA gene and its surrounding AT rich region is shown in figure 9. Although the tRNA sequence is not yet available for direct comparison, a cloverleaf model of this tRNA based on the DNA sequence is given in figure 12. There is no reason to believe that this tRNA gene contains nucleotides not found in the tRNA itself.

Glycine tRNA Gene. A Hpa II fragment from pYm318 that was subjected to DNA sequence analysis is 1029 base pairs long and contains a tRNA gene that specifies a tRNA with a glycine anticodon. The sequence of the $tRNA^{gly}$ gene and its surrounding region is shown in figure 10. Again, a cloverleaf model based on the DNA sequence (Figure 12) does not indicate an intervening sequence.

Although we have not done an experiment where the

recombinant plasmid containing this fragment has been digested with HpaII and mixed with digested uncloned mtDNA to prove it is present in the wild type genome, three lines of evidence suggest that it is. First, a band of about 1020 base pairs in a HpaII digest of uncloned DNA does hybridize tRNA. Second, three independently isolated recombinants contain a fragment of this size which hybridizes mitochondrial tRNA. Third, all three of these recombinants contained DNA that hybridized only to CEP-2 and C1 petite DNAs. This is consistent with the original location of the $tRNA^{gly}$ gene by deletion mapping (2).

```
1                              30                             61
GAGATTATATAATTATAGAATTATTATAATTATATATATATAAATAAATAAAATAATAATT

62                             90                             122
ATAAATAAATTAATAAGAGTTTGGATATATATCTGTGGAGTATATATTTTATAAAGGAGAT

AluI                                               TaqI
123                        150                                183
TAGCTTAATTGGTATAGCATTCGTTTTACACACGAAAGATTATAGGTTCGAACCCTATATT

184                            210                            244
TCCTAAATCTAGATATAATATTATATCTATCTTAATATAATAATATTTATTTATTTATTTA

                                          HinfI
245                            270                            305
TTAAATAAAAAAAAATAAATAATATTAATTAATATAAGATTCTTTTTAATTATAATAATAA

                                      HaeIII
306
ATAAATAAAAGAGTA...about 470 bp to...GGCC  3'
```

Figure 9. The nucleotide sequences of the mitochondrial $tRNA^{val}$ gene. The dots signify the tRNA structural gene.

```
HpaII
1                             30                              61
CCGG...about 540 bp to...AATAGTAATGATAGATGTAGGTTAATTGGTAAACTG

                         TaqI  HinfI
62                            90                              122
GATGTCTTCCAAACATTGAATGCGGGTTCGATTCCCGCTATCTATAATTAATATTAATATT

123                           150                             183
AACCAATATCCTATAATTAACTAAATACAAAATTATATTAAAACTTATATTATATTATAAT

                                      HpaII
184
ATTATATTAATATT...about 490 bp to...CCGG  3'
```

Figure 10. The nucleotide sequences of the mitochondrial $tRNA^{gly}$ gene. The dots signify the tRNA structural gene.

Arginine and Serine II Gene. Finally, we have sequenced a 420 base pair HpaII fragment isolated from the recombinant pYm 267. Figure 11 gives the nucleotide sequence of this fragment. It contains a second tRNA gene with a serine anticodon and we have designated this sequence as mitochondrial $tRNA^{serII}$. The same 420 base pair fragment also contains a tRNA gene with an arginine anticodon.

```
HpaII
1                            30                               61
CCGG...GGGAAAATTAATATAAAAATAATTAATAATTTATTAATAATTTATTAATTTATT
→
62                           90                              122
AATTTATTAATTTATTTATTAATTTATTAATTTATTTATTATTATATTTTTTTTAATAAAG
                                                               .
123                         150                              183
GAAAATTAACTATAGGTAAAGTGGATTATTTGCTAAGTAATTGAATTGTAAATTCTTATGA
.............................................................
184                         210                              244
GTTCGAATCTCATATTTTCCGTATATATCTTTAATTTAATGGTAAAATATTAGAATACGAA
....................     .....................................
245                         270                              305
TCTAATTATATAGGTTCAAATCCTATAAGATATTATATTATATTATATAATATTATATATT
.................................

306                         330                              366
AATAAATATTATTAATTAATTTATTTATTTATTTATTATTAAATAAAAATATTTAATAGTT

367  HpaII
CC...CCGG  3'
        ←
```

Figure 11. The nucleotide sequence of the mitochondrial $tRNA^{arg}$ and $tRNA^{serII}$ genes. The dots signify the tRNA structural genes.

The 420 bp fragment is the least well characterized fragment in terms of a comparison to the uncloned wild type genome. It is also potentially the most interesting with regard to tRNA transcription. Since the $tRNA^{ser}$ and $tRNA^{arg}$ genes are separated by only three nucleotides, the possibility is raised that these two tRNAs are transcribed as part of a multicistronic precursor. Such speculation is premature until a corresponding fragment in the uncloned wild type genome is identified. Experiments designed to identify such a fragment are currently in progress in our laboratory.

DISCUSSION

We have constructed recombinant plasmids containing yeast mtDNA and used them to isolate yeast mitochondrial tRNA genes for sequence analysis. The average size of the mtDNA inserts in the recombinant plasmids is about 2.5 kilobases. When the 347 recombinants were screened with petite mtDNA we found that only 98 of the clones were homologous to the CEP-2 and 0_I-3 petite DNAs which together represent 70% of the wild type genome (16). This figure predicts that a minimum of 140 of our recombinant plasmids must contain mitochondrial DNA. The statistical formulation described by Clarke and Carbon (18) gives a probability of greater than 99% that any given small sequence in the total mtDNA will be present in the 140 transformants.

There are several explanations for the unexpectedly low number of recombinants that hybridized to petite DNAs. First, some pBR322 DNA molecules may not contain inserts. Second, some plasmids could have nuclear inserts resulting from the cloning of low levels of nuclear DNA that may have contaminated the mtDNA used for cloning. If yeast nuclear DNA is more readily cloned, a disproportionate number of clones containing nuclear DNA could have resulted. Third, in practice it was more difficult to distinguish clones containing DNA that gave positive hybridization with nick translated petite DNA than it was to distinguish a positive hybridization in tRNA screenings. Therefore, transformants having small inserts homologous to petite DNAs may have been scored as negatives. Finally, it is possible that some regions of the mitochondrial genome of yeast are more difficult than others to maintain as cloned DNA fragments in bacterial plasmids. We cannot rule out, but have no evidence for, the possibility that regions of the wild type genome cannot be cloned.

There have been questions raised about the stability of yeast mtDNA - pBR322 recombinants in *E. coli*. Since it has been observed that highly repetitive DNA can be unstable in recombinant plasmids (19) and because mtDNA is very AT rich, some problems with the stability of our clones would not have been unexpected. Although we have not done chemostat experiments to determine the effect of long-term growth on the stability of the recombinant plasmids, we have repeatedly isolated DNA from several different clones in two different laboratories during the past year and have never observed a change in the restriction enzyme digest pattern of these DNAs.

Care must be taken when using recombinant plasmids for the study of gene organization and structure to be sure that the cloned DNA is unchanged with respect to its uncloned counterpart. We have analysed about 25 recombinants containing mitochondrial tRNA genes by restriction enzyme analysis followed by hybridization to mitochondrial tRNAs where uncloned mtDNA digests were run mixed with or next to the recombinant digests. The same size fragments hybridize tRNAs in these recombinants and in the uncloned mtDNA. Another argument against large rearrangements in fragments containing tRNA genes is that these fragments have been found in several independently isolated clones. Finally, where comparisons can be made with the tRNA sequences it is clear that rearrangements within the coding regions have not occurred.

The data obtained from the sequence analysis of the 420 base pair fragment is of particular interest because it identifies a new mitochondrial $tRNA^{ser}$ gene. Although the presence of a second mitochondrial $tRNA^{ser}$ has been suggested by column chromatography (20) and two dimensional electrophoresis (5) convincing hybridization data showing that a $tRNA^{ser}$ maps to a region of the genome other than the O_I region has not been obtained. Since pYm 267 DNA hybridized with petite DNAs from the CEP region of the genome it would appear that a second mitochondrial $tRNA^{ser}$ gene is located in that area.

The coding sequences of the $tRNA^{serI}$ and $tRNA^{phe}$ genes could be directly compared to their products since the tRNA sequences are known (figures 5 & 6, ref. 17). Neither gene encodes the CCA end. Although the tRNA sequences of the $tRNA^{arg}$, $tRNA^{val}$, $tRNA^{gly}$, and $tRNA^{serII}$ are not available, an examination of the cloverleaf models (Figure 12) and the nucleotides immediately following the 3' end demonstrate that the CCA end is not encoded in any of these tRNA genes. Some phage tRNA genes (21) and apparently all E. coli tRNA genes (22) do encode CCA ends. On the other hand, yeast nuclear encoded tRNAs appear to lack a transcriptionally encoded CCA end (23,24). Thus it appears that, with respect to the CCA end, yeast mitochondrial tRNA genes resemble eukaryotic tRNA genes rather than prokaryotic tRNA genes.

Of primary interest is a comparison of the tRNA sequences with their gene sequences to see if they contain the 14-19 base pair intervening sequence found in some yeast nuclear tRNAs genes and transcripts (23,24,25). $tRNA^{serI}$ and $tRNA^{phe}$ can be compared to their DNA sequences and the sequences are colinear. Again, although the tRNA sequences of the other genes we have sequenced are not available a consideration of the cloverleaf models (Figure 12) strongly indicates that these genes do not contain intervening sequences.

The regions directly surrounding the gene sequences are very A + T rich. There do not appear to be significant regions of either dyad symmetry, palindromes, tandem repeats, or G + C rich regions. Since the area is so A + T rich it is not suprising that sequences resembling the RNA polymerase binding site proposed by Pribnow (26) are seen near the 5' end of the genes. The tRNA^{phe} gene has a sequence 5' TATTATA 3' / 3' ATAATAT 5' nine base pairs from the 5' end of the gene, the tRNA^{serI} gene has the sequence 5' TATATTC 3' / 3' ATATAAG 5' seven bases from the 5' end of the gene, the tRNA^{val} gene has the sequence 5' TATATTT 3' / 3' ATATAAA 5' five base pairs from its 5' end and the tRNA^{arg} gene has the sequence 5'TATATTT 3' / 5'ATATAAA 3' eleven bases from its 5' end. All of these sequences are two base deviations from the prototype sequence. When the areas near the 5' end of the mitochondrial tRNA genes are compared with the sequence located before the nuclear tRNA^{phe} genes (24) and with the sequence located in front of the nuclear tRNA^{tyr} genes (23) no similar sequences are found. One mitochondrial sequence, 5' TTATATT 3' / 3' AATATAA 5', is found seven base pairs before the 5' end of the tRNA^{phe} gene, eight base pairs before the 3' end of the tRNA^{serI} gene, twelve base pairs before the tRNA^{arg} gene and with one nucleotide change, six base pairs before the tRNA^{val} gene. This sequence could be a signal for the initiation of transcription of mitochondrial tRNA genes.

An examination of the region following the 3' end of the mitochondrial gene sequences does not reveal the continuous string of seven or eight A's in the coding strands that were seen closely following the yeast nuclear tRNA genes (23,24). In fact, it does not appear that there are any significant areas of homology.

Prunell and Bernardi (27) have proposed that structural units exist within mtDNA which are composed of four sequence elements: (i) "site clusters" which are short G + C rich elements about 35 base pairs in length containing recognition sites for the restriction enzymes HaeIII and HpaII followed by (ii) the actual structural genes which are about 22-65% GC, (iii) A + T rich regions with an average size of 600 base pairs which are less than 5% G + C, (iv) G + C rich clusters which are about 35 base pairs in length but do not contain HaeIII or HpaII sites. Our results are relatively consistent with the predictions of this model. The fragments are defined at their termini by sites for HpaII and HaeIII. The tRNA genes themselves qualify as intermediately G + C rich regions. However, instead of following immediately after the "site cluster" as predicted, the genes are surrounded on both sides by the A + T rich regions. Also, none of the fragments contain regions which correspond to the G + C rich clusters that lack restriction sites for HpaI, HaeIII and HpaII. These differences not withstanding, our results do support the notion that the A + T rich regions are intergenic spacer sequences.

The sequences of yeast mitochondrial $tRNA^{phe}$ (17) and Neurospora mitochondrial $tRNA^{fmet}$ (28) and $tRNA^{tyr}$ (29) have been published recently. The sequences (Fig. 12) inferred from the yeast mitochondrial tRNA genes reported here provide further information on the structure of mitochondrial tRNAs. As a group, they are very A + U rich when compared to most other tRNAs. (Table II).

TABLE II
A + U CONTENT OF YEAST MITOCHONDRIAL tRNA

Mitochondrial tRNA	% A + U
Yeast $tRNA^{serI}$	61%
Yeast $tRNA^{gly}$	63%
Yeast $tRNA^{val}$	66%
Yeast $tRNA^{phe}$ (17)	67%
Yeast $tRNA^{serII}$	72%
Yeast $tRNA^{arg}$	82%

All can be formed into a cloverleaf with the standard number of base pairs in the stems. Both serine tRNAs have long variable arms as do their cytoplasmic counterparts. The mitochondrial tRNAs resemble all tRNAs in that they share certain invariant characteristics (30). All have a U in position 8 and all but $tRNA^{serII}$ have an A in position 14. The mitochondrial Neurospora $tRNA^{tyr}$ also lacks the invariant A_{14}. In contrast to Neurospora mitochondrial $tRNA^{fmet}$ (28) all of the yeast mitochondrial tRNAs have the GG sequence in the D loop so they can pair with $\psi_{55}C_{56}$ and form the teritiary structure dictated by this pairing. All of the mitochondrial tRNAs have a U at position 33, a G at position 53, a C at position 56 and have a TTC sequence in the TψC loop. It is likely that the tRNAs will have ψ at position 55 but direct sequencing of the tRNAs will be necessary to ascertain this.

A consideration of semi-invariant positions (30) again emphasizes the resemblance yeast mitochondrial tRNAs show to other tRNAs. For example, all have a purine at positions 9, 15 and 57. All have pyrimidines at the position before the TψC stem begins and at the end of the TψC loop. Of the mitochondrial tRNAs in figure 12, $tRNA^{gly}$ is the only one

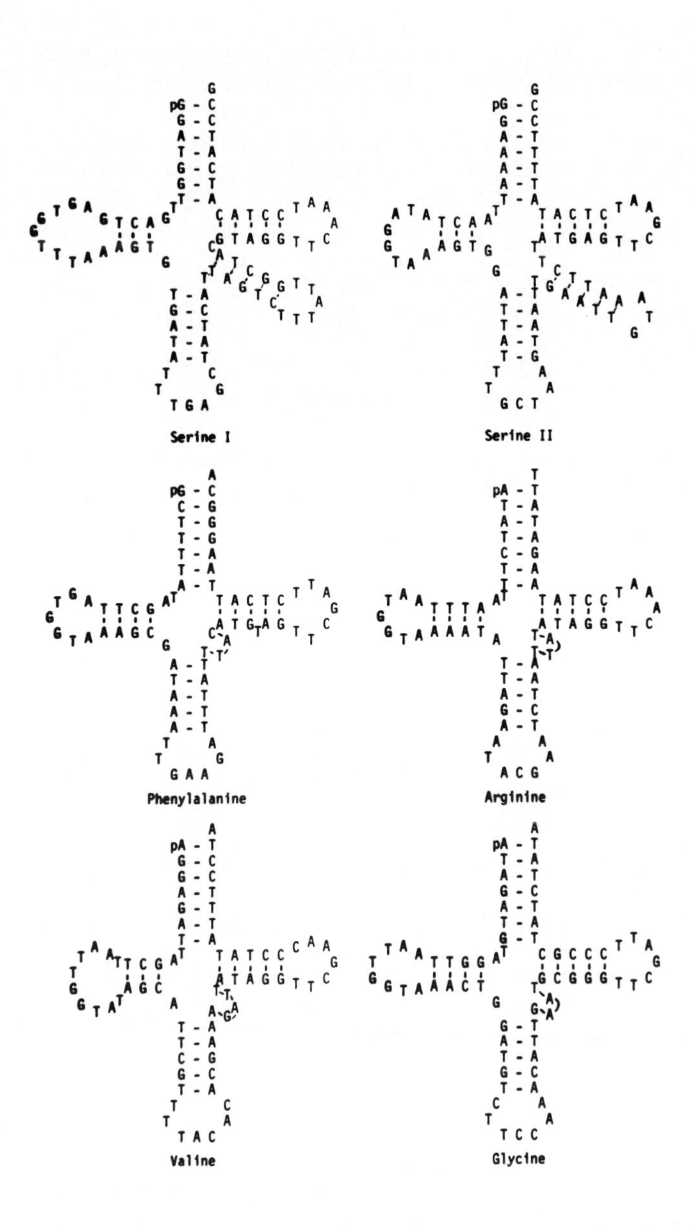

Figure 12. Cloverleaf models of yeast mitochondrial tRNAs based on their gene sequences.

that does not have a pyrimidine at position eleven, $tRNA^{val}$ does not have a purine preceeding the D stem and $tRNA^{arg}$ does not have a pyrimidine in the first position of the coding loop.

An examination of the sequences for some characteristic that might be considered "mitochondrial" reveals that all of the yeast mitochondrial tRNAs have an A-T base pair before the D loop. Most other tRNAs sequenced have a G-C pair at this position. It is interesting to note that this feature is also present in the mitochondrial $tRNA^{fmet}$ (28) and $tRNA^{tyr}$ (29) of *Neurospora crassa*.

ACKNOWLEDGMENTS

We would like to thank Connie Wandrey for her help in preparing the manuscript and Terry Jones for his help in preparing the figures.

REFERENCES

1. Borst, P., and Grivell, L.A. (1978). Cell 15, 705-723.
2. Fukuhara, H., Bolotin-Fukuhara, M., Hsu, H.J., and Rabinowitz, M. (1976). Mol. Gen. Genet. 145, 7.
3. Martin, N.C., Fukuhara, H., and Rabinowitz, M. (1977). Biochem. 16, 4672.
4. Wesolowski, M., and Fukuhara, H. In press.
5. Martin, N.C., and Rabinowitz, M. (1978). Biochem. 17, 1628.
6. Martin, R., Schneller, J.M., Stahl, A.J., and Dirheimer, G. (1977). Nuc. Acid. Res. 4, 3497.
7. Morimoto, R., Merten, S., Lewin, A., Martin, N.C., and Rabinowitz, M. (1978). Mol. Gen. Genet. 163, 241.
8. VanOmmen, G., Groot, G., and Borst, P. (1977). Mol. Gen. Genet. 154, 255.
9. Southern, E.M. (1975). J. Mol. Biol. 98, 503.
10. Livingston, D.M., and Klein, H.J. (1977). J. Bacteriol. 129, 472.
11. Schachat, F., and Hogness, D. (1973). Cold Spring Harbor SQB 38, 371.
12. Miller, D.L., Gubbins, E.J., Pegg, E.W., and Donelson, J. E. (1977). Biochem. 17, 1031.
13. Grunstein, M., and Hogness, D. (1975). Proc. Nat. Acad. Sci. 72, 3961.
14. Maxam, A., and Gilbert, W. (1977). Proc. Nat. Acad. Sci. 74, 560.

15. Simoncsits, A., Brownlee, G.G., Brown, R.S., Rubin, J.R., and Guilley, H. (1977). Nature 269, 833.
16. Lewin, A., Morimoto, R., Fukuhara, H., and Rabinowitz, M. (1978). Mol. Gen. Genet. 163, 257.
17. Martin, R.P., Sibler, A.P., Schneller, J.M., Keith, G., Stahl, A.J.C., and Dirheimer, G. (1978). Nuc. Acid Res. 5, 4579.
18. Clarke, L., and Carbon, J. (1976). Cell 9, 91.
19. Brutlog, D., Fry, R., and Nelson, T. (1977). Cell 10, 509.
20. Baldacci, G., Falcone, C., Frontali, L., Mancino, G., and Palleschi, C. (1976). "The Genetics and Biogenesis of Chloroplasts and Mitochondria." North Holland Publishing Co., Amsterdam. 759.
21. McClain, W.H., Seidman, J.G., and Schmidt, F. (1978). J. Mol. Biol. 119, 519-536.
22. Mazzara, G.P., and McClain, W.H. (1977). J. Mol. Biol. 117, 1061.
23. Goodman, H.M., Olson, M.U., and Hall, B.D. (1977). Proc. Nat. Acad. Sci. USA 74, 5453.
24. Valenzuela, P., Venegas, A., Weinberg, F., Bishop, R., and Rutter, W.J. (1978). Proc. Nat. Acad. Sci. USA 75, 190.
25. O'Farrell, P.Z., Cordell, B., Valenzuela, P., Rutter, W. J., and Goodman, H.M. (1978). Nature (London) 274, 438.
26. Pribnow, D. (1975). Proc. Nat. Acad. Sci. USA 72, 784.
27. Prunell, A., and Bernardi, G. (1977). J. Mol. Biol. 110, 53.
28. Heckman, J.E., Hecker, L.I., Schwartzback, S.D., Barnett, W.E., Baumstark, B., and RajBhandary, U.L. (1978). Cell 13, 83.
29. Heckman, J., Deweerd-Alzner, B., and RajBhandary, U.L. (1979). Proc. Nat. Acad. Sci. 76, 717.
30. Rich, A. and RajBhandary, U.L. (1976) Ann. Rev. Biochem. 45, 805.

MITOCHONDRIAL tRNAs AND rRNAs of NEUROSPORA CRASSA: SEQUENCE STUDIES, GENE MAPPING AND CLONING

U.L. RajBhandary, J.E. Heckman, S. Yin and B. Alzner-DeWeerd

Department of Biology, MIT, Cambridge, Massachusetts, USA 02139

ABSTRACT Using simple two step purification schemes, we have purified and sequenced four N.crassa mitochondrial tRNAs: the initiator methionine, tyrosine, alanine and valine tRNAs. Interestingly, every one of these tRNAs contains unusual structural features which differ from those of normal prokaryotic or eukaryotic tRNAs. We have obtained clone banks of HindIII and PstI fragments of N.crassa mitochondrial DNA cloned into pBR322, and have cloned a 16 kbp fragment which was produced by EcoRI + BamHI digestion of the mitochondrial DNA. Through analysis of these cloned DNA fragments we have derived a physical map for the region of the mitochondrial genome which encodes the ribosomal RNAs and virtually all of the tRNAs and have located tRNA and rRNA genes on this map by hybridization of purified ^{32}P-end-labeled RNA probes. A summary of our results is as follows:

(1) The mitochondrial DNA contains two genes for the initiator methionine tRNA. These two genes are not in tandem and are separated by at least 900 bp.
(2) The gene for tyrosine tRNA maps very close to that for the small ribosomal RNA.
(3) The genes for the small and large ribosomal RNAs are not adjacent as previously reported but are separated by about 5.5 kbp.
(4) The gene for the large ribosomal RNA contains an intervening sequence of approximately 2 kbp.
(5) Most of the tRNA genes are located within two clusters, one between the two ribosomal RNA genes, and the other immediately following the large ribosomal RNA gene. There is no evidence for tRNA genes in the intervening sequence of the large ribosomal RNA.
(6) Hybridization of labeled ribosomal and transfer RNAs to the separated strands of the cloned 16 kbp DNA fragment covering this region indicates that the two ribosomal RNAs and most, if not all, of the mitochondrial tRNAs are encoded on one strand of the mitochondrial DNA.

ISBN 0-12-198780-9

INTRODUCTION

The N.crassa mitochondrial genome consists of a circular double-stranded DNA molecule with a molecular weight of approximately 40×10^6 (1). It encodes approximately 25-30 tRNAs (2) and the mitochondrial ribosomal RNAs (3) as well as a number of proteins (4,5). Unlike the mitochondria of Saccharomyces cerevisiae and other facultative aerobes, the mitochondria of N.crassa and other obligate aerobes have not proven amenable to genetic analysis using classical genetic methods. Now, however, the techniques of DNA cloning and restriction enzyme analysis on the cloned DNA can be used for detailed analysis of these other mitochondrial genomes. This paper describes our work on the sequence analysis of several N.crassa mitochondrial tRNAs and on the gene mapping and cloning of tRNA and rRNA genes of N.crassa mitochondria. By restriction analysis of cloned DNA fragments we have derived a physical map of that region of DNA which codes for the ribosomal RNAs and virtually all of the tRNAs. Furthermore, using purified tRNAs as specific hybridization probes we have located precisely the genes for several of these tRNAs within this region.

RESULTS

SEQUENCES OF N.crassa MITOCHONDRIAL tRNAS

Using high pressure liquid chromatography on RPC-5 columns we have so far purified five N.crassa mitochondrial tRNAs and sequenced four of them. Figure 1 shows the nucleotide sequences of these four tRNAs(6-8). The most interesting and surprising finding from these sequences is that every one of these tRNAs contains novel features in its sequence which differ from normal tRNAs (9). These novel features are indicated as shadowed regions in Figure 1. Recent work of Dirheimer and his collaborators (10) and of Grivell, Tzagoloff, Martin, Drouin and coworkers (this volume) indicates that several, if not all, yeast and mammalian mitochondrial tRNAs also contain unusual structural features. These findings raise several very interesting questions.

(1) What is the effect of the novel features of tRNAs shown in Figure 1 on their secondary and tertiary structure?
(2) Does the presence of these novel features in the mitochondrial tRNAs imply that the mitochondrial protein synthesizing system in general is more flexible in what it requires of its tRNAs compared to prokaryotic and eukaryotic cytoplasmic protein synthesizing systems?

(3) Is there any functional significance associated with some of these unusual features in mitochondrial tRNAs? For example, is it possible that the presence of a pyrimidine next to the anticodon in the mitochondrial alanine tRNA instead of the usual purine--often a hypermodified purine--enables this tRNA to read all four alanine codons? If so, this would provide an explanation for the general observation (5) that the number of tRNA species in most mitochondria is less than the 32 needed as a minimum to read all the 61 codons of the genetic code using the rules of the Wobble Hypothesis (11).

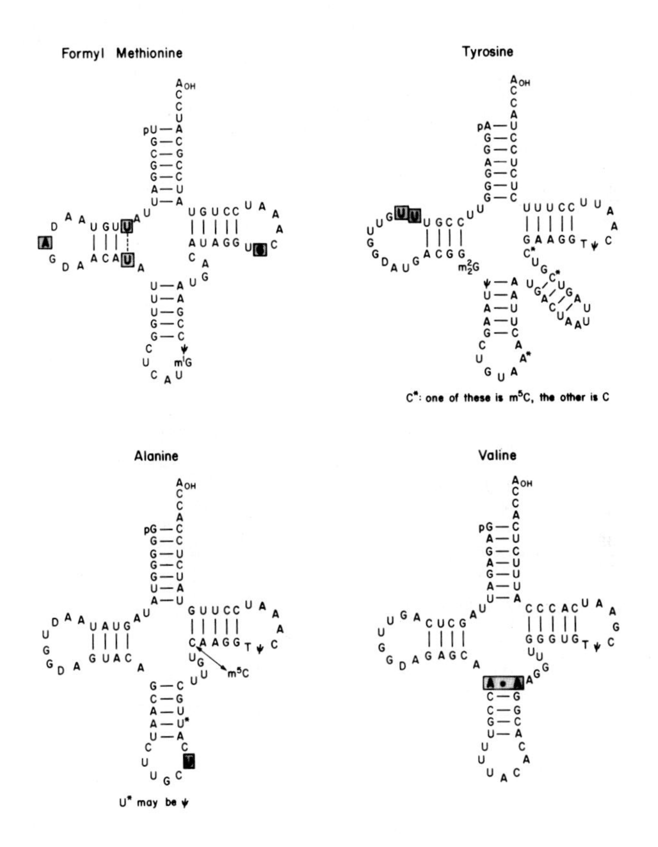

FIGURE 1. Nucleotide sequences of four N.crassa mitochondrial tRNAs, written in cloverleaf form.

Leaving aside these speculations, it is clear that answers to the questions raised can only come from detailed studies on the secondary and tertiary structure and functional studies on mitochondrial tRNAs. The amount of mitochondrial tRNAs that can be isolated and purified is unfortunately very limited and not enough for such detailed studies. Consequently, besides the immediate objective of mapping of tRNA and rRNA genes, an important reason for cloning tRNA genes is that the cloned tRNA genes, if they can be expressed in E.coli may be useful for producing relatively large amounts of mitochondrial tRNAs.

EVIDENCE FOR TWO GENES FOR THE INITIATOR METHIONINE tRNA

Bernard and Kuntzel (1) and Terpstra et al. (3) have reported that the largest EcoRI restriction fragment of the N.crassa mitochondrial DNA (12.7×10^6 MW) hybridizes both of the mitochondrial ribosomal RNAs and approximately 90% of the tRNAs. To determine the location of the gene for the initiator methionine tRNA, we used pure 5'-^{32}P-labeled $tRNA^{fMet}$ as a probe for Southern hybridizations to several restriction nuclease digests of N.crassa mitochondrial DNA. The results are shown in Figure 2. As expected from earlier work (3), the initiator methionine tRNA hybridizes to the largest DNA fragments in EcoRI and BamHI digests of the mitochondrial DNA; however, in HindIII digests, the $tRNA^{fMet}$ hybridizes to two distinct DNA bands approximately 2 and 3 kbp long. Since the known sequence of the $tRNA^{fMet}$ (see Figure 1) does not predict the presence of a HindIII site within the DNA coding it, this suggests that the mitochondrial DNA contains 2 genes for the initiator tRNA. Results in Figure 2 also show that the initiator tRNA hybridizes to one or more short DNA fragments about 350-450 base-pairs long in PstI digests of the mitochondrial DNA.

Further evidence that the two DNA bands in HindIII digests represent two initiator tRNA genes was obtained by cloning of the HindIII fragments onto pBR322 (12) and a study of the cloned DNAs which hybridize $tRNA^{fMet}$. Two classes of recombinant plasmids designated pHH67 and pHH100 which hybridized $tRNA^{fMet}$ were analyzed in detail. The plasmid DNAs were digested with HindIII restriction nuclease and the fragments were separated by electrophoresis. Mitochondrial DNA digested with HindIII nuclease was run on a parallel track to provide the necessary markers. Figure 3 shows the ethidium bromide staining pattern of the DNA fragments and the hybridization pattern with 5'-^{32}P-$tRNA^{fMet}$ as the probe. Besides the linear pBR322 vector DNA, HindIII digestion of pHH67 DNA yields a band which hybridizes to $tRNA^{fMet}$ and which

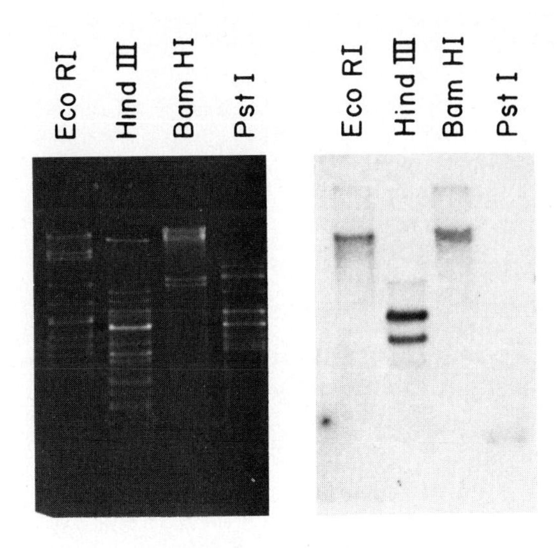

FIGURE 2. Southern hybridization of 5'-[^{32}P] tRNA^{fMet} to restriction nuclease digests of N.crassa mitochondrial DNA separated by electrophoresis on horizontal agarose gels. Left, ethidium bromide staining pattern of DNA fragments present in the various digests; right, hybridization pattern following transfer of DNA bands from agarose gel to nitrocellulose.

corresponds to the larger of the two HindIII fragments in a bulk mitochondrial DNA digest which hybridizes to tRNA^{fMet}. In a similar digest, pHH100 DNA yields linear pBR322 DNA plus two other fragments. One of these hybridizes to tRNA^{fMet} and this corresponds to the smaller of the two HindIII fragments present in total mitochondrial DNA digests which hybridize to tRNA^{fMet}. The other DNA band in HindIII digests of pHH100 DNA is a small fragment which apparently was co-cloned along with the 2 kbp long fragment in pHH100.

Our result indicating the presence of two gene copies for the initiator tRNA is somewhat surprising since most estimates of the number of tRNA genes in mitochondria are short of the 32 tRNAs needed to read all the codons of the genetic code (11). In the case of lower eukaryotes (13,14), current estimates of the number of tRNA genes lie around 25-30. Therefore, the initiator tRNA is most likely the only one or perhaps among the very few tRNAs which are coded for by more than one gene in N.crassa mitochondria. Fine mapping of the HindIII fragment present in pHH67 has shown (Heckman et al., in preparation) that these two $tRNA^{fMet}$ genes are separated by at least 900 bp on the mitochondrial DNA. Thus, the hybridization of $tRNA^{fMet}$ to two different HindIII fragments is not due to the presence of a HindIII restriction site within a short intervening sequence in the tRNA gene (15,16). Another possibility is that one of the two $tRNA^{fMet}$ genes repre-

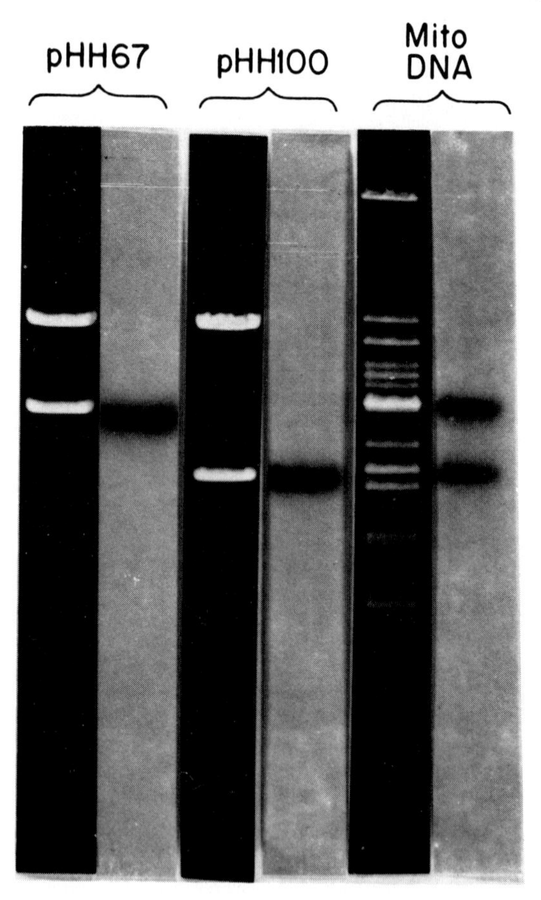

FIGURE 3. Southern hybridization of 5'-[^{32}P] $tRNA^{fMet}$ to HindIII restriction nuclease digests of recombinant plasmids pHH67 and pHH100 and of N.crassa mitochondrial DNA separated by electrophoresis on 1% agarose gels. The left track of each set is the ethidium bromide staining pattern and the right track, the hybridization pattern.

sents only a partial repeat and in this sense is "non-functional". We are investigating such a possibility and an unequivocal answer should be available when sequence studies on the cloned DNAs are completed. The observation (to be discussed later) that both of the $tRNA^{fMet}$ genes are part of a cluster of tRNA genes would, however, argue that they are functional.

GENES FOR THE SMALL 17S rRNA AND TYROSINE tRNA ARE IN CLOSE PROXIMITY AND ON THE SAME DNA STRAND

The location of the mitochondrial tyrosine tRNA gene was determined in the same way as that of the $tRNA^{fMet}$ genes - by hybridization of $5'-^{32}P-tRNA^{Tyr}$ to restriction patterns of the mitochondrial DNA. Figure 4 shows the results obtained.

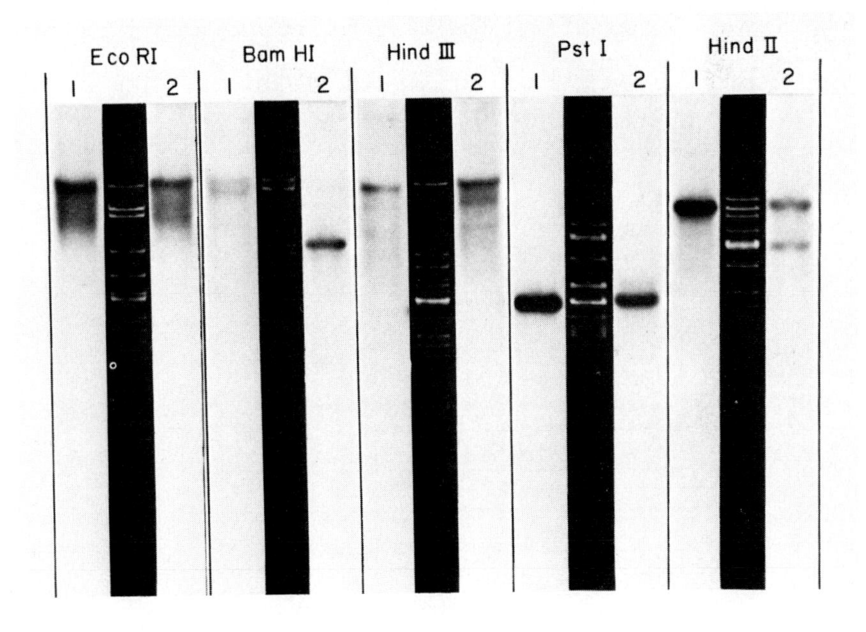

FIGURE 4. Southern hybridization of $5'-[^{32}P]$-labeled $tRNA^{Tyr}$ and 17S rRNA to various restriction nuclease digests of N.crassa mitochondrial DNA separated by horizontal electrophoresis on 1% agarose gels. The middle track in each set which is unnumbered is the ethidium bromide staining pattern; track 1 on the left, hybridization pattern with $tRNA^{Tyr}$ and track 2 on the right, hybridization pattern with 17S rRNA.

The hybridization pattern of tyrosine tRNA is shown in Track 1 to the left of each ethidium bromide staining pattern. It hybridizes, like most of the mitochondrial tRNAs, to the longest DNA fragment produced by EcoRI digestion, but it also hybridizes to the longest fragments from BamHI and HindIII digestions of the mitochondrial DNA. Since the region defined by the overlap of these three large restriction fragments has been implicated (3) as a "spacer" region between the two ribosomal RNA genes, it was of interest to determine whether the mitochondrial ribosomal RNAs would also hybridize to a selection of restriction fragments similar to those hybridized by $tRNA^{Tyr}$.

N.crassa mitochondrial 17S and 24S rRNAs were prepared from separated subunits of mitochondrial ribosomes. Figure 4 shows the hybridization pattern (in Track 2 to the right of each staining pattern) of $5'-^{32}P$-labeled 17S ribosomal RNA (and fragments thereof) to the restriction digests of the mitochondrial DNA. The 17S rRNA and $tRNA^{Tyr}$ hybridize to common fragments in all digests except BamHI, implying the presence of a BamHI cleavage site between the two genes. Hybridization of 17S rRNA to a HindII digest shows two bands, one of which is the same as that which also hybridizes $tRNA^{Tyr}$. This suggests that there is a HindII site within the 17S rRNA gene. In the PstI digest of the mitochondrial DNA, both $tRNA^{Tyr}$ and 17S rRNA hybridize to the same DNA band approximately 2700 bp long. Since this PstI band contained two or more DNA fragments of the same size, it was necessary to obtain a pure cloned fragment in order to analyze the proximity of these two genes in detail.

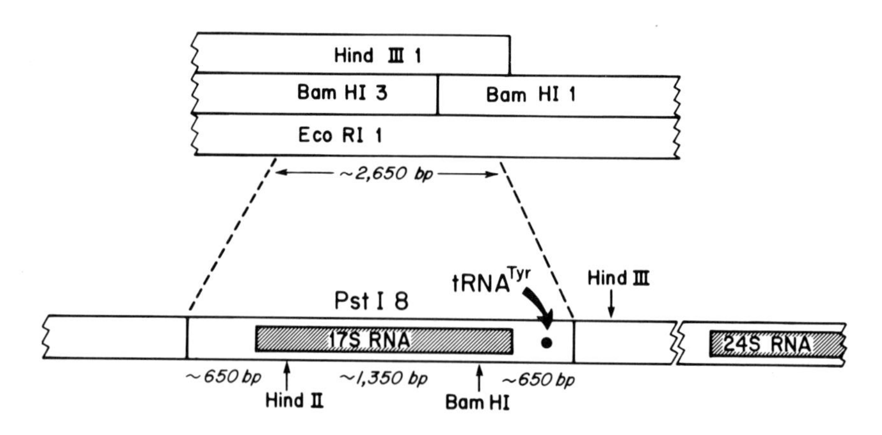

FIGURE 5. Partial restriction map of N.crassa mitochondrial DNA including the 17S rRNA and tyrosine tRNA genes.

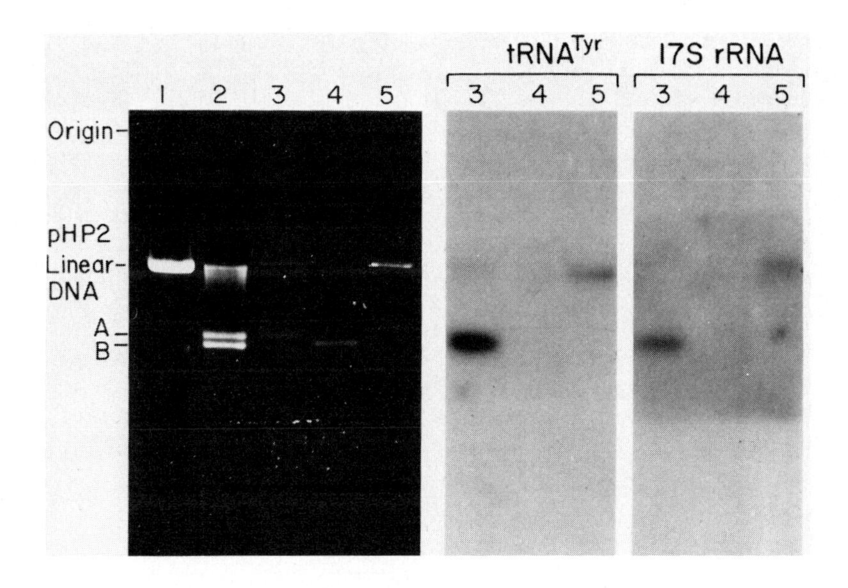

FIGURE 6. Separation of DNA strands of the recombinant plasmid pHP2 by electrophoresis on 1% agarose gels for Southern hybridization. Track 1, linear pHP2 DNA obtained by cleavage of covalently closed circular pHP2 DNA with EcoRI; track 2, linear pHP2 DNA as in 1 following denaturation in alkali; track 3, band A of track 2 following reannealing to remove any contaminating band B; track 4, band B of track 2 following reannealing to remove any contaminating band A and track 5, bands A and B of track 2 mixed, reannealed and then subjected to electrophoresis. The patterns on the right show the hybridization of 5'-^{32}P-labeled tRNA^{Tyr} and 17S rRNA to duplicate sets of tracks 3,4 and 5.

Plasmids containing the tRNA^{Tyr} gene were prepared from clones selected by colony filter hybridization of 5'-^{32}P-tRNA^{Tyr} to a clone bank of PstI fragments of mitochondrial DNA in pBR322. A detailed analysis of such a plasmid, designated pHP2, has yielded a partial map of this region of the mitochondrial genome as shown in Figure 5. The PstI fragment cloned contains within it the genes for both 17S rRNA and tyrosine tRNA and the tyrosine tRNA is located within 300-400 bp from the end of the 17S rRNA gene.

The close proximity between the 17S rRNA and tyrosine tRNA genes raised the interesting possibility that these RNAs may be synthesized by processing of a common precursor. If so both these RNAs must be encoded on the same DNA strand. To determine whether both $tRNA^{Tyr}$ and 17S rRNA hybridize to the same DNA strand in the recombinant plasmid pHP2, the covalently closed circular plasmid DNA was cleaved with EcoRI to yield linear DNA (Track 1 of Figure 6) and the DNA strands were separated by agarose gel electrophoresis following denaturation with alkali (Track 2). The separated DNA strands were freed of small amounts of contamination by the other strand by reannealing individually followed by agarose gel electrophoresis (Tracks 3 and 4). Hybridization of $tRNA^{Tyr}$ and 17S rRNA to the pattern obtained showed that both RNAs hybridized to the slower of the two DNA bands (labeled A in Tracks 2 and 3) but not to the faster band (labeled B in Tracks 2 and 4) of Figure 6. As expected, both RNAs also hybridized to the linear DNA duplex formed by annealing of the two DNA strands (Track 5).

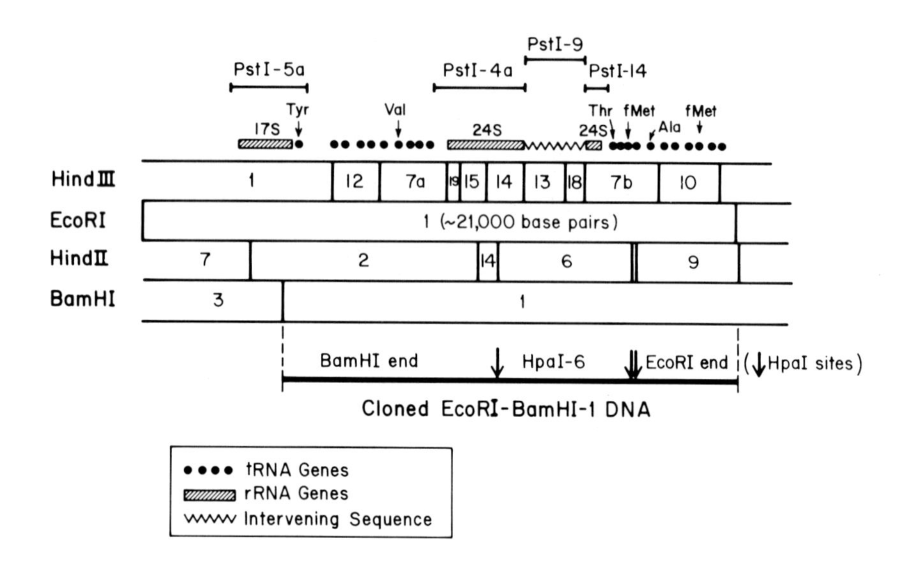

FIGURE 7. Map of tRNA and rRNA genes on the EcoRI-1 fragment of N.crassa mitochondrial DNA. Unspecified tRNA genes represent approximate locations in clusters detected by hybridization of total 3'-end labeled tRNA. Primary structures of all the specified tRNAs except Thr have been determined. This region of cloned DNA contains two short HindIII fragments 21 and 23 whose location within this map has not been precisely determined.

PHYSICAL MAP OF THE REGION OF THE N.CRASSA MITOCHONDRIAL GENOME ENCODING rRNAs AND tRNAs

To further map the tRNA and rRNA genes in N.crassa mitochondria, we cloned in pBR322 a 16 kbp DNA fragment obtained by double digestion of the mitochondrial DNA with EcoRI and BamHI. Analysis of this fragment and of other cloned PstI and HindIII fragments of the mitochondrial DNA (Heckman and RajBhandary, Cell, in the press) enabled us to construct the map shown in Figure 7, which covers about one-third of the mitochondrial genome. The relationship of PstI-5a, the DNA fragment bearing the 17S rRNA and $tRNA^{Tyr}$ genes, to this map was established by the finding that $5'$-^{32}P-$tRNA^{Tyr}$ hybridized to the junction fragment between the BamHI end of the cloned EcoRI-BamHI-1 fragment and the first HindIII site therein. The locations of the 24S rRNA gene and total tRNA genes were determined as follows:

MAPPING OF THE 24S RIBOSOMAL RNA GENE

Figure 8A shows the hybridization pattern of $5'$-^{32}P-labeled 24S rRNA (and large fragments thereof) to five different restriction digests of the mitochondrial DNA. The 24S rRNA hybridizes, as previously reported (2) to the largest DNA fragments in EcoRI and BamHI digests. In the HindII digest it hybridizes to three bands corresponding to the fragments HindII-2, 6, and 14 in Figure 7, but the decisive data for purposes of mapping are derived from the HindIII and PstI digestion patterns. The PstI track shows hybridization of the 24S rRNA to bands 4 and 14. The HindIII track of Figure 8 shows no detectable hybridization of 24S rRNA to HindIII-1, as would have been expected on the basis of previous reports (2), but it shows hybridization to bands 7, 14, 15, and 19. Evidence that the HindIII-7 which hybridizes 24S rRNA is HindIII-7b and not HindIII-7a was obtained by hybridization of the labeled rRNA to HindIII digests of plasmid DNAs containing the individual cloned HindIII-7a and 7b (Figure 8B). The first track shows that a HindIII digest of the cloned EcoRI-BamHI-1 fragment yields the same 24S rRNA hybridization pattern as that of the total mitochondrial DNA. The second track shows that the 24S rRNA hybridizes to HindIII digests of plasmid pHH67, which contains the cloned HindIII-7b, and the third track shows that it does not hybridize to the digest of pHH115, which contains HindIII-7a.

Based on the physical map of this region (Figure 7) which shows that HindIII-7b is separated from the other three 24S rRNA-hybridizing fragments by HindIII-13 and 18, which do not

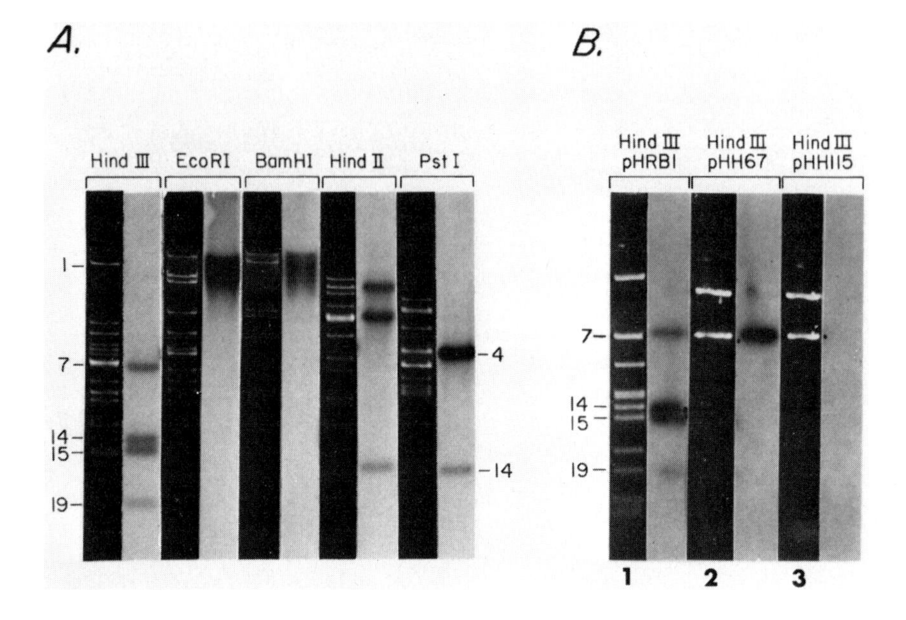

FIGURE 8. Southern hybridization of 5'-^{32}P-24S rRNA to (A) various restriction digests of N.crassa mitochondrial DNA and (B) HindIII digests of three recombinant plasmids. pHRB1, EcoRI-BamHI-1 cloned in pBR322; pHH67, HindIII-7b cloned in pBR322; pHH115, HindIII-7a cloned in pBR322. The slowest band in each ethidium bromide staining track in B is vector DNA (plus junction fragments in case of pHRB1). 24S rRNA hybridizes to HindIII-7b, but not to HindIII-7a.

hybridize the rRNA, these results suggest that the 24S rRNA gene is discontinuous. The other possible explanation, that the 24S rRNA is completely encoded by the contiguous HindIII fragments 19, 15, and 14 or PstI-4a (See Figure 7), and that HindIII-7b contains a sequence constituting a partial repeat of the information in the actual gene, was ruled out by experiments using the cloned PstI fragment 14 (representing the smaller segment of the 24S rRNA gene-see Figure 7) after labeling by nick-translation, as a hybridization probe. It was found that PstI-14 showed no hybridization to PstI-4a. The hybridization of the 24S rRNA to this small, nonadjacent region, therefore, represents hybridization to a unique segment of a discontinuous gene, and not to a partial repeat of some sequence in the larger segment. Thus, the N.crassa mitochondrial 24S rRNA gene contains an intervening sequence of approximately 2000 bp, similar to that found in the mitochondrial large ribosomal RNA gene in ω^+ strains of yeast (17).

IDENTIFICATION OF tRNA GENE CLUSTERS

The map positions of the genes for valine, alanine, and threonine tRNAs shown in Figure 7 were determined by hybridization studies with the purified end-labeled tRNAs. The unspecified tRNA genes on the maps represent approximate locations determined by hybridization of total 3'-end labeled tRNAs to mitochondrial DNA digests and cloned DNA fragments.

Mitochondrial tRNA was selectively labeled at the 3'-end with ^{32}P as described before (18). This procedure which utilizes tRNA nucleotidyl transferase labels selectively tRNAs but not small fragments of rRNAs or mRNAs which often contaminate total tRNA preparations. The mixture of labeled tRNAs was used as a hybridization probe and the results are shown in Figure 9. Track 1 shows that total mitochondrial 3'-^{32}P-tRNA hybridizes to only about six of the more than 20

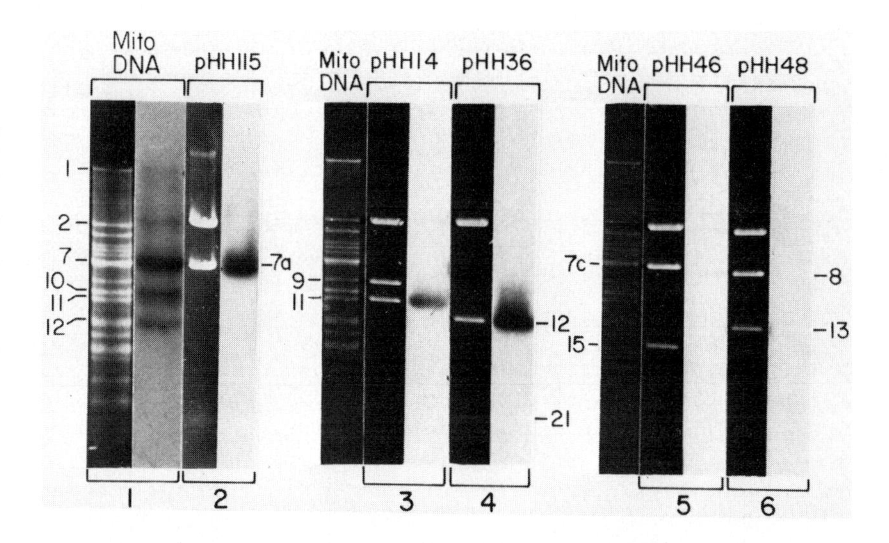

FIGURE 9. Hybridization of total mitochondrial 3'-^{32}P-tRNA to HindIII digests of mitochondrial DNA and of recombinant plasmids bearing specific HindIII fragments of the mitochondrial DNA.

DNA fragments produced by HindIII digestion of mitochondrial DNA. Most of the tRNAs hybridize to bands 7, 10 and 12, while weaker responses are seen for bands 1 ($tRNA^{Tyr}$, see Figure 5), 2, and 11. These findings are similar to those reported by Terpstra et al. (2) using ^{125}I-labeled 4S RNA as a probe, except that these investigators also detected weak hybridization to HindIII-3 and 9, but not to HindIII-11.

Further confirmation of these results was made possible by the availability of many of these HindIII fragments of interest in recombinant plasmids. Track 3 of Figure 9 shows that total tRNA hybridizes to HindIII-11, a response which was difficult to see in the pattern of total mitochondrial DNA because of the nearby strong hybridization of HindIII-10. The same track shows that HindIII-9 does not hybridize any detectable fraction of this labeled tRNA mixture. Similarly, tracks 4 and 6 of Figure 9 show that cloned HindIII-12 hybridizes total tRNA strongly, whereas HindIII-13 shows no detectable hybridization. The mixture of DNA fragments in HindIII bands 7 and 8 could also be investigated in this fashion. One of the three fragments in band 7, designated 7b, had been shown to hybridize fMet, alanine, and threonine tRNAs (see Figures 3 and 7). The other two components of HindIII band 7 were also isolated in clones and studied. Track 2 of Figure 9 shows that cloned HindIII-7a (which hybridizes $tRNA^{Val}$) responds strongly to total tRNA and thus probably contains several tRNA genes. Track 5 of Figure 9 shows that the third component of HindIII band 7, designated 7c, does not hybridize tRNA, and track 6 of Figure 9 shows that HindIII-8, which separates poorly from HindIII-7 on agarose gels, also fails to respond to tRNA.

These results allowed us to assign a minimal distribution of tRNA genes to the map shown in Figure 7. The mitochondrial genome codes for approximately 25-30 tRNAs (2) and all but a few of them (such as those located on HindIII-2 and 11) are encoded in the region included in this map. It can be seen from Figure 7 that the 20 or so tRNA genes in this region from two large clusters, one between the genes for 17S rRNA and 24S rRNA, and another closely following the smaller segment of the 24S rRNA gene. No evidence could be found for the presence of tRNA genes in the intervening sequence of the 24S rRNA gene.

ASYMMETRIC DISTRIBUTION OF tRNA AND rRNA GENES ON THE STRANDS OF THE MITOCHONDRIAL DNA

Are the ribosomal and transfer RNA genes located on the cloned EcoRI-BamHI-1 fragment (Figure 7) encoded upon one or both strands of the mitochondrial DNA? The approach used to answer this question was similar to that of the experiment shown in Figure 6. The 16 kbp EcoRI-BamHI-1 fragment was excised from plasmid pHRB1 by digestion with EcoRI plus BamHI. It was denatured in alkali and subjected to preparative agarose gel electrophoresis. The two denatured single strands, which separated slightly from one another, were excised, eluted, freed of contamination from each other by reannealing, and reelectrophoresed on an agarose gel. Figure 10A shows a pattern in which total 3'-^{32}P-labeled mitochondrial tRNA has been used as a hybridization probe against the separated DNA strands. Track 1 shows the hybridization of total tRNA to the duplex DNA from an EcoRI plus BamHI digest of plasmid pHRB1. Track 2 shows the hybridization of total tRNA to the purified slower moving single strand of the EcoRI-BamHI-1 fragment. Track 3 shows that no detectable portion of the total tRNA hybridizes to the purified faster moving strand of EcoRI-BamHI-1.

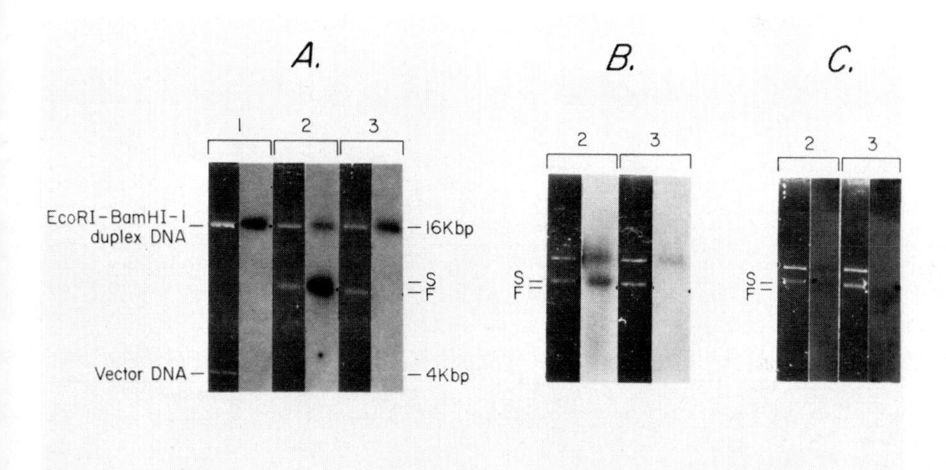

FIGURE 10. Southern hybridization of (A) total 3'-^{32}P-labeled mitochondrial tRNA, (B) 5'-^{32}P-24S rRNA, and (C) 5'-^{32}P-tRNA^{Tyr} to separated DNA strands of the EcoRI-BamHI-1 fragment. Track 1, duplex DNA derived from pHRB1 by EcoRI plus BamHI digestion. Track 2, isolated slower DNA strand (S) of EcoRI-BamHI-1, after reannealing to remove the contaminating faster strand (F) to the position of duplex DNA. Track 3, isolated faster strand of EcoRI-BamHI-1 after reannealing. Gels shown in B and C were subjected to electrophoresis for shorter times than that in A. All three probes hybridize exclusively to the slower DNA strand.

Figure 10B shows a similar analysis of EcoRI-BamHI-1, using 5'-^{32}P-labeled 24S rRNA as a hybridization probe against the purified slower strand (Track 2) and faster strand (Track 3) of the DNA. In each track, 24S rRNA hybridizes to the duplex DNA band, but it hybridizes only to the slower band of single stranded DNA. Thus the large ribosomal RNA is encoded upon the same DNA strand as all the mitochondrial tRNAs found in these studies.

Finally, since we had shown (Figure 6) that both 17S rRNA and $tRNA^{Tyr}$ were coded for by the same strand of the mitochondrial DNA, it was possible to extend the above analysis to the 17S rRNA gene by the use of $tRNA^{Tyr}$ as a hybridization probe. Figure 10C shows the results of such an analysis. 5'-^{32}P-$tRNA^{Tyr}$, like the 24S rRNA, hybridizes to the duplex DNA band and to the slower moving single strand (Track 2) of the mitochondrial DNA fragment. Thus all of the RNA genes shown in Figure 7 are coded for by the same strand of the mitochondrial DNA.

DISCUSSION

Besides the novel features found in the sequences of tRNAs from N.crassa mitochondria, one of the interesting results of this work is that the mitochondrial tRNA and rRNA gene organization of N.crassa is more similar to that of yeast and less similar to that of higher eukaryotes than had been previously believed (2,3). Several of the findings reported in here, some of which have also been independently confirmed by Kuntzel and coworkers (19), and Lambowitz and coworkers (20), differentiate the gene organization of N.crassa mitochondrial DNA from that of higher eukaryotes. These are (1) the presence of an intervening sequence in the 24S rRNA gene, (2) the separation of the 17S rRNA and 24S rRNA genes by a large (approximately 5.5 kbp) segment of DNA, (3) the clustering of most of the mitochondrial tRNA genes, and (4) the asymmetric distribution of rRNA and tRNA genes between the two DNA strands. All of these attributes resemble to some extent the organization of genes in yeast mitochondria (5), but they differ distinctly from the organization in mitochondria of mammals and amphibians (21-23), where the large and small ribosomal RNA genes are adjacent and lack intervening sequences, and where the tRNA genes are scattered over both strands of the mitochondrial DNA.

Perhaps the most important finding of this work is that virtually all of the tRNA genes are present in two major clusters, one between the rRNA genes and the other immediately following the large rRNA gene, and furthermore that all the tRNA and rRNA genes within this region are coded for by the same DNA strand. What are the possible implications of this genetic organization and the highly asymmetric distribution of tRNA and rRNA genes between the two DNA strands? First, it means that unlike the situation in HeLa mitochondria where tRNA genes are distributed fairly uniformly around the mitochondrial DNA and in both strands and where symmetrical transcription of both DNA strands is known to occur (24), there is *a priori* no need to invoke a similar symmetrical transcription within this region of *N.crassa* mitochondrial DNA just for the synthesis of tRNAs and rRNAs. Second, this highly asymmetric distribution of tRNA and rRNA genes leaves open the possibility that this entire region of DNA is part of a single transcriptional unit. It would clearly be of interest now to analyze the nature of the primary transcript(s), and examine whether the tRNA genes located between the rRNA genes are "transcribed" spacers in the rRNA cistron as in *E.coli* (25) or form part of a separate transcriptional unit(s).

ACKNOWLEDGEMENTS

We are grateful to Dr. W. Edgar Barnett for providing us with purified *N.crassa* mitochondria. This work was supported by grants GM17151 from NIH and NP 114 from the American Cancer Society.

REFERENCES

1. Bernard, U., and Kuntzel, H., (1976). In "The Genetic Function of Mitochondrial DNA", (C. Saccone and A.M. Kroon, eds.), pp. 105-109. North Holland, New York.
2. Terpstra, P., de Vries, H., and Kroon, A.M., (1977). In "Mitochondria 1977" (W. Bandlow, R.J. Schweyen, K. Wolf and F. Kaudewitz, eds.), pp. 291-302. De Gruyter, New York.
3. Terpstra, P., Holtrop, M., and Kroon, A.M., (1976). In "The Genetic Function of Mitochondrial DNA" (C. Saccone and A.M. Kroon, eds.), pp. 111-118. North-Holland, New York.
4. Tedeschi, H. (1976). "Mitochondria: Structure, Biogenesis and Transducing Functions". Springer-Verlag,New York.
5. Borst, P. and Grivell, L.A., (1978). Cell, 15, 705-723.
6. Heckman, J.E., Hecker, L.I., Schwartzbach, S.D., Barnett, W.E., Baumstark, B., and RajBhandary, U.L.(1978). Cell, 13, 83-95.
7. Heckman, J.E., Alzner-DeWeerd, B. and RajBhandary, U.L., (1979). Proc. Nat. Acad. Sci. USA 76, 717-721.

8. RajBhandary, U.L., Heckman, J.E., Yin, S.,Alzner-DeWeerd, B., and Ackerman, E. (1979). in press. Cold Spring Harbor Monograph on Transfer RNA. (J. Abelson, P. Schimmel and D. Soll, eds.).
9. Rich, A. and RajBhandary, U.L. (1976). Ann. Rev. Biochem. 45, 805-860.
10. Martin, R.P., Sibler, A.P., Schneller, J.M., Keith, G., Stahl, A.J.C. and Dirheimer, G. (1978). Nucl. Acids. Res. 5, 4579-4592.
11. Crick, F.H.C., (1966). J. Mol. Biol., 19, 548-555.
12. Boyer, H.W., Betlach, M., Bolivar, F., Rodriquez, R.L., Heyneker, H.L., Shine, J. and Goodman, J.M. (1977). In "Recombinant Molecules: Impact on Science and Society", (R.F. Beers, Jr., and E.F. Basset, eds.), pp. 9-20. Raven Press, New York.
13. Martin, N.C., Rabinowitz, M. and Fukuhara, H., (1977). Biochemistry, 16, 4672-4677.
14. Van Ommen, G.J.B., Groot, G.S.P. and Borst, P., (1977). Mol. Gen. Genet., 154, 255-262.
15. Goodman, H.M., Olson, M.V. and Hall, B.D., (1977). Proc. Nat. Acad. Sci. USA, 74, 5453-5457.
16. Valenzuela, P., Venegas, A., Weinberg, G., Bishop, R. and Rutter, W.J., (1978). Proc. Nat. Acad. Sci. USA 75, 190-194.
17. Bos, J.L., Heyting, C., Borst, P., Arnberg, A.C. and VanBruggen, E.F.J., (1978). Nature, 275, 336-338.
18. Silberklang, M., Gillum, A.M., and RajBhandary, U.L., (1977). Nucleic Acids Res., 4, 4091-4108.
19. Hahn, U., Lazarus, C.M., Lunsdorf, H. and Kuntzel, H., (1979). Cell, in press.
20. Manella, C., Collins, R.A., Green, M.R., and Lambowitz, A.M., (1979). Proc. Nat. Acad. Sci. USA, in press.
21. Angerer, L., Davidson, N., Murphy, W., Lynch, D., and Attardi, G., (1976). Cell, 9, 81-90.
22. Dawid, I.B., Klukas, C.K., Ohi, S., Ramirez, J.L. and Upholt, W.B., (1976). In "The Genetic Function of Mitochondrial DNA" (C. Saccone and A.M. Kroon, eds.) North-Holland, New York, pp. 3-14.
23. Saccone, C., Pepe, G., Bakker, H., and Kroon, A., (1977). In "Mitochondria 1977" (W. Bandlow, R.J. Schweyen, K. Wolf and F. Kaudewitz, eds.), De Gruyter, New York, pp. 303-316.
24. Murphy, W.I., Attardi, B., Tu, C., and Attardi, G., (1975). J. Mol. Biol., 99, 809-814.
25. Lund, E., Dahlberg, J.E., Lindahl, L., Jaskunas, S.R., Dennis, P.P. and Nomura, M. (1976). Cell, 7, 165-177.

STRUCTURE AND EVOLUTION OF ANIMAL MITOCHONDRIAL DNA

Igor B. Dawid and Eva Rastl

Laboratory of Biochemistry, National Cancer Institute,
Bethesda, Maryland 20205

ABSTRACT This article summarizes our knowledge on the function and evolution of animal mitochondrial DNA (mtDNA) with emphasis on our results with *Xenopus laevis*. In this frog and in several other animal species the map locations of the regions coding for the mitochondrial ribosomal RNA molecules have been determined. In human cells and in *X. laevis* sites coding for 4S RNA have been located. In both species about 20 sites were found distributed widely over both strands of the mtDNA. Poly(A) containing RNA molecules have been detected in several animal species; these RNAs are likely to be mitochondrial mRNAs. We have mapped relatively abundant poly(A) RNA molecules onto the mtDNA of *X. laevis*. Most of these RNAs are encoded by the H strand and essentially the entire H strand is expressed into these discrete RNA molecules. Electron microscopy and mapping with the aid of the single strand specific nuclease S1 and exonuclease VII suggest that no intervening sequences occur in the major mitochondrial RNAs in *Xenopus*. Rare RNA molecules were detected which overlap the coding regions of major RNAs and which might represent precursor RNAs. At the end of the article we discuss some ideas regarding the evolution and transmission of mtDNA.

THE MAIN FUNCTIONS OF mtDNA

Studies in many organisms have shown that mtDNA is a small genome that codes for various functions that are encoded in cellular genomes, including rRNA, tRNA and mRNA. However, mtDNA does not code for all components that make up a mitochondrion and most of these components depend on nuclear genes (1,2). In every organism studied including higher animals and yeast, mtDNA codes for the two large structural RNAs in the mitochondrial ribosome, a limited set of tRNAs, and a limited set of mRNAs which code for about 9 mitochondrial polypeptides. It appears likely that

ISBN 0-12-198780-9

mtDNAs of groups of organisms like protozoa and plants, whose coding functions have not yet been studied in detail, will prove to have a similar range of products. It remains to be explained why mtDNAs of different organisms differ greatly in overall size. While the mtDNAs of metazoan animals have a very similar size of 15 to 18 kilobase pairs (kb), yeast mtDNA is about five times larger, protozoan mtDNAs are intermediate and plant mtDNAs even larger (3). In the well-studied case of yeast mtDNA it appears that large portions of the DNA are composed of regions very rich in A and T which appear not to code for any products (2). There is no evidence for similar regions in protozoan or plant mtDNAs but it may be hypothesized that these DNAs contain long segments of noncoding sequences. In contrast, the small animal mtDNAs appear to utilize essentially all of their coding capacity as will be discussed later in this article.

THE CODING REGIONS FOR MITOCHONDRIAL rRNA

All mitochondria studied so far have a specific ribosome which differs in physical and functional properties from the cytoplasmic ribosome of the same organisms (4,5,6,7). The two large rRNAs in this particle are encoded in the mtDNA while most or all of the ribosomal proteins are encoded in nuclear genes. Mitochondrial ribosomes of animals and fungi do not contain 5S RNA (8,9), but it remains an open question whether they contain a small structural RNA of about 4S RNA size. Mitochondrial ribosomes have several functional similarities to prokaryotic ribosomes but the physical properties of both the ribosomes and the rRNAs of mitochondria differ greatly from those in prokaryotes (4,6,7).

Mitochondrial rRNA in metazoan animals is composed of two molecules of about 900 and 1600 nucleotides. The size varies little among the organisms studied. The two RNAs are encoded by stretches of mtDNA which are positioned close to each other and separated by a distance of about 150 base pairs (bp). The two RNAs are encoded by the same strand of mtDNA which in the case of vertebrate animals is always the strand that is heavy in alkaline cesium chloride gradients. Transcription proceeds in the direction from the smaller to the larger rRNA. Most or all of these points have been established for the following animal species: human (10); *Xenopus laevis* (11); *Drosophila melanogaster* (12); mouse (13). While the sample is heavily weighted towards vertebrate animals the presence of an insect in it makes it probable that this structural arrangement is general among metazoan animals, all of which share 15 kb circular mtDNA. The proximity of the rRNA coding regions, the conservation of this arrangement in evolution and the example of nuclear rRNA genes may suggest that the two rRNAs could be transcribed together in the form of a precursor molecule. However, no evidence for a specific rRNA precursor in animal mitochondria has been presented.

Mitochondrial rRNA molecules in other organisms are larger than in animals. The arrangement of coding regions also differs. Without going into too much detail here we mention that in yeast (2) and _Tetrahymena_ (14) the coding regions of the two rRNAs are widely separated. Further, in some strains of yeast (15) the coding region for the large rRNA is interrupted by an "intervening sequence". No such interruptions have been found in the rRNA genes of higher animals (see also below).

MITOCHONDRIAL tRNA

In all organisms studied mtDNA codes for specific tRNAs. Acceptor species for many amino acids have been identified in yeast (2), rat (16), _Tetrahymena_ (17) and human (18). In some cases isoacceptor tRNAs have been found. In Hela cells tRNAs accepting all but four amino acids were detected (19).

The study of mitochondrial tRNA concerns identification of specific tRNAs, the counting of tRNAs in the mitochondrial set and mapping of the coding regions in the mtDNA. The latter problem has been attacked in three organisms only. In yeast a number of specific tRNAs were mapped by a variety of hybridization methods often involving _petite_ mutant mtDNA (2). In Hela and _Xenopus_ mtDNA coding regions for the total set of mitochondrial 4S RNAs were mapped by electron microscopy (10, 20,21). The results for _Xenopus_ are summarized in Figure 1. The first point that emerges from these studies is the total number of sites coding for mitochondrial 4S RNA. Nineteen (Hela cells) or 21 (Xenopus) such sites were found, and the work in _Xenopus_ suggested that 5 additional sites may exist. In yeast, the total number of identified tRNA coding sites in mtDNA stands at 26 (2). These values, together with the fact that no mitochondria-specific tRNAs for some amino acids could be found in Hela cells (19) may suggest that the mitochondrial set is incomplete and that perhaps some tRNAs are imported from the cytoplasm. This suggestion has support from the work in _Tetrahymena_ which provided some direct evidence for tRNA import (22). Even if the few "missing" mitochondrial tRNAs in Hela cells could still be found we would be left with a total number of coding sites in the mtDNA very much lower than required for a complete set of isoaccepting tRNAs. One possibility would be that the usage of codons in mitochondria might be quite restricted. This possibility is suggested by recent work of Macino and Tzagoloff (23) showing that codon usage in a section of the yeast gene for subunit 9 of ATPase is quite limited. Further sequencing studies will show whether this phenomenon is general in mtDNA.

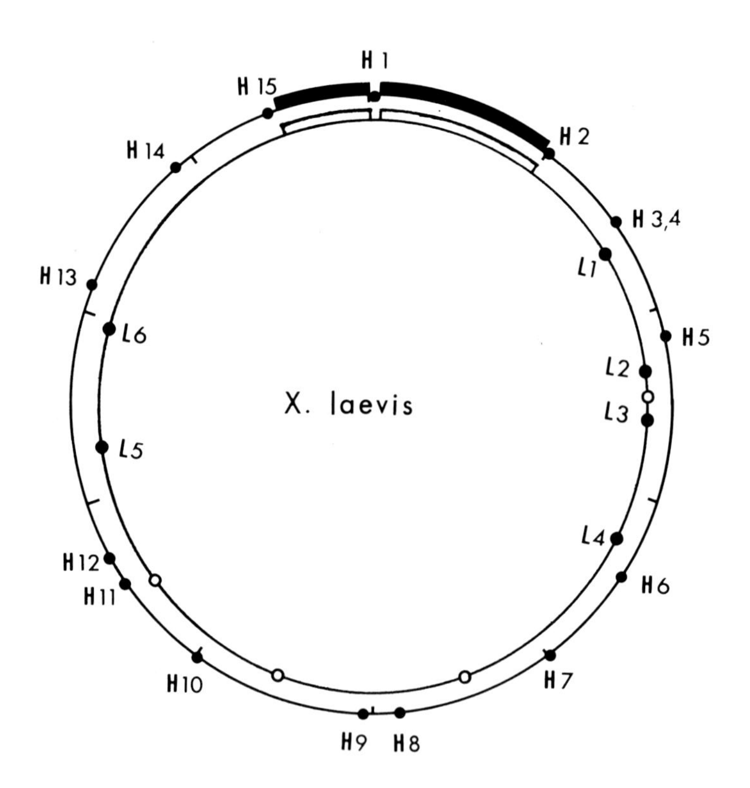

FIGURE 1. Map of 4S RNA coding sites on mtDNA of X. laevis. The heavy lines indicate the rRNA coding regions and dots show 4S RNA sites. The sites are numbered with H and L referring to the strand of the mtDNA. The two EcoRI sites are indicated by short arrows.

Another aspect of interest is the arrangement of tRNA coding sites in mtDNA. The map for Xenopus (Figure 1) is remarkably similar to that for Hela (20). In particular, both animal mtDNAs carry sites on both strands, the sites are distributed over the entire strand, and three sites are present between and surrounding the rRNA regions. These facts point to a considerable evolutionary conservatism in the arrangement of mtDNA sequences. In yeast mtDNA, most tRNA coding sites are found in about one fourth the total genome. This has sometimes been referred to as clustering of tRNA sites. However, the region in which yeast tRNA sites are found is larger than the entire mitochondrial genome of higher animals. Further, it is not known whether all tRNAs in yeast are coded on one strand of the mtDNA or whether both strands are involved as in animal mtDNA.

MITOCHONDRIAL mRNA

In yeast a combination of genetic and biochemical studies has demonstrated that mtDNA codes for several polypeptides (2). While the particular mRNA species have not always been identified their presence may be inferred. In higher animals the evidence for the presence of mitochondrial mRNA is indirect. There is good evidence that Xenopus mitochondria synthesize several polypeptides including subunits of oligomycin-sensitive ATPase and cytochrome c oxidase (24). These are the analogous polypeptides also synthesized in yeast mitochondria and encoded in yeast mtDNA. Further, animal mitochondria contain a series of non-ribosomal RNA molecules, most of which contain poly(A) tails. It is reasonable to infer that these RNA molecules, or at least some of them, are mRNAs directing the synthesis of mitochondrial polypeptides.

A number of non-ribosomal RNA species have been identified in Hela cell mitochondria (18,25). The total length of these RNAs far exceeds the coding capacity of the mtDNA so that these RNAs must overlap on the DNA. All these RNAs are coded by the H strand except one short RNA which is coded by the L strand.

In mouse L cells several poly(A)-containing RNAs have been identified and mapped onto the mtDNA (13). These RNAs represent about half the coding capacity of the H strand. It is likely that they represent mitochondrial mRNAs but their specific coding capacities have not been reported.

We have carried out experiments to identify and map non-ribosomal RNA molecules derived from mitochondria in the ovary of X. laevis. Total mitochondrial RNA was isolated and separated into poly(A)-containing and poly(A)-lacking fractions. In the first approach we prepared R-loops (26) between mtDNA and the poly(A)-containing RNA. Mitochondrial rRNA was added to provide for R-loops that could be used as reference for orientation of the map. In this way we identified six abundant mitochondrial RNAs (27). These RNA molecules are spread over the mtDNA length without apparent overlaps.

For a more sensitive detection of mitochondrial RNAs we employed the method of Alwyne, Kemp and Stark (28). In this procedure RNAs are separated by electrophoresis on a denaturing gel, in our case using glyoxal treatment of the RNA followed by electrophoresis in agarose (29). The RNA is then transferred from the gel to activated paper where it binds covalently and is available for hybridization with labeled DNA. Such transfers were carried out with poly(A)-containing and poly(A)-lacking mitochondrial RNA, and the paper strips were hybridized with labeled restriction fragments of mtDNA. This allowed the visualization of various RNAs and their mapping relative to the restriction map of the DNA. It is a great advantage of this procedure that different RNA molecules of similar size and therefore similar mobility can nevertheless be distinguished by virtue of their hybridization with distinct restriction fragments. However, the procedure does not actually map the endpoints of the various RNAs but only determines whether a particular RNA molecule overlaps a certain restriction fragment. We have been able to place the more abundant non-ribosomal RNA molecules on the mtDNA map by combining the information obtained from the R-loop mapping and the gel transfer analysis. The results of these experiments are summarized in Figure 2. The major RNA molecules we identified appear to be encoded in the H strand of the mtDNA. Only a single short L strand transcript has been identified sofar (RNA 9 in Figure 2). The RNAs shown in Figure 2 almost completely occupy the H strand, suggesting that most or all of that strand is transcribed into RNAs which accumulate to a fairly high concentration in ovarian mitochondria. Further, two apparent overlaps of RNA coding regions are observed. RNAs 4 and 10 and RNAs 13 and 14 are encoded in the same regions of the mtDNA. While we have no direct evidence on this point it is possible that two of these overlapping RNAs are encoded by the L strand. Alternatively, the overlapping pairs could have precursor-product relationships or they could be independent transcripts from the same regions. At present we cannot distinguish between these possibilities.

Are the non-ribosomal RNAs shown in Figure 2 mRNAs? We have no evidence on this point but it appears likely that the answer will be affirmative. One reason for this suggestion is the number of these RNAs which agrees quite closely with the number of polypeptides known to be synthesized in mitochondria

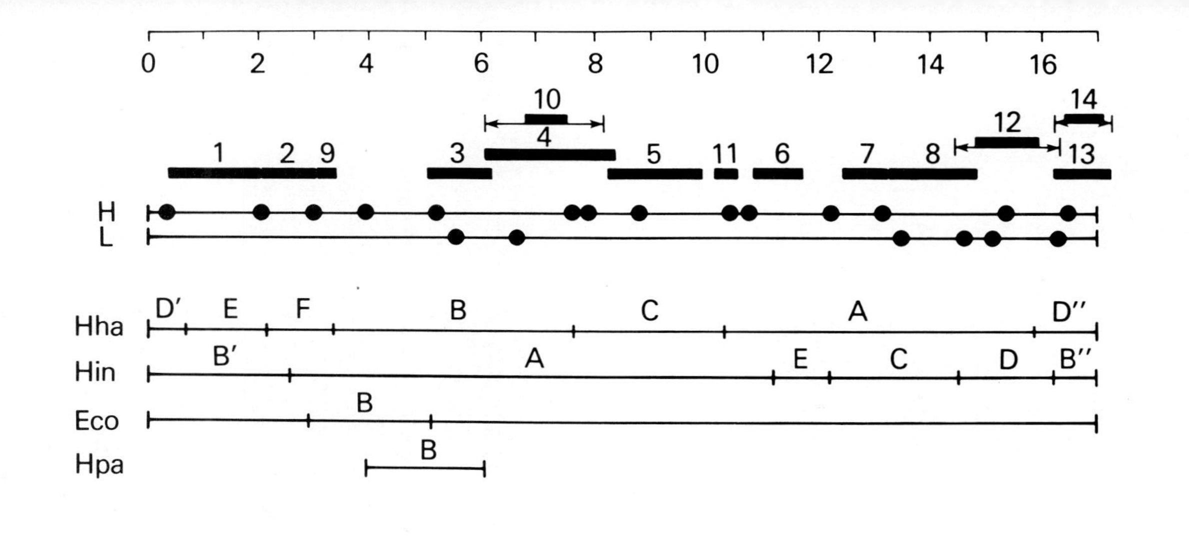

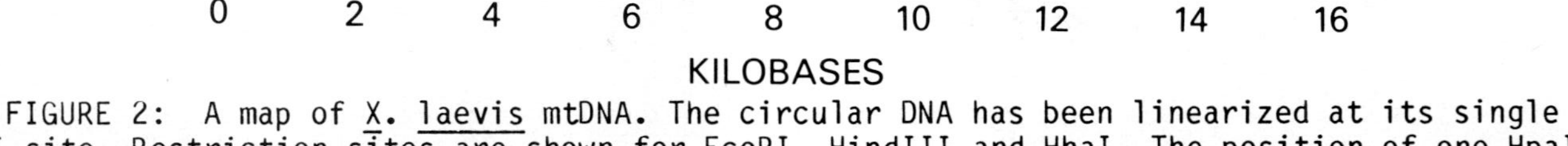

FIGURE 2: A map of *X. laevis* mtDNA. The circular DNA has been linearized at its single BamHI site. Restriction sites are shown for EcoRI, HindIII and HhaI. The position of one HpaII fragment is shown since it was used in the analysis. Sites for 4S RNA are shown from Figure 1 on the H and L strand, and are indicated by dots. Major RNA molecules that were mapped are indicated by heavy lines. Molecules 1 and 2 are the rRNAs. The other mapped RNA molecules are all poly(A) containing. When the map location of an RNA is not certain the possible range is indicated by arrows while the heavy line represents the length of the RNA.

(see above). All of these RNAs contain a poly(A) segment, but this in itself does not necessarily indicate their mRNA nature.

Figure 2 also shows the positions of 4S RNA coding sites on the mtDNA taken from Figure 1. It is not easy to decide in every case whether or not putative mRNA coding regions overlap with 4S RNA sites since the maps for the two kinds of RNA were derived in independent and rather different ways. Thus, superposition of the two maps might lead to significant distortion. Nevertheless it would appear that many of the 4S RNA sites are in fact located between adjoining coding regions for the larger RNAs. However, even when taking into account the possible inaccuracy in the maps it seems certain that at least some of the 4S RNA sites overlap other coding regions or anticoding regions.

In our gel transfer analysis we noticed a number of RNA bands of lower intensity and higher molecular weight than the bands leading to the map assignments in Figure 2. We believe that these bands are due to RNA molecules that are precursors or intermediate processing products leading to the major RNAs, the rRNAs, tRNAs and putative mRNAs. All these putative precursor RNAs map in regions that overlap the coding regions of major RNAs. In some cases more than one precursor occurs in a particular region. It is possible that these RNAs are not precursors but simply minor transcription products which overlap the coding regions of other products. We cannot distinguish between these interpretations on the basis of existing evidence but we hold the precursor hypothesis to be the more likely one. Several of these RNAs overlap one or more 4S RNA sites, and due to the greater length of the putative precursors there can be no doubt about the overlap even when maximum map distortion is allowed for.

Mitochondrial RNA in yeast and various nuclear RNAs are known to be transcribed from interrupted genes. We are therefore asking whether "intervening sequences" occur in mtDNA. We have tested this question in two ways. First, the R-loop analysis failed to show the characteristic pattern of hybrids between RNA and interrupted genes that has been described first by White and Hogness (26). This absence of such structures suggests that intervening sequences have to be short or occur in genes for rare RNAs, if they exist at all. Second, we have carried out an analysis of the type introduced by Berk and Sharp (30,31). In this analysis RNA/DNA hybrids are treated with S1 nuclease and the protected DNA fragments are visualized by gel electrophoresis. This gives the length of DNA regions that continuously code for RNA. In the case of an interrupted gene two (or more) DNA fragments arise whose lengths add up to the length of the RNA. In parallel RNA/DNA hybrids are treated with exonuclease VII. This enzyme digests single stranded DNA exonucleolytically (32). As a result the hybridized coding regions and any intervening sequences that interrupt the gene are protected from digestion. In hybrids between mtDNA and RNA from *Xenopus* all major bands visualized by S1 digestion (corresponding to all RNAs described in Fig. 2) were also present in the exo VII digest. We conclude that the mitochondrial genes in *Xenopus* coding for major RNAs lack intervening sequences.

THE LEVEL OF mtDNA TRANSCRIPTION

The question what fraction of mtDNA sequences is transcribed into RNA is not completely resolved. Earlier work in Hela cells indicated that both strands are transcribed fully into RNA some of which is long enough to be a complete transcript of each strand (33). In the analysis of individual RNAs from Hela mitochondria all but one was found to be encoded by the H strand; the only L strand products identified were one small RNA (7S RNA), and six 4S RNAs (18,20). However, polydisperse L strand transcripts were detected (25). In mouse L cells Battey and Clayton (13) detected only H strand transcripts that accounted for about half the coding capacity of this strand. Some additional H strand transcripts of lower abundance were found but no L strand transcripts.

In *X. laevis* we had earlier suggested that less than one complete strand is transcribed into stable RNA (34). In our present mapping studies we have identified a number of RNAs coded by the H strand which account for about the entire length of this strand (Figure 2). Thus, the earlier value must have been an underestimate, and in *Xenopus* as in Hela cells, all or nearly all the H strand is transcribed. We tried to find very long transcripts, possibly accounting for the entire strand. These attempts were negative and we do not know the size of the primary H strand transcripts in *Xenopus*.

As in the case of Hela cells we identified five 4S RNA sites on the L strand of *Xenopus* mtDNA (21). We also detected an additional small RNA of about 300 nucleotides derived from this strand (27). However, other dicrete L strand transcripts seem to be rare since we have not detected any sofar. These experiments are still in progress and some L strand transcripts may eventually be found. It is already clear, however, that by far the majority of stable transcripts in *Xenopus* mitochondria derive from the H strand.

SOME ASPECTS OF mtDNA EVOLUTION

The size and shape of mtDNA is quite conservative within

broad groups of organisms. All metazoan animal mtDNAs are similar in this respect. However, great differences occur when more diverse organisms are considered. Nevertheless, the functions of mtDNA appear to be largely conserved in that rRNA, tRNA and the mRNAs for a few specific proteins are encoded in all mtDNAs. In general, the polypeptides encoded in mtDNA appear to be analogous in different organisms. Considering this similar coding capacity the size differences between mtDNA require an explanation.

The study of mtDNA evolution is more profitably carried out by considering quite closely related organisms. In such cases the DNAs are very similar and relatively small differences in primary sequence can be utilized in detailed comparisons. This has been done in an early study comparing the mtDNAs of two frogs X. laevis and X. borealis (35), using crosshybridization and heteroduplex melting studies. The main conclusions from this study were shown by later work to be generally true: The mtDNAs of even closely related species differ considerably in their sequences. Sequence divergence varies along the length of the DNA in that the rRNA coding regions are more conserved than the average sequence. In this latter respect the evolutionary behavior of mtDNA is similar to that of total nuclear DNA (36). It is of interest that the rRNA coding regions in mtDNA are more extensively changed during evolution than the nuclear rDNA sequences (35,37).

More detailed comparisons of mtDNA became possible by the introduction of restriction endonucleases. The first report of the application of these enzymes to mtDNA comparison (38) led to very interesting conclusions. In particular, these authors noted that different individuals from the same species of animals could contain mtDNA molecules that differed in their restriction pattern and thus in their primary sequence. DNAs from closely related animal species differed greatly when examined in this way. In our work with the mtDNAs of sheep and goats we made a quantitative study of sequence differences within and between these species (39). Restriction maps were derived from the separate study of three individual sheep and two goats using the enzymes EcoRI and HindIII. Great similarity was seen but some restriction sites differ between the species. In addition, a few site differences emerged when individuals of the same species were contrasted. A particularly sensitive way to probe differences between similar DNAs involves the use of endonucleases with frequent cleavage sites. This was done by comparing the cleavage products with HaeIII without mapping the sites. Band differences were detected and could be evaluated quantitatively. When making certain assumptions such differences can be interpreted in terms of nucleotide sequence substitution (40). The application of these methods lead to the following conclusions. The mtDNA from each single individual is homogenous as far as can be deduced from the restriction analysis. The mtDNAs from all five individual sheep and goats which we tested differed from each other. Intraspecific differences in restriction pattern suggested a sequence divergence of 0.5 to 2%. In contrast, the interspecific sequence divergence is approximately 6 to 11%.

This type of restriction endonuclease analysis of related mtDNAs has since been applied to additional examples, and this symposium volume contains some articles on the subject. There are two general applications of these kinds of analysis. First, one may use the relatedness of mtDNA as an evolutionary marker. Small differences could be utilized to distinguish subspecies or strains of organisms and different lines of cultered cells. The degree of sequence divergence in mtDNA could be used as an indication of taxonomic relations between species. In the latter context, however, it must be clear that mtDNA sequences alone cannot be a sufficient guide in setting up evolutionary trees and can only be helpful in conjunction with other evidence.

The second point was raised in our work on sheep and goat mtDNA (39). We would like to understand the mechanisms that lead to structural homogeneity of mtDNA in a single individual but nevertheless allow significant sequence differences between members of a species, i.e., within an interbreeding population of animals. We have speculated that the pattern of maternal inheritance that is followed by mtDNA (41) may be causing the observed sequence relations. Reduction of the number of mtDNA molecules to a small population during the early development of the oocyte and subsequent "amplification" to the very large number in the mature egg could act as a correction mechanism that maintains homogeneity in the mtDNA population. Because of the maternal inheritance it would appear that the sequences of mtDNA in different individuals could not exchange, and therefore any changes that did become fixed in one female would be transmitted to her progeny without the chance of mixing or recombination with sequences from other individuals. The possibility must be considered whether mitochondria from the sperm do in some cases become incorporated into the zygote and eventually contribute to an exchange of sequences within a breeding population. But in general the mtDNA would be transmitted in "maternal clones". These clones would assure homogeneity of mtDNA within an individual and also between all members of the same clone, e.g. in an inbred population of mice. This model would explain the observations on individual homogeneity and intraspecific heterogeneity of mtDNA. Further studies of mtDNA sequences in individual animals will provide more data that may allow a test of our speculative models of mtDNA evolution.

REFERENCES

1. Bucher, Th., Neupert, W., Sebald, W., and Werner, S. Eds. (1976). p. 895. North-Holland: Amsterdam.
2. Borst, P., and Grivell, L. A. (1978). Cell 15, 705.
3. Borst, P., and Flavell, R. A. (1976). In "Handbook of Biochemistry and Molecular Biology" (G. D. Fasman, ed.), pp. 363-374. CRC Press, Cleveland.
4. Dawid, I. B. (1972). In "Mitochondria Biomembranes" (S. G. van den Bergh et al., eds), pp. 35-51. Amsterdam: North-Holland.
5. Leister, D. E., and Dawid, I. B. (1974). J. Biol. Chem. 249, 5108.
6. O'Brien, T. W. (1976). In "Protein Synthesis, Vol. 2". (E. H. McConkey, ed.), pp. 249-307. Marcel Dekker, New York.
7. O'Brien, T. W., and Matthews, D. E. (1976). In "Handbook of Genetics, Vol. 5", (R. C. King, ed.), pp. 535-580. Plenum Publishing Co.
8. Lizardi, P. M., and Luck, D. J. L. (1971). Nature New Biology 229, 140.
9. Dawid, I. B., Chase, J. W., and Tartof, K. D. (1972). Carnegie Inst. Year Book 71, 30.
10. Wu, M., Davidson, N., Attardi, G., and Aloni, Y. (1972). J. Mol. Biol. 71, 81.
11. Ramirez, J. L., and Dawid, I. B. (1978). J. Mol. Biol. 119, 133.
12. Klukas, C. K., and Dawid, I. B. (1976). Cell 9, 615.
13. Battey, J., and Clayton, D. A. (1978). Cell 14, 143.
14. Goldbach, R. W., Borst, P., Bollen-De Boer, J. E., and Van Bruggen, E. F. J. (1978). Biochim. Biophys. Acta 521, 169.
15. Bos, J. L., Heyting, C., and Borst, P. (1978). Nature 275, 336.
16. Nass, M. M. K., and Buck, C. A. (1970). J. Mol. Biol. 54, 187.
17. Chiu, N., Chiu, A. O. S., and Suyama, Y. (1974). J. Mol. Biol. 82, 441.
18. Attardi, G., Albring, M., Amalric, F., Gelfand, R., Griffith, J., Lynch, D., Merkel, C., Murphy, W., and Ojala, D. (1976). In "Genetics and Biogenesis of Chloroplasts and Mitochondria" (Th. Bucher et al., eds.) pp. 573-585. North Holland: Amsterdam.

19. Lynch D. C., and Attardi, G. (1976). J. Mol. Biol. 102, 125.
20. Angerer, L., Davidson, N., Murphy, W., Lynch, D., and Attardi, G. (1976). Cell 9, 81.
21. Ohi, S., Ramirez, J. L., Upholt, W. D., and Dawid, I. B. (1978). J. Mol. Biol. 121, 299.
22. Chiu, N., Chiu, A., and Suyama, Y. (1975). J. Mol. Biol. 99, 37.
23. Macino, G., and Tzagoloff, A. (1979). Proc. Nat. Acad. Sci. USA 76, 131.
24. Koch, G. (1976). J. Biol. Chem. 251, 6097.
25. Amalric, F., Merkel, C., Gelfand, R., and Attardi, G. (1978). J. Mol. Biol. 118, 1.
26. White, R. L., and Hogness, D. S. (1977). Cell 10, 177.
27. Rastl, E. (1978). Carnegie Inst. Year Book 77, 95.
28. Alwyne, J. C., Kemp, D. J., and Stark, G. R. (1977). Proc. Nat. Acad. Sci. USA 74, 5350.
29. McMaster, G. K., and Carmichael, G. G. (1977). Proc. Nat. Acad. Sci. USA 74, 4835.
30. Berk, A. J., and Sharp, P. A. (1977). Cell 12, 721.
31. Berk, A. J., and Sharp, P. A. (1978). Proc. Nat. Acad. Sci. USA 75, 1274.
32. Chase, J. W., and Richardson, C. C. (1974). J. Biol. Chem. 249, 4545.
33. Murphy, W. I., Attardi, B., and Attardi, G. (1975). J. Mol. Biol. 99, 809.
34. Dawid, I. B. (1972). J. Mol. Biol. 63, 201.
35. Dawid, I. B. (1972). Develop. Biol. 29, 139.
36. Sinclair, J. H., and Brown, D. D. (1971). Biochemistry 10, 2761.
37. Brown, D. D., Wensink, P. C., and Jordan, E. (1972). J. Mol. Biol. 63, 57.
38. Potter, S. S., Newbold, J. E., Hutchison, C. A. III, Edgell, M. H. (1975). Proc. Nat. Acad. Sci. USA 72, 4496.
39. Upholt, W. B., and Dawid, I. B. (1977). Cell 11, 571.
40. Upholt, W. B. (1977). Nucl. Acids Res. 4, 1257.
41. Dawid, I. B., and Blackler, A. W. (1972). Develop. Biol. 29, 152.

STRUCTURE AND REPLICATION OF MITOCHONDRIAL DNA FROM THE GENUS *DROSOPHILA*[1]

David R. Wolstenholme, Judy M. Goddard and Christiane M.-R. Fauron

Department of Biology, University of Utah, Salt Lake City, Utah 84112

ABSTRACT Mitochondrial DNA (mtDNA) molecules from different species of the melanogaster group of the genus *Drosophila* differ in size from 15.7 to 19.5 kilobase pairs (kb). These differences appear to be accounted for by differences in size (1.0 to 5.4 kb) of a single adenine and thymine (A+T)-rich region in each molecule. The sizes of the mtDNA molecule of other *Drosophila* species are within the narrow range 15.7 to 16.8 kb, and contain an A+T-rich region of approximately 1.0 kb. Restriction enzyme mapping results indicate the A+T-rich region of *D. melanogaster*, *D. simulans*, *D. mauritiana*, *D. takahashi*, and *D. virilis* to be at homologous positions on the mtDNA molecules. We have studied the various structural forms of partially replicated mtDNA molecules from the above mentioned species, and concluded that in each species most molecules are replicated by a highly asymmetrical mode in which synthesis on one strand can be between 60% and 100% complete before synthesis on the other strand is initiated. Using the A+T-rich regions and *Eco*RI cleavage sites as markers, we have determined that in all species studied replication of mtDNA molecules is initiated in the A+T-rich region and proceeds unidirectionally around the molecule towards the nearest *Eco*RI site common to the mtDNAs of all six species. We have found that the A+T-rich regions of the different species have little or no sequence homologies.

[1]This work was supported by National Institute of Health Grant Nos. GM 18375 and K4-GM-70104, American Cancer Society Grant No. NP-41B and National Science Foundation Grant No. BMS-74-21955.

ISBN 0-12-198780-9

INTRODUCTION

A most remarkable feature of metazoan mtDNAs is that while they may differ extensively in sequence content the size of the molecules has been highly conserved. All metazoan mtDNAs have been found to be in the form of circular molecules and the majority are between 14.5 kilobase pairs (kb) and 16.5 kb in size (1). Two notable exceptions which have been well studied are the mtDNAs of the anuran amphibia, (*Xenopus* and *Rana*) which comprise circular molecules of approximately 19 kb (2, 3), and mtDNAs of species of the melanogaster group of the genus *Drosophila* which comprise circular molecules with species specific sizes ranging up to 19.5 kb (4). While the greater size of mtDNA molecules of anuran amphibia has not been explained in regard to sequence content, the range in sizes of *Drosophila* mtDNA molecules has been clearly shown to result from the presence of a region rich in adenine and thymine (A+T), which varies in size between the mtDNA molecules of different species (4).

The presence of A+T-rich DNA in mtDNA molecules of *D. melanogaster* was discovered independently by three groups (5, 6, 7), from heat denaturation studies of this DNA. Using electron microscope denaturation mapping (8) Peacock et al (7) demonstrated that much of the A+T-rich DNA was confined to a single region of the molecule. The constancy in the size of this region in different *D. melanogaster* mtDNA molecules was demonstrated by Fauron and Wolstenholme (4) and Klukas and Dawid (9), and the former workers further showed that a region with similar denaturation properties to the A+T-rich region of *D. melanogaster* mtDNA molecules, but of a lesser, variable size is present in mtDNA molecules of a number of *Drosophila* species. While A+T-rich regions have been found in yeast mtDNA (10), the presence of an A+T-rich region in *Drosophila* mtDNAs appears to be unique among metazoans mtDNAs.

We present here a review of our studies of the structure and replication of mtDNA molecules from a number of *Drosophila* species.

MATERIALS AND METHODS

The origins of all species of *Drosophila* used in these investigations are given in detail in reference 4. Methodological details including culturing of flies; preparation of mitochondria from eggs and from ovaries, and isolation of DNA from mitochondria; analytical neutral and alkali cesium chloride equilibrium buoyant density centrifugation; preparative cesium chloride and cesium chloride-ethidium bromide equilibrium buoyant density centrifugation; restriction enzyme digestions and mapping of restriction enzyme sites on mtDNA molecules; preparation of DNA for electron microscopy by the aqueous and formamide protein monolayer techniques; estimations of sizes of whole molecules and single-stranded and double-stranded regions of molecules either using bacteriophage fd single-stranded or double-strand RF DNA (6,408 base pairs [11]) as internal standard, or by using a correction factor for the greater mass per unit length of single-strand DNA molecules under the standard conditions used, are given in references 4, 12 and 13.

RESULTS AND DISCUSSION

We have studied the size and denaturation properties of *Drosophila* mtDNA, using electron microscopy (4). In electron microscope preparations, *D. melanogaster* mtDNA appeared as a collection of circular molecules with a uniform contour size of 19.54 kb. When these molecules were heated at 40°C in 0.05 M sodium phosphate and 10% formaldehyde (pH 7.8), a single continuous region of denaturation, representing approximately 25% of the molecule was observed (Figure 1). Other small regions of denaturation occurred at this temperature but collectively they represented no more than 10% of the molecule contour length. Similar observations have been made by Klukas and Dawid (9) following alkali denaturation. As it would be expected that the large region of denaturation observed when *D. melanogaster* molecules are heated at 40°C is rich in adenine and thymine, we termed this region, and a region of varying size in mtDNA molecules heated at 40°C from other *Drosophila* species, the A+T-rich region. There is evidence from buoyant density studies (7, 14, 15) supporting this interpretation.

Mitochondrial DNA molecules of *Drosophila simulans* (which forms genetic hybrids with *D. melanogaster*) were found to have similar properties in regard to whole molecule size and size of the A+T-rich region, to the mtDNA molecules of *D. melanogaster*. In contrast, we found that the mtDNA molecules of *D. virilis* were circular molecules of only 16.0 kb and that the A+T-rich region was only 0.5 kb. (The correct value for the size of the A+T-rich region of *D. virilis* mtDNA appears to be 1.0 kb[16]). In view of this latter finding, we determined the sizes of the mtDNA molecules of 36 other species of *Drosophila*. The results are summarized in Figures 2 and 3. Only a single length class of circular molecule was observed for each species examined. The 39 species examined

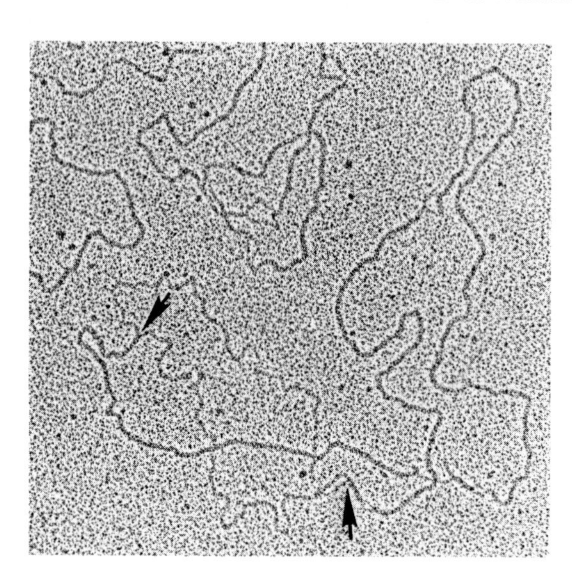

Figure 1. Electron micrograph of a D. melanogaster mtDNA molecule heated at 40°C for 10 min. in 0.05 M sodium phosphate (pH 7.8) and 10% formaldehyde. The arrows show the limits of the denatured region. x 62,500.

represent 13 groups of 5 subgenera of the genus Drosophila. Mitochondrial DNA molecules of all these species other than members of the melanogaster group had, with one exception, sizes in the rather narrow range of 15.7 to 16.4 kb. The one exception was D. robusta mtDNA which was 16.8 kb. In contrast, mtDNA molecules from 19 species of the melanogaster group (Figure 3) had sizes covering the considerably greater range of 15.7 to 19.5 kb. The results of denaturation mapping studies of mtDNA molecules from selected melanogaster group species are shown in Figure 4. It was clear that each mtDNA molecule studied contained a distinct A+T-rich region, and a plot of the size of the whole mtDNA molecule against the size of the corresponding A+T-rich region for each of the 8 melanogaster group species indicated that most of the variation in total size of the mtDNA molecules of these species could be accounted for by differences in size of the A+T-rich region.

It was noted from these results (4) that the distribution of sizes of the A+T-rich region of mtDNA molecules of the different melanogaster group species does not follow a simple taxonomic pattern (Figure 3). Of the six species of the melanogaster subgroup, three (D. melanogaster, D. simulans, and D. mauritiana) have A+T-rich regions of 4.2 to 5.4 kb, while at least one (D. yakuba) has an A+T-rich region of 1.5 kb. The mtDNA molecules of D. takahashi, D. birchii

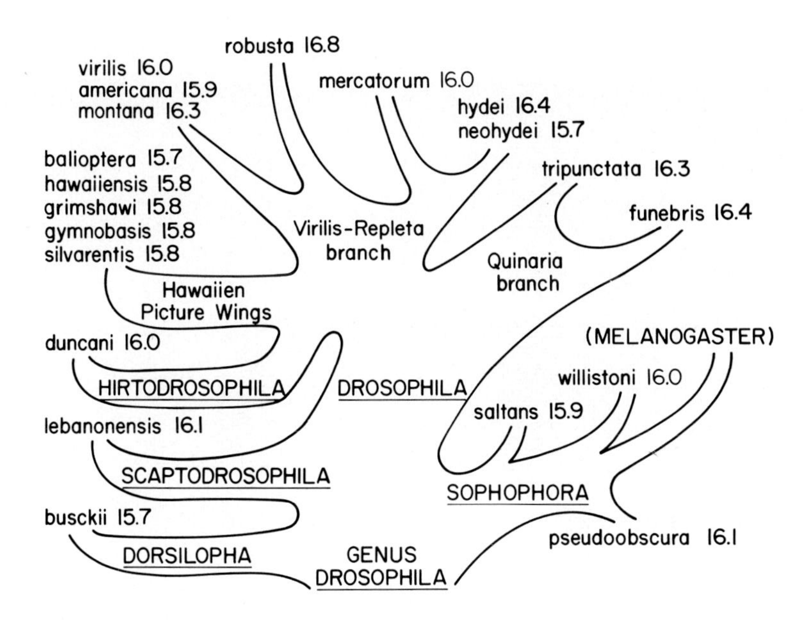

Figure 2. The size of mtDNA molecules (kb) from 20 species representing 5 subgenera of the genus Drosophila. Species classification is given in reference 4.

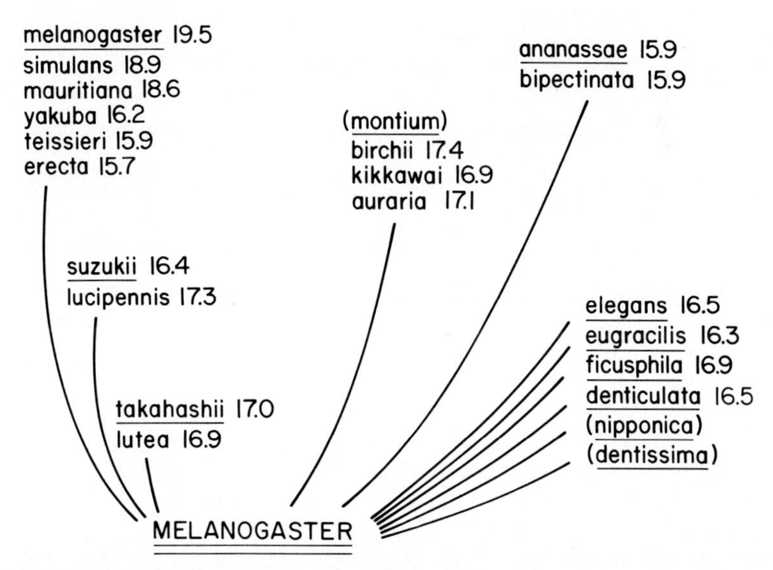

Figure 3. The size of mtDNA molecules (kb) from 19 species representing 9 of the 11 subgroups (names underlined) of the melanogaster species group. Species classification is given in reference 4.

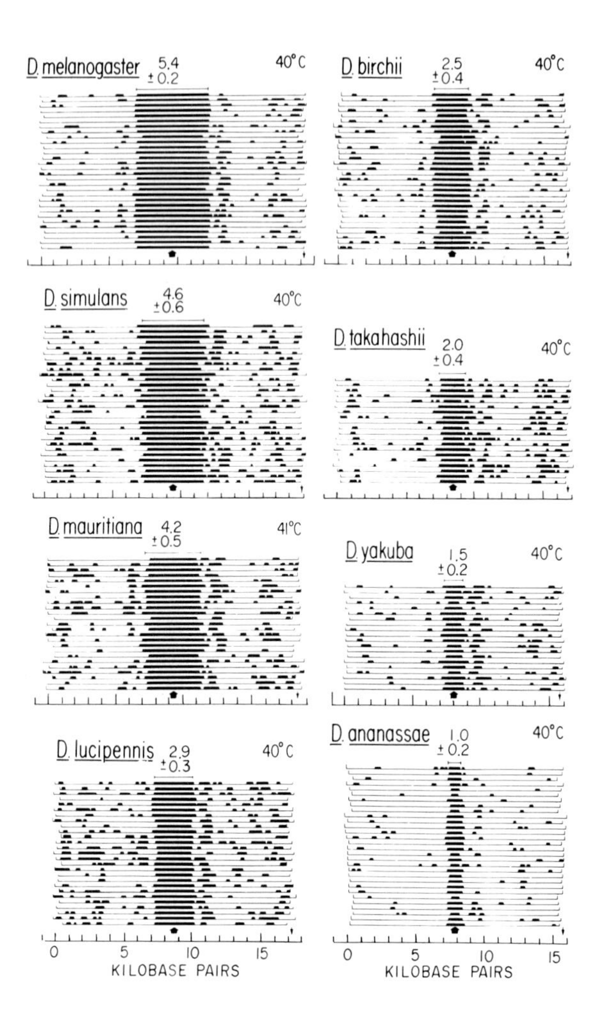

Figure 4. Denaturation maps of nicked circular mtDNA molecules of the species indicated, produced by heating at 40°C for 10 min. in 0.05 M of sodium phosphate (pH 7.8) and 10% formaldehyde. For the purpose of comparison each circular molecule was converted to a linear rod by opening it at one-half the circumference length away from the midpoint of the main region of denaturation. Denatured and undenatured regions are represented by thick and thin lines respectively. The midpoint of the largest region of denaturation was taken as the common point by which the molecules were aligned (large arrow in the center of the abscissae). The mean size (in kb ± SD) of the A+T-rich region is indicated above each set of molecules. (Modified from Figure 3 of reference 4).

and *D. lucipennis*, which represent three different subgroups, have A+T-rich regions with sizes intermediate to those of *D. yakuba* and the other three melanogaster subgroup species. If we assume that the melanogaster group species have indeed evolved along the lines indicated by the taxonomic classification adopted (Figures 2 and 3) then the results indicate either that increase in size of the A+T-rich region of mtDNA molecules has occurred independently in separate evolutionary paths, or that decrease in size of the A+T-rich region has accompanied evolution along at least one path.

Mapping of the A+T-rich region relative to sites sensitive to cleavage by the restriction enzyme *EcoRI* on mtDNA molecules of *D. melanogaster* (12) and mtDNA molecules of *D. simulans*, *D. mauritiana*, *D. yakuba*, *D. takahashi* and *D. virilis* (16) confirm that the A+T-rich regions are located at homologous positions in these mtDNA molecules.

Using electron microscopy we studied the replicative intermediates present in *D. melanogaster* mtDNA obtained from embryonated eggs (13). Approximately 1% of the molecules in the sample had the characteristics of structure and size which indicated them to be partially replicated molecules (17-19). All of these molecules (n = 59) had one totally double-stranded segment (the first daughter segment) in the replicated region. In the majority (49) of these (with the replicated region ranging from 29 to 99-100% of the genome length) the second daughter segment was totally single-stranded (Figure 5B and C). A distinct excess of D loop containing molecules (20) was not observed. In 6 of the replicative intermediates observed, the second daughter segment was also double-stranded and the replicated region was 90 to 96% of the genome length (Figure 5A). In 6 of the replicative intermediates one daughter segment was single-stranded at both forks but double-stranded in a central region. Replicative intermediates have since been found in which the second daughter segment is single-stranded at one fork and double-stranded at the other (Figure 5D). A separate sample of *D. melanogaster* embryo mtDNA was scored only for replicative intermediates in which the second daugher segment was either totally or partially double-stranded. Of 156 such molecules observed, 120 were totally double-stranded. Although the replicated region of such molecules ranged from 5 to 97% of the genome length, in 99 (83%) of the molecules the daughter segment lengths fell within the narrow range of 87 to 97% of the genome length. In all preparations of mtDNA, we have observed simple (nonforked) circular molecules which were either totally single-stranded (Figure 5E) or in which a single region measuring between 2 and 77% of the genome length appeared to be single-stranded (Figure 5F).

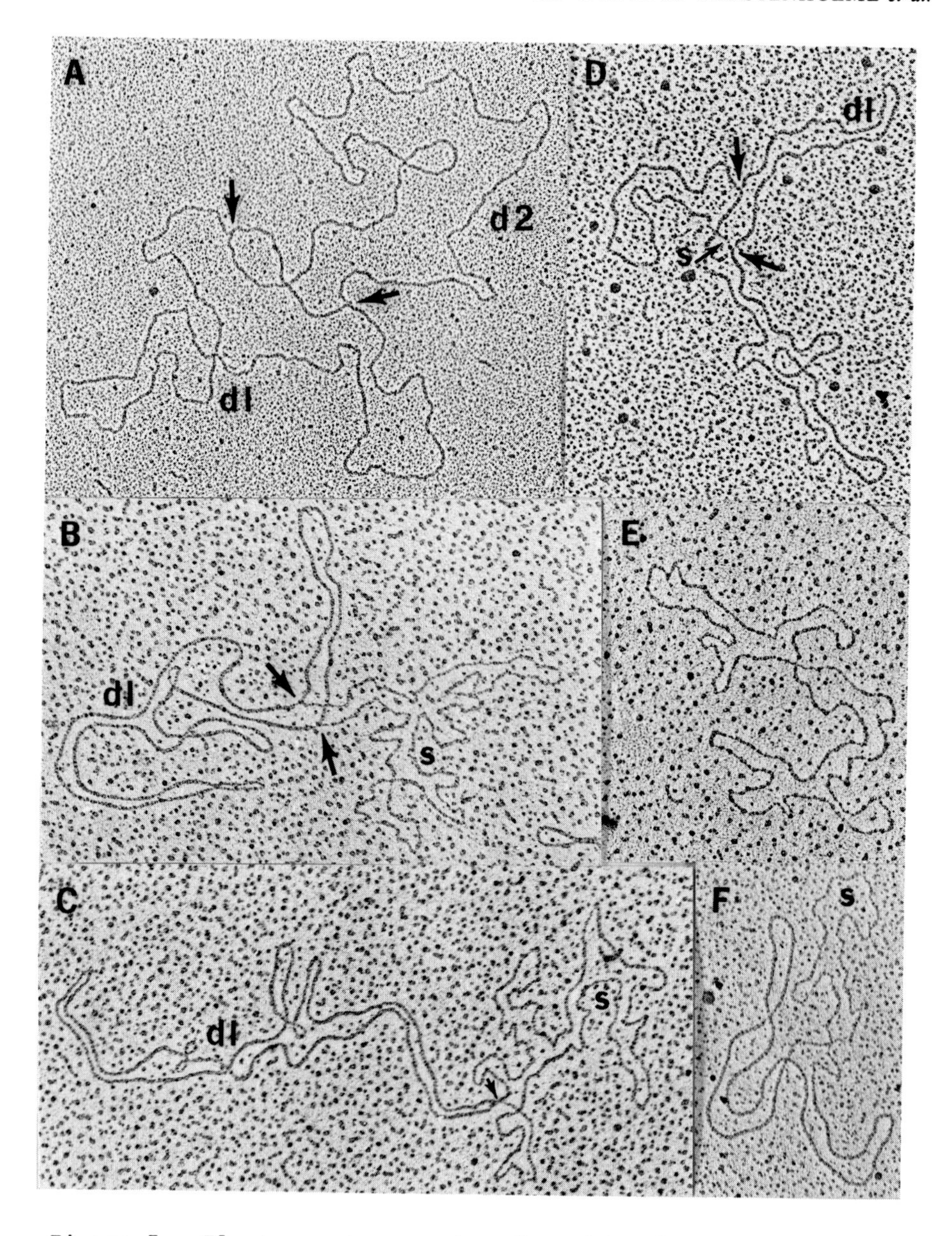

Figure 5. Electron micrographs of replicative intermediates of D. melanogaster mtDNA · A-D, Double forked (arrows) circular molecules in which one daughter segment (dl) is totally dougle-stranded and the second daughter segment is either totoally double-stranded (d2; a), totally single-stranded (s, B and C), or double-stranded (d) at one fork and single-stranded (s) at the other fork (D). The replicated regions in A, B,

We have used the relative locations of the A+T-rich regions and *EcoRI* sites in *D. melanogaster* mtDNA molecules to locate the site at which replication originates and to study the direction in which replication proceeds around these molecules (13). First, we examined samples of mtDNA in the electron microscope following heating at 40°C for 10 min. in 0.05 M of sodium phosphate. Approximately 99% of the molecules appeared to be simple circular molecules in which the A+T-rich region was fully denatured (Figure 1). Ten molecules were observed, however, which had a more complex structure. These molecules were interpreted as shown in Figure 6, and from a consideration of the lengths and relative arrangement of the various single-stranded and double-stranded segments of these molecules, it was concluded that replication can proceed unidirectionally from a unique origin lying approximately in the center of the A+T-rich region, along one strand for at least 58% of the genome length. From this data it is not possible to say whether or not the direction of replication is the same for individual molecules.

Next, we examined in the electron microscope samples of mtDNA following digestion with *EcoRI*. Linear fragments containing two forks were located. These fragments could be placed into three structural classes as shown in Figure 7 a"-c". Fragments of the three classes were interpreted as being derived from partially replicated molecules in which replication had proceeded from a unique point 39% of the genome length from one end of the A fragment, to various points on the A fragment short of the nearest *EcoRI* site (Figure 7a"), or through the nearest *EcoRI* site, to various positions on the B fragment (Figure 7b"), or around the molecule back to the A fragment (Figure 7c"). As well as confirming that replication originates from a unique origin situated on the A fragment, these data indicate that replication always proceeds in the same direction around the molecule. Because one of the two locations for a point 39% of the genome length from one end of the A fragment is the center of the A+T-rich region, these observations are in agreement with the conclusion drawn from observations on partially denatured molecules, that the origin of replication lies in the center of the A+T-rich region.

Finally, we digested samples of the mtDNA with *EcoRI*

C, and D are 90%, 77%, 99-100% and 34% of the genome length respectively. E and F, simple (nonforked) circular molecules. E is completely single-stranded and F contains a single-stranded region(s) accounting for 40% of the genome length. All micrographs x 39,000.

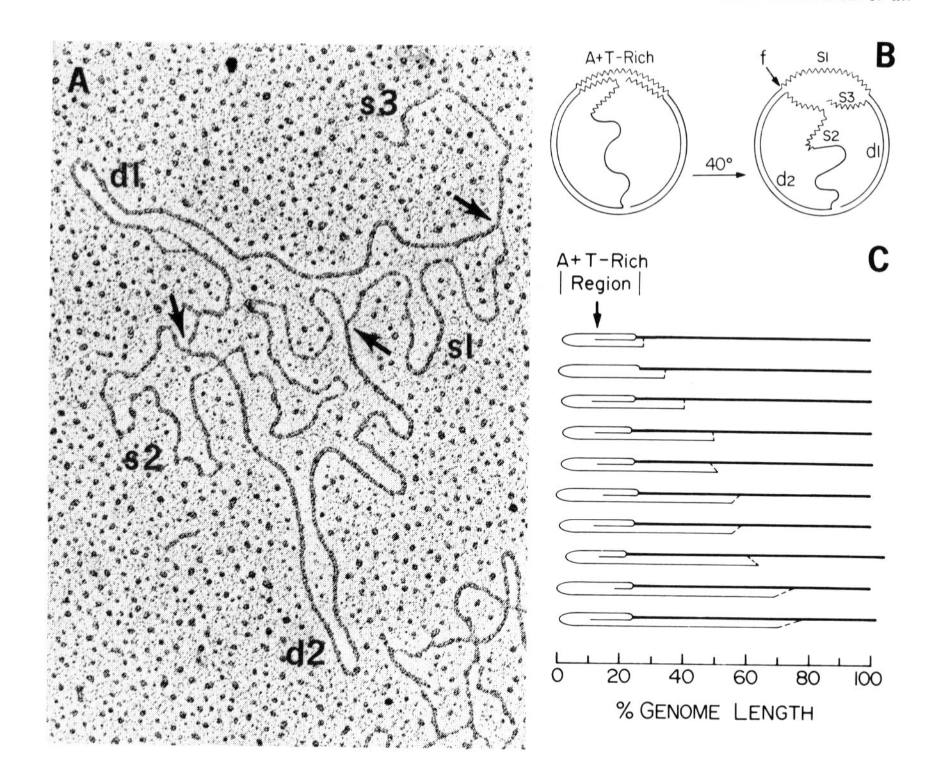

Figure 6. Partially replicated mtDNA molecules in which the A+T-rich region has been denatured by heating at 40°C for 10 min. in 0.05 M sodium phosphate and 10% formaldehyde. A, an example electron micrograph. Three forks are indicated by arrows. B, diagrammatic interpretation of the molecule shown in A. Corresponding segments of double-stranded DNA (d) and single-stranded DNA (s) in A and B are indicated by numbers. C, a comparison of ten molecules. For this purpose each is shown as a linear rod produced by cutting the circular molecule at the fork marked f in B and placing this fork to the left. Double-stranded and single-stranded segments are indicated by thick and thin lines respectively. (Modified from Figure 1 of reference 13.) Micrograph, x 55,500.

and following partial denaturation examined the products in the electron microscope. A number of complex forms were found which could in each case be interpreted as having arisen from replicative intermediates in which replication originated in the center of the A+T-rich region and proceeded unidirectionally around the molecule in the direction of the nearest EcoRI site.

Our conclusions concerning the modes of replication of

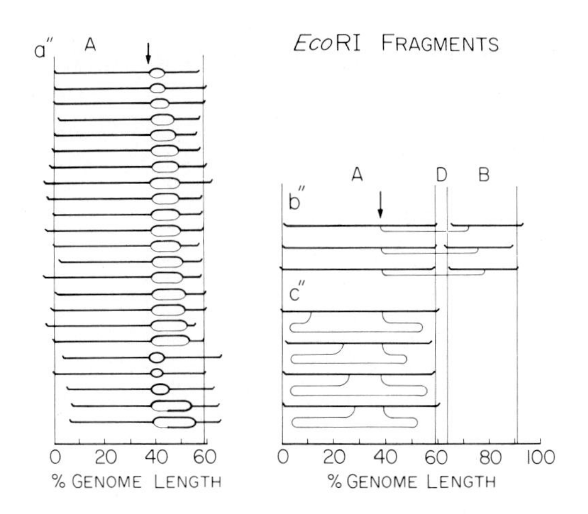

Figure 7. EcoRI fragments of D. melanogaster mtDNA which are interpreted as resulting from partially replicated molecules. A-D refer to the 4 EcoRI fragments of D. melanogaster mtDNA which are equal to 59%, 27.5%, 9% and 4.5% of the genome length respectively (12). a", b" and c", comparisons of the molecules of the three structural classes. In a" and b" each fragment is aligned (arrow) by the fork associated with the larger double-stranded segment and this double-stranded segment is placed to the left. In b", the second fork of each fragment is placed to the right as shown. In c", each fragment is aligned (arrow) by the fork associated with the double-stranded segment which measures approximately 20% of the genome length and this double-stranded segment is placed to the right. Double-stranded and single-stranded segments are represented by thick and thin lines respectively. (Modified from Figure 2 of reference 13.)

D. melanogaster mtDNA are summarized in Figure 8. As in all other metazoan mtDNAs studied (21-25) replication begins at a unique site and proceeds unidirectionally around the molecule. As the majority of replicative intermediates are molecules in which one daughter segment is totally single stranded, it seems clear that this mtDNA is replicated by a highly asymmetrical mode. Such an asymmetrical mode of replication was suggested by Zakian (26) for D. virilis mtDNA and seems to be operative in mtDNA of tissues of various organisms (18, 19, 27). As noted here for D. melanogaster mtDNA (Figure 5F) simple circular molecules containing a single-stranded region (gapped molecules) have been found in other animal mtDNAs and evidence has been obtained (28) that they are daughter mole-

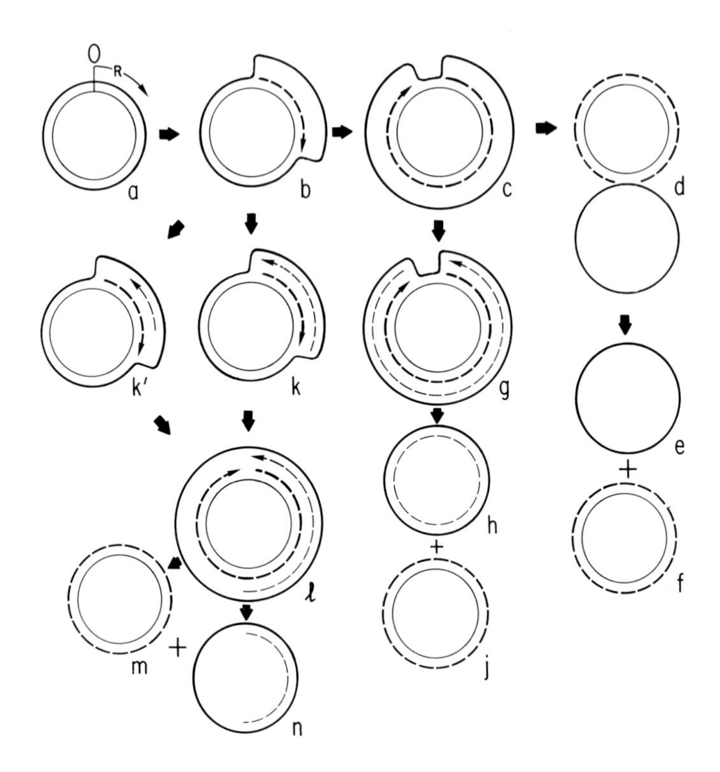

Figure 8. A scheme for the replication of Drosophila mtDNA which accounts for all of the molecular forms observed in electron microscope preparations. Thick and thin continuous lines represent the complementary parental strands, and thick and thin broken lines represent the corresponding complementary progeny strands. The arrows on progeny strands show the direction of synthesis (assuming antiparallel synthesis of the two complementary strands). In a, the origin of replication (O) and direction of replication (R) around the molecule are indicated. The scheme a-f accounts for the molecular forms shown in Figure 5 B, C and E. The scheme a-c-j accounts for the molecular forms shown in Figure 5 A and B. The scheme a-b-k(k')-n accounts for the molecular forms shown in Figure 5 D and F.

cules which have separated before synthesis is complete (Figure 8n).

In contrast to what has been found in other animal tissues, it appears from observation of molecules like that shown in Figure 5C, that in D. melanogaster mtDNA, DNA synthesis of one strand can be virtually (99-100%) complete before synthesis of the complementary strand is initiated (Figure 8a-c).

The observation of totally single-stranded simple circular molecules (Figure 5E) suggests that, in fact, daughter strand separation may occur before synthesis commences on the second strand (Figure 8a-f). However, the findings of molecules in which the replicated region measured 10% to 98% of the genome length (Figure 5A and C) and in which the second daughter segment was either partially or totally double-stranded, indicates that synthesis of the second daughter strand may be initiated when synthesis on the first daughter strand is as little as 10% complete (Figure 8 a-b-k(k')-n). It was observed that of molecules in which synthesis had occured in the second daughter strand, those in which synthesis was 87-98% completed predominated. This suggests that there is a temporal pause in replication when synthesis is completed on both daughter strands to an extent greater than 87% (Figure 8a-g).

Replication of mouse L cell mtDNA appears to follow a more defined mode (17, 28) than mtDNA taken directly from animal tissues. Synthesis on the second daughter strand is initiated only when synthesis on the first daughter strand is at least 60% complete.

D loop-containing molecules have been found in mtDNA from many metazoans, and the D loop has been identified as the first step in replication of mouse L-cell mtDNA (20, 28). Failure to find a distinct excess of D loop containing *Drosophila* mtDNA molecules by ourselves (12, 13, 29) and others (9, 30) is therefore of particular interest as it may indicate a real difference between the mechanisms of replication operative in mtDNA of this organism and of mtDNAs of other metazoans studied. However, it has not been ruled out that the lower resistance to denaturation of the A+T-rich region contributes to loss of D loops by branch migration (31) at some time in the isolation procedure of the DNA.

In Figure 9 the position of the origin of replication and the direction of replication around the molecule are shown relative to the location of the rRNA genes. Assuming that, as in all other cases studied, synthesis of DNA and RNA both occur 5'→3', then the same strand of the parental mtDNA molecule must serve as template both for the initial DNA synthesis of DNA replication (synthesis of the lead strand) and for rRNA transcription. In mouse mtDNA (32) rat mtDNA (33) human mtDNA (34) and *Xenopus* mtDNA (35), the two rRNA genes have been shown to occupy similar positions on the molecule relative to the origin of replication as shown in Figure 9 for *Drosophila* mtDNA, except that the smaller rRNA gene is separated from the origin of replication by 0.3 to 1.0 kb. However, in all four cases, in contrast to *Drosophila* mtDNA, replication proceeds around the molecule in the direction *away from* the rRNA genes. Further, it has been clearly shown in

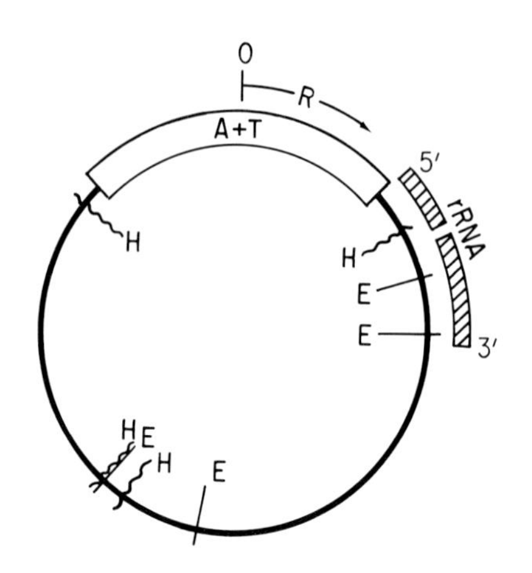

Figure 9. Map of the D. melanogaster mtDNA molecule showing the origin (O) and direction (R) of replication relative to the A+T-rich region, and to the EcoRI (E) and HindIII (H) sites derived from our data given in references 12, 13, and 16. The position and polarity of the rRNA molecules transcribed from this mtDNA molecule, relative to the HindIII sites were determined by Klukas and Dawid (9). The two sets of data were aligned by the HindIII site in the smaller rRNA gene.

mouse mtDNA (32) that the rRNAs are transcribed from the opposite (H) strand to that strand (L) which acts as template for synthesis of the lead strand in DNA replication (20).

Similar studies to those described above for D. melanogaster mtDNA, have been carried out with D. simulans, D. mauritiana, D. yakuba, D. takahashi and D. virilis (29). In the mtDNAs of all of these species it appears that similar modes of replication are followed, that the origin of replication occurs in the A+T-rich region and that replication proceeds unidirectionally around the molecule towards the nearest EcoRI site common to each of these mtDNAs.

The distribution of sizes among A+T-rich regions of the mtDNA molecules shown in Figure 4 suggested the possibility that they represent integral multiples of a 0.5 kb unit (4). The possibility that the A+T-rich region of D. melanogaster mtDNA comprises tandemly repeated sequences has been studied by examining the products of denaturation and renaturation of EcoRI and AluI restriction fragments which contain the entire A+T-rich region (16). The predominant molecular form observed in each case was a perfect duplex the size of the original restriction fragment. Molecular forms expected if the A+T-rich region contains repeated sequences were not observed.

The relationship between the A+T-rich regions of mtDNA molecules of different *Drosophila* species (*D. simulans*, *D. mauritiana*, *D. yakuba*, *D. takahashi*, and *D. virilis*) has been studied by examining heteroduplexes formed between restriction fragments of the mtDNA molecules of the different species (16). While extensive or complete pairing of regions of molecules outside the A+T-rich regions was found in all species combinations examined, A+T-rich regions of the different species completely failed to pair in all but two combinations. In heteroduplexes between *D. melanogaster* and *D. simulans* mtDNAs, and between *D. melanogaster* and *D. mauritiana* mtDNA, up to approximately 40% of the A+T regions appeared duplexed.

At the present time little is known about the actual sequence structure of the A+T-rich regions of *Drosophila* mtDNA molecules. Available data (5-7, 14-16) concerning *D. melanogaster* mtDNA indicate that if guanine and cytosine occur in the A+T-rich region they account for no more than 5% of this DNA. It also appears that there is little or no base bias between the complementary strands of this A+T-rich region (15). A region with similar properties to the A+T-rich region has not been found in the mtDNAs of other metazoans, although regions rich in adenine and thymine have been demonstrated and well studied in yeast mtDNA (10).

The function of the A+T-rich regions of *Drosophila* mtDNAs is unknown. Any function which may be predicted for the A+T-rich regions must take into account the apparently extensive divergences of sequences which have occurred in these regions relative to other regions of the mtDNA molecule, as well as the differences in size of the A+T-rich regions in different *Drosophila* mtDNAs.

ACKNOWLEDGMENTS

We are grateful to Lawrence M. Okun for discussions and for constructive criticism of the text.

REFERENCES

1. Altman, P. L., and Katz, D. D. (1977) ed. Biological Handbooks I. Cell Biology. Fed. Am. Soc. Expl. Biol. Bethesda Md. pp. 217.
2. Wolstenholme, D. R., and Dawid, I. B. (1968). J. Cell Biol. 39, 222.
3. Jones, S. S., and Wolstenholme, D. R., manuscript in preparation.

4. Fauron, C. M.-R., and Wolstenholme, D. R. (1976). Proc. Natl. Acad. Sci. USA. 73, 3623.
5. Polan, M. L., Friedman, S., Gall, J. G., and Gehring, W. (1973). J. Cell Biol. 56, 580.
6. Bultman, H., and Laird, C. D. (1973). Biochem. Biophys. Acta 299, 196.
7. Peacock, W. J., Brutlag, D., Goldring, E., Appels, R., Hinton, C., and Lindsley, D. C. (1974). Cold Spring Harbor Symp. Quant. Biol. 38, 405.
8. Inman, R. B. (1966). J. Mol. Biol. 18, 464.
9. Klukas, C. K., and Dawid, I. B. (1976). Cell 9, 615.
10. Christiansen, C., Christiansen, G., and Bak, A. L. (1974). J. Mol. Biol. 84, 65.
11. Beck, E., Sommer, R., Auerswald, E. A., Kurz, Ch., Zink, B., Osterburg, G., Schaller, H., Sugimoto, K., Sugisaki, H., Okamoto, T., and Takanami, M., (1978). Nucleic Acids Res. 5, 4495.
12. Wolstenholme, D. R., and Fauron, C. M.-R. (1976). J. Cell Biol. 71, 434.
13. Goddard, J. M., and Wolstenholme, D. R. (1978). Proc. Natl. Acad. Sci. USA. 75, 3886.
14. Goldring, E. S., and Peacock, W. J. (1977). J. Cell Biol. 73, 279.
15. Fauron, C. M.-R., and Wolstenholme, D. R. (1977). J. Cell Biol. 75, 311a.
16. Fauron, C. M.-R., and Wolstenholme, D. R., Manuscript in preparation.
17. Robberson, D. L., Kasamatsu, H., and Vinograd, J. (1972). Proc. Natl. Acad. Sci. USA. 69, 737.
18. Wolstenholme, D. R., Koike, K., and Cochram-Fouts, P. (1973). J. Cell Biol. 56, 230.
19. Wolstenholme, D. R., Koike, K., and Cochran-Fouts, P. (1974). Cold Spring Harbor Symp. Quant. Biol 38, 267.
20. Kasamatsu, H., Robberson, D. L., and Vinograd, J. (1971). Proc. Natl. Acad. Sci. USA. 68, 2252.
21. Robberson, D. L., and Clayton, D. A. (1972). Proc. Natl. Acad. Sci. USA. 69, 3810.
22. Kasamatsu, H., and Vinograd, J. (1973). Nature (London) New Biology 241, 103.
23. Brown, W. M., and Vinograd, J. (1974). Proc. Natl. Acad. Sci. USA. 71, 4447.
24. Robberson, D. L., Clayton, D.A., and Morrow, J. F. (1974). Proc. Natl. Acad. Sci USA. 71, 4447.
25. Buzzo, K., Fouts, D. L., and Wolstenholme, D. R. (1978) Proc. Natl. Acad. Sci. USA. 75, 909.
26. Zakian, V. A., (1970). J. Mol. Biol. 108, 305.
27. Matsumoto, L., Kasamatsu, H., Piko, L., and Vinograd, J. (1974). J. Cell Biol. 63, 146.

28. Berk, A. J., and Clayton, D. A. (1974). J. Mol. Biol. 86, 801.
29. Goddard, J. M., and Wolstenholme, D. R., Manuscript in preparation.
30. Rubenstein, J. L. R., Brutlag, D., and Clayton, D. A. (1977). Cell 12, 471.
31. Lee, C. S., Davis, R., and Davidson, N. (1970). J. Mol. Biol. 48, 1.
32. Battey,J., and Clayton, D. A. (1978). Cell 14, 143.
33. Kroon, A. M., Pepe, A.,Bukker, H., Holtrop, M.,Bollen, J. E., van Bruggen, E. F. J., Cantatore, P., Terpstra, P., and Saccone, C. (1977). Biochem. Biophys. Acta 478, 128.
34. Ojala, D., and Attardi, G. (1978). J. Mol. Biol. 122, 301.
35. Ramirez, J. L., and Dawid, I. B. (1978). J. Mol. Biol. 119, 133.

TRANSCRIPTION PATTERN OF DROSOPHILA MELANOGASTER MITOCHONDRIAL DNA[1]

Jim Battey, John L.R. Rubenstein and David A. Clayton

Department of Pathology
Stanford University School of Medicine
Stanford, California 94305

ABSTRACT The expression of Drosophila melanogaster mitochondrial DNA has been examined by mapping and sizing a set of mitochondrial transcription products present in both tissue culture and larval salivary gland cells. We have utilized five recombinant plasmids containing cDNA complementary to polyadenylated Drosophila salivary gland mitochondrial transcripts as hybridization probes to identify and localize regions of the mitochondrial DNA genome which are transcribed. The sizes of the mitochondrial transcripts generating the cDNA-containing plasmids were determined. The overall pattern of mitochondrial transcription is not remarkably different during the first 21 hours of embryogenesis.

INTRODUCTION

The mitochondrial DNA (mtDNA) genome of Drosophila melanogaster has been structurally characterized (1-5) and many of its properties are similar to those of other animal mtDNAs. One of the more remarkable distinctions is the fact that ∿ 25% of the closed circular mtDNA sequence exists as a contiguous region of predominantly AT base pairs (bp) and different species of this genus exhibit a range of mtDNA molecular weights that can be rationalized on the basis of variability in the size of the AT-rich region (6-8). The functional significance of this variability is not known, but it

[1]This investigation was supported by grant no. NP-9E from the American Cancer Society and grant no. CA-12312-08 from the National Cancer Institute. J.B. and J.L.R.R. are Medical Scientist Training Program Fellows of the National Institute of General Medical Sciences (GM-07365-03) and D.A.C. is a Faculty Research Awardee of the American Cancer Society (FRA-136).

ISBN 0-12-198780-9

is notable that this portion of the genome is at or near the origin of mtDNA replication (9) and the map position of the two most abundant transcripts, the rRNA genes (3). These two mitochondrial rRNAs were positioned relative to HaeIII and HindIII restriction endonuclease cleavage sites and map within 160 bases of each other adjacent to the AT-rich region (3). Spradling et al. (10) have isolated 12 poly A^+ and one poly A^- RNA species from Drosophila mitochondria. The poly A^- RNA was identified as the small rRNA and is thus further distinguishable from the larger, poly A^+ rRNA. The map positions of these putative transcripts have recently been determined (11).

We have previously examined the conformational properties of Drosophila mtDNA during different stages of embryogenesis (5) and have recently determined the basic transcription map of mouse L-cell mtDNA (12). In order to identify transcriptionally active regions of the Drosophila mtDNA genome, we isolated and characterized five recombinant DNA clones, each containing a cDNA copy of a discrete poly A^+ transcript complementary to mtDNA. We have determined the map positions of the cDNA clones as well as the sizes of the transcripts which generated each of the individual cDNA clones. The basic pattern of Drosophila mtDNA transcription appears unchanged during embryogenesis.

MATERIALS AND METHODS

Mapping of HpaI and PstI Restriction Endonuclease Cleavage Sites. 1 μg of Drosophila mtDNA isolated from eggs was cleaved with combinations of the enzymes HaeIII (Biolabs), HindIII (Biolabs), PstI (Bethesda Research Labs), or HpaI (Biolabs). All digestions were performed in 6 mM Tris (pH 7.5), 6 mM NaCl, 6 mM $MgCl_2$, 100 μg/ml gelatin at 37° C. Electrophoretic analyses utilized 1% agarose gels in 89 mM Tris (pH 8.5), 89 mM sodium borate, 1 mM EDTA at 1 V/cm. Gels were stained with 0.5 μg/ml ethidium bromide (EB) in distilled water, illuminated with UV light, and the resultant fluorescent DNA bands were photographed for analysis.

Preparation and Cloning of Mitochondrial cDNA Plasmids. Preparation and cloning of cDNA segments reverse transcribed from total poly A^+ *Drosophila melanogaster* mRNA was similar to the method used to clone rabbit globin mRNA (13). cDNA clones complementary to *Drosophila* mtDNA were selected by colony hybridization (14) using ^{32}P-labeled, nick translated *Drosophila* mt-DNA (15) as a hybridization probe.

Isolation of Mitochondrial cDNA Plasmids from E. coli. Closed circular recombinant plasmid DNA was isolated from *E. coli* as described by Battey and Clayton (12). Host cells were grown and plasmid DNA was isolated under P2-EK1 conditions.

Mapping of Mitochondrial cDNA Clones to Drosophila mtDNA Restriction Fragments. 10 μg of *Drosophila* mtDNA was cleaved with both *Hind*III and *Hae*III restriction endonucleases producing seven restriction fragments (3) and two prominent partial digestion products. The cleaved mtDNA was electrophoresed on a native agarose slab gel in 89 mM Tris (pH 8.5), 89 mM sodium borate, 1 mM EDTA at 1 V/cm. The gel was composed of two concentrations of agarose, to allow resolution of all seven restriction fragments. The 7 cm proximal to the origin was 0.7% agarose, and the remaining 7 cm distal from the origin was 2% agarose. The gel was stained with 1 μg/ml EB and the relative mobility of the fragments was assessed from photographs of the illuminated gels.

The slab gel was transferred to nitrocellulose (16) and five lanes of the nitrocellulose blot were used for subsequent hybridization. The five nitrocellulose strips were pretreated with solution A (4 x SSC, 100 mM Tris (pH 8.0), and a mixture of 0.02% bovine serum albumin, 0.02% polyvinylpyrollidine, 0.02% Ficoll (17)) at 65°C for 4 h prior to hybridization.

The five DNA probes for filter hybridization were labeled by nick-translation of 1 μg of each of the five mitochondrial cDNA clones with [α-^{32}P] GTP (15) to a specific activity of approximately

10^6 cpm/μg. The five hybridizations on nitrocellulose filters were performed in solution A, 50 μg/ml denatured salmon sperm DNA carrier, and ∿10^6 cpm denatured, nick-translated mitochondrial cDNA plasmids. The hybridization reaction proceeded for 40 h at 65°C. After hybridization, the strips were washed successively for 1 h at 65°C in [1] solution A, [2] 100 mM KH_2PO_4 (pH 7.0), 2 x SSC, and [3] 2 x SSC. The washed nitrocellulose strips were autoradiographed using Kodak RP-X-Omat film and, in some cases, a Dupont Cronex Lightning-Plus intensifying screen.

Sizing of Cloned mtDNA Transcripts.

(A) Preparation of Drosophila nucleic acid. A 500 ml spinner culture of KC cells (18) was grown to mid-logarithmic density (5-8 x 10^6 cells/ml). The cells were pelleted, resuspended in 2 ml 100 mM NaCl, 10 mM Tris (pH 8.0), 0.1% Sarkosyl, 1 mM EDTA, and immediately extracted with 5 ml phenol-chloroform (50:50 v/v, pH 8). The aqueous phase was then extracted with chloroform alone. The phenol-chloroform and chloroform extractions were repeated twice, and two volumes of ethanol at -20°C was added to the aqueous phase to precipitate nucleic acid. The resulting Drosophila nucleic acid was used in the preparation of mitochondrial RNA-DNA hybrids.

(B) Preparation of RNA-DNA hybrids. 10 μg of mtDNA was cleaved with HpaI restriction endonuclease, which makes a single cleavage in the circular genome. The restricted mtDNA was hybridized to 15 mg bulk Drosophila nucleic acid in a 500 μl reaction mixture containing 70% recrystallized formamide, 50 mM PIPES (pH 7.0), 0.3 M NaCl and 2 mM EDTA for 5 h at 41.5°C in a Haake water bath. The calculated T_m (19) for the most GC-rich regions of double-stranded Drosophila mtDNA in this reaction mixture is 36.5°C. The reaction temperature is therefore at least 5°C above the reannealing temperature of mtDNA, allowing only the formation of RNA-DNA heteroduplexes, and excluding the reannealing of Drosophila mtDNA (20). The 500 μl reaction mixture was diluted into 5 ml of S_1 nuclease reaction buffer (190 mM NaCl, 30 mM sodium acetate (pH 4.5), 3 mM $ZnCl_2$) and digested with 500 units *Aspergillus oryzae* S_1 (Sigma) for 30 min at 37°C. All single-stranded mtDNA sequences not protected by complementary RNA are digested to oligonucleotides, leaving intact (as RNA-DNA hybrids) DNA regions the same size as the protecting complementary transcripts in the *Drosophila* nucleic acid population. 50 μg yeast sRNA was added, and nucleic acid was ethanol precipitated and resuspended in 200 μl 30 mM NaOH, 2 mM EDTA. The S_1-resistant single-stranded mtDNA regions were resolved on five slots (33 μl/slot) of a 2% alkaline agarose slab gel (21) after electrophoresis at 1 V/cm overnight. Restriction digests producing fragments of known molecular weights were used as size standards to allow a calibration of migration distance vs. single-stranded DNA size on the alkaline agarose gels. The denaturing alkaline gel was neutralized with two 30-min immersions in 3 M NaCl, 500 mM Tris (pH 7.2). DNAs from the five lanes were transferred to nitrocellulose paper and the resultant nitrocellulose strips were pretreated for 4 h at 42°C in solution A and 100 μg/ml denatured salmon sperm DNA. Hybridization probes for each of the nitrocellulose strips were labeled by nick-translation of 1 μg of each of the mitochondrial cDNA recombinant plasmids to a specific activity of approximately 10^7 cpm/ g. Hybridizations to nitrocellulose strips were performed in solution A plus 1 mg/ml denatured salmon sperm DNA and 10^7 cpm nick-translated mitochondrial cDNA plasmid. Hybridization proceeded for 40 h at 42°C. After hybridization, the nitrocellulose strips were washed as described previously, and autoradiographed.

Detection of mtRNA transcripts found in developing Drosophila embryos.

(A) *Isolation of Drosophila nucleic acid from embryos.* *Drosophila* eggs were collected over a 2 h window bracketing the specified time points. Eggs were dechorioniated (5), weighed and split into six 500 mg cultures and incubated in yeast and cornstarch vials for the developmental periods indicated. Nucleic acid was isolated by resuspending the eggs in 1 ml 200 mM NaCl. Eggs were immediately dounced in 5 ml of phenol:chloroform (50:50, pH 7.5) and 1 ml Sarkosyl lysis buffer (1.0% Sarkosyl, 50 mM Tris (pH 7.5), 2 mM EDTA). The lysate was extracted with chloroform alone. The aqueous phase was removed and extracted with phenol:-chloroform, and then chloroform alone. The aqueous phase was made 0.3 M in sodium acetate, and ethanol precipitated. The precipitate was rinsed with cold ethanol to remove residual phenol, and the pellet was dessicated. The dry pellet was resuspended in 0.5 ml 10 mM Tris (pH 7.5), 1 mM EDTA, and stored at -20°C.

(B) Single-nick translation of *Drosophila* mtDNA. 2 μg of *Drosophila* mtDNA was singly-nicked with DNase I and EB (22). The mtDNA was phenol extracted and ethanol precipitated. The singly-nicked mtDNA circles were nick-translated with (α-^{32}P)dGTP and DNA polymerase I. The specific activity of the DNA was ∿ 2.5×10^6 cpm/ g mtDNA.

(C) Hybridization reaction. Four percent of the *Drosophila* nucleic acid isolated from each time point was hybridized to 300 ng (8×10^5 cpm) of singly nick-translated mtDNA in 80% formamide, 0.3 M NaCl, 60 mM PIPES (pH 7.0), and 1.5 mM EDTA. The reaction mix was heated to 60°C (T_m + 25 for *Drosophila* mtDNA reassociation) for 3 min to denature the mtDNA. The mixture was then incubated at 41.5°C (T_m + 5 for mtDNA reassociation) allowing the formation of mitochondrial RNA-DNA heteroduplexes and preventing the reassociation of mtDNA-mtDNA duplex (20). The heteroduplexed material was then treated with S_1 nuclease, which cleaves the single-stranded DNA not protected by RNA-DNA duplex formation into oligonucleotides. This digestion leaves intact a ^{32}P-labeled single-stranded mtDNA replica complementary to homologous mitochondrial transcripts in the *Drosophila* nucleic acid. The heteroduplexed material was ethanol precipitated and resuspended in 50 μl 30 mM NaOH, 2 mM EDTA. The alkaline solution

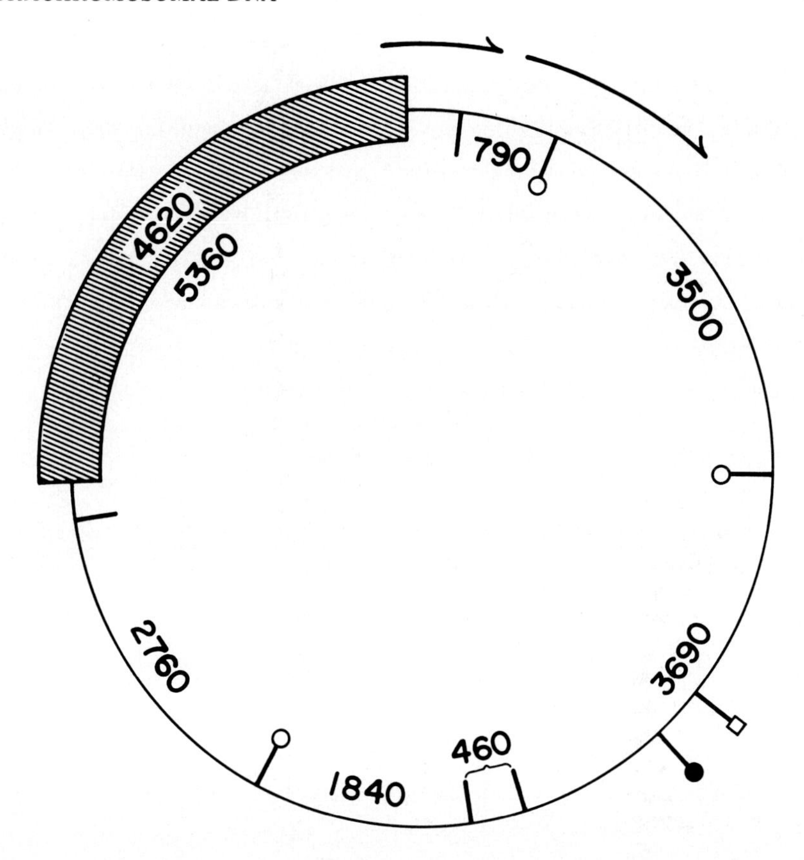

FIGURE 1. Circular map of Drosophila mtDNA. The previously determined map locations of the 4620 bp AT-rich region and the small and large rRNA cistrons with their 5'→3' orientation are indicated (the half-arrow denotes the 3'-end). The location of HaeIII, HindIII, HpaI and PstI restriction sites, and the sizes of restriction fragments produced by combined HaeIII-HindIII digestion are also indicated. Mapping of the single HpaI and PstI sites was by restriction fragment size comparison to known restriction enzyme sites (data not shown). Mapping of HaeIII sites, HindIII sites, rRNAs and their polarity, and the AT-rich region are from Klukas and Dawid (3). | , HaeIII; HindIII; HpaI; PstI; AT-rich region; ⟶, rRNAs.

denatures the heteroduplex and hydrolyzes the RNA, leaving a single-stranded, ^{32}P-labeled mtDNA replica of all homologous transcripts in the Drosophila nucleic acid being analyzed. The single-stranded DNA replicas were resolved and sized relative to restriction fragment markers by electrophoresis on alkaline agarose gels. Electrophoresis was overnight at 1 V/cm in 30 mM NaOH, 2 mM EDTA running solution. The alkaline gel was neutralized with two 30-min immersions in 0.5 M Tris (pH 7.0), and dried. The dried gel was autoradiographed using Kodak RP-X-Omat film.

RESULTS

Mapping of cDNA Clones to the HaeIII and HindIII Restriction Fragments of Drosophila mtDNA. MtDNA-specific plasmid DNAs were isolated from the total population of plasmids containing cloned cDNA copies of polyadenylated Drosophila salivary gland transcripts. These plasmids were constructed by "A:T tailing" and insertion into pSC105 as described in MATERIALS AND METHODS. Five colonies each specific for an actively transcribed region of mtDNA were identified by hybridization to mtDNA sequence using the gel transfer technique of Southern (16).

Closed circular Drosophila mtDNA was digested with HaeIII and HindIII restriction endonucleases and the resultant fragments were electrophoresed on a native agarose slab gel. The slab gel was blotted to nitrocellulose and five strips of this nitrocellulose blot, containing all seven HaeIII and HindIII restriction fragments (Figure 1), were used as filters for hybridization. Approximately 1 μg of each of the five mitochondrial cDNA clones was nick translated with DNase I and DNA polymerase I in the presence of [α-^{32}P] GTP (15). 10^6 cpm of each of the five ^{32}P-labeled, nick-translated cDNA clones were hybridized to a strip of the nitrocellulose blot containing the seven Drosophila mtDNA HaeIII-HindIII restriction fragments. An autoradiogram of the five strips is shown in Figure 2, along with the positions on the strips of five restriction fragments and

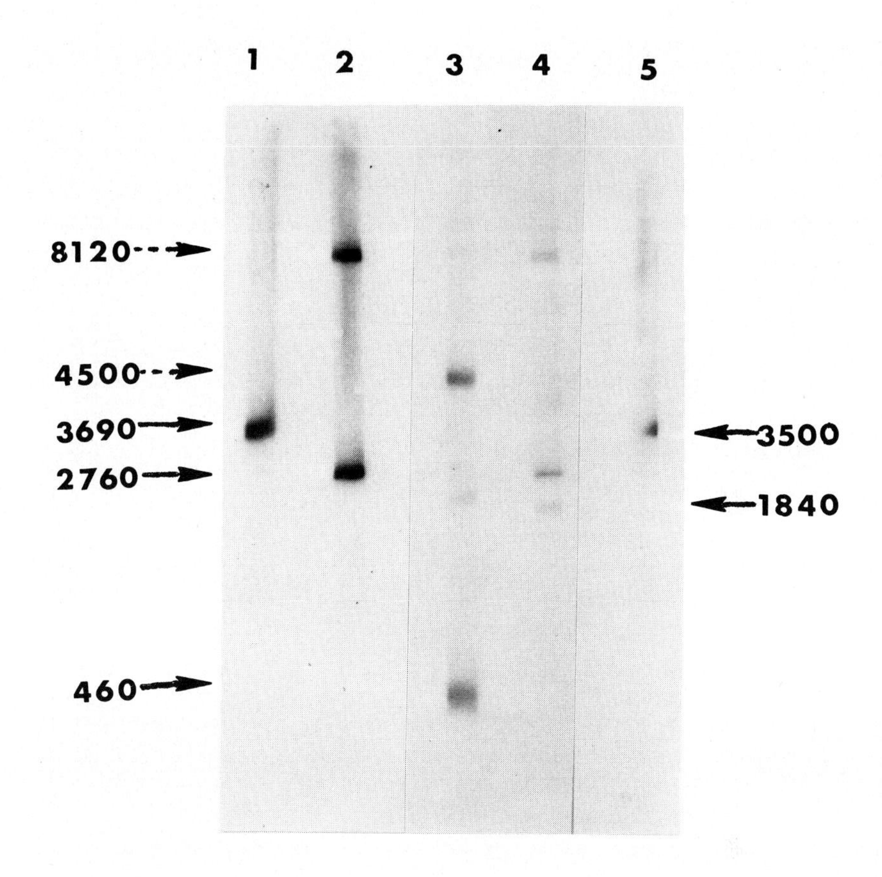

FIGURE 2. Autoradiogram showing homology between specific HaeIII-HindIII restriction fragments of Drosophila mtDNA and the five ^{32}P-labeled mitochondrial cDNA plasmids. The five ^{32}P-labeled cDNA plasmids were each hybridized to a nitrocellulose blot of an agarose gel resolving five Drosophila mtDNA HaeIII-HindIII restriction fragments and two prominent partials. The positions and sizes of the five HaeIII-HindIII restriction fragments and partials are indicated by arrows. ——, fragment; — — —, partial.

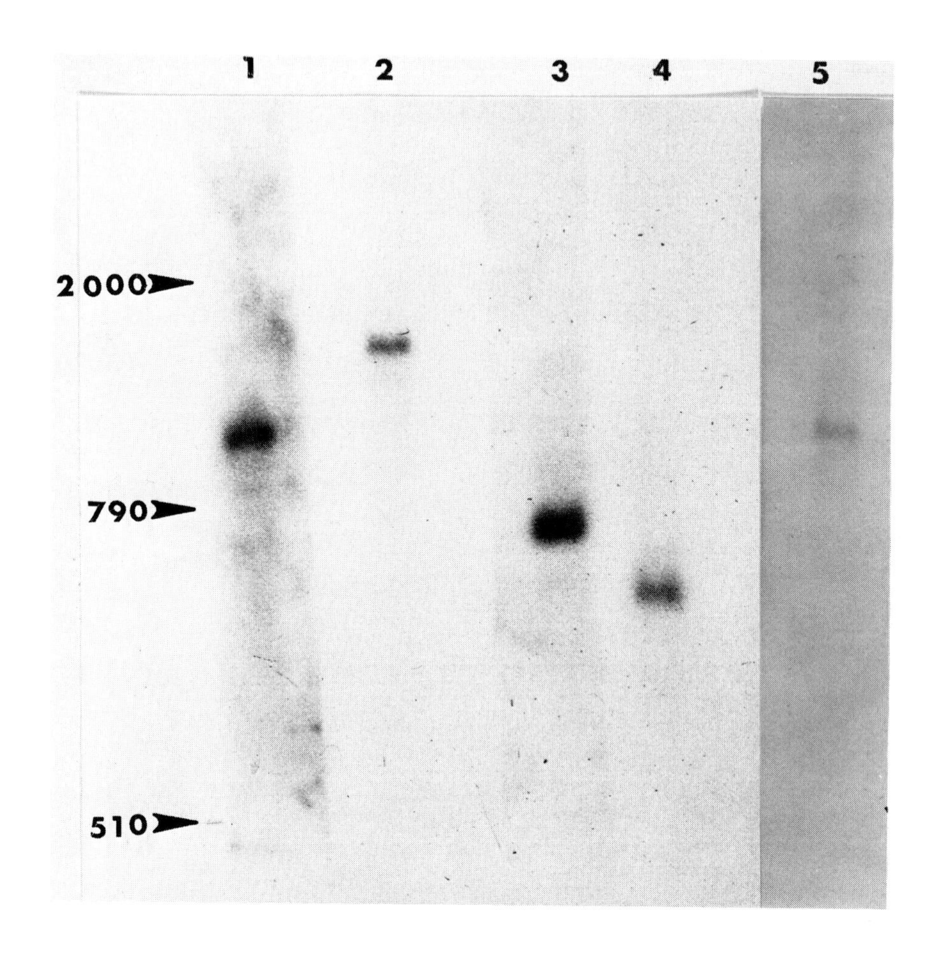

FIGURE 3. Autoradiogram showing homology between S_1-protected DNA copies of mitochondrial transcripts and the five ^{32}P-labeled mitochondrial cDNA plasmids. Unlabeled DNA regions homologous to mitochondrial transcripts from Drosophila tissue culture cells are generated by hybridization and S_1 digestion, subsequently resolved and sized on an alkaline agarose gel, and then blotted onto nitrocellulose. The five ^{32}P-labeled cDNA plasmids were each hybridized to a strip of the nitrocellulose blot. Positions of DNA restriction fragments serving as molecular weight size standards are indicated by arrows.

two prominent partials as determined by EB staining of the slab gel before blotting. The map positions are based on determining which of the mtDNA fragments hybridize to each of the five labeled cDNA clones. The cDNA clones are numbered 1-5 and are displayed in the composite Figure 4 of the next section.

Determination of the molecular weights of the cloned m-RNAs. We determined the molecular weights of the transcripts which were the templates for our cDNA clones by utilizing a modification of the S_1 nuclease protection technique developed by Berk and Sharp (23) to map and size viral DNA transcripts. Crude nucleic acid from Drosophila tissue culture cells was hybridized to full-length mtDNA cleaved by HpaI restriction endonuclease. Hybridization reactions were in high formamide, 5° C above the melting temperature for a DNA-DNA duplex, allowing only DNA-RNA heteroduplex formation (20). Under these conditions, all mitochondrial transcripts present in the nucleic acid population hybridize to their complementary mtDNA gene, forming an RNA-DNA heteroduplex. The resultant hybrid molecules were then treated with S_1 nuclease, digesting all single-stranded mtDNA not protected by mitochondrial transcripts. The remaining S_1-resistant DNA copies of all mitochondrial transcripts were sized and resolved by electrophoresis in alkaline agarose gels, and blotted to nitrocellulose filters. The resulting blot contains nonradioactive DNA copies of all mitochondrial transcripts resolved and sized by their electrophoretic mobilities in the alkaline gel. To identify and size the transcripts which have sequence homology to a particular cDNA clone, we hybridized nick-translated ^{32}P-cDNA from each particular clone to a nitrocellulose strip containing all of the single-stranded DNA sequences complementary to mitochondrial RNA found in tissue culture cells (Figure 3). The autoradiograph reveals one sequence with homology to each individual cloned DNA. With the use of restriction fragment size standards co-electrophoresed on the alkaline gel, we determined the molecular weights of the primary transcripts showing homology

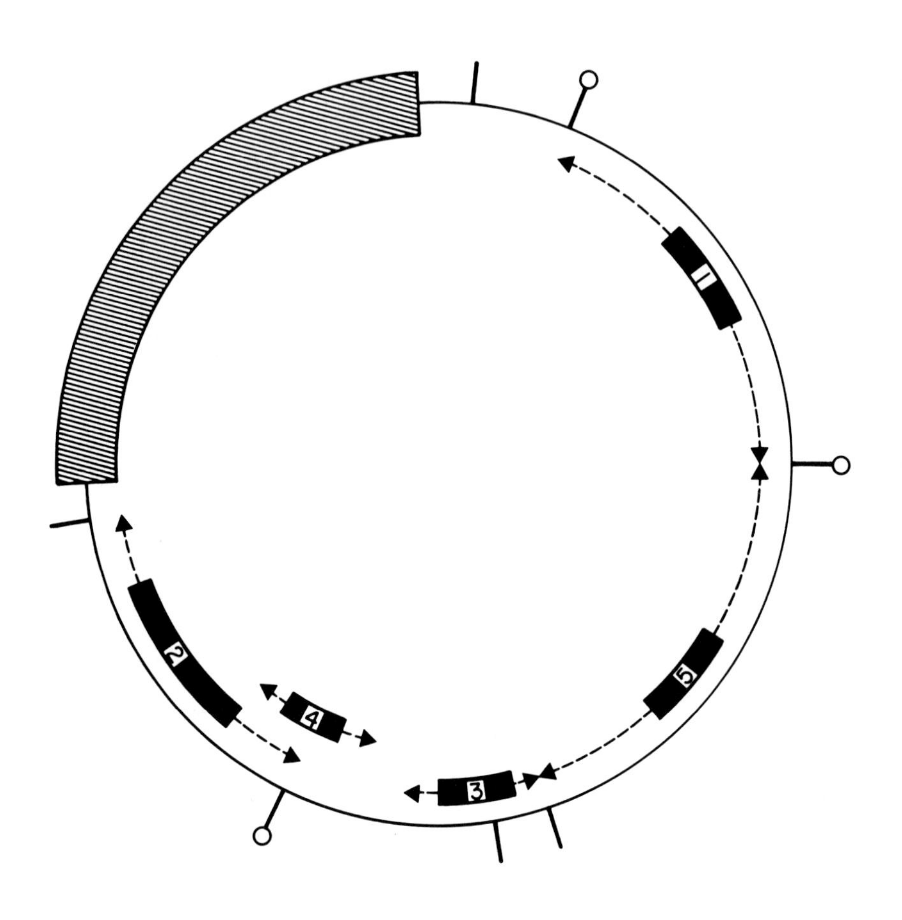

FIGURE 4. Circular map of Drosophila mtDNA. The sizes of transcripts generating cDNA plasmids 1-5 are drawn to scale on the 18.4 kb circular map and are indicated by heavy bars. The range of map positions for each transcripts consistent with the data is shown by dashed arrows. The position of the AT-rich region as well as HaeIII and HindIII restriction sites are shown. [hatched bar], AT-rich region; [black bar], transcript; |, HaeIII; [lollipop], HindIII.

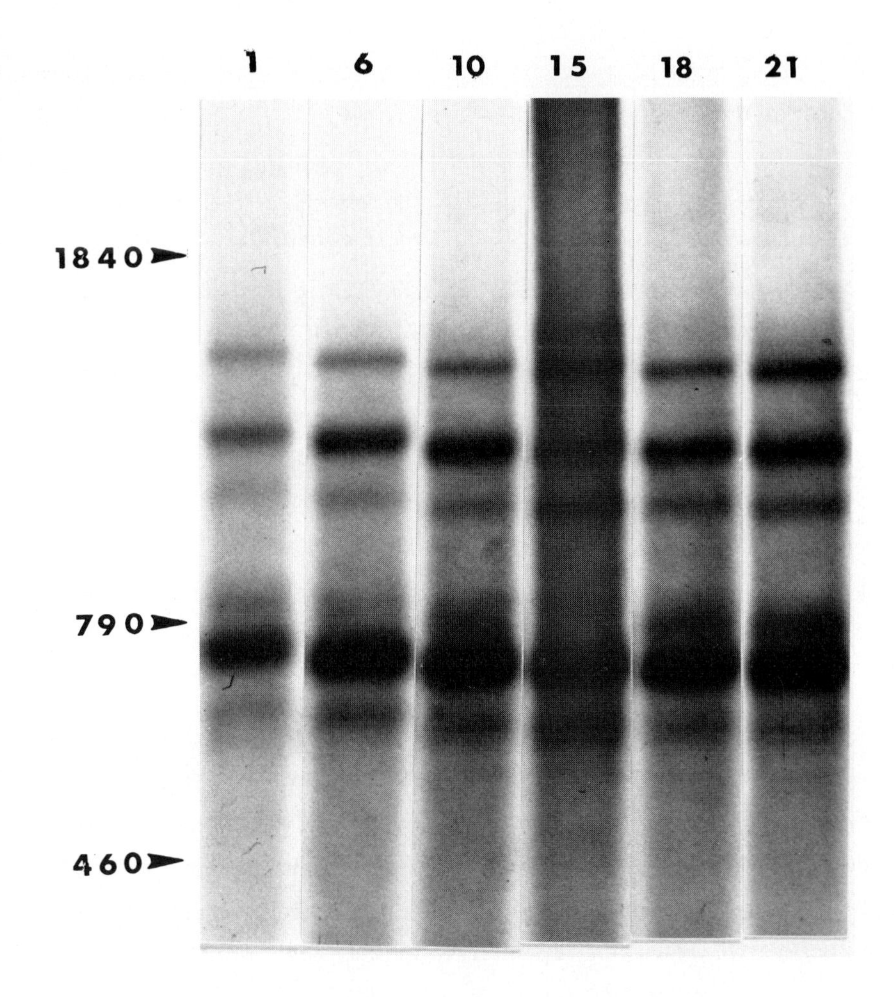

FIGURE 5. Autoradiogram of an agarose gel resolving ^{32}P-labeled mtDNA regions protected from S_1-digestion by transcripts present at different times during Drosophila embryogenesis. The positions and sizes of HaeIII-HindIII size-standard restriction fragments of Drosophila mtDNA co-electrophoresed on the slab gel are indicated by arrows. The post-fertilization time (in hours) of mitochondrial RNA isolation is noted at the top of each lane.

with each clone, as depicted in Figure 4.

Transcription pattern during embryogenesis. Similar techniques to those used to resolve regions of mtDNA protected from S_1 digestion by mitochondrial transcripts found in tissue culture can be used to investigate changes in the mitochondrial transcription pattern during embryogenesis. We therefore isolated nucleic acid from developing embryos at 1, 6, 10, 15, 18 and 21 hours after fertilization (see MATERIALS AND METHODS). These nucleic acid populations were hybridized to full-length, single-nick translated ^{32}P-labeled mtDNA in ^{32}P-mtDNA excess as detailed in MATERIALS AND METHODS. The resultant hybrids were digested with S_1 nuclease and analyzed on an alkaline agarose gel (Figure 5), thus comparing DNA regions protected from S_1 digestion by transcripts present at the different times in embryogenesis. There is a clear correspondence of major transcripts present at all time points.

DISCUSSION

The expression of Drosophila mitochondrial genes has been investigated by mapping and sizing a set of mitochondrial transcription products present in tissue culture cells and larval salivary gland cells. Five cloned DNAs were utilized for mapping which, by definition, represent specific localized regions of the mtDNA genome transcribed in larval salivary glands. From an analysis of the size of the RNAs from which the clones were constructed, we determined that transcription occurs over at least 35% of the 18.4 kb mitochondrial genome and none of the clones examined mapped in the 4.6 kb AT-rich region. The general distribution of transcriptionally active regions is similar to that reported for mouse L-cell mtDNA (12) and represents a subset of the recently determined transcripts of Drosophila mtDNA (11).

We have measured steady-state levels of mitochondrial RNA transcripts present at different times in development. We found that all of the transcripts present at one hour after fertilization were also present at 6, 10, 15, 18 and 21 hours in the same relative amounts. In general, the steady-state population of transcripts does not change significantly during the first 21 hours of development. In contrast to the constitutive expression of these mtDNA transcripts, mtDNA replication is maximal between 10 and 18 hours after fertilization (5).

ACKNOWLEDGEMENTS

We thank Douglas Brutlag for use of the Drosophila growing facility and David Kemp and David S. Hogness for gifts of cloned cDNAs.

REFERENCES

1. Bultmann, H., and Laird, C.D. (1973). Biochem. Biophys. Acta 299, 196.
2. Polan, M.L., Friedman, S., Gall, J.G., and Gehring, W. (1973). J. Cell. Biol. 56, 580.
3. Klukas, C.K., and Dawid, I.B. (1976). Cell 9, 615.
4. Wolstenholme, D.R., and Fauron, C.M.R. (1976). J. Cell. Biol. 71, 434.
5. Rubenstein, J.L.R., Brutlag, D., and Clayton, D.A. (1977). Cell 12, 471.
6. Bultmann, H., Zakour, R.A., and Sosland, M.A. (1976). Biochim. Biophys. Acta 454, 21.
7. Fauron, C.M.R., and Wolstenholme, D.R. (1976). Proc. Nat. Acad. Sci., U.S.A. 75, 3623.
8. Shah, D.M. and Langley, C.H. (1979). Plasmid 2, 69.
9. Goddard, J.M., and Wolstenholme, D.R. (1978). Proc. Nat. Acad. Sci., U.S.A. 75, 3886.
10. Spradling, A., Pardue, M.L., and Penman, S. (1977). J. Mol. Biol. 109, 559.
11. Bonner, J.J., Berninger, M., and Pardue, M.L. (1978). Cold Spring Harbor Symp. on Quant. Biol. 42, 803.
12. Battey, J., and Clayton, D.A. (1978). Cell 14, 143.
13. Maniatis, T., Kee, S.G., Efstratiadis, A., and Kafatos, F.C. (1976). Cell 8, 163.
14. Grunstein, M., and Hogness, D.S. (1975). Proc. Nat. Acad. Sci., U.S.A. 72, 3961.
15. Rigby, P.W.J., Dieckmann, M., Rhodes, C., and Berg, P. (1977). J. Mol. Biol. 113, 237.
16. Southern, E. (1975). J. Mol. Biol. 98, 503.

17. Denhardt, D.T. (1963). Biochem. Biophys. Res. Comm. 23, 641.
18. Echalier, G. (1971). Curr. Topics Microbiol. Immunol. 55, 220.
19. Schildkraut, C., and Lifson, S. (1965). Biopolymers 3, 195.
20. Casey, J., and Davidson, N. (1977). Nucl. Acids Res. 4, 1539.
21. McDonell, M.W., Simon, M.D., and Studier, R.W. (1977). J. Mol. Biol. 110, 119.
22. Greenfield, L., Simpson, L., and Kaplan, D. (1975). Biochem. Biophys. Acta 407, 365.
23. Berk, A.J., and Sharp, P.A. (1977). Cell 12, 721.

THE ORGANIZATION OF THE GENES IN THE HUMAN MITOCHONDRIAL GENOME AND THEIR MODE OF TRANSCRIPTION[1]

Giuseppe Attardi, Palmiro Cantatore, Edwin Ching, Stephen Crews, Robert Gelfand, Christian Merkel and Deanna Ojala

Division of Biology, California Institute of Technology, Pasadena, California 91125

ABSTRACT The organization of the genes in the human mitochondrial genome and their mode of transcription are being investigated by a detailed mapping, structural and metabolic analysis of the various discrete poly(A)-containing and non-poly(A)-containing RNA species coded for by mitochondrial DNA (mit-DNA). Several approaches, involving different RNA-DNA hybridization techniques, RNA and DNA sequencing methods, and analysis of kinetic behavior, are being used in this work. A fine structural study of the ribosomal RNA region of mit-DNA has failed to reveal the presence of inserts in the main body of the 12S and 16S rRNA cistrons. The sequence of 71 nucleotides at the 5'-end of 12S rRNA and that of 32 nucleotides at the 5'-end of 16S rRNA have been determined. Experiments of hybridization of nascent RNA chains with different restriction fragments of mitochondrial DNA have led to the identification in this DNA of a region containing a point of initiation of L strand transcription. Several giant poly(A)-containing and non-poly(A)-containing L strand transcripts have been identified and characterized in their map location and metabolic behavior.

INTRODUCTION

In recent years, work in several laboratories has revealed that the mitochondrial genome of animal cells possesses unique features of gene organization (1,2), replication (3-6), and transcription (1). The replication of mit-DNA has been reviewed elsewhere in this Symposium. In this talk I would like to discuss the recent progress we have made in under-

[1]Supported by grants GM-11726 and T32 GM-07616-01 from the National Institutes of Health.

ISBN 0-12-198780-9

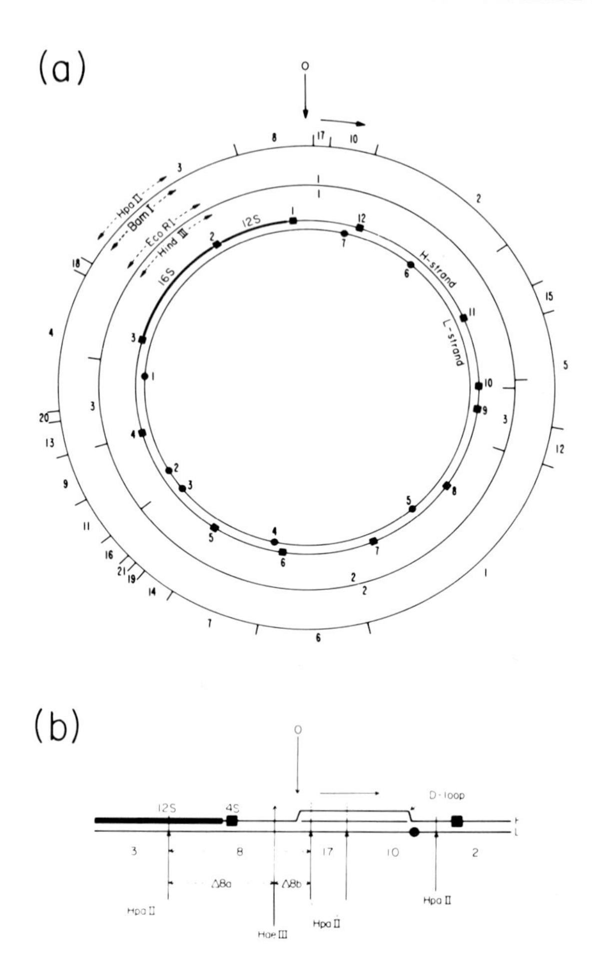

FIGURE 1. (a) Physical and genetic maps of HeLa cell mit-DNA. (b) Region of the physical map expanded to show the precise positions of the origin of replication (vertical arrow [marked O]) and of the D-loop. The rightward arrow indicates the direction of H strand synthesis (modified from [4]).

standing the organization of the genes in HeLa cell mit-DNA and their mode of expression.

Figure 1a shows the physical and genetic maps of HeLa cell mit-DNA, as have been constructed in the last few years by a combination of biochemical and electron microscopic approaches (7,8). The positions of the genes for the two rRNA species, 16S and 12S RNA, on the heavy (H) strand and those of the 4S RNA sites on the H and light (L) strands are shown. At least 17 tRNA species coded for by mit-DNA have been identified by RNA-DNA hybridization involving AA-tRNA complexes labeled in the amino acid moiety (9). It is likely that all, or almost all, of these 4S RNA sites are tRNA genes. It may be worthwhile to mention that, recently, the position in the map of some of these tRNA sites has been confirmed by DNA sequencing in F. Sanger's laboratory in Cambridge, and their amino acid specificity determined (F. Sanger, personal communication). In Figure 1b the region of the map surrounding the origin of replication has been enlarged to show its details. By a variety of *in vitro* DNA synthesis experiments utilizing the H strand initiation sequence (7S DNA) as a primer or template, it has been possible to localize the position of the 5'-end of 7S DNA at about 80 base pairs from the border between *Hpa* II fragments 3 and 8 (10). Furthermore, a subfragment of *Hpa* II-8, about 220 base pairs long and which contains the origin of H strand synthesis, has been generated by *Hae* III digestion ($\Delta 8b^{Hae}$). More recently, this fragment, as well as the 5'-proximal portion of 7S DNA have been sequenced and aligned with each other (11). The results have allowed us to localize the origin of 7S DNA synthesis precisely at 87 nucleotides from the border between *Hpa* II fragments 3 and 8.

GENERAL ORGANIZATION OF MITOCHONDRIAL DNA TRANSCRIPTS

Symmetrical Transcription of Mitochondrial DNA. One striking feature of the transcription process of HeLa cell mit-DNA, discovered in our laboratory about eight years ago (12), is its symmetry. Both strands are transcribed completely (13); furthermore, pulse-labeling experiments have indicated that the two strands are transcribed at an equal or almost equal rate. However, the transcripts of the H and L strands have a quite different half-life. In particular, the transcripts of the L strand have a very fast turnover and do not accumulate to any great extent (12). The symmetry of transcription of mit-DNA has to be contrasted with the great difference in informational content of the two strands. The H strand codes for the two rRNAs (14), for most of the tRNAs (7,9) and, as we will discuss below, for most of the poly(A)-containing RNAs (15). The L strand codes apparently for 7 tRNAs (7,9) and one small poly(A)-containing RNA (15). In the light of the low informational content of the L strand it is reasonable to ask whether the complete transcription of this strand is related exclusively to the expression of the few genes contained in this strand. It is clear that the occurrence of symmetrical transcription of mit-DNA has added one degree of complexity to the RNA metabolism in HeLa cell mitochondria. Further complexity to the picture has

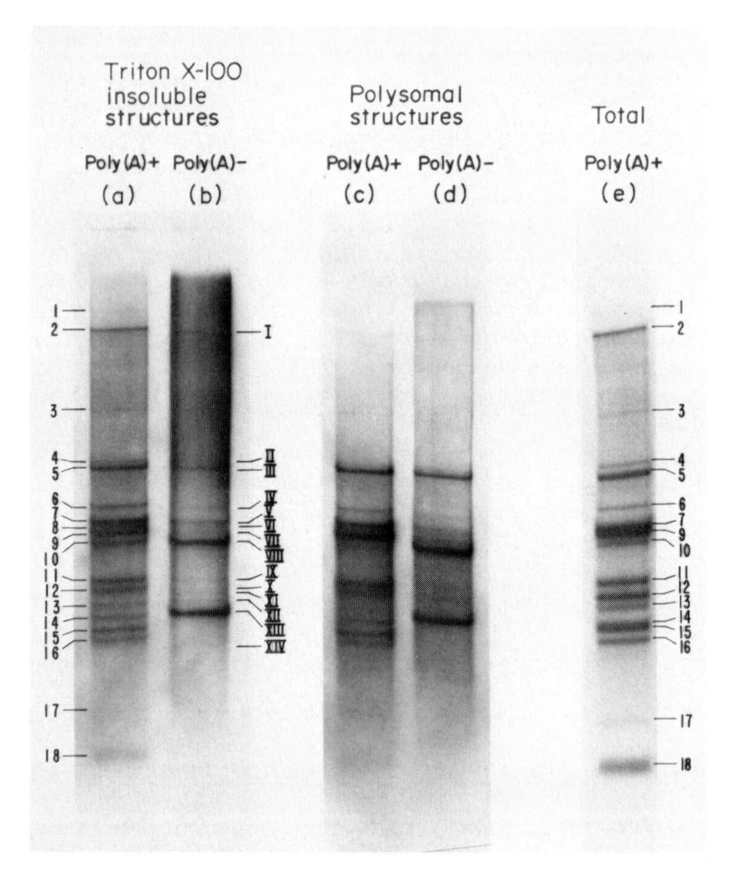

FIGURE 2. Autoradiograms, after electrophoresis through agarose-CH_3HgOH slab gels, of mitochondrial poly(A)-containing and non-poly(A)-containing RNAs. See text for details.

come from the discovery of the large number of poly(A)-containing RNAs and non-poly(A)-containing RNAs, distinct from rRNAs and tRNAs, which are coded for by HeLa mit-DNA (15).

Poly(A)-Containing and Non-Poly(A)-Containing RNAs. Figure 2 shows the fractionation by high resolution agarose-CH_3HgOH electrophoresis of mitochondrial RNA from HeLa cells labeled with ^{32}P-orthophosphate under different conditions. Lanes (a) and (b) show the oligo(dT)-bound and, respectively, unbound RNA fractions from the Triton X-100 insoluble mitochondrial structures (which contain most of the newly synthesized RNA species and a portion of mitochondrial polysomes) of cells labeled for 3 hr with ^{32}P-orthophosphate in the presence of 20 μg/ml camptothecin (to block all high mol. wt.

nuclear RNA synthesis) (16). At least 18 bands of varying intensity can be recognized in the autoradiogram of the oligo-(dT)-bound RNA fraction and 14 bands in that of the non-bound RNA fraction. Most prominent among the latter are the two rRNA species, 16S (VIII) and 12S RNA (XIII); the 4S RNA, under the conditions of electrophoresis used here, has run out of the gel. Many bands in this autoradiogram correspond roughly in position to bands in Figure 2a; in particular, bands II and IX correspond to components 4 to 11, and may be precursors of them. Direct analysis by pancreatic and T1 RNase digestion followed by polyacrylamide gel electrophoresis has shown that the oligo(dT)-bound material contains poly(A) stretches corresponding to about $4S_E$ (∽55 nucleotides). No such stretches were found in the unbound fraction. Heterogeneous material can be seen as a fairly uniform background throughout the major portion of the gel of the oligo(dT)-bound RNA fraction, and, in greater abundance, in the upper third of the gel of the non-bound RNA fraction.

Lanes (c) and (d) show the autoradiograms of the oligo-(dT)-bound and, respectively, unbound RNA fractions from the mitochondrial polysomal structures, separated from the mitochondrial Triton X-100 lysate by sucrose gradient centrifugation. One can observe in general the same components as in the RNA fractions from the Triton X-100 insoluble structures, though with some differences in the relative intensity of the bands. In particular, the high mol. wt. poly(A)-containing RNA components 1 to 4 appear to be present in much lower amounts in the polysomal structures, suggesting that they are not intrinsically associated with such structures, but represent, presumably, contaminants.

The components shown in Figure 2a-d are very reproducible, and are also observed if actinomycin D is used instead of camptothecin to block high mol. wt. nuclear RNA synthesis; more significantly, the same components are found if the labeling of the cells is carried out in the absence of drugs, using treatment of the mitochondrial fraction with RNase A or micrococcal nuclease to destroy the contaminating extra-mitochondrial RNA. Figure 2e shows the electrophoretic pattern obtained for the poly(A)-containing RNA fraction extracted from the micrococcal nuclease treated mitochondrial fraction of HeLa cells labeled for 2.5 hr with ^{32}P-orthophosphate in the absence of inhibitors. The identity in pattern to that shown in Figure 2a is striking. These observations clearly indicate that the RNA species detected here are products of the normal cell metabolism and do not result from drug-induced artifacts. The fact that they become labeled in the presence of inhibitors of nuclear RNA synthesis strongly points to their intramitochondrial origin. Such origin has been more directly confirmed by experiments which have

TABLE 1
MOLECULAR WEIGHT AND STRAND SPECIFICITY OF MITOCHONDRIAL DNA CODED POLY(A)-CONTAINING RNAs

Poly(A)-containing RNA component	Molecular weight x 10^{-5}	Coding strand
1	34	L
2	28	L
3	14	L
4	9.0	H
5	8.6	H
6	6.7	H
7	6.2	H
8	6.0	H
9	5.8	H
10	5.4	H
11	4.2	H
12	4.0	H
13	3.6	H
14	3.3	H
15	3.2	H
16	2.9	H
17	1.4	H
18	0.93	L

shown the complete sensitivity to 1 μg/ml ethidium bromide of the labeling of these RNA species (15) and, more significantly, their sequence homology to mit-DNA (15), as will be discussed below. The apparent absence, in the RNA pattern from nuclease treated mitochondrial fraction, of components distinct from those coded for by mit-DNA has an important implication, namely, that the great majority, if not all of the presumptive mRNA components present in HeLa cell mitochondria are mit-DNA transcripts.

Table 1 shows the molecular weights of the poly(A)-containing RNA species isolated from HeLa cell mitochondria, as estimated by comparison with the electrophoretic mobility of known standards (mitochondrial 16S, 12S, and 7S RNA). Also shown in Table 1 is the sequence complementarity of these RNA species to separated mit-DNA strands. With the exception of the largest species, #1, 2, and 3, and the smallest component, #18 (7S RNA), all the discrete poly(A)-containing RNA species are complementary to the H strand. Here again appears the asymmetry in informational content of the two strands, which is strongly biased in favor of the H strand. If one sums up the mol. wt. of all of the RNA species coded for by the H strand, one obtains a total length which is

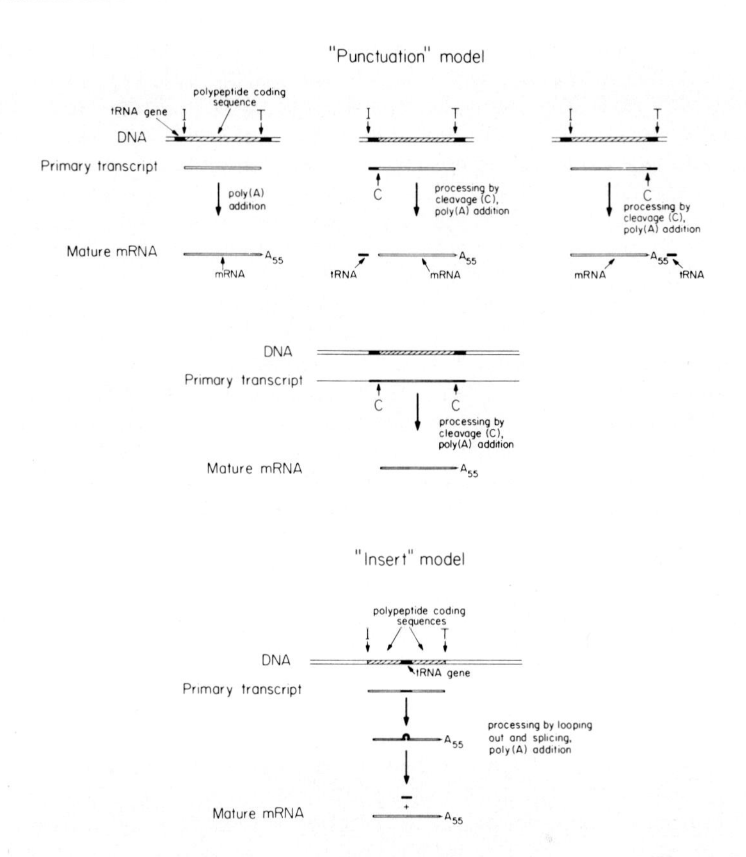

FIGURE 3. Possible mapping relationships of tRNA and mRNA sequences. The "punctuation" model and the "insert" model.

greater by a factor of more than 1.7 than the portion of the H strand which is not occupied by rRNA and 4S RNA genes. This suggests that some of the H strand specified poly(A)-containing RNA species may be precursors of other, and/or that there may be overlapping transcription of the same segments of the DNA.

Mapping Relationship of tRNA and mRNA Sequences - the "Punctuation" Model and the "Insert" Model. Further constraints in the organization of the mitochondrial DNA transcripts on the mit-DNA map come from the arrangement of the 4S RNA genes. One of the most striking features of the organization of the animal cell mitochondrial genome is the fairly uniform distribution of the tRNA genes along the genome. Barring the possibility, which is by no means excluded, of overlapping tRNA and mRNA sequences, the inter-tRNA sequences represent the obvious candidates for being the coding sequences for mit-DNA-specified polypeptides. There are two extreme models which can be envisaged as concerns the organization of tRNA and mRNA coding sequences and their expression in animal cell mit-DNA. The first model, which can be called the "punctuation" model, postulates that the tRNA coding sequences separate the sequences corresponding to individual mRNAs. In this model, the tRNA genes provide the punctuation in the reading of the mitochondrial DNA information, and this function may explain their scattered arrangement in the genome as has emerged in the course of evolution. The tRNA sequences may be even the recognition sites for initiation or termination of transcription or for processing of mRNA sequences (Figure 3). The second model, which can be called the "insert" model, assumes that one or more tRNA coding sequences can interrupt the continuity of a given mRNA coding sequence. According to this model, which has been suggested by the discovery of discontinuous genes in eukaryotic cells, the "inserted" tRNA sequences are presumably extruded by a looping out and splicing process, in the course of processing of the primary transcripts (Figure 3). The two above mentioned models are not mutually exclusive, and may indeed apply in different sections of the genome.

Mitochondrial Translation Products. Information concerning the possible mapping relationship of the tRNA and mRNA sequences can be derived from a comparison of number and size of the mitochondrial translation products (which, for the arguments discussed above, are in most part, if not all, coded for by mit-DNA) with the number and size of the inter-tRNA tracts. Figure 4 shows an autoradiogram of the products of HeLa cell mitochondrial protein synthesis labeled *in vivo* for 2 hr with ^{35}S-methionine in the presence of the cytoplasmic protein synthesis inhibitor emetine, and fractionated either on an SDS-polyacrylamide gradient slab gel (I) or on an SDS-polyacrylamide/8 M urea slab gel (II). At least 19 discrete components (in the mol. wt. range between 3500 and 51,000 daltons) can be recognized on the SDS-polyacrylamide gradient slab gel, and at least 18 components on the SDS-polyacrylamide-urea slab gel. Control experiments utilizing chloramphenicol have shown that all the above mentioned discrete components are synthesized in mitochondria. The same components were observed if phenylmethylsulfonyl fluoride was present throughout cell homogenization and preparation of the mitochondrial proteins, or if a total HeLa cell SDS lysate was directly analyzed on either of the two slab gel systems. These observations strongly suggest that degradation phenomena were not involved in generating the patterns

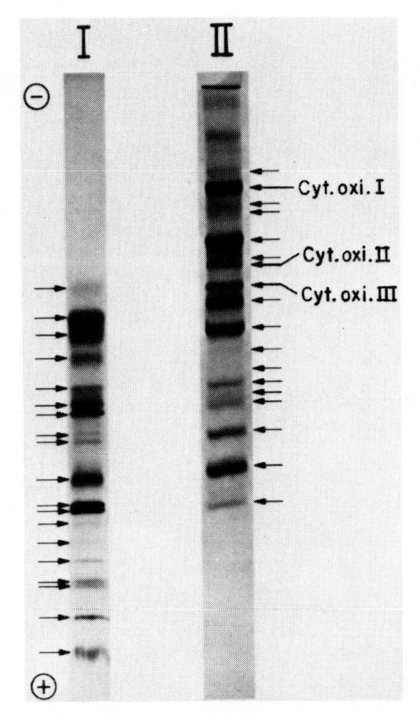

FIGURE 4. Fluorographs, after electrophoresis through an SDS-polyacrylamide 15 to 25% gradient slab gel (I) or an SDS-polyacrylamide/8 M urea slab gel (II), of the proteins from the mitochondrial fraction of HeLa cells labeled for 2 hr with ^{35}S-methionine in the presence of 100 μg/ml emetine.

shown in Figure 4. Experiments of fractionation by two-dimensional polyacrylamide/8 M urea and SDS-polyacrylamide gradient slab gel have clearly indicated that some of the discrete bands shown in Figure 4 consist of more than one component. Thus, at least 20 discrete translation products have been identified in HeLa cell mitochondria. Three of these polypeptides have been shown to be the three largest subunits of human cytochrome c oxidase (Figure 4), which has been recently isolated and characterized in our laboratory (17).

Although we do not know whether all the polypeptides shown in Figure 4 are independent translation products, i.e., products which are not related to each other by precursor to product relationship or by secondary modifications, it is interesting to note that their number is appreciably greater than previously reported in a mitochondrial system (18). However, the information required for the synthesis of these

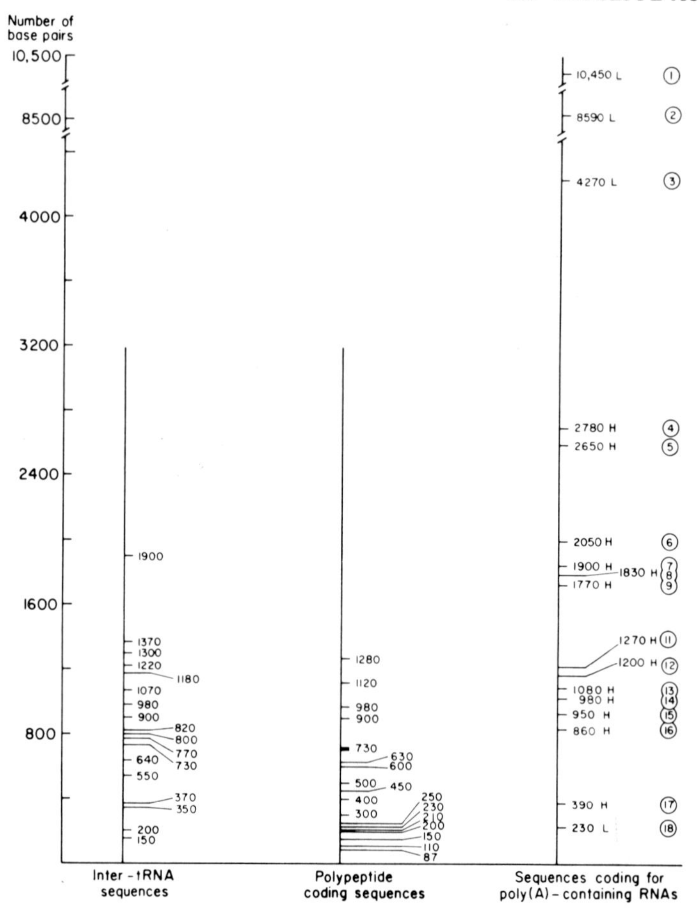

FIGURE 5. Size distributions of "inter-tRNA tracts," polypeptide coding sequences and poly(A)-containing RNAs.

20 polypeptides need not exceed that available in HeLa mit-DNA, even if one excludes the occurrence of overlapping genes. In fact, it can be calculated that the total number of base pairs coding for such a set of polypeptides (∽10,500), assuming that they are independent translation products and coded for by different sequences in mit-DNA, is still lower than the total length of the portion of mit-DNA not occupied by the rRNA genes and by the 4S RNA genes detected so far (∽12,000 base pairs).

Size Distributions of "Inter-tRNA Tracts," Polypeptide Coding Sequences and Poly(A)-Containing RNAs. Figure 5 shows a comparison of the length distribution of the "inter-tRNA tracts" with that of the mRNA sequences coding for the 20

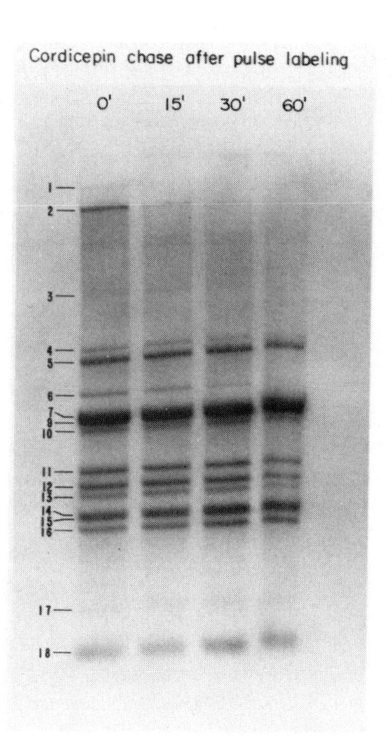

FIGURE 6. Autoradiogram, after electrophoresis through agarose-CH_3HgOH, of mitochondrial poly(A)-containing RNAs in a cordycepin chase experiment. See text for details.

mitochondrially synthesized polypeptides. It is obvious that both the number and the size of the inter-tRNA tracts would be compatible with the "punctuation" model. Also shown in Figure 5 is the size distribution of the poly(A)-containing RNAs coded for by HeLa cell mit-DNA. One sees that one half of the poly(A)-containing RNA species are larger than the largest inter-tRNA tract (excluding that harboring the 16S rRNA cistron). That some of these larger RNAs, in particular #1, 2, 3, 4, and 6, are not mature species, but either precursors or intermediates in the pathway of maturation of the functional species is suggested by the observations mentioned above which indicate that they are not in polysomes. Further evidence in this respect comes from an analysis of their metabolic behavior, discussed below. There are, however, other relatively large RNA species like the species 5, 7, and 9, which for their relative abundance in long-term labeling experiments and their occurrence in polysomal structures are very probably mature species. The absence of any inter-tRNA tracts in which the sequences of these species

TABLE 2
METABOLIC STABILITY OF MITOCHONDRIAL POLY(A)-CONTAINING RNA COMPONENTS AFTER CORDYCEPIN BLOCK

Poly(A)-containing RNA component	Half-life (minutes)
2	7
4	39
5	87
6	16
7	112
9	116
11	56
12	51
14+15	141
16	191
18	118

would fit (Figure 5) strongly suggests that the extrusion of tRNA sequences may be an important mechanism operating in the processing of mitochondrial RNA.

Metabolic Stability of the Poly(A)-Containing RNAs. Figure 6 shows the results of a chase experiment utilizing the nucleoside analogue, cordycepin (3'-deoxyadenosine), a drug which blocks mitochondrial RNA synthesis (19). In this experiment, the labeling of mitochondrial poly(A)-containing RNA with ^{32}P-orthophosphate was measured after a 2.25 hr pulse (a), and after a pulse followed by a 15 min (b), or 30 min (c) or 60 min (d) cordycepin chase. The very rapid decay of species 1, 2, and 3, and the somewhat less rapid decay of species 6 is clearly apparent. Table 2 shows the half-lives of the main poly(A)-containing RNAs, as estimated in this experiment by measuring the ^{32}P-radioactivity in each band and normalizing it for the recovery of 5-^{3}H-uridine-labeled RNA from cells added to the ^{32}P-labeled cells as an internal standard. RNA species 2 has an estimated half-life of about 7 min; species 3 has a similar half-life, though the high background in this region of the gel prevents an accurate measurement; species 1 is the fastest turning over, disappearing completely after a 15 min chase. With the exception of species 6, all other species have a considerably longer half-life, ranging between 40 and 190 min. Although caution should be used in interpreting the results of metabolic experiments involving the use of drugs, there are reasons to believe that the half-life estimates made

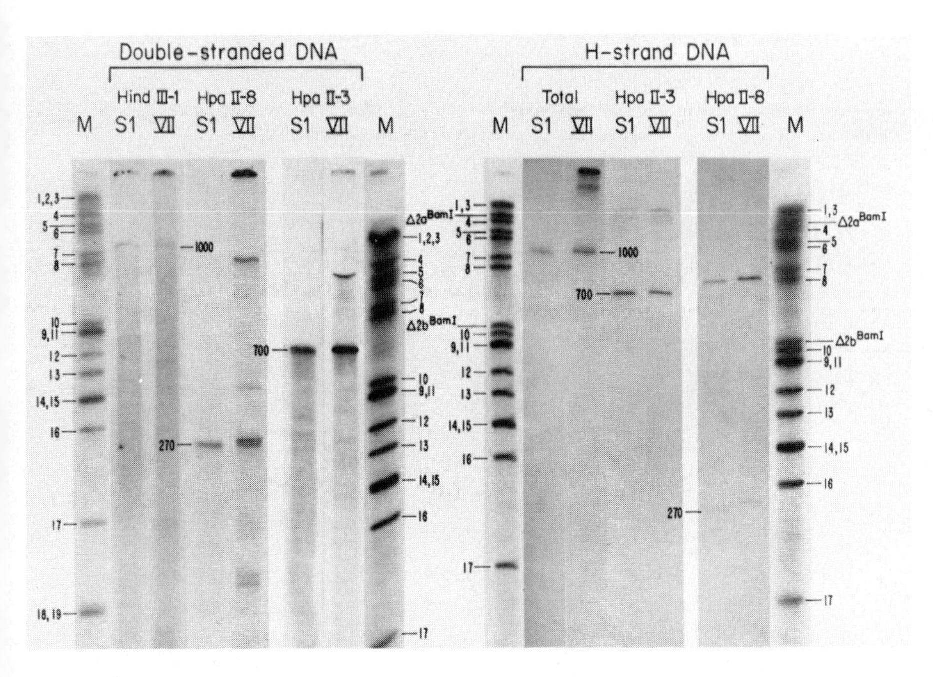

FIGURE 7. Nuclease S1 and exonuclease VII sizing and mapping of mitochondrial 12S rRNA. See text for details.

in this cordycepin chase experiment reflect the real metabolic behavior of these species. In fact, very similar half-lives have been determined for most of the species 7 to 18 by measuring their kinetics of labeling in the absence of drugs, with correction for changes in pool-specific activity; furthermore, the labeling of species 1 to 3 and of species 6 appears to be greatly reduced relative to that of the other species in long-term labeling experiments in the absence of drugs, an observation which again points to their greater metabolic instability.

FINE MAPPING OF THE RIBOSOMAL RNA GENES

Recent work in several laboratories has revealed that the gene for the large rRNA species in mit-DNA from yeast (20) and Neurospora (21) contains an intervening sequence 1.1-1.2 and 2.0-2.5 kb long, respectively. These surprising observations have indicated that the mitochondrial system shares with other eukaryotic systems the occurrence of discontinuous genes and the involvement of splicing mechanisms in the processing of the primary transcripts. In animal cell mit-DNA, previous EM studies of RNA-DNA hybrids involving

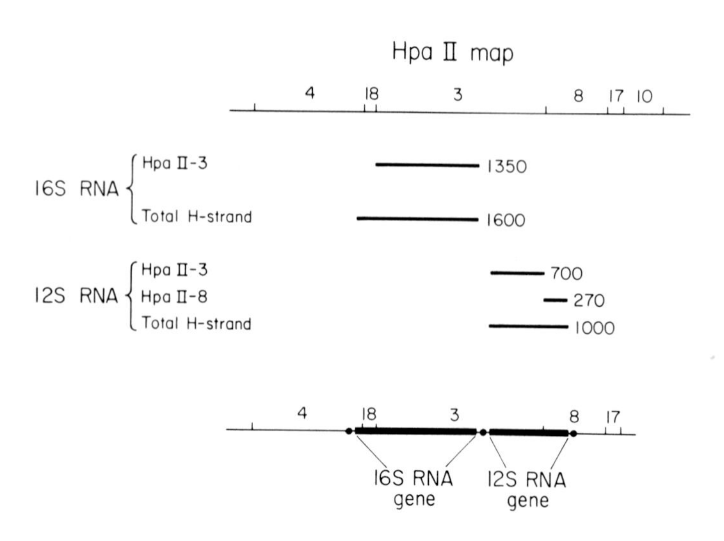

FIGURE 8. Diagrammatic representation of the results shown in Figures 7 and 9.

rRNA species (2,13,22) had not revealed the presence of obvious discontinuities in the rRNA genes. However, the resolution of this analysis was not such that would have detected discontinuities of 100 base pairs or less. And, indeed, there is evidence that some of the known intervening sequences in eukaryotic genes are only 15-20 base pairs long (23,24). We have therefore carried out experiments aimed at obtaining information on the possible existence of inserts in the rRNA genes of HeLa cell mit-DNA. For this purpose, we have used the methodology developed by Berk and Sharp (25): this involves trimming of RNA-DNA hybrids (formed with ^{32}P-labeled DNA) with the single-strand specific *Aspergillus* nuclease S1 and *E. coli* exonuclease VII and size analysis of the dissociated hybridized DNA sequences in denaturing gels. The S1 enzyme is known to have endonuclease activity and is therefore expected to digest any terminal or internal single-stranded DNA segment in the hybrid (26); in contrast, the exonuclease VII lacks endonuclease activity and thus would only trim nonhybridized DNA segments at the 5'- and 3'-ends of the hybrid (27).

Figure 7 shows the autoradiograms after polyacrylamide/urea gel electrophoresis of mit-DNA segments protected by mitochondrial 12S rRNA in the hybrids against digestion with either S1 nuclease or exonuclease VII. In one set of experiments, 12S rRNA was hybridized either to *Hind* III-1 fragment (see map in Figure 1) at 49°C in high formamide (conditions

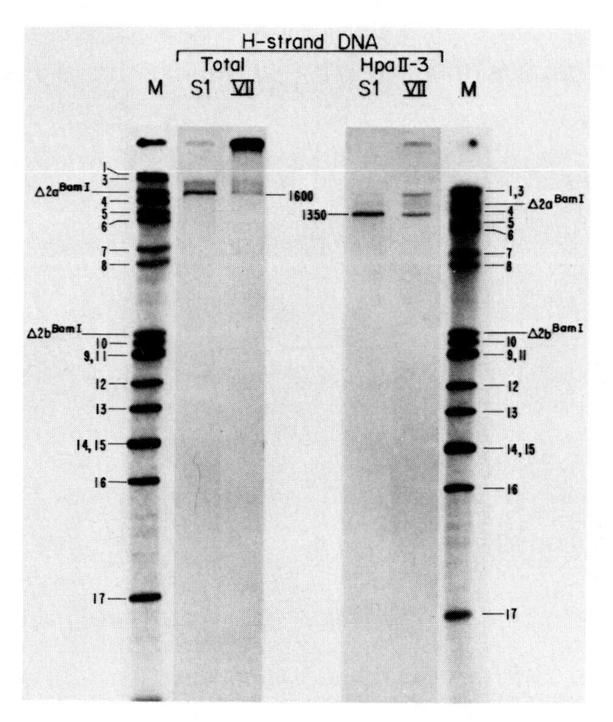

FIGURE 9. Nuclease S1 and exonuclease VII sizing and mapping of mitochondrial 16S rRNA. See text for details.

favoring RNA-DNA hybridization over DNA-DNA reassociation), or to total mit-DNA H strand at 66°C in 0.4 M salt. Under the two experimental conditions, the hybridized 12S rRNA protects from both S1 and exonuclease VII digestion a DNA segment about 1000 nucleotides long. Since the size of 12S rRNA previously estimated by EM is about 1100 nucleotides (28), these results indicated that the main body of the 12S rRNA gene lacks intervening sequences susceptible to S1 digestion. Analysis of the protected DNA segments after hybridization of 12S rRNA to either the whole Hpa II fragment 3 (under the high formamide conditions) or to the H strand of this fragment (at high temperature) showed a band corresponding to a 700 nucleotide long segment after both S1 or exo VII digestion (Figure 7). A similarity in behavior with respect to S1 and exo VII digestion was likewise observed with hybrids formed, under the high formamide conditions, between 12S rRNA and Hpa II fragment 8. As one can see in Figure 7, after S1 digestion, there is a band corresponding to a 270 nucleotide long segment; after exo VII digestion, in addition to this band, there is another one, somewhat more pronounced,

12S rRNA

```
5' (N)A U A G G P P U G G U C C P A G C P P P U C U A C
      1                 10                  20

    U A G C U P P P A G P A A G A P P A P A P A P G P
            30                  40                   50

   (A)A G P A P P P P P G P P P P A G P G A G----
                      60                  70
```

16S rRNA

```
5'  G C U A A A P P P A G C C P P A A A C C P A C U P
    1                 10                  20

    P A C C C P A----
            30

    P = pyrimidine
```

FIGURE 10. Nucleotide sequence of the 5'-end regions of mitochondrial 12S and 16S rRNAs.

moving slightly slower and corresponding to an approximately 275 nucleotide long segment. It seems likely that the latter band represents the protected DNA segment incompletely trimmed by the exo VII at one or both ends due to some structural peculiarities. When the same 12S rRNA preparation was hybridized at high temperature to the separated H strand of the Hpa II fragment 8, a segment of about 270 nucleotides was protected from both S1 and exo VII digestion, with a hint that the protected segment was slightly longer in the case of exo VII.

The results described above are represented diagrammatically in Figure 8. While these results tend to exclude the occurrence of intervening sequences in the main body of 12S rRNA, the experimental approach used is not adequate to reveal what occurs near the ends. The proximity of the 5'-end of the 12S rRNA gene to a 4S RNA site in DNA (13,22) raises the question of whether the promoter for 12S rRNA synthesis is located between the 12S rRNA gene and the 4S RNA gene or on the 5'-end side of the latter. An interesting possibility is that a short 5'-end proximal segment of the 12S rRNA coding sequence, which would not form a stable hybrid with the complementary 12S rRNA sequence, thus allowing exo VII

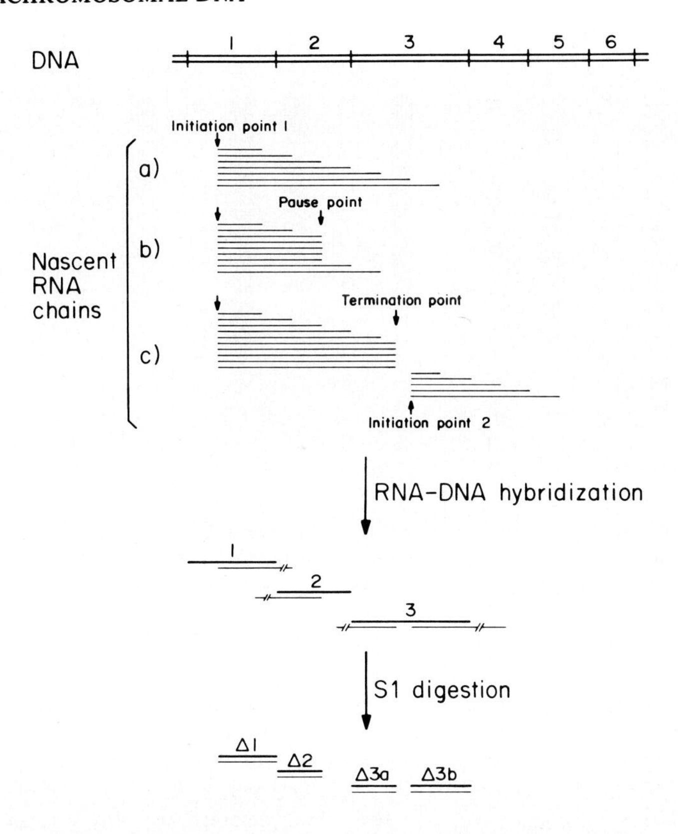

FIGURE 11. Diagram illustrating the rationale of an experiment designed to map nascent RNA chains from mit-DNA transcription complexes. See text for details.

attack, is located on the 5'-end side of the 4S RNA coding sequence. Synthesis of the 12S rRNA would thus involve processing of the primary transcript by looping out and excision of a segment containing the 4S RNA sequence, followed by splicing. Direct sequence analysis of the 5'-end-proximal segment of 12S rRNA and of the corresponding region of the DNA will be necessary to localize precisely the 5'-end of 12S rRNA relative to the 4S RNA site.

A similar analysis to that described above for 12S rRNA was performed with 16S rRNA. As shown in Figure 9, after hybridization with the H strand of total mit-DNA, a fragment about 1600 nucleotides long appears to be protected from both S1 and exo VII digestion. This size is slightly smaller than that estimated by EM for 16S rRNA (about 1700 nucleotides) (28). Hybridization of 16S rRNA to the H strand of

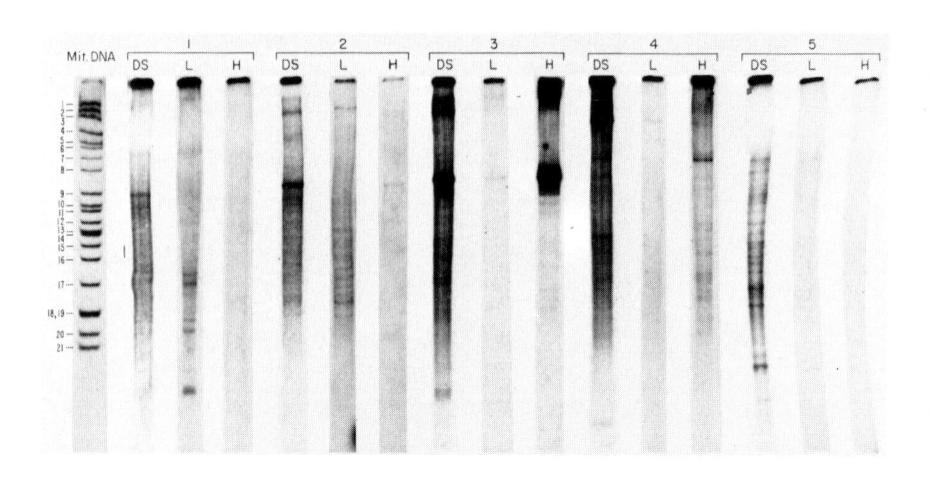

FIGURE 12. Protection from nuclease S1 digestion of mit-DNA from Hpa II restriction fragments 1 to 5 hybridized to nascent RNA chains from mit-DNA transcription complexes. See text for details.

Hpa II fragment 3 protects a 1350 base long fragment from both S1 and exo VII digestion. By contrast, no protected fragment was reproducibly observed after hybridization of 16S rRNA to the H strand of Hpa II fragment 4. This result suggests that the portion of 16S rRNA extending into fragment 4 is considerably shorter than previously estimated. Indeed, from the length of the segment of 16S rRNA corresponding to Hpa II fragment 3 (about 1350 nucleotides) and from the sizes of Hpa II fragment 18 (about 150 nucleotides) (2) and of 16S rRNA (1600-1700 nucleotides) one can estimate that a stretch of only 100-200 nucleotides of 16S rRNA could have its coding sequence in fragment 4.

The above described results are summarized in Figure 8. As in the case of 12S rRNA, the main conclusion of these experiments is that there are no obvious intervening sequences in the main body of the 16S rRNA gene. In this respect, the situation appears to be different from that described for the large mitochondrial rRNA of yeast and Neurospora. However, here too, one cannot exclude the occurrence of sub-terminal intervening sequences, possibly related to the two 4S RNA genes flanking the 16S rRNA gene.

As mentioned above, the only unambiguous approach to analyze the relationship of the 5'-end and 3'-end of the 12S and 16S rRNAs to the flanking 4S RNA coding sequences is the sequencing approach. We have made a start in this direction by applying the methodology developed by

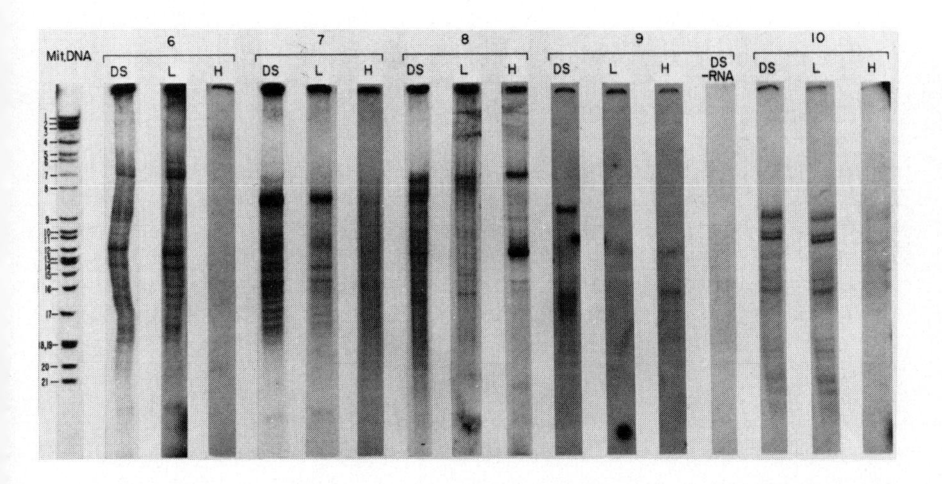

FIGURE 13. Protection from nuclease S1 digestion of mit-DNA from *Hpa* II restriction fragments 6 to 10 hybridized to nascent RNA chains from mit-DNA transcription complexes. See text for details.

Donis-Keller *et al.* (29) and Simoncsits *et al.* (30) to determine the 5'-end-proximal sequence of 12S and 16S rRNA. These sequences are shown in Figure 10.

There is nothing striking in these sequences except the presence, in both, of several stretches of pyrimidines and the almost complete absence of G in the first 32 nucleotide segments of the 16S rRNA. The nucleotide sequence of the fragment $\Delta 8a^{Hae}$ (Figure 1) is at present being determined in order to align it with the 5'-end-proximal 71 nucleotides of 12S rRNA and to identify possibly the position of a tRNA sequence corresponding to the 4S RNA site recognized by EM (13,22).

SEARCH FOR INITIATION SITE(S) OF TRANSCRIPTION

The presumptive mature transcripts of the L strand identified so far, i.e., seven 4S RNA and one poly(A)-containing RNA, account all together for less than 5% of the length of the L strand. This makes the complete transcription of this strand very intriguing. Understanding of the significance of such a phenomenon would be facilitated if one knew whether and how the complete transcription of the L strand is related to the expression of the few genes situated in this strand. In order to obtain information on these questions, an investigation has been carried out of the location of the initiation sites of L strand transcription. Previous evidence indicating that the nascent L transcripts consist mainly of large molecules (14,15) suggested two approaches for the identification of their initiation sites.

The first approach is based on an application of the methods developed by Berk and Sharp (25) for mapping RNA species, and is illustrated schematically in Figure 11. In a steady state population of nascent RNA molecules one can in principle recognize subsets characterized by a common 5'- and/or a common 3'-end. A common 5'-end is represented by a fixed initiation point, while a common 3'-end is either a nonrandom pause point in the progression of the RNA polymerase or a fixed termination point. The occurrence of such common ends in the nascent molecules makes it possible to use the Berk and Sharp methods for the analysis of the RNA-DNA hybrids formed between different restriction fragments and the nascent molecules. If the restriction fragments are small relatively to the average length of the nascent chains, it should be possible to identify from the length of the protected DNA segments the positions of the initiation points, pause points or termination points in the nascent RNA molecules.

The availability in our laboratory of a procedure for the isolation from HeLa cells of transcription complexes of mit-DNA (31) has provided a convenient source of nascent RNA molecules. Figures 12 and 13 show an electrophoretic analysis in polyacrylamide gel, under nondenaturing conditions, of the S1-resistant hybrids formed, in high formamide, between nascent RNA molecules and either *Hpa* II fragments 1 to 10 or separated strands of these fragments. In each case, a cascade of discrete bands is observed, one of which in general corresponds to the entire length of the fragment. Hybridization of nascent RNA molecules to separated strands of the above mentioned *Hpa* II fragments shows that the majority of the S1-resistant duplexes involve the L strand in the case of fragments 10, 2, 5, 1, 6, and 7, and, by contrast, the H strand, in the case of fragments 3, 4, and 9. Fragment 8 is unique in that it produces two different sets of S1-resistant duplexes with its L and its H strand. This distribution of the hybrids formed with the L or H strands reflects clearly the position, in the physical map of mit-DNA, of the fragments involved. This is illustrated in Figure 14, which summarizes diagrammatically the results of these experiments. It is clear that the hybrids formed with the L strand correspond mainly to the right half of the genome (defined relative to the origin of replication set at the noon position, assuming a clockwise direction of L strand transcription), while those formed with the H strand correspond to the left half of the genome. The discrete hybrid bands not coinciding with the entire length of the *Hpa* II

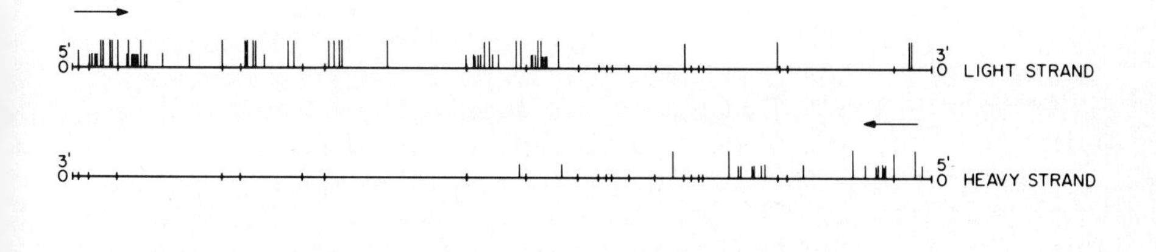

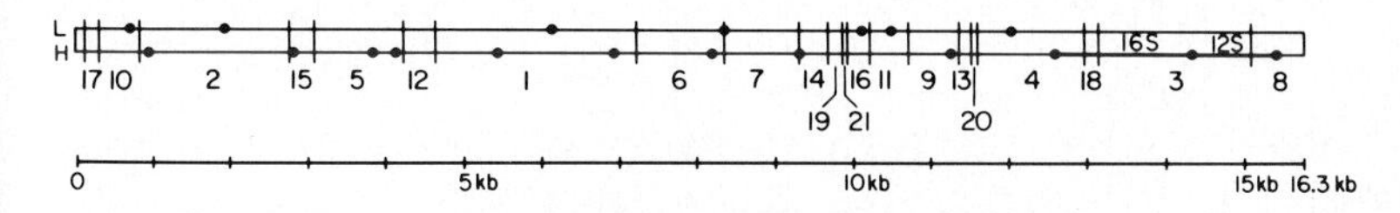

FIGURE 14. Diagrammatic representation of the results shown in Figures 12 and 13. In the upper part of the figure, the positions in the *Hpa* II physical map of mit-DNA corresponding to the discrete bands in the gels (determined from the length of the protected DNA segments, assuming arbitrarily the origin-proximal end of each fragment in the direction of transcription as a starting point) are indicated by ticks. Long and short ticks refer to relatively intense and faint bands, respectively. In the lower part of the figure the linearized *Hpa* II map of mit-DNA is shown. O: origin of replication.

fragments could in principle mark the position of initiation points, pause points or termination points in the transcription process. However, it does not seem likely that terminated nascent chains would normally accumulate in transcription complexes, from which these chains were derived. On the other hand, not all these bands could correspond to initiation points, since one finds many of these in the middle of the 12S rRNA or 16S rRNA cistron. A plausible interpretation is that many, and perhaps the majority of these bands correspond to nonrandom pause points in the progression of the RNA polymerase. Such nonrandom pauses in chain elongation, possibly related to potential secondary structures in the product or template strand, have been reported to occur with most, if not all, nucleic acid polymerases (32). Whatever the correct interpretation is, the distribution in the mit-DNA physical map of the hybrids formed with nascent chains should reflect the distribution of the RNA sequences involved, and thus may tell us something about the location of the main initiation sites for transcription. An examination of Figure 14 suggests a gradient of decreasing density of these bands, starting from the origin of replication in *Hpa* II fragment 8 and proceeding clockwise, with a secondary cluster of bands in fragments 6 and 7, about 180° away from the origin. The bands are very rare in the left half of the genome. This is what would be expected for a population of nascent L transcripts growing clockwise and having their initiation points in the region close to the origin; the clustering of bands in fragments 6 and 7 suggests a possible second initiation point for L strand transcription in this region.

As concerns the hybrid duplexes involving the H strand, their position in the map suggests at least one initation point for H strand transcription on the left side of the origin of replication, in Hpa II fragment 8. The nascent H strand transcripts hybridizing with Hpa II fragments 8 and 3 are presumably nascent rRNA chains.

Corroboration of the results described above as concerns the location of the initiation sites for L strand transcription has come from a different approach, based on that used by Bachenheimer and Darnell (33) to map the main promoter of adenovirus DNA late transcription. For this approach, nascent transcripts of the L strand, isolated from cells labeled with a very short pulse of 5-^{3}H-uridine, were fractionated in a sucrose gradient under denaturing conditions, and the different size classes were purified by affinity chromatography on H strand mit-DNA-cellulose, reduced to a uniformly small size by controlled alkali treatment and finally hybridized in solution with isolated Hpa II restriction fragments of mit-DNA. Under these conditions, the fragments closer to a transcription initiation point should be recognized for the higher hybridization level with the labeled segment of the shortest RNA chains. The results of these experiments pointed to Hpa II fragment 8, and probably to fragments 6 and 7, as the main sites of initiation of L strand transcription.

The work that I have just described has thus strongly suggested that the region of mit-DNA close to the origin of replication, contains one promoter for L strand transcription and one promoter for H strand transcription, probably involved in rRNA synthesis. This organization is reminiscent of that of the SV40 genome, where the initiation sites for transcription of both early and late strands appear to be situated close to the origin of replication.

IDENTIFICATION AND CHARACTERIZATION OF GIANT, DISCRETE L STRAND TRANSCRIPTS

Evidence in agreement with the above proposed location of a major site of initiation of L strand transcription in the region close to the origin of replication has come from the identification and characterization of discrete, high mol. wt. L strand transcripts. As mentioned above, the largest discrete poly(A)-containing RNAs detected in HeLa cell mitochondria (#1, 2, and 3) are coded for by the L strand (Table

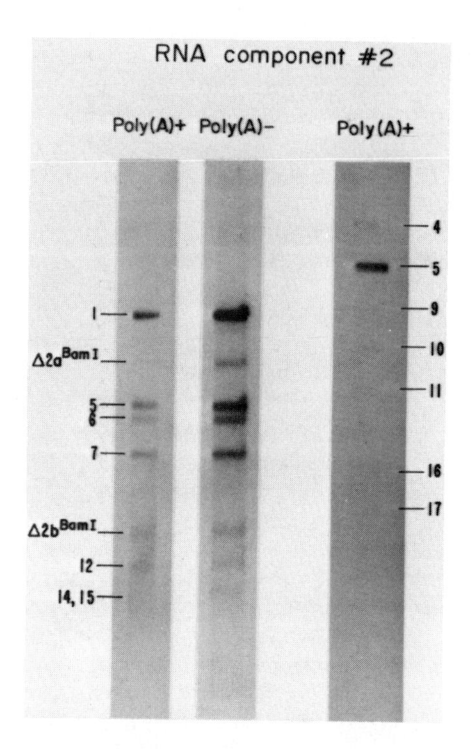

FIGURE 15. Hybridization of ^{32}P-labeled poly(A)-containing and non-poly(A)-containing RNA species 2 to Hpa II + Bam I digests of mit-DNA or to isolated Hpa II fragments transferred to nitrocellulose filters. See text for details.

1). More recently, the map location of the sequences coding for the most abundant of these species, poly(A)-containing RNA component 2 (Figures 2 and 6), has been investigated by hybridizing this species, isolated from cells labeled for 2 to 3 hr with ^{32}P-orthophosphate, to total Hpa II + Bam I digests of HeLa mit-DNA transferred to a nitrocellulose filter by the Southern technique (34). As shown in Figure 15, there is clear hybridization of the species 2 to Hpa II fragments 1, $\Delta 2a^{\text{Bam I}}$, 5, 6, 7, 12, and 14 and/or 15. These results allowed the approximate localization of the sequences coding for the poly(A)-containing RNA species 2 in the right half of the genome. In order to better define the location in the map of the 5'-end and 3'-ends of this RNA species, hybridization experiments were carried out with isolated fragments 9, 10, 11, 16, and 17, and with fragment 4 as a negative control, and fragment 5 as a positive control, each transferred to a small nitrocellulose filter. As one can see in Figure 15, all fragments failed to show any significant radioactivity, with the expected exception of fragment 5. In other experiments (not shown), a definite hybridization to Hpa II fragment 14 was observed. Assuming that there are no segments spliced out of species 2, as seems likely, in view of its kinetic characteristics of early transcript, the above results allow a localization of the sequences coding for RNA species 2 in the region of the mit-DNA map defined by coordinates 8.5 ± 2 to 61.5 ± 2, as shown in Figure 16.

Also shown in Figure 15 are the results of hybridization with a Hpa II + Bam I digest of the non-poly(A)-containing RNA species I (Figure 2), which has an electrophoretic mobility identical to that of species 2. The two species gave an identical pattern of hybridization, thus indicating that they are transcribed from the same region of mit-DNA. Similar experiments (not shown) carried out with poly(A)-containing RNA species 1 and the corresponding non-poly(A)-containing RNA species gave patterns of hybridization with the whole Hpa II + Bam I digest very close to that observed with RNA species 2; however, the hybridization with fragment $\Delta 2a^{\underline{Bam\ I}}$ relative to that with fragments 1 and 5 was definitely and reproducibly stronger with these RNA species than with species 2. This suggested that the coding sequences of RNA species 1 extend further towards the origin of replication as compared to species 2 (see map in Figure 16). Experiments are in progress to define more precisely the location in the map of the 5'- and 3'-ends of species 1.

The above described results have several interesting implications. First, they agree with the mapping data of nascent RNA chains from transcription complexes in pointing to the region close to the origin of replication as a major initiation site for L strand transcription. It seems possible that the RNA species 2 is not the primary transcript, but derives from the primary transcript by processing at the 5'-end. The observation of the occurrence of large L strand transcripts lacking poly(A) but with the same mapping specificities as the poly(A)-containing species 1 and 2 suggests that poly(A) addition in mit-DNA coded RNA is not an event necessarily coupled with termination of transcription or with cleavage of larger transcripts. Finally, there is no evidence in our work of any poly(A)-containing RNA coded for by the L strand and present in polysomal structures, of which the giant species 2 could be a precursor; on the other hand it seems likely that this poly(A)-containing RNA species contains the sequences of at least the 4S RNAs complementary to the sites 4L and 5L. Therefore, it would appear

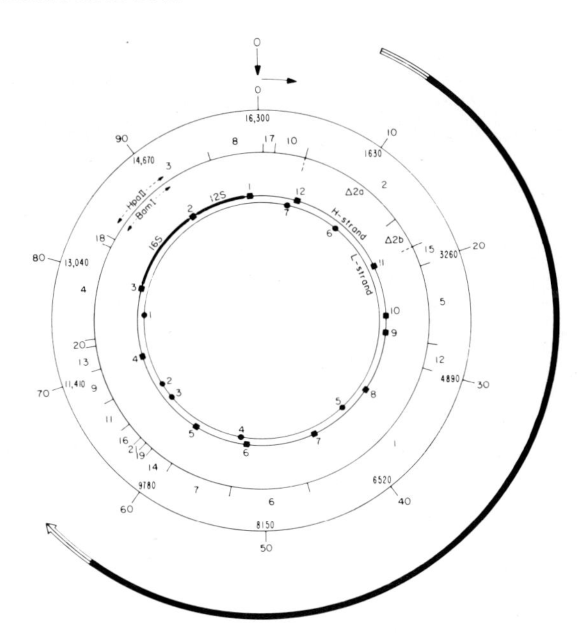

FIGURE 16. Location on the Hpa II map of mit-DNA of the sequences coding for the poly(A)-containing and non-poly(A)-containing RNA species 2. The open bar represents the region of uncertainty in the map assignment at each end of the RNA.

that poly(A) presence, at least in mitochondria, is not an exclusive feature of mRNA or its precursors.

ACKNOWLEDGMENTS

A generous gift of exonuclease VII from Dr. P. Berg is thankfully acknowledged. We are grateful to Dr. M. Albring for his suggestion to use micrococcal nuclease.

REFERENCES

1. Attardi, G., Albring, M., Amalric, F., Gelfand, R., Griffith, J., Lynch, D., Merkel, C., Murphy, W., and Ojala, D. (1976). In "Genetics and Biogenesis of Chloroplasts and Mitochondria" (Th. Bücher et al., eds.), pp. 749-758. North-Holland, Amsterdam.
2. Dawid, I. B., Klukas, C. K., Ohi, S., Ramirez, J. L., and Upholt, W. B. (1976). In "The Genetic Function of Mitochondrial DNA" (A. Kroon and C. Saccone, eds.), pp. 3-13. North-Holland, Amsterdam.
3. Kasamatsu, H., Grossman, L. I., Robberson, D. L., Watson, R., and Vinograd, J. (1973). Cold Spring Harbor Symp. Quant. Biol. 38, 281.

4. Arnberg, A. C., Van Bruggen, E. F. J., Flavell, R. A., and Borst, P. (1973). Biochim. Biophys. Acta 308, 276.
5. Wolstenholme, D. R., Koike, K., and Cochran-Fouts, P. (1973). J. Cell Biol. 56, 230.
6. Bogenhagen, D., Gillum, A. M., Martens, P. A., and Clayton, D. A. (1978). Cold Spring Harbor Symp. Quant. Biol. 43, in press.
7. Angerer, L., Davidson, N., Murphy, W., Lynch, D. C., and Attardi, G. (1976). Cell 9, 81.
8. Ojala, D., and Attardi, G. (1977). Plasmid 1, 78.
9. Lynch, D. C., and Attardi, G. (1976). J. Mol. Biol. 102, 125.
10. Ojala, D., and Attardi, G. (1978). J. Mol. Biol. 122, 301.
11. Crews, S., Ojala, D., Posakony, J., Nishiguchi, J., and Attardi, G. (1979). Nature 277, 192.
12. Aloni, Y., and Attardi, G. (1971). Proc. Nat. Acad. Sci. U.S.A. 68, 1957.
13. Murphy, W., Attardi, B., Tu, C., and Attardi, G. (1975). J. Mol. Biol. 99, 809.
14. Aloni, Y., and Attardi, G. (1971). J. Mol. Biol. 55, 271.
15. Amalric, F., Merkel, C., Gelfand, R., and Attardi, G. (1978). J. Mol. Biol. 118, 1.
16. Abelson, H. T., and Penman, S. (1970). Nature New Biol. 237, 144.
17. Hare, J., Ching, E., and Attardi, G. (1979). Submitted for publication.
18. Schatz, G., and Mason, T. L. (1974). Ann. Rev. Biochem. 43, 51.
19. Penman, S., Fan, H., Perlman, S., Rosbash, M., Weinberg, R., and Zylber, H. (1970). Cold Spring Harbor Symp. Quant. Biol. 35, 561.
20. Bos, J. L., Heyting, C., Borst, P., Arnberg, A. C., and Van Bruggen, E. F. J. (1978). Nature 275, 336.
21. Mannella, C. A., Collins, R. A., Green, M. R., and Lambowitz, A. M. (1979). Proc. Nat. Acad. Sci. U.S.A., in press.
22. Wu, M., Davidson, N., Attardi, G., and Aloni, Y. (1972). J. Mol. Biol. 71, 81.
23. Goodman, H. M., Olson, M. V., and Hall, B. D. (1977). Proc. Nat. Acad. Sci. U.S.A. 74, 5453.
24. Valenzuela, P., Venegas, A., Weinberg, F., and Bishop, R. (1978). Proc. Nat. Acad. Sci. U.S.A. 75, 190.
25. Berk, A. J., and Sharp, P. A. (1978). Proc. Nat. Acad. Sci. U.S.A. 75, 1274.
26. Vogt, V. M. (1973). Eur. J. Biochem. 33, 192.
27. Chase, J. W., and Richardson, C. C. (1974). J. Biol. Chem. 249, 4553.

28. Robberson, D., Aloni, Y., Attardi, G., and Davidson, N. (1971). J. Mol. Biol. 60, 473.
29. Donis-Keller, H., Maxam, A. M., and Gilbert, W. (1977). Nucleic Acids Res. 4, 2527.
30. Simoncsits, A., Brownlee, G. G., Brown, R. S., Rubin, J. R., and Guilley, H. (1977). Nature 269, 833.
31. Carré, D., and Attardi, G. (1978). Biochemistry 17, 3263.
32. Mills, D. R., Dobkin, C., and Kramer, F. R. (1978). Cell 15, 541.
33. Bachenheimer, S., and Darnell, J. E. (1975). Proc. Nat. Acad. Sci. U.S.A. 72, 4445.
34. Southern, E. M. (1975). J. Mol. Biol. 98, 503.

CLONING OF HUMAN MITOCHONDRIAL DNA

Jacques Drouin[1] and Robert H. Symons[2]

MRC Laboratory of Molecular Biology
Cambridge, England

ABSTRACT In order to determine its nucleotide sequence, human mitochondrial DNA (mtDNA) purified from term placentae was cloned in *E. coli* using the plasmid vector pBR322. The products of an mtDNA MboI digestion (23 fragments ranging in size from 2800 to 25 base pairs, bp) were ligated with BamHl-cut pBR322. The ampicillin-resistant tetracycline-sensitive colonies obtained upon transformation of *E. coli* χ1776 were screened by agarose gel electrophoresis of colony lysates, colony hybridization and restriction analysis. All but MboI fragment 2 were obtained in this way. MboI fragments 5 and 8 were each found only once among the 705 clones screened. All other MboI fragments were approximately equally represented in the population of clones except for a slight bias towards smaller fragments. MboI fragment 2 overlaps with the mtDNA BamHl/HindIII (1.75 kb) or BamHl/EcoRI (1.7 kb) and the 0.9 kb HindIII fragments. These were cloned in similarly restricted pBR322 to provide a set of clones covering the entire mtDNA. Clones representative of each MboI fragments were shown to be complementary to mtDNA by hybridisation to Southern blots of mtDNA digests and were thereby partially mapped. Further mapping was obtained by restriction analysis of mtDNA sequentially degraded by exonuclease III. A collection of recombinant clones has thus been obtained using the mtDNA isolated from a single placenta and is now being used to obtain a complete nucleotide sequence of human mtDNA.

[1]Present address: Department of Biochemistry and Biophysics, University of California, San Francisco, California 94143

[2]Present address: Department of Biochemistry, University of Adelaide, Box 498 GPO, Adelaide, South Australia 5001

ISBN 0-12-198780-9

INTRODUCTION

The human mitochondrion, like that of other eucaryotes, contains a genome of its own. Human mitochondrial DNA (mtDNA) can be isolated as a supercoiled circular molecule with a contour length of approximately 5 microns (1). Animal mtDNA is replicated unidirectionally from its origin of replication, which can be easily located by the presence of a single stranded replication intermediate ("7s DNA") hydrogen-bonded to the L-strand (2,3). This replication intermediate has been characterised (4,5,6) and the HeLa mtDNA region in which it was precisely located has been sequenced (6).

The products of human mtDNA transcription are still ill-defined except for the ribosomal and transfer RNAs. HeLa cell mtDNA was shown to contain sequences complementary to the two ribosomal RNAs (12S and 16S) and to at least 19 mitochondrial "4S" RNAs, presumably tRNAs (7). A fraction of mitochondrial RNA contains poly(A) tails (8,9). There is evidence that the mature mitochondrial RNA species are processed from symmetrically transcribed mitochondrial RNA that is complementary to nearly all of the two mtDNA strands (10). Animal mitochondria-encoded proteins are even less well defined but, indirect evidence suggests that they may be highly hydrophobic proteins associated with the inner mitochondrial membrane (1).

Although mitochondrial function is dependent on a large number of nuclear-encoded proteins, the mitochondrial genome represents a well integrated genetic entity that is amenable to complete structural investigation given the recently developed rapid DNA sequencing techniques (11,12). Mammalian mtDNA has been shown to be replicated with fidelity when cloned in *E. coli* (13). Human mtDNA was therefore cloned in *E. coli* to provide a reliable and abundant source of mtDNA for determination of its complete DNA sequence.

RESULTS

Mitochondrial DNA (mtDNA) was extensively purified from human placentae obtained at term. Placental mitochondria prepared by differential centrifugation were incubated with pancreatic DNAase I before lysis with SDS. Supercoiled mtDNA was then purified twice by equilibrium sedimentation in ethidium bromide-caesium chloride. The mtDNA obtained from one placenta was used for all the experiments described herein. Digestion of this mtDNA with EcoRI and HindIII generated fragments similar to those reported for HeLa mtDNA (14,15).

XbaI digestion products were also similar to those reported by Brown et al. (16). BamHI cuts human mtDNA at a single site while both PstI and KpnI generate a big and a small fragment, the small fragments being of 2.25 and 3.1 kb respectively. A map of the sites cut by these restriction endonucleases was generated by double digests and found to be similar to those reported previously (14,15,16,17).

Cloning of EcoRI/BamHI and HindIII/BamHI mtDNA Fragments
Preliminary attempts to clone mtDNA were done by ligating mtDNA cut with BamHI or with both BamHI plus HindIII or BamHI plus EcoRI with the similarly cut plasmid vector pBR322 (18). The ligated DNA was introduced by transformation into *E. coli* HB101 or χ1776 (19) and ampicillin-resistant tetracycline-sensitive (amp^r-tet^s) colonies were screened by agarose gel electrophoresis of colony lysates (20). Clones containing the small EcoRI (1.15 kb) and HindIII (0.9 kb) fragments were obtained as well as the small overlapping fragments generated by BamHI and EcoRI (1.7 kb) or HindIII (1.75 kb). Two clones shown by restriction analysis to contain the 1.15 kb EcoRI fragment and the 1.7 kb BamHI-EcoRI fragment (pmt1) and the 0.9 kb HindIII fragment (pmt2) were further characterised by hybridisation to human mtDNA (Figure 2). No clones were found to contain the bigger fragments or the complete BamHI-cut mtDNA and the remainder of the genome was cloned as the MboI fragments. Since the purpose of the mtDNA cloning was the determination of a complete DNA sequence, cloning of a set of restriction fragments of mtDNA rather than the entire molecule would only simplify the sequencing task.

MboI Digestion of mtDNA. An MboI digest of mtDNA was chosen for this purpose since MboI generates the same 5' over-hanging single stranded ends as BamHI. The MboI fragments could therefore be ligated *in vitro* with BamHI-cut pBR322. At least 23 fragments are generated by MboI cleavage of mtDNA. They were resolved on thin polyacrylamide gels (21) after labelling with $\alpha[^{32}P]$-dGTP and the Klenow fragment of *E. coli* DNA polymerase I (as in Figures 1 and 3). The sizes of the mtDNA MboI fragments as well as those generated by double digestion are compiled in Table 1. Restriction fragments of known sequence from φχ174 (22) and pBR322 (23) were used as size markers. EcoRI cuts MboI fragment 1 twice, generating the 1.5 kb EcoRI fragment and two new fragments; it also cuts MboI fragment 2 once. One HindIII site is present in each of MboI fragments 1, 2 and 3b. Fragment 3b was first defined as possessing an HindIII site while fragment 3a is cut by KpnI but it was later found on better resolved gels that fragment 3b migrates slightly slower than fragment 3a. KpnI cuts fragment 5 as well. Only one PstI site, in MboI fragment 4, was detected. Since two PstI sites are evident from digestion of

TABLE I

PRODUCTS OF mtDNA MboI DIGESTION

Fragment	Size (bp)	Double Digests* Second Enzyme	New Fragments Sizes (bp)
1	2800	EcoRI	1250+1150+460 = 2860
		HindIII	2600 + 240 = 2840
2	2400	EcoRI	1700 + 720 = 2420
		HindIII	1750 + 625 = 2375
3b	1830	HindIII	1575 + 255 = 1830
3a	1800	KpnI	1470 + 325 = 1800
4	1650	PstI	1400 + 295 = 1695
5	1000	KpnI	490 + 460 = 950
6a	800	XbaI	575 + 230 = 805
6b	780	XbaI	460 + 310 = 770
7	730		
8	620		
9	600		
10	465		
11	300		
12	280	XbaI	250 +
13	240		
14	215		
15	205		
16	195		
17	115		
18	100	XbaI	55 + 39 = 94
19	67		
20	33		
21	25		
TOTAL =	17200		

*The assignment of most of the products of double digestion (except for subfragments of MboI fragments 12 and 18) was later verified by digestion of cloned fragments.

mtDNA with PstI alone, this suggests that the second PstI site is too close to an MboI site for a change in fragment size to be measured or for a new fragment to be detected (it would have to be smaller than 15 bp). A similar situation is found in the MboI + XbaI digest where four MboI fragments seem to be cut by XbaI and only seven new fragments can be detected in the double digest, by comparison with the five fragments generated by XbaI alone. Both MboI fragments 6a and 6b are cut by XbaI and clones containing these were later identified by XbaI digestion, since the size difference between the two is not very large. The assignment of the products of double digestion of MboI fragments shown in Table 1 was later verified by similar digestion of the cloned MboI fragments, except for the fragments produced by XbaI digestion of MboI fragments 12 and 18.

Cloning of mtDNA MboI Fragments. mtDNA restricted with MboI was ligated with BamHI-cut pBR322. The ligated DNA was transformed in the disabled *E. coli* strain, χ1776, and $amp^r tet^s$ transformants were picked. These were first screened by the colony lysate method (20) and DNA was prepared for restriction analysis from a collection of clones that would hopefully represent the complete MboI digest of mtDNA. Most MboI fragments were found in the first 45 clones examined but only one clone out of 161 was found to contain MboI fragment 8. The frequency of occurence of the MboI fragments in this set of clones was approximately equal, except for MboI fragments 2, 5 and 8. Fragments bigger than fragment 11 were each found in approximately 4% of the clones whereas fragments 11, 12 and 13 were found twice as frequently. Fragments smaller than fragment 13 were probably present at a similar or higher frequency but the smaller recombinants (as estimated by colony lysate electrophoresis) were not systematically screened by restriction analysis and therefore estimates of the frequency of the presence of these fragments in the 161 clones considered cannot be calculated.

In an effort to find the missing MboI fragments, 700 independent clones were screened by filter hybridisation (24,25) for sequences complementary to MboI fragments 2, 5 and 8. Only one positive clone was found. This clone (pmt172) was later shown by restriction analysis to contain MboI fragments 5,9,16 and 18. A control hybridisation with MboI fragments 3a, 3b, 4 and 10 of a filter made from one of the plates used in the previous search revealed 16 positives out of 92 clones. These were all shown by restriction analysis to contain at least one of the four MboI fragments used as probes. MboI fragment 2 was the only fragment that did not appear in the 700 clones screened, but since the mtDNA region covered by

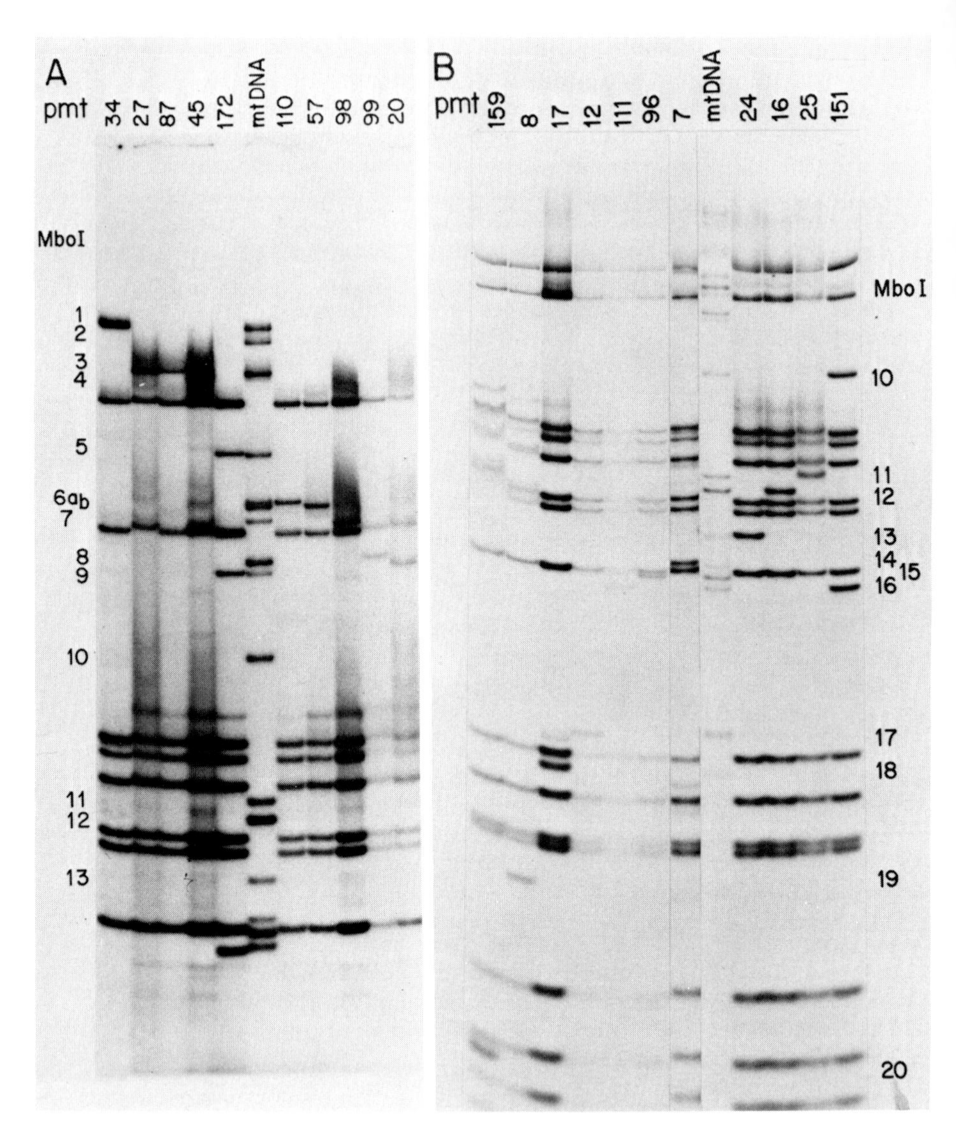

FIGURE 1. Acrylamide gel (A: 3.5% and B: 5%) electrophoresis of FnuE (or Sau3A)-digested pBR322 plasmids containing at least one mtDNA MboI fragment. MboI digestion products were labelled by filling-in with α[^{32}P]-dGTP using the Klenow fragment of *E. coli* DNA polymerase I. The MboI fragment contained in pmt27 can be cut by KpnI while that contained in pmt87 can be cut by HindIII (not shown). pmt17, pmt98 and pmt172 each contain another mtDNA MboI fragment that is not visible on the gels shown but is listed in Figure 2.

this fragment is included in the previously obtained clones, pmtl and pmt2, it was not necessary to obtain a separate clone for this fragment.

A set of clones representing all the mtDNA MboI products of digestion (except for fragment 2) was chosen for further characterisation. Since the A residue of the MboI recognition sequence is methylated in DNA isolated from χ1776, isoschizomers of MboI (FnuE or Sau3A) not affected by this methylation were used in the analysis of the clones (27). DNA isolated from small cultures (26) of each clone was restricted with FnuE or Sau3A and labelled with $\alpha[^{32}P]$-dGTP. The digests were resolved by electrophoresis on 3.5%, 6% or 8% polyacrylamide gels (Figure 1). The bands common to all clones are the pBR322 MboI fragments and each clone is shown to contain one or, in some cases, a few mtDNA MboI fragment(s). Although MboI fragment 3b is slightly bigger than fragment 3a (not obvious on all gels), clones containing these were digested with KpnI and HindIII together with MboI to differentiate them. Accordingly, the pmt27 cloned insert is cut by KpnI while the pmt87 insert is cut by HindIII. Similarly, clones containing the two MboI fragments 6a and 6b were differentiated using double digests with MboI and XbaI. pmt172 contains four MboI fragments (5,9,16 and 18) while pmt98 (MboI 7 + 19), pmt151 (MboI 10 + 16) and pmt17 (MboI 18 + 7) each contain two. Clones containing MboI fragment 21 were not sought since such a short sequence will be easily determined by priming with neighborring fragments on the original mtDNA.

The mtDNA clones were further characterised by hybridisation (28) to a nitrocellulose filter blot (29) of an mtDNA digest with the four enzymes BamHI, HindIII, PstI and KpnI (Figure 2). This demonstrated complementarity of the cloned fragments to mtDNA and provided partial MboI restriction mapping information. It can be seen from Fig. 2 that pmtl DNA hybridises with the 1.75 kb BamHI/HindIII fragment (A) and with the HindIII/KpnI fragment (F) within which all of the 1.15 kb EcoRI fragment is contained. pmt2 DNA hybridises to the 0.9 kb HindIII fragment (B) that it contains (over-exposure showed hybridisation to partial digestion products). Since the hybridisation data only mapped the MboI fragments to the resolution of the eight fragments in the digest used, Figure 2, further restriction data was necessary.

Restriction Mapping with Exonuclease III A restriction site map for the enzyme MboI was deduced from the sequence of dissapearance of MboI fragments in exonuclease III-digested mtDNA. An example of such a mapping experiment is shown in Figure 3, where BamHl-cut HeLa mtDNA is digested with exonuclease III for various periods of time, restricted with

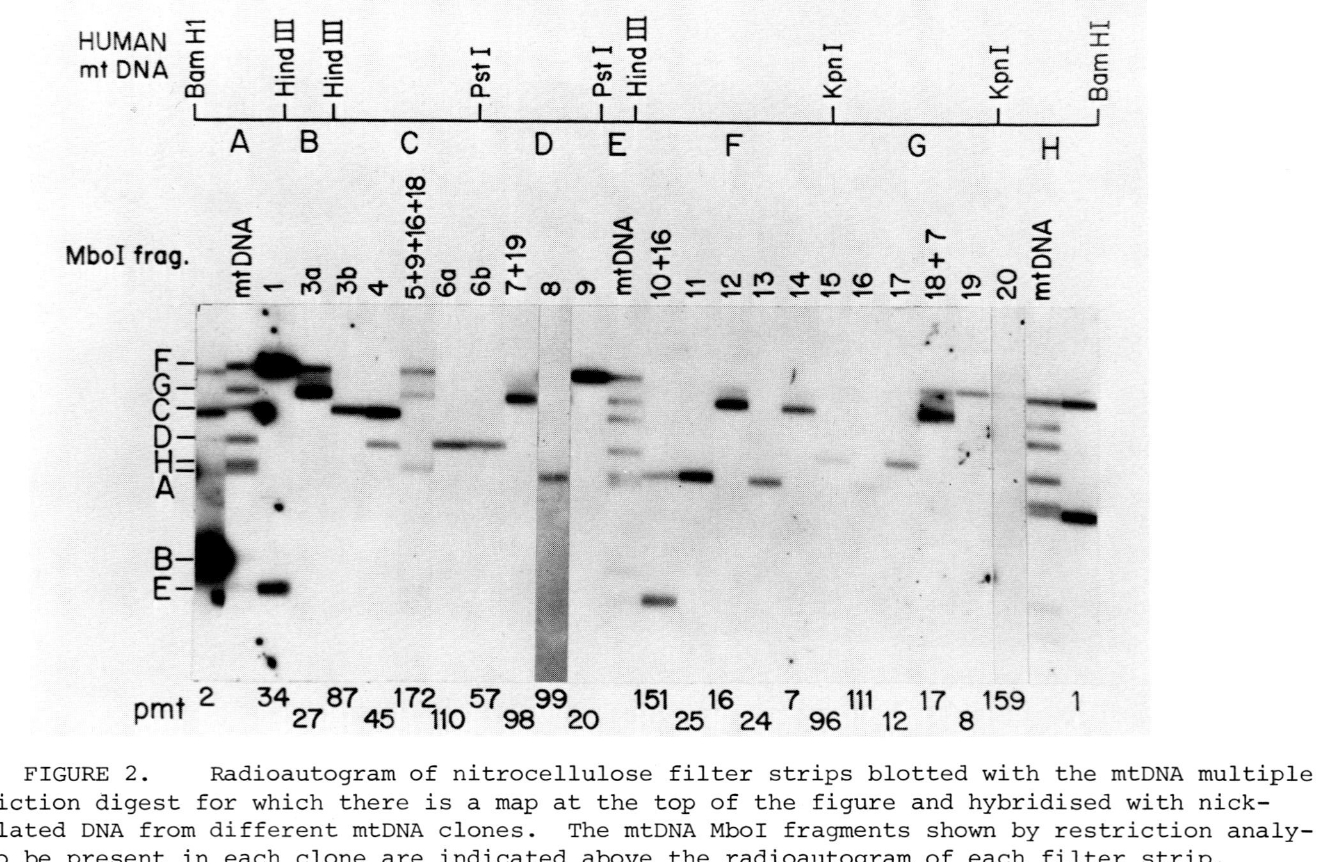

FIGURE 2. Radioautogram of nitrocellulose filter strips blotted with the mtDNA multiple restriction digest for which there is a map at the top of the figure and hybridised with nick-translated DNA from different mtDNA clones. The mtDNA MboI fragments shown by restriction analysis to be present in each clone are indicated above the radioautogram of each filter strip.

MboI and labelled with $\alpha[^{32}P]$-dGTP as previously. One minute after the addition of exonuclease III, MboI fragments 2 and 8 are the only two missing from the digest indicating that they are located on either side of the single BamHI site (which is also an MboI site). The following fragments can then be seen to sequentially dissapear from the digest: fragment 16 at 5 min., fragment 11 at 7.5 min., fragment 13 at 10 min., fragment 5 at 10-20 min., fragment 3b and 7 at 20 min. The sequence of the fragments is easily deduced from these data when combined with the hybridisation data (Figure 2) that shows on which side of the BamHI site each fragment lies. The samples at 20 and 30 min. have less DNA because exonuclease III had to be added at every 10 min. of incubation (the preparation of exonuclease III used in this experiment was not active after 10 min. at 37°C). Similar experiments were conducted on mtDNA cut with HindIII, EcoRI, KpnI and PstI (not shown) in order to deduce the restriction map shown in Figure 4. The 3'-single stranded ends produced by KpnI and PstI were digested away so that these would become substrates for exonuclease III by incubation of the restricted DNA for 5 min. at room temperature with the Klenow fragment of *E. coli* DNA polymerase I in the presence of one dNTP at 25 μM. The relative order of fragments 18, 19 and 21 shown in Figure 4 is the most likely one but the data supporting it is not completely unambiguous. The MboI restriction map shown here was found to be the same in both HeLa mtDNA and in the placental mtDNA preparation used for cloning.

DISCUSSION

In view of the early successes reported in cloning mouse mtDNA (13,30), we first attempted to clone the complete human BamHI-cut mtDNA and its large HindIII and EcoRI fragments but these attempts were not very successful. Only the small HindIII and EcoRI fragments were obtained, together with the small BamHI/EcoRI fragment. In similar attempts made elsewhere with mtDNA the 0.9 and 5.5 kb HindIII fragments were obtained but not the biggest HindIII fragment (Wesley M. Brown, personal comm.). Since subcloning of a large mtDNA recombinant was already envisaged to simplify sequencing, cloning of an mtDNA MboI digest into the BamHI site of pBR322 seemed a straightforward way of doing so. When this was done, most MboI fragments were found at a close to random frequency among the clones obtained except for fragments 2, 5 and 8. These fragments were found in one (5 and 8) or none (2) out of 700 independent clones whereas other MboI fragments were each present in 5-10% of the clones. The data indicate a slight preference for smaller inserts but the biggest MboI fragment

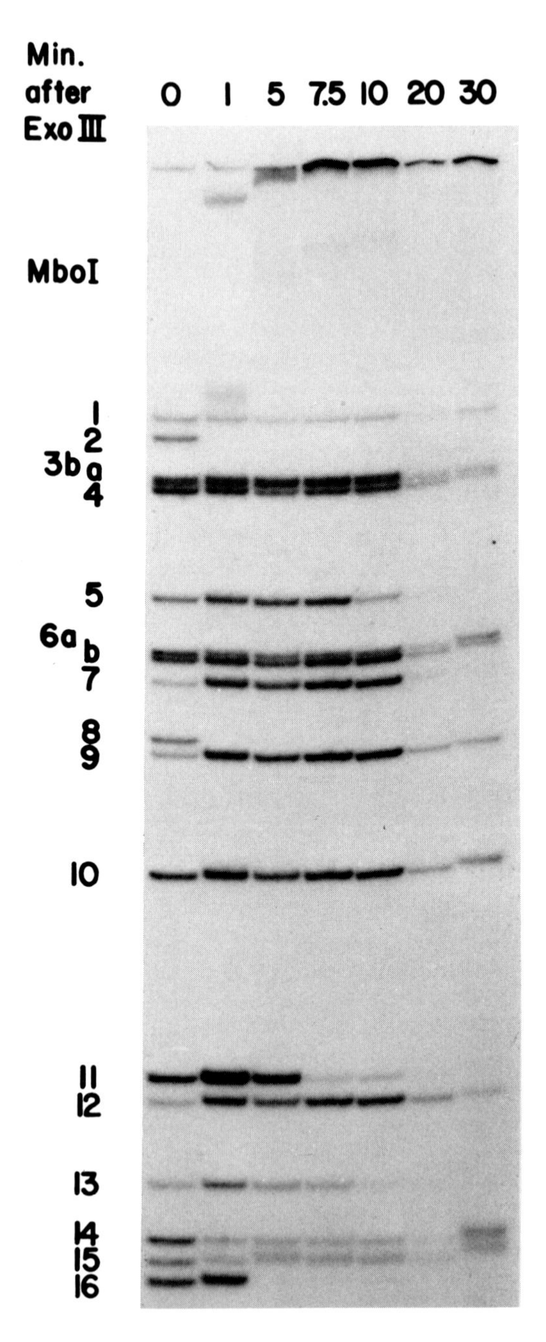

FIGURE 3. Exonuclease III mapping of the mtDNA MboI fragments. MboI fragments 17 to 21 were not digested by exonuclease III in this experiment (not shown).

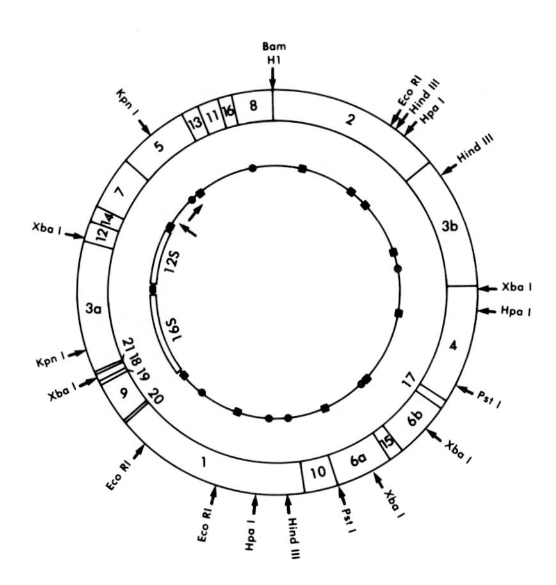

FIGURE 4. Human mtDNA MboI restriction map derived from the data presented. The map was aligned with the previously determined restriction sites for the enzymes EcoRI and HindIII (14,15) and with a functional map previously derived for HeLa mtDNA (7). The origin and direction of H strand synthesis are indicated, respectively, by the radial and tangential arrows inside the innermost circle (14,6). The position of the ribosomal (complementary to H-strand) and "4S" (■, complementary to H-strand, ●, complementary to L-strand) RNAs determined by Angerer et al., (7) are also shown. Fragments 18, 19 and 21 are shown in their likely relative order but more data is needed to confirm it.

(2,8 kb) was only cloned at half the frequency of the smaller ones. It therefore seems that some feature(s) present in MboI fragments 2, 5 and 8 hamper(s) either their *in vitro* recombination with pBR322 DNA, the ability of the recombinants to replicate in *E. coli*, or the viability of *E. coli* once such recombinants are introduced into the bacterium. It is unlikely that these fragments could not form circular recombinants *in vitro* since the other MboI fragments serve as internal controls in this respect. On the other hand, sequences within these three fragments could contain modified (or otherwise altered) nucleotides such that they would not be compatible with replication in *E. coli* or would be degraded within the bacterium before they could be replicated.

Arguing against this is the observation that the whole of MboI fragment 2 could be cloned in two pieces, i.e. the 1.7 kb BamHI/EcoRI (or BamHI/HindIII) and the 0.9 kb HindIII fragment. Bacteria transformed with recombinants containing these fragments could also have decreased plating efficiencies because of expression (in a large sense) of sequences they contain. This would be compatible with the observation that MboI fragment 2 could be cloned in two pieces but not whole.

The object of this work was to obtain a set of recombinant clones containing the entire human mtDNA sequence for its determination. Such a set was obtained with the exception of MboI fragment 21 (25 bp) and possibly fragments smaller than 15 bp which would have eluded analysis. The identity of the fragments cloned was verified by restriction analysis and hybridisation to human mtDNA. The precise orientation of each of these cloned fragments with respect to its neighbors will be firmly established by primed synthesis sequencing on an intact mtDNA template. Such analysis will also determine the position and sequence of MboI fragment 21 and any other that would not have been detected in the present work. It is hoped that this work will yield a complete DNA sequence for human mtDNA and that this sequence can be a basis for a functional understanding of the human mitochondrial genome.

ACKNOWLEDGMENTS

This work was done in the laboratory of Dr. Fred Sanger whom we thank for offering the facilities, many helpful discussions and above all a stimulating presence. The help of Jack L. Coote in preparing placental mtDNA was very useful. J.D. was supported by a Fellowship of the Medical Research Council of Canada.

REFERENCES

1. Borst, P. (1972). Ann. Rev. Biochem. 41, 333.
2. Kasamatsu, H., Robberson, D.L., and Vinograd, J. (1971). Proc. Natl. Acad. Sci. USA 68, 2252.
3. Robberson, D.L., and Clayton, D.A. (1972). Proc. Natl. Acad. Sci. USA 69, 3810.
4. Brown, W.M., Shine, J., and Goodman, H.M. (1978). Proc. Natl. Acad. Sci. USA 75, 735.
5. Gillum, A.M., and Clayton, D.A. (1978). Proc. Natl. Acad. Sci. USA 75, 677.
6. Crews, S., Ojala, D., Posakany, J., Nishiguchi, J., and Attardi, G. (1979). Nature 277, 192.
7. Angerer, L., Davidson, N., Murphy, W., Lynch, D., and Attardi, G. (1976). Cell 9, 81.

8. Perlman, S., Abelson, H.T., and Penman, S. (1973). Proc. Natl. Acad. Sci. USA 70, 350.
9. Hirsh, M. and Penman, S. (1973). J. Mol. Biol. 80, 379.
10. Amalric, F., Merkel, C., Gelfand, R., and Attardi, G. (1978). J. Mol. Biol. 118, 1.
11. Sanger, F., and Coulson, A.R. (1975). J. Mol. Biol. 94, 441.
12. Sanger, F., Nicklen, S., and Coulson, A.R. (1977). Proc. Natl. Acad. Sci. USA 74, 5463.
13. Brown, W.M., Watson, R.M., Vinograd, J., Tait, K.M., Boyer, H.W., and Goodman, H.M. (1976). Cell 7, 517.
14. Brown, W.M., and Vinograd, J. (1974). Proc. Natl. Acad. Sci. USA 71, 4617.
15. Robberson, D.L., Clayton, D.A., and Morrow, J.F. (1974). Proc. Natl. Acad. Sci. USA 71, 4447.
16. Brown, W.M., George, M. Jr., and Wilson, A.C. (1979). Proc. Natl. Acad. Sci. USA, in press.
17. Ojala, D., and Attardi, G. (1977). Plasmids 1, 78.
18. Bolivar, F., Rodriguez, R.L., Greene, P.J., Betlach, M.C., Heyneker, H.L., and Boyer, H.W. (1977). Gene 2, 95.
19. Norgard, M.V., Keem, K., and Monahan, J.J. (1978). Gene 3, 279.
20. Barnes, W.M. (1977). Science 195, 393.
21. Sanger, F., and Coulson, A.R. (1978). FEBS Letters 87, 107.
22. Sanger, F., Coulson, A.R., Friedmann, T., Air, G.M., Barrell, B.G., Brown, N.L., Fiddes, J.C., Hutchison, C. A., Slocombe, P.M., and Smith, M. (1978). J. Mol. Biol. 125, 225.
23. Sutcliffe, J.G. (1978). Nucleic Acids Res. 5, 2721.
24. Grunstein, M., and Hogness, D.S. (1975). Proc. Natl. Acad. Sci. USA 72, 3961.
25. Cami, B., and Kourilsky, P. (1978). Nucleic Acids Res. 5, 2381.
26. Ohlsson, R., Hentschel, C.C., and Williams, J.G. (1978). Nucleic Acids Res. 5, 583.
27. Lui, A.C.P., McBride, B.C., Vovis, G.F., and Smith, M. (1979). Nucleic Acids Res. 6, 1.
28. Jeffreys, A.J., and Flavell, R.A. (1977). Cell 12, 429.
29. Southern, E.M. (1975). J. Mol. Biol. 98, 503.
30. Chang, A.C.Y., Lansman, R.A., Clayton, D.A., and Cohen, S.N. (1975). Cell 6, 231.

QUANTITATION OF INTRAPOPULATION VARIATION BY RESTRICTION ENDONUCLEASE ANALYSIS OF HUMAN MITOCHONDRIAL DNA[1]

Wesley M. Brown[2] and Howard M. Goodman

Department of Biochemistry and Biophysics
University of California
San Francisco, California 94143

ABSTRACT Mitochondrial DNA can be obtained with relative ease from individuals. Multiple samples can be quickly analysed using gel electrophoretic analysis of restriction endonuclease digests of the DNA. Mitochondrial DNA samples from 21 humans of diverse racial background were analysed using 8 restriction endonucleases. The mitochondrial DNA population from each individual appeared to be highly homogeneous. By contrast, mitochondrial DNA from some of the individuals were observed to differ from one another in their nucleotide sequence. Fourteen of the 21 samples appeared identical to one another in the 64 restriction sites examined. The remaining seven samples differed at one or more restriction sites from the wild type and from each other. We calculate that mitochondrial DNA from randomly chosen pairs of humans will differ on an average at 65 base pairs per mitochondrial genome, or at one base pair out of 250. This degree of difference exceeds that observed in studies of vertebrate single copy nuclear DNA and is consistent with other evidence indicating that the rate of evolution of mitochondrial DNA is rapid. Mitochondrial DNA analyses will thus be especially valuable for studies of genetic variability within and among populations and among closely related species.

[1]This work was supported by National Science Foundation Grants BMS75-10468 (to HMG) and DEB76-20599 (to WMB).
[2]Present address: Department of Biochemistry, University of California, Berkeley, California 94720.

ISBN 0-12-198780-9

INTRODUCTION

The mitochondrial genome of humans is a closed circular duplex DNA molecule of approximately 16,500 nucleotide pairs (1). Among multicellular animals the genome size, relative to the size of Øχ174 RFDNA, ranges from 15,000 to 18,000 nucleotide pairs (1-3). The relative positions of the mitochondrial origin of replication and the large and small mitochondrial ribosomal genes have been shown to be the same in animals as evolutionarily diverged as amphibians (4), and mammals, including humans (5,6). It is likely that the relative order of the remaining mitochondrial genes, whose functions are presently unknown, will also prove to be highly conserved, at least among the vertebrates (7). However, despite the relative uniformity of genome size and possible conservation of gene order, mitochondrial DNA (mtDNA) has been shown to be a rapidly evolving molecule with regard to its nucleotide sequence, having an average rate of nucleotide substitution 5 to 10 times as fast as that of single copy nuclear DNA (3). Because of this rapid rate of evolution and because of the relative ease of its preparation in highly purified form from individuals, we regard mtDNA as a useful molecule to employ in assessing the degree of genetic variability within a population. The human species was chosen for study because it is genetically outbreeding and because large amounts of mtDNA could be obtained from each individual (from placental tissue) thus enabling many restriction endonuclease analyses to be performed per sample. However, the same techniques have been successfully applied to individual samples from much smaller animals (7), and the use of radioactive labeling methods can make multiple analyses on even the smallest animals possible (W. Brown, manuscript in preparation).

In order to assess the amount of variation present among mtDNA samples we have digested them with class II restriction endonucleases, which cleave DNA molecules at specific base sequences (8,9). The locations of these sequences in the mitochondrial genome can be determined and a physical map of the genome constructed. The maps can be related to each other by multiple enzyme digests or by orienting each map relative to any recognizable feature of the genome that is uniquely located. Some animal mtDNAs contain a structure, the D-loop, that is present at the origin of replication and can be seen with the electron microscope (10). This structure has been used to align the cleavage maps of human (HeLa Cell) mtDNA for a number of restriction endonucleases (1,11; W. Brown, M. George, S. Ferris, H.M. Goodman and A. Wilson, manuscript in preparation).

Human mtDNA samples have been obtained from 21 individuals of diverse racial background. The restriction endonuclease fragment patterns yielded by digestion with eight different enzymes have been analyzed by gel electrophoresis. The patterns with each enzyme have been compared among the samples. An estimate of the average number of differences between individuals has been obtained. From the cleavage maps of these enzymes for human mtDNA the variant positions have been determined. The results indicate that restriction endonuclease analyses of mtDNA will be a precise and valuable method to employ in studies of genetic variability within and among populations and among closely related species.

MATERIALS AND METHODS

DNAs. Closed circular mtDNAs from HeLa cells (strain S3) and from human placentas were prepared as described (1,2). The placentas were from individuals of the negroid, mongoloid and caucasoid races as indicated in Table 1. Bacteriophage λ DNA was prepared according to the method of Hedgepeth et al. (12). Bacteriophage PM2 DNA was a gift from R.M. Watson.

Enzymes and Enzyme Digestion. Abbreviations for the names of the restriction endonucleases are those used by Roberts (9). The enzyme *Eco*RI was a gift from Drs. Pat Greene and Herbert Boyer. The enzymes *Hinc*II, *Hind*III, *Pst*I and *Xba*I were purified from bacterial strains supplied by Dr. Richard Roberts. The identities of the enzymes obtained were verified using DNA substrates whose digestion patterns with the enzymes are known. The enzymes *Bam*HI, *Hha*I, and *Hpa*II were purchased from New England Biolabs (Beverly, Mass.). Digestion conditions for the enzymes have been described (9,13).

Electrophoresis of DNA. The electrophoretic analysis of DNA fragments in vertical slab gels of 1.2% and 2% agarose (Seakem, ME) and of 5% polyacrylamide, respectively, was performed as described by Brown et al. (14) and by Maniatis et al. (15). Fragment sizes were estimated using the *Eco*RI fragments of λ DNA (16) and the *Hind*III fragments of PM2 DNA (7,17,18) as size standards. A Joyce-Loebl densitometer was used to scan the negatives of gel photographs. No attempt was made to detect fragments smaller than ∿50 base pairs.

RESULTS

Gel Analysis of Restriction Endonuclease Cleaved Human mtDNAs. The patterns produced by agarose gel electrophoresis of digests of the mtDNAs with *Hinc*II endonuclease were identical for twenty of the twenty-one samples and consisted of a series of eleven bands (A through K, Figure 1). One

TABLE I

RACE AND COUNTRY OF ORIGIN OF MOTHER[a]

mtDNA Sample #	Race[b]	Birthplace
1	W	U.S.
2	W	Phillipines
3	O	China
4	W	Phillipines
5	O	China
6	W	U.S.
7	B	U.S.
8	B	U.S.
9	W	U.S.
10	O	China
11	W	U.S.
12	W	U.S.
13	W	Egypt
14	O	China
15	B	U.S.
16 - 20	W	U.S.
21	B	U.S.

[a]Data obtained from medical admissions forms.

[b]W = white, B = black, O = oriental

sample (#3) lacked two of the bands (B and H) characteristic of the remaining samples and showed one new band that ran slightly ahead of band A. The estimated size of the DNA fragment in this band was 3000 base pairs, close to the value expected for a fragment that would be produced by the fusion of the fragments in bands B (2240 base pairs) and H (770 base pairs). Some of the lanes in figure 1 exhibit faint bands in addition to the major bands. Redigestion of the correspon-

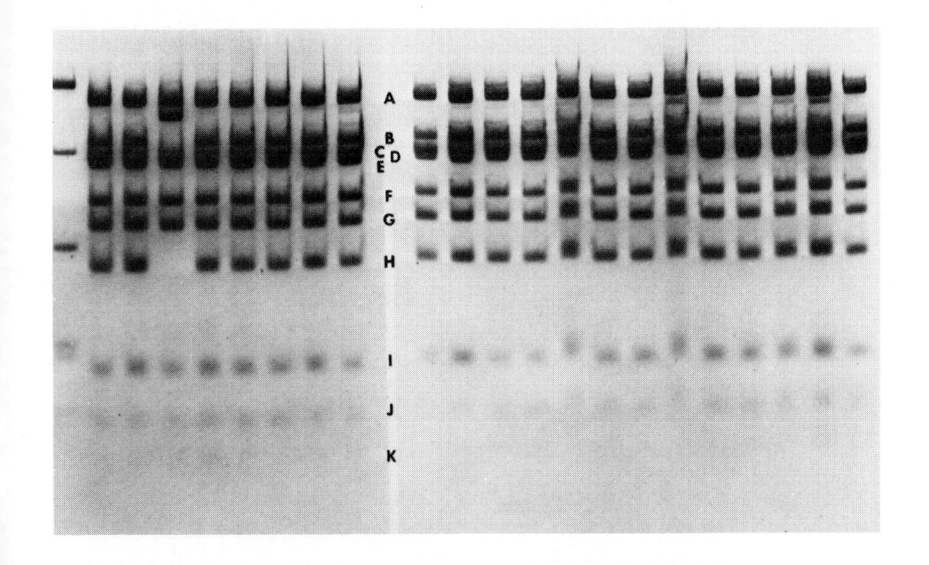

FIGURE 1. Gel electrophoresis patterns of HincII digested human mtDNAs in 2% agarose. Lane 1 contains HindIII digested PM2 DNA as a size standard. Lanes 2-22 contain mtDNA samples 1-21, respectively. Lanes are numbered from left to right. The K band, not seen in this reproduction, was clearly visible in the gel and on the negative of the gel photograph. Fragments C, D, and E, not well resolved in this gel, are usually quite distinct (data not shown).

ding samples under more exhaustive conditions resulted in the reduction of the faint bands below the limits of detectability (data not shown).

The patterns produced by agarose gel electrophoresis of each mtDNA sample after combined digestion with EcoRI, BamHI and PstI showed identity among 19 of the 21 samples, figure 2. The "type" pattern (e.g. sample #1, figure 2) consisted of five bands (A through E). The fourth band (D) was especially intense, indicating the presence of more than one fragment as shown by the densitometer trace in figure 3. Two samples

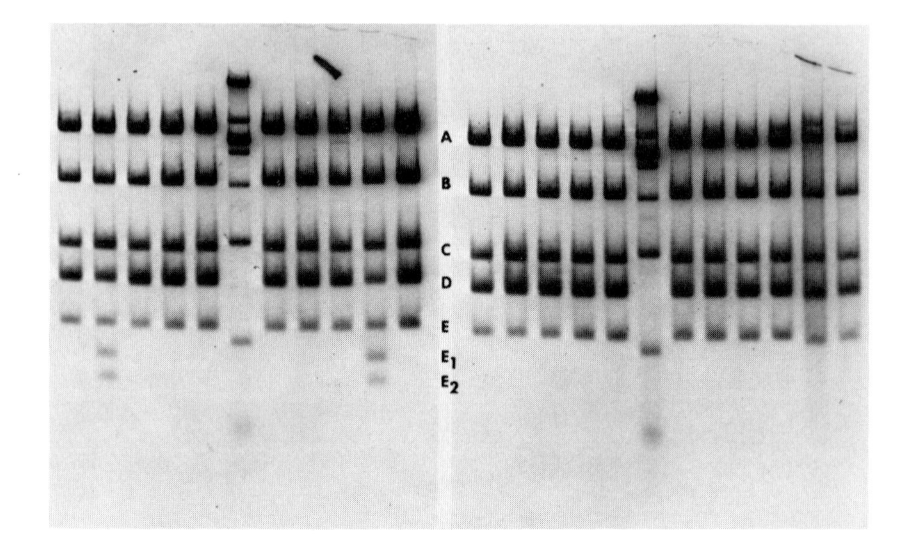

FIGURE 2. Gel electrophoresis patterns of EcoRI, BamHI, PstI digested human mtDNAs in 1.2% agarose. Lanes 6 and 17 contain HindIII digested PM2 marker DNA. Lanes 1-5, 7-16 and 18-23 contain mtDNA samples 1-5, 6-15, and 16-21, respectively. Lanes are numbered from left to right.

(#2 and #9, figure 2) showed a loss of one of the two fragments comprising band D and a gain of two new fragments, E_1 and E_2. The loss of one of the fragments from band D, suggested by visual inspection of the gel photograph, figure 2, was confirmed by densitometric analysis, figure 3. A comparison of the electrophoretic mobilities of the fragments relative to those of the size standards yielded size estimates of 830 base pairs for E_1 and 680 base pairs for E_2. The combined size of E_1 and E_2 (1510 base pairs), is the same as the estimated size for D and indicates that E_1 and E_2 were derived

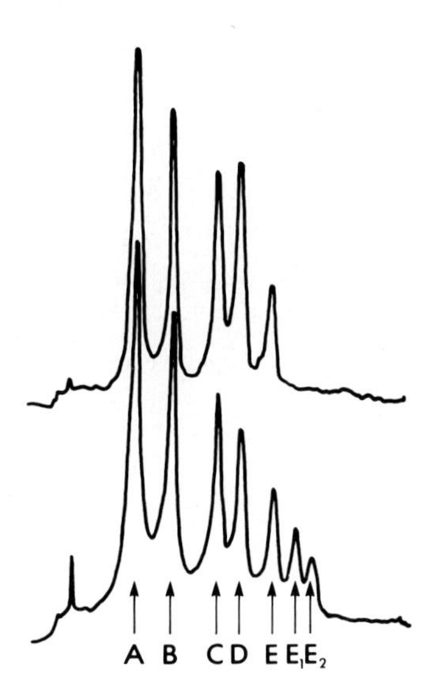

FIGURE 3. Densitometer scans of EcoRI + BamHI + PstI digests of human mtDNAs. The lanes scanned were those containing samples #1 (upper trace) and #9 (lower trace) of the gel negative shown in figure 2.

from D. Single enzyme digests of samples #2 and #9 with EcoRI and with PstI gave patterns identical to the respective "type" patterns, indicating that these enzymes were not responsible for the additional cleavage. Digestion with BamHI, which typically produces only one band, produced two bands in each of the samples #2 and #9, figure 4. The estimated size of the faster band, identical in samples #2 and #9, was 840 base pairs, corresponding in size to band E_1 of figure 2. Combined digestion of the samples with BamHI and PstI produced three bands with HeLa mtDNA and four with samples #2 and #9. Analysis of these patterns, figure 4, indicated that the additional BamHI cleavage site was contained in the B fragment (5060 base pairs), producing fragments of 4150 (B_1) and 840 (C_1) base pairs.

No atypical patterns were observed in the digests with the endonucleases EcoRI, HindIII, PstI and XbaI. Two samples (#2 and #7) provided atypical patterns when digested with HpaII, which typically produced 22 fragments in human mtDNA.

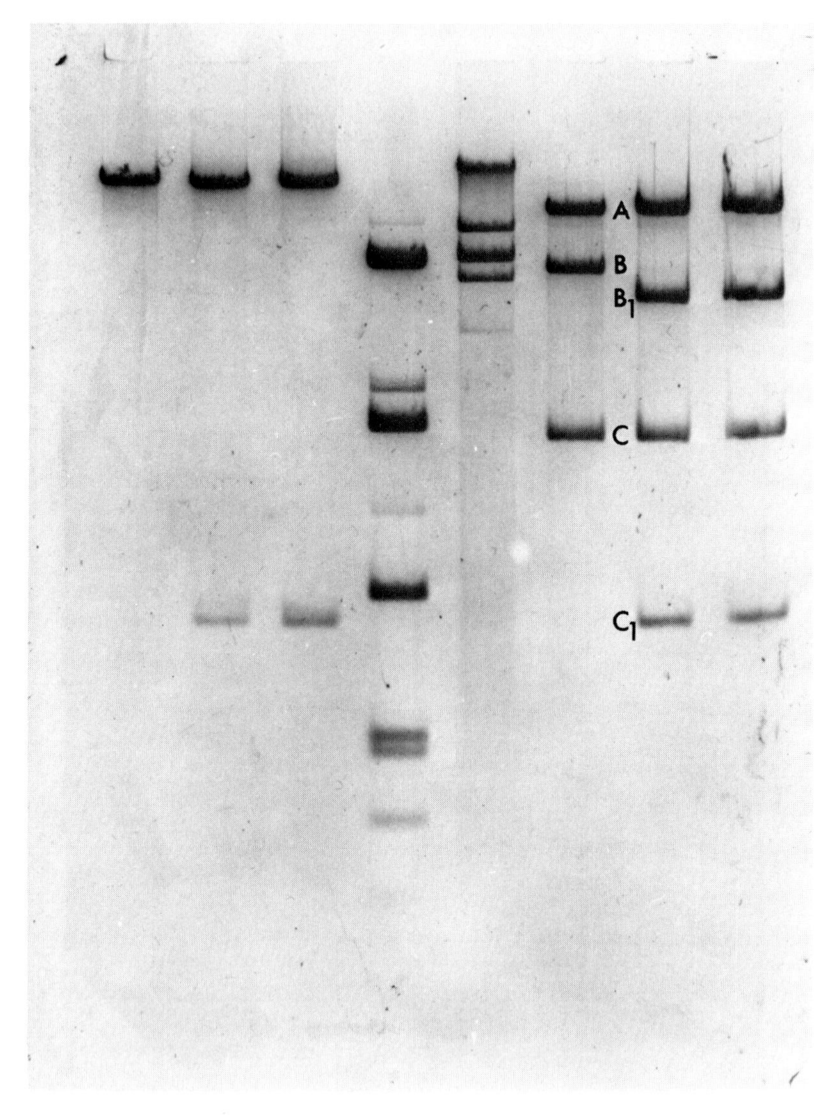

FIGURE 4. Gel electrophoresis patterns of BamHI and BamHI + PstI digested human mtDNAs in 1.2% agarose. In order, lanes 1-3 contain BamHI digests of mtDNAs from HeLa and samples #2 and #9; lanes 4 and 5 contain PM2, HindIII and λ, EcoRI digests; lanes 6-8 contain BamHI + PstI digests of mtDNAs from HeLa and samples #2 and #9.

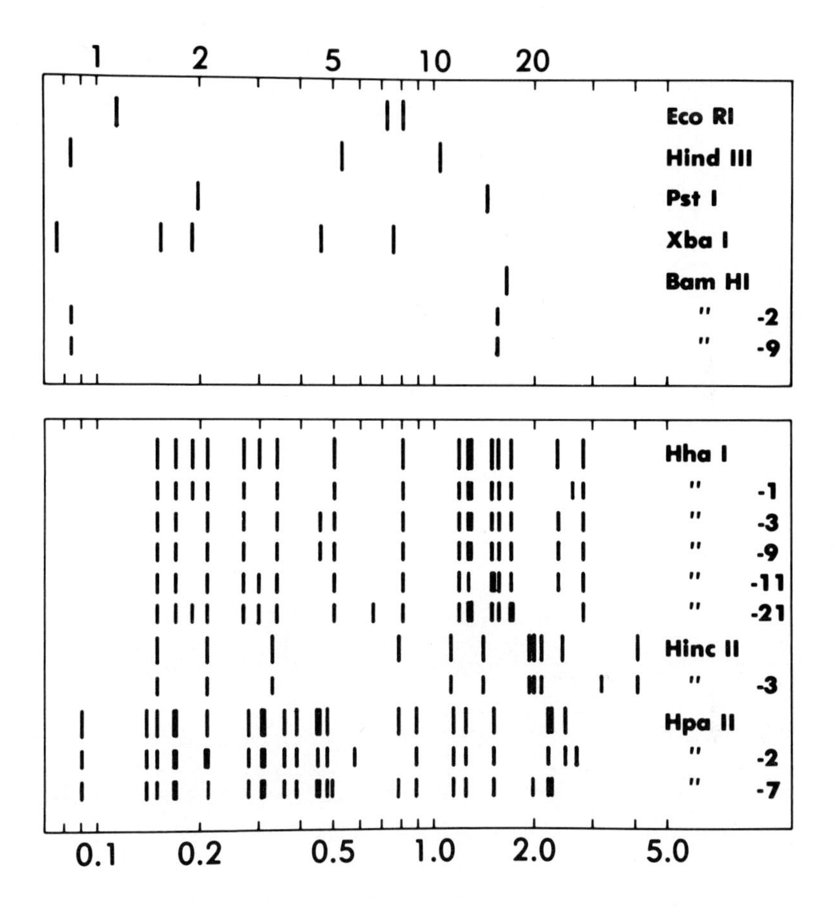

FIGURE 5. Diagramatic representation of gel electrophoresis patterns obtained by restriction endonuclease digestions of human mtDNAs. The enzymes which gave rise to the patterns are indicated at the top of the figure; lanes headed with the name of an enzyme show the most common, or "type" patterns; lanes headed by ditto marks followed by a sample number show the atypical patterns obtained with those samples. The ordinates are in base pairs x 10^{-3}. The fragment sizes are based on data from multiple electrophoresis runs in 1.2% and 2% agarose gels and (for the HpaII fragments) in 5% polyacrylamide gels.

Sample #2 showed two changes, one a site loss and the other a site gain. Sample #7 showed a site gain. Five samples (numbers 1,3,9,11 and 21) exhibited atypical patterns when digested with HhaI. This enzyme produced 17 cleavages in the majority of the samples. Four of the five atypical patterns involved site losses, the same site being lost in two of the samples (#3 and #9). The fifth was a site gain, in HeLa mtDNA (#21). This was the only enzyme digest in which HeLa mtDNA gave an atypical pattern. These results are summarized diagrammatically in figure 5.

DISCUSSION

Locations of the Altered Sites in the Mitochondrial Genome. The positions of the altered sites have been determined from cleavage maps for the restriction endonuclease recognition sites, figure 6 (1,11; W. Brown, M. George, S. Ferris, H. Goodman and A. Wilson, manuscript in preparation). In all cases, the interpretations of the variant patterns with site losses based on size data from gels (see Results) are consistent with the independently determined locations of the cleavage sites. For example, the fusion of the fragments in the HincII bands B and H by loss of a common site is consistent with the adjacent location of these two fragments in the HincII cleavage map, figure 6. In all but two cases, the positions of the variant site gains have been mapped. For example, data from the BamHI, PstI double digests of mtDNA samples #2 and #9 (figure 4, lanes 7 and 8) combined with the map positions of the "type" BamHI and PstI sites indicates that the position of the additional BamHI site in both of these samples is at 0.20 in the map. In two cases, both involving HpaII site gains, the positions of the variant sites were not determined. As a consequence, two locations for each site are equally possible, since any new asymmetric site that appears within a restriction fragment has two possible locations that cannot be resolved without more information.

The clustering of altered sites in the genome (from ∿0.96 G to ∿0.46 G) does not reflect clustering of the typical restriction recognition sites. For the enzymes used in this survey there are typically 64 restriction sites in the human mtDNA genome, 27 of which occur in the map interval 0.96 to 0.46 and 37 in the interval 0.46 to 0.96. The significance, if any, of the clustering of the altered sites cannot be assessed on the basis of these data.

Estimate of the Heterogeneity of mtDNA among Individuals. We have observed individual variation in the mtDNA sites cleaved by restriction endonucleases, in agreement with

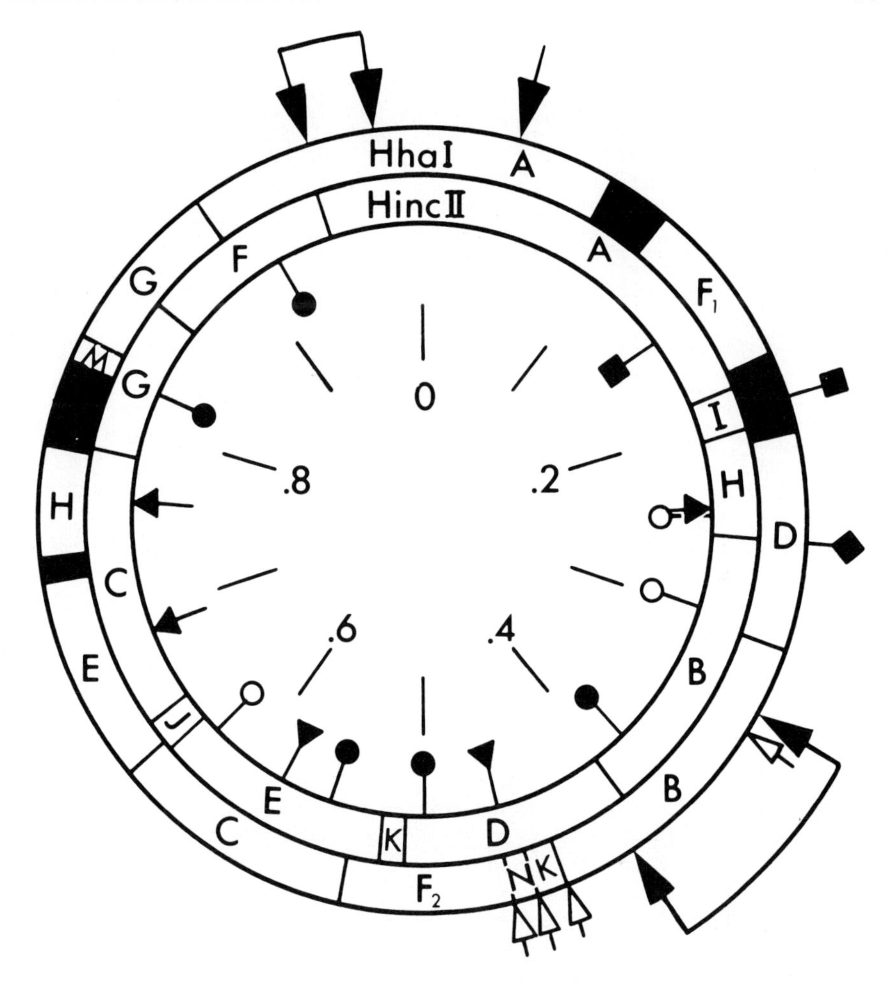

FIGURE 6. The normal and altered restriction endonuclease cleavage sites in human mtDNA. The normally occurring sites for BamHI (■), EcoRI (↓), XbaI (●), HindIII (○), and PstI (▼) are shown inside the inner circle. The fragment maps for HincII and HhaI (Brown and Goodman, unpublished) are shown within the concentric circles. Fragments A through K of the HincII map correspond to the similarly labeled bands of figure 1. The altered sites for BamHI (■), HincII (◆), HhaI (▽), and HpaII are shown outside the circles. The ambiguity of the positions of two of the altered HpaII sites (see text) is indicated by the linked arrows. The mtDNA origin of replication is located at map position zero, with the fraction of the genome increasing in the direction of mtDNA replication. The filled in portions of the HhaI map represent the locations of fragments I,J,L,O and P, whose exact positions have not been determined.

published reports (19-22). Our results indicate that random pairs of individuals differ, on an average, at 65 of the 16,500 nucleotide positions in their respective mtDNAs. However, the amount of variation that we have observed in humans is significantly less than the amounts reported by Upholt and Dawid (1977) in a comparison among 3 sheep and between 2 goats. One possible explanation for this difference is that the amount of variation observed may be related to the length of time since the occurence of the most recent population constriction ("bottleneck") in the species. If such an event occurred at different times for different species, the species would exhibit different degrees of heterogeneity.

Thus far, the observed variation has been attributed to sequence changes resulting in either the loss of recognition sites or the creation of new ones. While this interpretation is probably correct, it should be noted that the methylation of a base within a site could also cause a site loss and the lack of a usual methylation could cause a site gain. The maximum amount of base methylation in mtDNA from mouse and hamster cells has been estimated to be 0.2% and 0.6%, or 12 and 36 bases per mtDNA, respectively (23). It is reasonable to expect to find a similarly low magnitude of mtDNA methylation in humans. The sequence specificity of nucleotide methylation among vertebrates, including humans, has been found to be highly conserved (24) and a single enzymatic species may be responsible for the methylation (25). Comparative studies, each using several methods of analysis, have demonstrated that the rate of evolutionary divergence of mtDNA sequences among vertebrate species is very rapid (2,3,26). One could reasonably expect this to be reflected within a species. These factors suggest that variation in the methylation of mtDNA that is not caused by a sequence change is unlikely and that the method of restriction endonuclease analysis furnishes data that are directly related to nucleotide substitutions.

No genome rearrangements (inversions, translocations) were detected among the samples. All of the atypical patterns observed can be explained by single base changes. Taking the 64 "type" sites and 4 additional atypical site gains into account we have examined 313 out of 16,500 base pairs-1.9% of the mitochondrial genome in each of the 21 samples. We found 11 recognition site changes out of 313 x 21 = 6573 bases, or an average of 1.7×10^{-3} changes per base pair, assuming one base change per recognition site loss or gain. Of the 21 mtDNA samples examined, 14 showed no site alterations, 5 showed one and 2 showed two. In two of the digests, two samples showed alterations at the same site (though not necessarily at the same base within the site). In a total of eleven site alterations, six were losses and five were gains. These results are summarized in Table 2.

TABLE 2

ANALYSIS OF HUMAN mtDNAs WITH EIGHT RESTRICTION ENDONUCLEASES

Enzyme	Recognition Sites Sequence[a]	No./mtDNA[b]	Atypical Sample Nos.	Site Gain (+) or Loss (-)
EcoRI	GAATTC	3	-	
HindIII	AAGCTT	3	-	
XbaI	TCTAGA	5	-	
PstI	CTGCAG	2	-	
BamHI	GGATCC	1	2,9	+,+
HincII	GTPyPuAC[c]	11	3	-
HpaII	CCGG	22	2,7	+-,+
HhaI	GCGC	17	1,3,9,11,21	-,-,-,-,+

[a] Recognition sequences from Table 2 of Roberts (9) and, for XbaI, from Zain & Roberts (13).

[b] This is the "type" number, seen in the majority of the samples.

[c] Because of the Py-Pu ambiguity, the HincII site is treated as a pentanucleotide site for quantitative purposes.

Because of the maternal mode of inheritance of mtDNA (22,27,28) each restriction pattern difference in the present sample probably represents a different matrilineal clone. If a larger sample were analyzed it is possible that group specific patterns would emerge. No such grouping is apparent in this study. In two cases (samples #2 and #9 with BamHI and #3 and #9 with HhaI) the same site has been altered in two different samples. The occurence of the same sample, #9, in both cases suggests the possibility of a genetic relationship among the three samples. However, since #2 lacks the altered HhaI site common to #3 and #9, and #3 lacks the BamHI site common to #2 and #9, the samples can only be related by invoking a back mutation of one of the sites in sample #2 and #3 after the mutation of the second site. Since the exact base altered in either site is unknown, and possibly different among the three samples in one or both cases, the present data does not distinguish between a polyphyletic and a monophyletic origin of the alteration.

Two BamHI sites occur at the same relative locations in the cleavage map of chimpanzee mtDNA as the two BamHI sites in human mtDNA samples #2 and #9 (W. Brown, unpublished data). From the close phylogenetic relationship between chimpanzees and humans (29) such a correlation is not surprising. Comparative restriction site data has been used to estimate the overall rate of evolution of the mitochondrial genome (3) and may provide information about the relative rates of evolution of different portions of the genome. It may also contribute to a better quantitative understanding of the relationships among closely related species of animals.

These results demonstrate the usefulness of restriction endonuclease analysis of mtDNA in estimating the amount of genetic variation present in a population. The analyses are both rapid and quantitative. The method is sensitive enough to allow the detection of heterogeneity within, as well as among, samples. We observed no heterogeneity within samples, but the possible ocurrence of this has been reported by Potter et al. (19). If such heterogeneous individuals can be found in a species with which breeding experiments can be performed, the type of analysis we have employed might also provide insights into the details of cytoplasmic inheritance among multicellular organisms.

ACKNOWLEDGMENTS

We wish to thank Allan C. Wilson for many helpful comments during the preparation of this manuscript. The technical assistance of Fran DeNoto and Dave Russel is gratefully acknowledged, as is the cooperation of the staff in the delivery room of the U.C. Medical Center.

REFERENCES

1. Brown, W.M., and Vinograd, J. (1974). Proc. Nat. Acad. Sci. USA 71, 4617-4621.
2. Brown, W.M. (1976). Ph.D. Thesis. California Institute of Technology.
3. Brown, W.M., George, M., Jr., and Wilson, A.C. (1979). Proc. Nat. Acad. Sci. USA 76, no. 4.
4. Dawid, I.B., Klukas, C.K., Ohi, S., Ramirez, J.L., and Upholt, W.B. (1976). In "The Genetic Function of Mitochondrial DNA" (Saccone, C., and Kroon, A.M., eds), pp. 3-13. North Holland: Amsterdam.

5. Attardi, G., Albring, M., Amalric, F., Gelfand, R., Griffith, J., Lynch, D., Merkel, C., Murphy, W., and Ojala, D. (1976). In "Genetics and Biogenesis of Chloroplasts and Mitochondria" (Bucher, T., et al., eds.), pp. 573-585. Elsevier/North Holland: Amsterdam.
6. Battey, J., and Clayton, D.A. (1978). Cell 14, 143-156.
7. Brown, W.M., and Wright, J.W. (1979). Science 203, 1247-1249.
8. Nathans, D., and Smith, H.O. (1973). Annu. Rev. Biochem. 44, 273-293.
9. Roberts, R.J. (1976). C.R.C. Crit. Rev. Biochem. 3, 123-164.
10. Kasamatsu, H., Robberson, D.L., and Vinograd, J. (1971). Proc. Nat. Acad. Sci. USA 68, 2252-2256.
11. Ojala, D., and Attardi, G. (1977). Plasmid 1, 78-105.
12. Hedgepeth, J., Goodman, H.M., and Boyer, H.W. (1972). Proc. Nat. Acad. Sci. USA 69, 3443-3452.
13. Zain, B.S., and Roberts, R.J. (1977). J. Mol. Biol. 115, 249-255.
14. Brown, W.M., Watson, R.M., Vinograd, J., Tait, K.M., Boyer, H.W., and Goodman, H.M. (1976). Cell 7, 517-530.
15. Maniatis, T., Jeffrey, A., and Van deSande, H. (1975). Biochem. 14, 3787-3794.
16. Thomas, M., and Davis, R.W. (1975). J. Mol. Biol. 91, 315-328.
17. Brack, C., Eberle, H., Bickle, T.A., and Yuan, R. (1976). J. Mol. Biol. 104, 305-309.
18. Parker, R.C., Watson, R.M., and Vinograd, J. (1977). Proc. Nat. Acad. Sci. USA 74, 851-855.
19. Potter, S.S., Newbold, J.E., Hutchinson, C.A., III, and Edgel, M.H. (1975). Proc. Nat. Acad. Sci. USA 72, 4496-4500.
20. Upholt, W.B., and Dawid, I.B. (1977). Cell 11, 571-583.
21. Francisco, J.F., and Simpson, M.V. (1977). FEBS Letters 79, 291-294.
22. Buzzo, K., Fouts, D.L., and Wolstenholme, D.R. (1978). Proc. Nat. Acad. Sci. USA 75, 909-913.
23. Nass, M.M.K. (1973). J. Mol. Biol. 80, 155-175.
24. Browne, M.J., and Burdon, R.H. (1977). Nucleic Acids Res. 4, 1025-1037.
25. Browne, M.J., Turnbull, J.F., McKay, E.L., Adams, R.L.P., and Burdon, R.H. (1977). Nucleic Acids Res. 4, 1039-1045.
26. Dawid, I.B. (1972). Devel. Biol. 29, 139-151.
27. Dawid, I.B., and Blackler, A.W. (1972). Devel. Biol. 29, 152-161.
28. Hutchinson, C.A., III, Newbold, J.E., Potter, S.S., and Edgell, M.H. (L974). Nature 251, 536-538.
29. King, M.-C., and Wilson, A.C. (1975). Science 188, 107-116.

RESTRICTED MITOCHONDRIAL DNA FRAGMENTS AS GENETIC MARKERS IN CYTOPLASMIC HYBRIDS

Ching Ho and Hayden G. Coon

Laboratory of Cell Biology, National Cancer Institute
National Institutes of Health, Bethesda, Md. 20205

ABSTRACT Cytoplasmic hybrids (cybrids) were made by fusing the chloramphenicol resistant (CAP-r) cytoplast of one mouse species with the chloramphenicol sensitive (CAP-s) whole cell of another mouse species or subspecies. The mitochondrial DNA (mt-DNA) of parents and cybrids was analysed by restriction endonuclease digestion and electrophoresis. Two types of cybrids were found: 1) those that maintained a balanced combination of endogenous, CAP-s and alien, CAP-r mt-DNA, and 2) those that showed repopulation by the alien CAP-r mt-DNA. When cybrids of the first type were returned to medium without CAP, the CAP-r phenotype segregated concordantly with the mt-DNA of the CAP-r subspecies, indicating that they are linked. In cybrids that showed repopulation, no segregation was observed, indicating that substitution was probably complete. No evidence of recombination was found. The repopulated heterospecific cybrids may be useful in determining which polypeptides are encoded by the mt-DNA. When CAP-s cells were treated with purified mt-DNA from CAP-r strains, CAP-s cells were efficiently "transformed" to CAP-r. Only the endogenous CAP-s mt-DNA was detected in these transformants. Fine structure restriction analysis suggests that small regions of the CAP-s mt-DNA may be altered, perhaps by incorporation of some CAP-r mt-DNA.

INTRODUCTION

Mitochondrial genetics has received considerable attention in recent years, since the organelle contains what is presumably the simplest mammalian chromosome and a pattern of inheritance distinguishable from and at least partially independent of the nucleus (1-4). Some basic features of mammalian mitochondria have emerged. They represent about 0.1% of the DNA in mammalian cells. The 15 Kb DNA is packaged as a 5 μm supercoiled circle and contains both GC and AT rich

ISBN 0-12-198780-9

regions but no repetitive sequences. These mitochondrial chromosomes replicate within the organelle and initiate at a fixed origin using enzymes which are uniquely associated with the organelle, but are probably encoded in the nucleus (5,6). Mitochondria contain their own molecular apparatus for the transcription of the genome and translation of messages. Genes for about 20 t-RNAs, ribosomal RNA and 9-12 proteins have been localized on the yeast mitochondrial DNA (mt-DNA), which is 5 times larger than mammalian cell mt-DNA. The maps of t-RNA, ribosomal RNA, and a few mRNA locations have been determined in yeast and in certain mammalian cells by heteroduplex analysis and electron microscopy (7-11). Mitochondria rely on up to 90 different nuclear proteins for the continuity of the organelle. Thus, the organelle genome and the nuclear genome must interact extensively during the biogenesis of the organelle. For example, the nucleus provides 4 of the 7 cytochrome oxidase subunits, 6 out of 7 cytochrome bc_1 subunits, and subunit number 7 of 10 subunits of ATP-ase, all of which are proteins found in the inner membrane of yeast mitochondria (12,13). Although precise information is not yet available for mammalian mitochondria, there are experiments which suggest that they have also retained a similarly interdependent relationship with the cell's nucleus (14,15,16).

A major problem that has confronted investigators interested in mammalian mitochondrial biogenesis is that obtaining stable cytoplasmic mutants has proven to be technically difficult. Eisenstadt and coworkers (17) first demonstrated chloramphenicol resistant (CAP-r), cytoplasmically transmissible mutants in mammalian cells. Since their success, others have been able to get CAP-r markers, but additional mutants useful in cell selection techniques have remained especially difficult to obtain (18,19). Perhaps the perculiar dynamics of CAP sele tion make the CAP-r markers especially well suited to the task of cell selection. Considering that the average cell contains 50-150 mitochondria and 1000-2000 copies of mt-DNA, a single mutation in one organelle chromosome might not be expected to sustain the cell through the crisis of intense immediate selection. Moreover, if th selective pressureinhibits mitochrondrial biogenesis, the cells ma not survive in culture until the mutant can expand sufficiently in the mitochrondrial population. However, CAP sensitivity (CAP-s) is cell density dependent and possibly related to glucose metabolism as well (20,21). Furthermore, mouse cells that are CAP-s may divid 8-10 times in CAP medium before dying. These attributes interact to ease the burden of most CAP selection regimes and perhaps they help to explain why the CAP-r mutants are the ones most used today

We have established a set of CAP-r mutants in tissue culture cells from different species and subspecies of mouse. The subspeci have mt-DNAs that are easily distinguished by the endonuclease digestion products listed in Table 1. The CAP mutants that we have

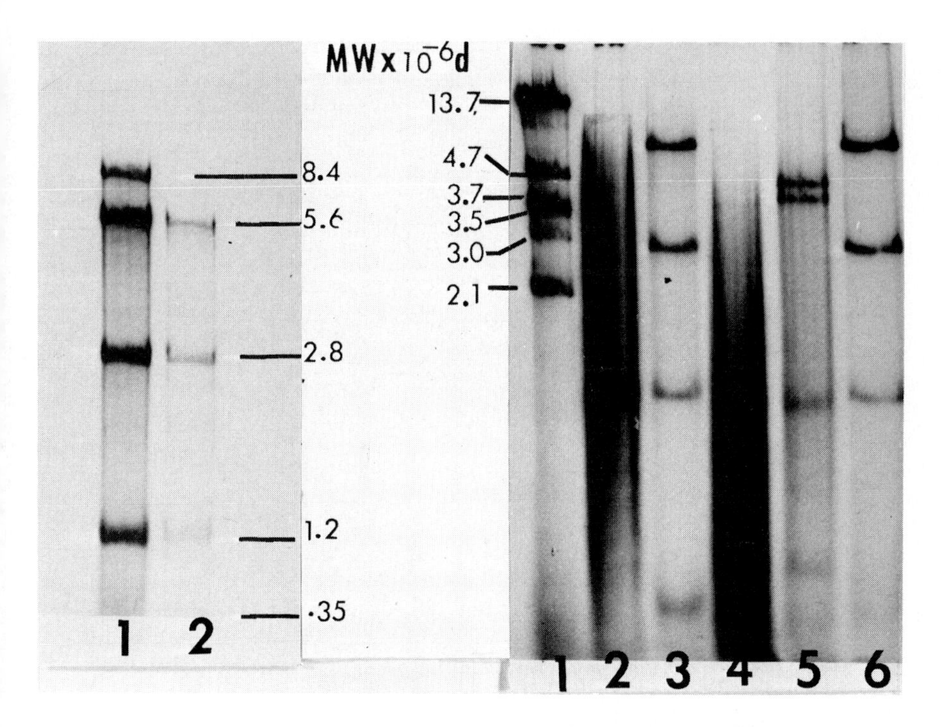

FIGURE 1. Mitochondrial DNAs from KYK, MCH-C18, MTOC10 and KYK(MTOC10)-1 were prepared by purification of component I from Eth-Br CsCl gradients. The samples were nick translated (ref 29), digested with EcoR1 (from BRL), separated on 1% agarose gels by electrophoresis, and then autoradiographed. λ phage DNA was treated identically and used as a molecular weight standard. The results show concordant segregation of MTOC10 mt-DNA and the CAP-r phenotype. The left panel contains KYK(MTOC10)A-1 that were selected with CAP (lane 1) and without CAP (lane 2). The right panel shows the parental controls: λ(lane 1), MTOC10 (lane 2), KYK (lanes 3,6) and MCH (lanes 4,5).

characterized are: 1) genetically stable at least over months of culture in medium without CAP and 2) are capable of cytoplasmically transferring CAP-r to CAP-s cells via fusion with the cytoplasmic fragment remaining after cytochalasin B treatment (cytoplast). Table 2 summarizes the mutant cell strains used in the experiments described in this communication.

IDENTIFICATION OF mt-DNA IN HETEROSUBSPECIFIC AND HETEROSPECIFIC CYBRIDS

The cytoplasmic hybrids (cybrids) are genetically identical to the whole cell parent in karyotype, retention of nuclear markers, and isoenzyme patterns. When the mt-DNA was analyzed from the CAP-r heterosubspecific cybrids using

TABLE 1
SUMMARY OF RESTRICTION DIGEST ANALYSES

	Estimated molecular weight (x 10^6)		
	Mus musculus musculus	*M. m. molossinus*	*Mus caroli*
	DBA BALB/c (MTOC10,BU8) C3H (A9, Cl 1 D)	KYK KYK-CAP KYK6TG	MCH MCH-C18
EcoR1	8.4 ± 0.2 1.2 ± 0.01 0.35	5.6 ± 0.13 2.8 ± 0.05 1.2 ± 0.07 0.38 ± 0.03	4.6 4.1 1.2 0.45, 0.34
HindIII	8.1 ± 0.5 1.21 ± 0.01 0.52 ± 0.02	10 ± 0.1 0.49 ± 0.01 0.3	7.8 2.4 2.0 0.43

TABLE 2
SUMMARY OF CELL LINES

	Species	Tissue	Nuclear marker	Cytoplasmic marker
MTOC10	*Mus m. musculus*	mammary tumor	Ouabain-r	CAP-r
MTBU8	"	"	TK^-	no
A9	"	skin	$HGPRT^-$	no
Cl 1 D	"	"	TK^-	no
A9B3	"	"	$HGPRT^-$	CAP-r
Cl 1 DA1,2,3	"	"	TK^-	CAP-r
KYK	*Mus m. molossinus*	kidney	no	no
KYK-CAP	"	"	no	CAP-r
KYK6TG	"	"	$HGPRT^-$	no
MCH	*Mus caroli*	heart	no	no
MCH-C18	"	"	no	CAP-r

TABLE 3
PROPOSED NOTATION FOR CELL COMBINATIONS

Type	Symbol	Example	Definition
Hybrids:	/ or x	CELL/cell or CELLxcell	somatic cell hybrid strain made by fusing whole cells of strain "CELL" with whole cells of strain "cell"
Cybrids:	( )	CELL(cell)	enucleated "cytoplast" donor designated within parentheses - strain "CELL" to which cytoplasts of strain "cell" were fused
Transformants:	*	CELL*cell	strain "CELL" transformant made by exposure to DNA prepared from strain "cell"

restriction endonuclease digestion, the cybrids with KYK (*Mus musculus molossinus*) as whole cell recipient and MTOC10 (*Mus musculus musculus*) as CAP-r cytoplast donor were found to contain a nearly equal mixture of complete mt-DNA molecules as in KYK(MTOC10)-1 (Figure 1). (The letters before the parenthesis designate the whole cell parent and the letters within the parentheses designate the cytoplasmic donor - see our porposed notation scheme and nomenclature in Tables 2,3.) Three independently arising clones from the same fusion experiment, KYK(MTOC10)-1, -2, -3, and one subclone KYK(MTOC10)-1a, gave identical results, indicating that the mt-DNA of both species must cohabit in the same cell. Table 4 shows a summary of the results we have obtained in this way for the cybrid cell strains that we have studied.

TABLE 4
SUMMARY OF mt- DNA IN CYBRIDS AND TRANSFORMANTS

Code	Species	mt-DNA in CAP medium
Cybrids:		
KYK(MTOC10)	*molossinus(musculus)*	*molossinus* + *musculus*
MTBU8(KYK)	*musculus(molossinus)*	*molossinus*
L-A9(MCH)	*musculus(caroli)*	*caroli*
MCH(KYK)	*caroli(molossinus)*	not done
L-A9(KYK)	*musculus(molossinus)*	no cybrids
Transformants:		
KYK*MTOC10	*molossinus* * *musculus*	*musculus*
MCH*MTOC10	*caroli* * *musculus*	*caroli*

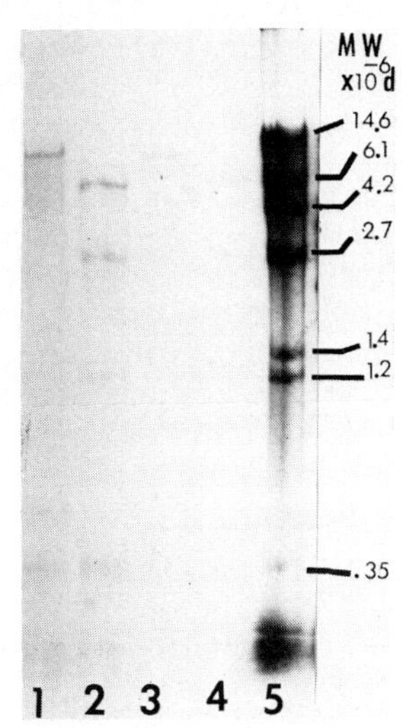

FIGURE 2. The EcoR1 and HindIII digest patterns of mt-DNA from MTBU8(KYK) is shown. The methods were the same as in Figure 1. Complete repopulation of the CAP-s cell's mt-DNA by alien CAP-r mt-DNA is illustrated. Lanes 1 and 3 are HindIII digests and lanes 2 and 4 are EcoR1 digests. HindIII fragments of λ DNA were used as molecular weight standards (lane 5).

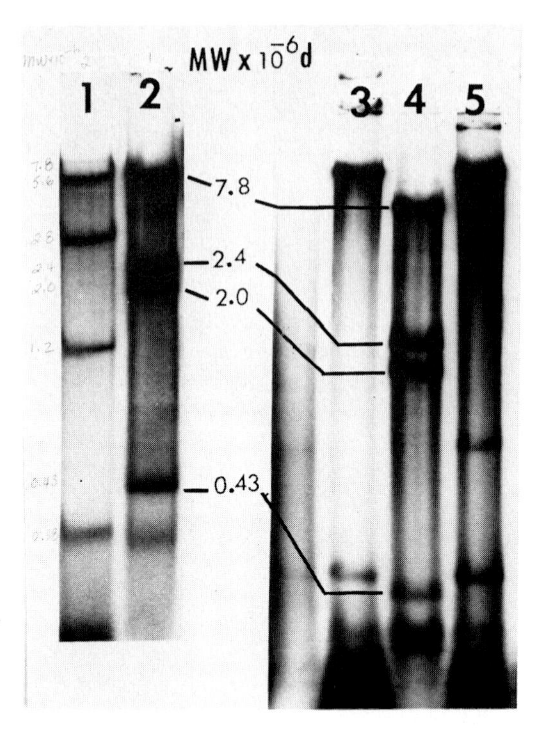

FIGURE 3. Complete repopulation of CAP-s L-A9 mt-DNA (*Mus m. musculus*) by CAP-r, mt-DNA of a distinct species (*M. caroli*) is shown. The HindIII digests of mt-DNAs from L-A9(MCH) with CAP (lane 2), KYK (lane 3), MCH-C18 (lane 4), MTOC10 (lane 1) are compared. Methods were the same as in Figure 1.

If, on the other hand, the cybrid cross is made in the opposite direction, *molossinus* to *musculus*, MTBU8(KYK-CAP) then only KYK mt-DNA was found in the CAP-r cybrids Figure 2. Apparently *molossinus* mt-DNA can survive and substantially repopulate a cell with a *musculus* nucleus but the cell with a *molossinus* nucleus does not tolerate a total substitution of *musculus* mt-DNA; instead it requires about 50% representation of its own mt-DNA in addition.

Still another example was found in the heterospecific combination of L-A9 (*M. m. musculus*) with MCH-C18 (*Mus caroli*) CAP-r cytoplasts. In this situation the MCH-C18 mt-DNA substitutes for the endogenous *musculus* mt-DNA (Figure 3). These cybrids may be the first examples of cells that have shown unequivocally the nuclear genes of one species and at least the great majority of their mt-DNA from another (albeit closely related) species. We do not yet have data on the reciprocal heterospecific transfer.

CONCORDANT SEGREGATION OF THE CAP-rPHENOTYPE AND THE mt-DNA OF THE CAP-r SPECIES

Loss of the CAP-r phenotype from cybrid cells during culture in medium lacking CAP was first studied by Eisenstadt's group (22,23). Their results support the assumption that in cybrid populations the cells are heterogeneous for many unlinked CAP-r and CAP-s determinants and that the CAP-r ones are gradually lost in some strains (and are stable in others) after the selective pressure is removed. We have examined the segregation of different mt-DNAs in similar experiments.

After about 150 population doublings in CAP-medium, CAP selective pressure was removed from the KYK(MTOC10) cybrids that had a 50-50 mixture of the mt-DNAs of both species and within 50-100 population doublings (depending on the cybrid clone) the MTOC10 mt-DNA was lost (Figure 1). In each case the CAP-r phenotype was lost concordantly. This result demonstrates that 1) a tight association between the CAP-r phenotype and the presence of the MTOC10 mt-DNA exists and 2) that KYK mt-DNA must replicate faster than MTOC10 mt-DNA in the presence of the KYK nuclear genes. The preference of a whole cell (nucleus?) for its own mt-DNA was further confirmed in a "second generation" transfer by fusing KYK(MTOC10) CAP-r cytoplasts with CAP-s MTBU8 cells. The resulting cybrids, MTBU8(KYK(MTOC10)), which were selected in CAP and BrdU were found to contain only the MTOC10-BU8 mt-DNA (*musculus*) and no KYK mt-DNA (*molossinus*). The two cybrid crosses that had shown apparently complete substitution of donor mt-DNA for the endogenous mt-DNA, L-A9(MCH) and MTBU8(KYK-CAP), have not shown any segregation after a comparable time (50-100 population doublings) in medium minus CAP. This result is tentatively taken to mean that these cybrids were totally substituted by the foreign CAP-r mt-DNA. We estimate, however, that as much as 2% of the endogenous mt-DNA could still be present in the cells and not be detected in our assays.

TRANSFORMATION OF CAP-s CELLS BY PURIFIED CAP-r mt-DNA

We have previously reported that treatment of CAP-s cell strains by the $CaPO_4$ coprecipitates (24) with the purified mt-DNA from CAP-r cells yields "transformation" to the CAP-r phenotype with frequencies in the range of 10^{-4} to 10^{-3} treated cells (20,25). While most of the data came from homospecific combinations of *musculus* CAP-s cells and CAP-r mt-DNA, heterosubspecific and recently heterospecific transformations have been made successfully - see Table 5.

TABLE 5
CAP-r mt-DNA TRANSFORMATION EXPERIMENTS[a]

DNA source (μg DNA/10^6 cells)	Treated cells	Cells per plate 3×10^3	1×10^3
		Colonies/Plates	
MTBU8 (CAP-s) mt-DNA (0.35μg)	3T34EF(1)	0/6	
	N186TG-2	0/2	
	MTBU8	0/7	
MTOC10 (CAP-r) mt-DNA (0.35μg)	3T34EF(1)	9/7	
	3T34EF(2)	120/5	14/7
	N186TG-2	6/6	
MCH-C18 (CAP-r) mt-DNA (0.5 μg)	3T34EF	72/5	27/9
MCH (CAP-s) mt-DNA (0.4 μg)	3T34EF		0/8
$CaPO_4$ (0 - DNA)	3T34EF(1)	0/8	
	3T34EF(2)		0/7
E. coli (CAP-s) DNA (10 μg)	3T34EF	0/2	
	MTBU8	0/8	

[a]Results of treating various CAP-s cell strains with different purified DNA followed by selection in CAP (50μg/ml). Control DNA purified from *E. coli* (a CAP-s strain) and purified mt-DNA from CAP-r and CAP-s cell strains were used to treat the 3 CAP-s cell strains indicated. Petri plates containing 1-3 x 10^6 CAP-s cells were treated with $CaPO_4$ -coprecipitates of DNA from the sources indicated (ref 24). After 18 hours the treated cells were trypsinized, counted, and distributed into 100mm Petri plates at 1 or 3 x 10^3 cells per plate and grown in CAP medium. Viable colonies were counted after at least 6 weeks of selection. The data are presented as the number of colonies found in the total number of plates and are not corrected for plating efficiency. Postscripts (1) and (2) indicate independent experiments. 3T34EF is a TK^- 3T3 cell strain and N186TG-2 is a 6-thioguanine resistant variant of the C-1300 mouse neuroblastoma (see refs 20,25 for details).

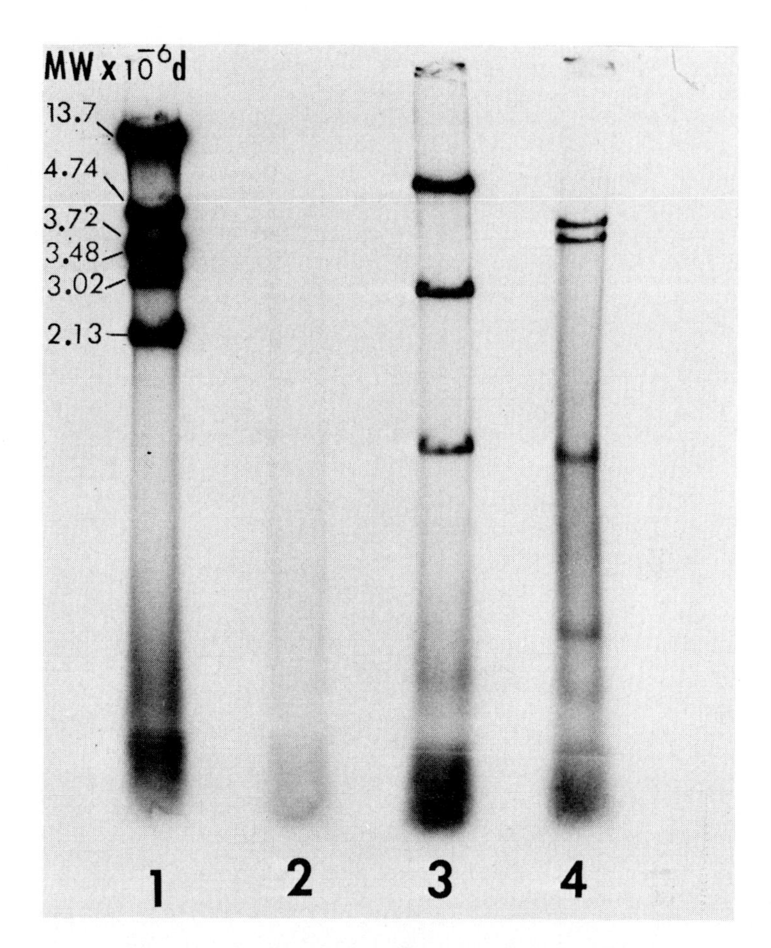

FIGURE 4. The EcoRl digest patterns of three colonies from CAP-s cell lines transformed to CAP-r by CAP-r MTOC10 mt-DNA are shown: KYK*MTOC10-7 (lane 1), KYK*MTOC10-2 (lane 3) and MCH*MTOC10-10 (lane 4). In transformants no mt-DNA from the CAP-r DNA donor was detected. The experimental procedures were the same as for Figure 1.

The restriction digests of mt-DNA from 2 heterosubspecific and 3 heterospecific transformants have been examined and show that the mt-DNA was like the original CAP-s whole cell parent and not like that of the CAP-r donor from which the mt-DNA was extracted.(Figure 4). Moreover, using EcoRl, HindIII and HpaII no indication of recombination could be found, i.e., no novel DNA bands were detected in the mt-DNA of the putative transformants. The possibility, therefore, had to be considered that the transformation to CAP-r had occurred as a

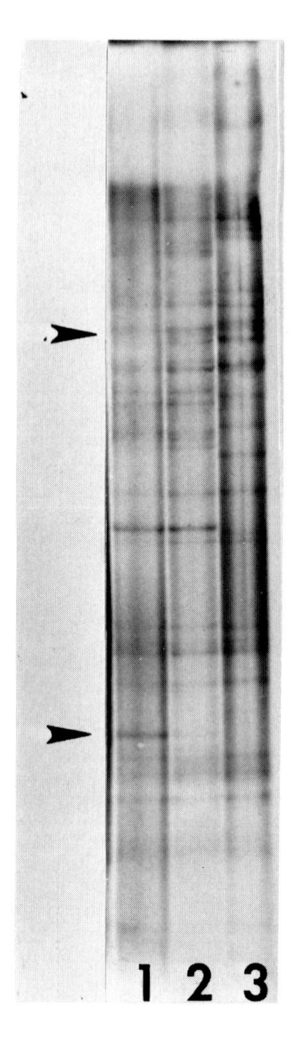

FIGURE 5. Comparison of a CAP-r transformant, KYK*MTOC10-2 (lane 2) with the KYK recipient cell (lane 1) and MTOC10 CAP-r mt-DNA donor (lane 3). The arrows indicate possible band changes found in the transformant. Purified mt-DNAs were labelled by nick translation (ref 29), digested with HaeIII (from BRL), separated on a 6% acrylamide gel and exposed 18hrs.

result of a change in the nucleus and not at the level of the mt-DNA. However, "second generation" cybridization experiments using the transformed cells as CAP-r cytoplast donor showed efficient cytoplasmic transfer of the CAP-r phenotype in every case. Since the first set of restriction digests were made using enzymes that produced only 4-6 fragments from mt-DNA, a further analysis was performed using HaeIII on KYK*MTOC10 mt-DNA (Figure 5). Thirty five fragments were clearly visible on the gels indicating that the mouse mt-DNA, as demonstrated in yeast (26), retains many GGCC sites. Secondly, the parent cell, KYK, and the CAP-r donor, MTOC10, differ appreciably, so that any recombination that might occur between them would appear to offer a good chance of removing or inserting a restriction site and thus to produce a novel band(s). The mt-DNA of one of the five transformants studied, KYK*MTOC10-1, did turn out to differ from the KYK parent by having a new high molecular weight HaeIII band and at the same time one of the low molecular weight bands was reduced - see Figure 5. This probably means that the cells contained a mixed population of mt-DNAs, some of which, the putative transformants, had deleted one restriction site. The high background we have experienced with our nick translation technique precludes a firm interpretation of this observation as yet and future work will be required. Heteroduplex analysis should allow us to distinguish between insertions of transforming mt-DNA and deletions from the host mt-DNA.

DISCUSSION

Our results so far demonstrate that: 1) CAP-r is associated with the presence of a particular marked species of mt-DNA; 2) heterosubspecific "transformation" may be the result of an alteration within a limited region of the mt-DNA rather than the incorporation of a large fragment or retention of the whole CAP-r mt-DNA molecule, or possibly it may result from a mechanism that encourages the appearance of "spontaneous" CAP-r mutants in the CAP-s mt-DNA; and 3) cybridization, by contrast, appears to result in the successful recruitment of complete mt-DNA molecules. Of course these results are not surprising since much of the CAP-r mt-DNA used in "transformation" experiments must be imagined to be fragmented while in transit to the mitochondrion, whereas a large number of intact mitochondria are incorporated directly by cytoplast fusion.

Since observing apparent recombinant human/mouse mt-DNA molecules in hybrid cells (27,28) we have been interested in applying cytoplast fusion procedures to look for more evidence of genetic recombination in mammalian mt-DNA. Of course, a more nearly ideal system would combine these techniques with multiple, independently selectable markers. However, just as the species specific mt-DNA sequences recognized by cRNA probes provided useful markers to search for evidence of recombination in human/mouse hybrids, so the different restriction sites in the mt-DNA of the mouse subspecies may serve as convenient physical markers to trace the mt-DNA associated with CAP selection. The KYK(MTOC10) cybrids, with a stable and nearly equal representation of both species' mt-DNA molecules, should present an optimal opportunity for observing mt-DNA recombination. It appeared to us that if recombination did occur we might select for recombinants in the KYK(MTOC10) cybrids. Because they maintain a 50-50 balance of both molecules at equilibrium (after many population doublings), it may be inferred that the cells require functions supplied by both molecules - probably CAP-r from the MTOC10 mt-DNA and some second function like a nucleus or MTOC10 mt-DNA compatibility factor supplied by the KYK mt-DNA. It seemed that we might detect examples of all three of the principal recombinant types: 1) catenanes that covalently bound both mt-DNAs into a single molecule would usually show at least one novel restriction fragment, and 2) KYK mt-DNA incorporating the CAP-r marker or 3) MTOC10 mt-DNA incorporating the requisite compatibility factor would both appear to offer a reasonable chance to see a novel restriction band. But these were not found with the endonucleases studied - and the analysis will have to proceed to a higher level of resolution provided by sequential digestion and enzymes that produce more fragments. At present our results suggest either that recombinants extensive enough to be likely to involve restriction sites are rare or that CAP-r and the second compatibility factor are so closely associated that recombinants usually destroy one or the other function.

One cybrid scheme derived from what we already know about the KYK(MTOC10) crosses may offer still a better hope of selecting for recombinant mt-DNAs. That possibility would be first to fuse CAP-r KYK, KYK-CAP, cytoplasts to CAP-s MCH whole cells to make the MCH(KYK-CAP) cybrid. In a second generation transfer these cybrids would be enucleated and then fused to L-A9 or L Cl 1 D cells to form the compound cybrid, L(MCH(KYK-CAP)). We have never been able to obtain cybrids between L-cells and KYK-CAP cytoplasts; presumably the KYK mitochondria and the L-cell genes are incompatible. However, MCH-C18 (CAP-r *Mus caroli*) mt-DNA has repeatedly been observed to repopulate CAP-s L-cells. Therefore, if recombinants were formed transferring the CAP-r marker of KYK to predominantly MCH mt-DNAs in the first cybridization, then these should be selectively amplified and expand in the L-cell second generation cybrids. The competing KYK-CAP mt-DNAs we anticipate would be lost because of the apparent incompatibility noted above.

The heterosubspecific and heterospecific cybrids like L-A9(MCH) present a unique opportunity to determine which organelle proteins are encoded in the mitochondrial genome.

Comparison of the protein composition of purified mitochondria of MCH and L-A9(MCH) or of KYK and MTOC10(KYK) and MTOC10 should produce variations best explained by the different mt-DNA. Those polypeptides that are shifted on 2-D gels and presumably encoded by the alien mt-DNA can then be related to the homologous subunits of the various functional complexes of mitochondria.

REFERENCES

1. Borst, P., and Grivell, L. A. (1978). Cell 15, 705-723.
2. Wallace, D., Bunn, C., Eisenstadt, J. M. (1975). J. Cell Biol. 67, 174.
3. Bunn C., Wallace, D., Eisenstadt, J. M. (1974). Proc. Natl. Acad. Sci. (Wash.), 71, 1681.
4. Mitchell, C. H., and Attardi, G. (1978). Som. Cell Genet. 4, 737.
5. Borst, P. (1972). In "Annual Review of Biochemistry", 41, pp 333-376.
6. Bogenhagen, D. and Clayton, D. A. (1978). J. Mol. Biol. 119, 49.
7. Borst, P. (1976). In "International Cell Biology 1976-77" (B. Brinkley and K. Porter, eds.), Rockefeller Univ. Press.
8. Moorman, A. F. M., Van Ommen, G. J. B., and Grivell, L. A. (1978). Molec. Genet. 160, 13.
9. Battey, J. and Clayton, D. A. (1978). Cell 14, 143.
10. Angerer, L., Davison, N., Murphy, W., Lynch, D., and Attardi, G. (1976). Cell 9, 81.
11. Klukas, C. K. and Dawid, I. B. (1976). Cell 9, 615.
12. Schatz, G. and Mason, T. L. (1974). Annual Review of Genetics 8, 51.
13. Tzagoloff, A. (1977). BioScience 27, 18.
14. Costantino, P., and Attardi, G. (1975). J. Mol. Biol. 96, 296.
15. Jeffreys, A.J. and Craig, I. W. (1976). Eur. J. Biochem. 68, 301.
16. Yatscoff, R. W., Freeman, K. B., and Vail, W. J. (1977). FEBS Letters 81, 7.
17. Spolsky, C. M., and Eisenstadt, J. M. (1972). FEBS Letters 25, 319.
18. Harris, M. (1978). Proc. Natl. Acad. Sci. (Wash.) 75, 5604.
19. Lichton, T. and Getz, G. C. (1978). Proc. Natl. Acad. Sci. (Wash.) 75, 324.
20. Coon, H. G., and Ho, C. (1977). In Brookhaven Symposia in Biology No 29, "Genetic Interaction and Gene Transfer" (C. W. Anderson, ed.), pp 166-177. Brookhaven National Laboratory, Upton, New York.
21. Ziegler, M.L., and Davidson, R. L. (1978). J. Cell Biol. 79, Part 2, 5G2514.

22. Bunn, C., Wallace, D., Eisenstadt, J. M. (1977). Som. Cell Genet. 3, 71.
23. Wallace, D. C., Bunn, C. 1., and Eisenstadt, J. M. (1977). Som. Cell Genet. 3, 93.
24. Graham, F. L., and Van Der Eb, A. J. (1973). Virology 54, 536.
25. Coon, H. G. (1978). Natl. Cancer Inst. Monogr. 48, 45.
26. Prunell, A., Kopecka, H., Strauss, F., and Bernardi, G. (1977). J. Mol. Biol. 110, 17.
27. Coon, H. G., Horak, I., and Dawid, I. B. (1973). J. Mol. Biol. 81,285.
28. Horak, I., Coon, H. G., and Dawid, I. B. (1974). Proc. Natl. Acad. Sci. (Wash.) 71, 1828.
29. Maniatis, T., Jeffrey, A., and Kleid, D. G. (1976). Proc. Natl. Acad. Sci. (Wash.) 72, 1184.

STRUCTURE AND FUNCTION OF KINETOPLAST DNA OF THE AFRICAN TRYPANOSOMES

P.Borst and J.H.J.Hoeijmakers

Section for Medical Enzymology and Molecular Biology,
Laboratory of Biochemistry, University of Amsterdam,
Jan Swammerdam Institute, P.O.Box 60.000, 1005 GA Amsterdam
(The Netherlands)

ABSTRACT The kinetoplast DNA from Trypanosoma brucei consists of networks containing two types of circles, maxi-circles and mini-circles. The maxi-circles are 20 kilo-base pairs, homogeneous in sequence and they represent about 10% of the network. Maxi-circles are transcribed; the transcripts are enriched in poly-$(A)^+$ RNA and cross-hybridize to maxi-circles from Crithidia. The maxi-circle restriction fragments of different T. brucei strains are largely identical.

We have not detected mini-circle transcripts in total cell RNA of T. brucei. Restriction enzyme digestion shows marked size and sequence heterogeneity of mini-circles (but less than inferred from renaturation studies). TaqI digests of mini-circles from different T. brucei strains have hardly a band in common.

In African trypanosomes, unable to make functional mitochondria, we find all sorts of kDNA: 'normal' kDNA (one T. brucei), 'normal' mini-circle networks without maxi-circles (one T. equiperdum, two T. evansi), no kDNA at all (one T. brucei, two T. evansi). In the two strains without maxi-circles we find hardly any size or sequence heterogeneity in the mini-circles.

We conclude that the maxi-circles are the equivalent of mtDNA in other organisms, that the ribosomal RNAs of trypanosome mitochondria are very small, that mini-circles are not transcribed and that recombination between mini-circles and maxi-circles may contribute to mini-circle sequence heterogeneity.

INTRODUCTION

Trypanosomes are unicellular flagellates with a single mitochondrion, which contains a mass of DNA large enough to be visible in the light microscope (Fig. 1). Extraction of this kinetoplast DNA (kDNA) yields compact networks of catenated circles, that can be isolated intact and visualized in the electron microscope by the protein monolayer technique

ISBN 0-12-1

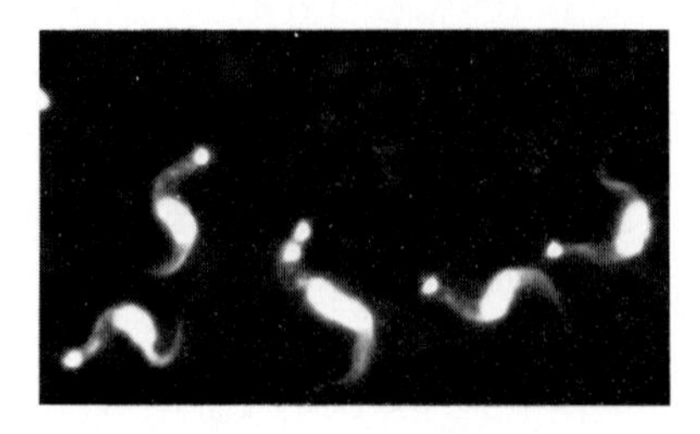

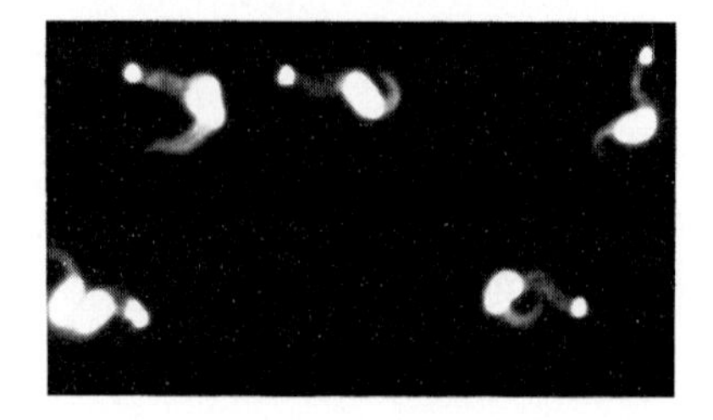

Fig. 1. DAPI-stained T. brucei under the fluorescence microscope. Bloodstream trypanosomes in blood smears were fixed with ethanol/ether (50/50, 30 min), stained for 10 min with either 10 μg DAPI (4,6-diamino-2-phenylindole, see ref. 1) per ml (left) or 2.5 μg DAPI/ml (right) and photographed with incident light. Final magnification, 936 X. The preparations were deliberately over-stained to show the outline of the trypanosome. (Photograph of Dr. J.James.)

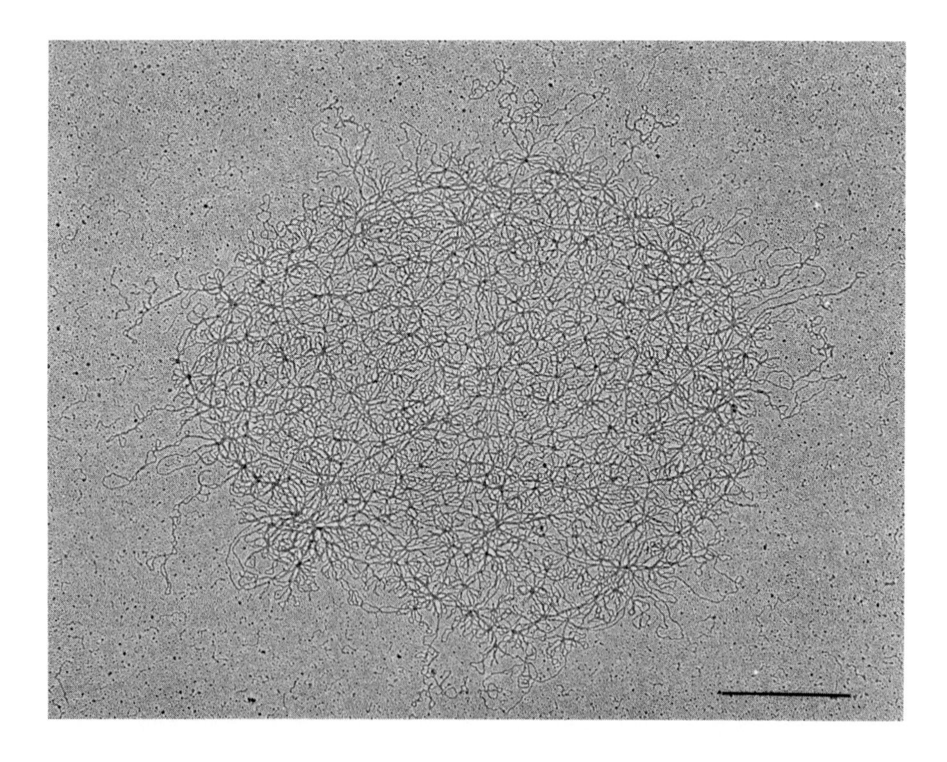

Fig. 2. Electron micrograph of a kDNA network, isolated from T. brucei (strain 31), spread in a protein monolayer according to a modified Lang and Mitani method (see ref. 2). The bar is 1 μm.

(Fig. 2). Most of the circles in this *T. brucei* network are 0.3-μm 'mini-circles'; a minority consists of 6-μm 'maxi-circles', visible as long edge loops in Fig. 2. kDNA networks from other trypanosome genera are essentially similar to those of *T. brucei*, as shown by the data presented in Table I.

We have recently reviewed our experiments on kDNA up to 1978. This paper summarizes our recent attempts to further define the function of kDNA components and the fate of these components in African trypanosomes of the *brucei* group that have lost the ability to make functional mitochondria.

SEQUENCE EVOLUTION OF MINI-CIRCLES AND MAXI-CIRCLES IN *Crithidia*

When mini-circles from *Crithidia luciliae* are digested with restriction endonuclease HapII (CCGG), a complex pattern of bands is obtained on gels. The bands are present in non-stoicheiometric amounts and their added molecular weights are at least 10-fold higher than that of the mini-circle. Several lines of evidence indicate that this is not due to partial methylation of recognition sites (4). One of the most direct experiments is the demonstration (unpublished) that the fragmentation patterns obtained with endonucleases HapII and MspI are the same. MspI also cuts at the CCGG sequence, but scission is not blocked by the methylation that prevents scission by HapII (5). Double digestion analysis with two restriction enzymes has suggested that rearrangements of sequence blocks rather than point mutations play a major role in mini-circle heterogeneity and electron micrographs of renatured mini-circles, which show denaturation bubbles usually around 0.12 μm long, support this interpretation (unpublished).

Comparison of the HapII restriction fragments of mini-circles from *C. luciliae* and *C. fasciculata* shows that these have hardly a fragment in common. Sequence evolution in the mini-circles is so fast in fact that the fragment patterns of *C. fasciculata* kDNA obtained with several endonucleases have detectably changed in a 2-year culture period in our lab in the absence of selective conditions. This confirms the marked sequence difference between the two types of mini-circles already inferred from cross-hybridization with complementary RNAs (6). On the other hand, the fragment maps of the maxi-circles of these two organisms are largely identical. From a comparison of fragments generated by seven hexamer and two tetramer-recognizing enzymes we calculate with Upholt's procedure (7) a sequence divergence of less than 5% (unpublished).

These results show that mini-circle sequence evolution in *Crithidia* is much faster than maxi-circle evolution.

TABLE I

PROPERTIES OF kDNA NETWORKS FROM VARIOUS TRYPANOSOME GENERA[a)]

Organism	Spread networks	Mini-circles			Maxi-circles[d)]		Mini/Maxi
	Diameter (μm)	Size (μm)	Mass[b)] (daltons x 10^{-6})	Sequence[c)] hetero-geneity	Size (μm)	Mass[b)] (daltons x 10^{-6})	mass ratio in networks[e)]
Crithidia							
C. luciliae		0.76		++	11.3	23	20
C. fasciculata	15-20	0.79	1.59	++	∿12	23	20
Herpetomonas							
H. muscarum[f)]			0.7	++		21[g)]	>20
Leishmania							
L. tarentolae	8-10	0.29	0.55	+		∿20?	20
Schizotrypanum							
T. cruzi	10-15	0.49	0.94	+		26[h)]	>20
Trypanosoma							
T. brucei	4-6	0.32	0.56	+++	6.3	13	10
T. mega[i)]	12-15	0.74	1.49	+++	9.0	16.1	>20
Phytomonas							
P. davidi (3)		0.37	0.70	++		∿24	20

SEQUENCE EVOLUTION OF MINI-CIRCLES AND MAXI-CIRCLES IN *T. brucei* kDNA

Optical DNA.DNA renaturation experiments by Steinert *et al*. (8) have shown that the reassociation rate of *T. brucei* kDNA is less than 1% of that expected for a single DNA sequence of mini-circle size. Renatured DNA was hardly mismatched as shown by remelting. These results suggested that *T. brucei* kDNA contains at least 100 sequence classes that show so little homology that they do not cross-hybridize.

Initial attempts to verify this surprising result with restriction enzyme analysis were inconclusive: the mini--circles of *T. brucei* contain about 30 mole percent G+C and 1000 base pairs (bp) and only restriction enzymes that recognize a tetranucleotide sequence containing two AT pairs can be expected to cut such molecules more than once. Indeed, in initial experiments with the hexanucleotide-recognizing restriction endonucleases HapII (CCGG) and HaeIII (GGCC), only part of the mini-circles were cut at all and the patterns obtained were uninformative, although clearly showing the size heterogeneity of the once-cut mini-circles (see Fig. 3A, slot 2). Endonucleases AluI (AGCT) and TaqI (TCGA), however,

---

Table I - continued.

a) Modified from Table I of ref. 2. References are given in brackets; data without references are documented in ref. 2.

b) Calculated from size in electron micrographs or mobility in gels relative to reference DNA (fragments) of known molecular weight.

c) +, micro-heterogeneity, more than one sequence class present; ++, micro-heterogeneity, more than 10 sequence classes present; +++, extensive heterogeneity present.

d) No sequence heterogeneity has been found in maxi-circles.

e) These values are very approximate and estimated by us from the data available; since maxi-circles are preferentially lost during isolation if networks are damaged, the ratio's given could be higher than those of intact networks *in vivo*.

f) Hajduk, S.L., Vickerman, K., Hoeijmakers, J.H.J. and Borst, P., unpublished experiments.

g) Circles with a mass of 15 x 10^6, micro-heterogeneous in sequence and not cross-hybridizing to mini-circles or maxi-circles are present as minor components in kDNA preparations from this organism[f)].

h) Leon, W., Hoeijmakers, J.H.J., Fase-Fowler, F. and Borst, P., unpublished experiments.

i) Genus, *Trypanosoma*; sub-genus, unclassified.

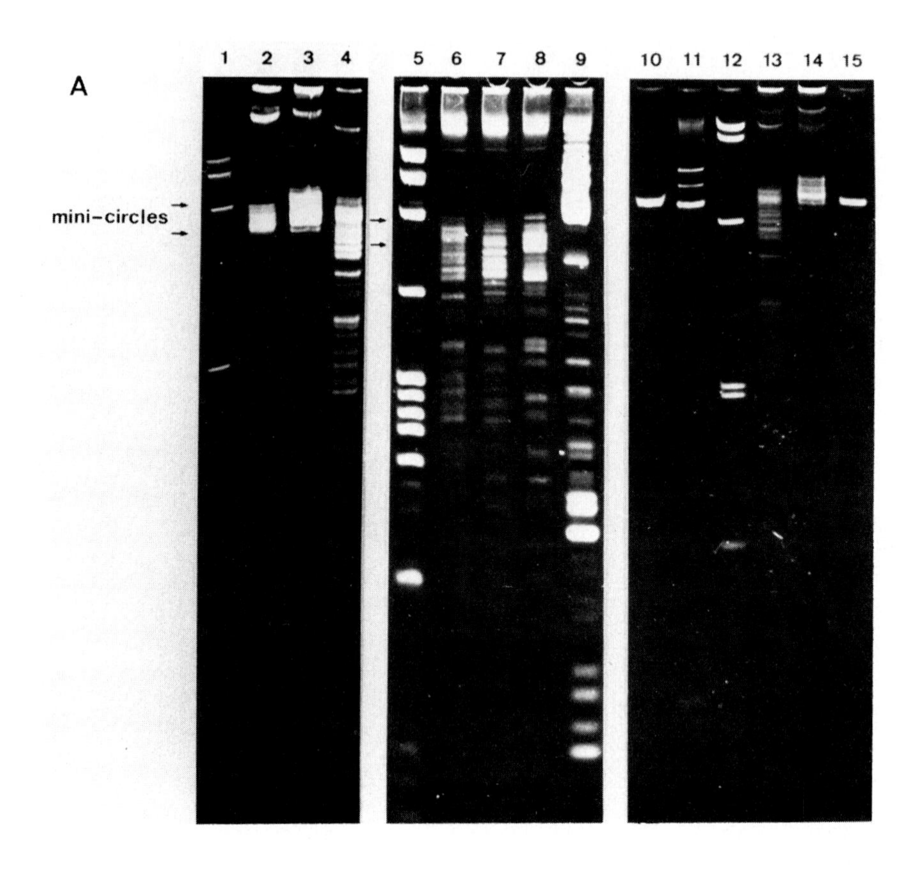

Fig. 3. Restriction digests of T. brucei, T. evansi (Zwart, +) and T. equiperdum (Zwart, +) kDNAs. Panel A shows polyacrylamide (PAA)-0.5% agarose gels, 2.5% PAA for lanes 1-9, 3% PAA for lanes 10-15. Panel B is a 1.5% agarose gel. Panel A: Lanes 1 and 5, phage PM2 DNA x HindII+III; lane 12, phage PM2 DNA x HindIII; lane 2, T. brucei (427) kDNA x HapII; lanes 3 and 14, T. brucei (427) kDNA x AluI; lanes 4, 6 and 13, T. brucei (427) kDNA x TaqI; lane 7, T. brucei (31) kDNA x TaqI; lane 8, T. brucei (1125) kDNA x TaqI; lane 9, C. fasciculata kDNA x HapII; lane 10, T. evansi kDNA x TaqI; lane 11, T. evansi kDNA x AluI; lane 15, T. equiperdum kDNA x TaqI. Panel B: Lanes 1 and 8, phage PM2 DNA x HindII+III; lanes 2-7 are TaqI digests of kDNA from T. brucei 427, 31, 1125, 839, T. equiperdum and T. evansi, respectively. See refs 4 and 10 for details of gel electrophoresis and photography.

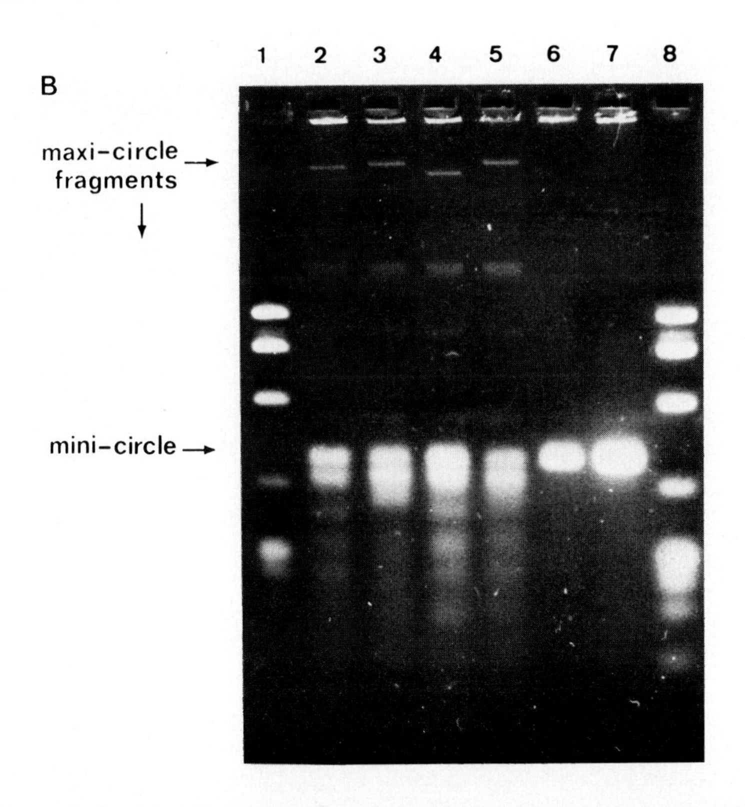

cut (nearly) all mini-circles and the latter results in highly informative digests in acrylamide gels, as shown in Fig. 3A, slot 4). In 1.5% or 2.0% agarose gels the TaqI digests show a series of distinct bands smaller than once-cut mini-circles (Fig. 3A, slots 2-5). We have also observed this with endonucleases MboI, MboII and HinfI (not shown). In acrylamide gels these bands resolve in clusters of multiple bands, like the once-cut mini-circles (Fig. 3A, slot 4). These results show that the distance between restriction sites is more conserved than would be expected from the presence of at least 100 sequence classes that do not cross-hybridize. It is, furthermore, clear from Fig. 3A that the mini-circle sequences are strikingly unstable in evolution, as judged from the fact that the TaqI digests of kDNAs from three different T. brucei strains (Fig. 3A, slots 6-8) have hardly a fragment in common, confirming again earlier RNA.DNA hybridization experiments with complementary RNA (8).

On the other hand, the maxi-circle sequence is relatively conserved in these same T. brucei strains. The differences observed between strains are nearly exclusively in a 6.5 bp 'silent' segment of the maxi-circle (Fig. 3B; ref. 9), which lacks recognition sites for 20 of the 22 restriction enzymes tested (Fig. 4). Since the maxi-circle has a very low density in CsCl (1.682 g/cm³) this 'silent' segment could be analogous to the AT-rich region in *Drosophila* mtDNA, which also varies in size in different species.

In general, these results confirm the conclusions from the analysis of *Crithidial* networks, although mini-circle sequence heterogeneity is even more pronounced in *T. brucei* than in *Crithidia*.

TRANSCRIPTION OF kDNA

To get more information on the function of maxi-circles and mini-circles, we have analysed *Crithidia* for the presence of kDNA transcripts. To avoid loss of RNA components we have isolated total cell RNA by a drastic hot phenol procedure and labelled it *in vitro* with polynucleotide kinase. This RNA was hybridized with Southern blots of restriction digests of kDNA. Whereas maxi-circle transcripts were readily detected, no mini-circle transcripts were found at all, even though the concentration of mini-circle DNA sequences on the filter was at least a factor 100 higher than of maxi-circle sequences (10).

We have recently repeated these experiments with *T. brucei* and with essentially similar results (Fig. 5). Obviously, if mini-circles are ever transcribed at all in the cell, the transcripts are either very short-lived or not detectable by standard procedures for other reasons. When the total cell RNA is first enriched in poly(A)$^+$ RNA by two cycles of (dT)-cellulose chromatography, the fraction of the RNA that hybridizes with maxi-circle fragments increases. Although this is compatible with poly(A) tails on (part of the) maxi-circle transcripts, this requires verification.

An interesting aspect of the hybridization experiments is that the predominant hybridization is with a rather small segment of the maxi-circle. In *Crithidia* this segment is only 1500 bp (unpublished), in *T. brucei* we do not yet have enough restriction enzyme recognition sites in this area of the map for precise localization and the smallest segment obtained is a HaeIII x HindIII fragment of about 2000 bp (Fig. 5; see also Fig. 4). The sequence of this segment is conserved in trypanosome evolution, because *T. brucei* RNA cross-hybridizes to the predominantly transcribed segment of *Crithidial* maxi-circle DNA (Fig. 5). We think that this segment codes for the two major RNAs, observed in mitochondrial preparations of *Leishmania* (9S and 12S; ref. 11) and *Crithidia* (11.5S and 14.6S; ref. 12).

Simpson has suggested that these RNAs are messenger RNAs, we that they are unusually small ribosomal RNAs (rRNAs) (2). In all mitochondria studied, the rRNAs are always the predominant RNA species and we see no reason why this should

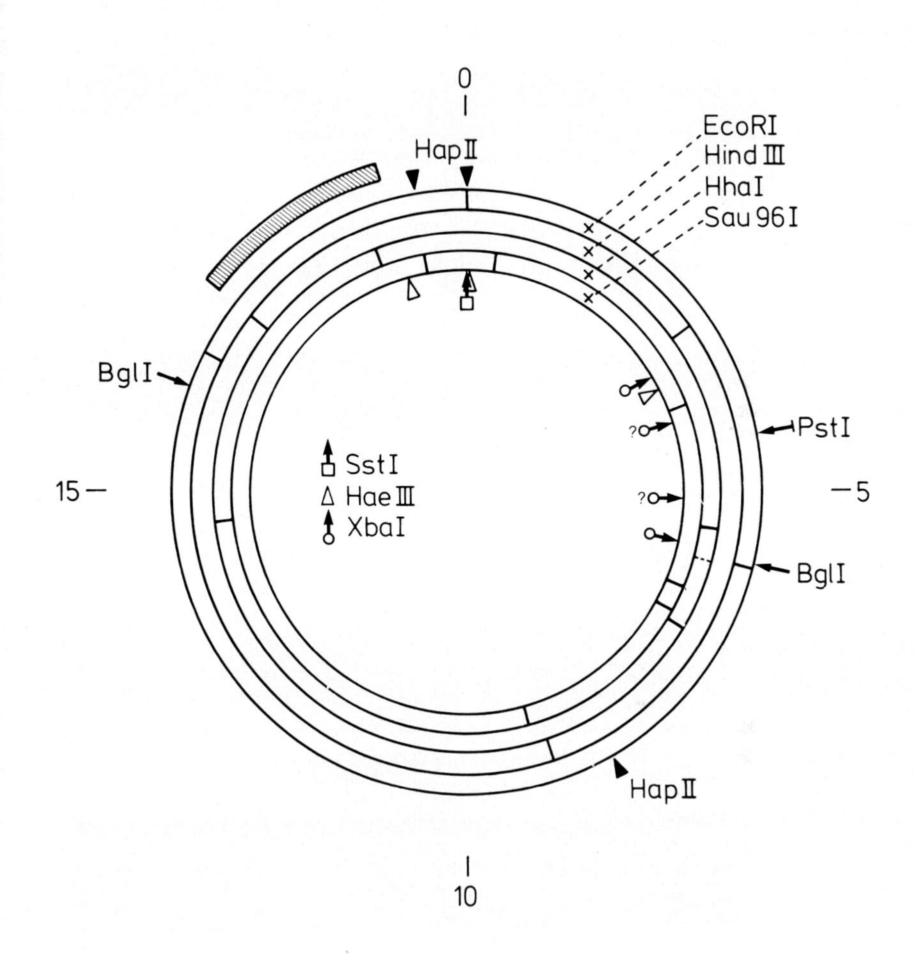

Fig. 4. A restriction fragment map of the maxi-circle of kDNA from T. brucei (EATRO 427). The scale is in kilobase pairs. The black bar indicates the HindIII x HaeIII fragment that gives preferential hybridization with total cell RNA.

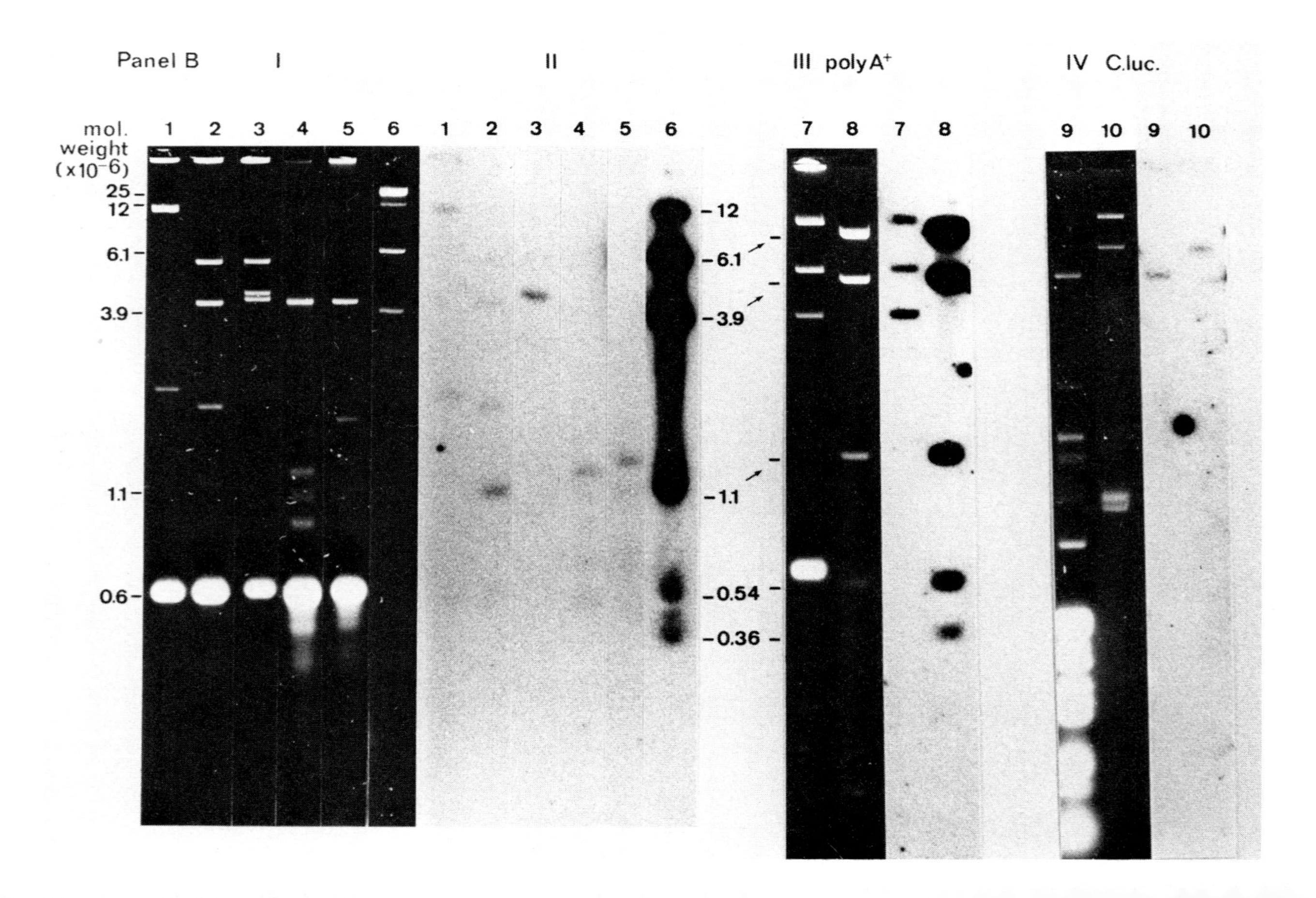
Panel B
I
II
III polyA+
IV C.luc.
mol. weight (x10-6)
1 2 3 4 5 6 1 2 3 4 5 6 7 8 7 8 9 10 9 10
25
12
6.1
3.9
1.1
0.6
-12
-6.1
-3.9
-1.1
-0.54
-0.36

Fig. 5

be different in trypanosomes. Although rRNAs could be lost or fragmented during isolation, our hybridization experiments indicate that the predominant RNA species are complementary to a segment of about 1500 bp, which fits the estimated size of 1540 nucleotides for the two major kinetoplast RNA species in *Leishmania* (11). The combined size of the two rRNAs of animal mitochondria is about 2400 nucleotides, already much smaller than the rRNA of *Escherichia coli*, and there is no compelling reason why it should not be possible to make ribosomes with even smaller RNAs. A 70S ribosome was observed in *T. brucei* extracts by Hanas *et al*. (13) and tentatively identified as the mitochondrial ribosomes of this organism. Although it seems a little difficult to fit 15S and 11S RNAs into a 70S ribosome, this is not impossible. Further analysis of mitochondrial ribosomes and mitochondrial RNA from trypanosomes is obviously required to settle this point.

THE STRUCTURE OF kDNA IN AFRICAN TRYPANOSOMES THAT HAVE LOST THE ABILITY TO MAKE FUNCTIONAL MITOCHONDRIA

In the vertebrate host *T. brucei* suppresses mitochondrial biogenesis and the resulting rudimentary 'pro-mitochondrion' lacks a functional respiratory chain and Krebs cycle. Under these conditions the parasite lives by glycolysis alone and all known products of the mitochondrial genome should be

---

Fig. 5. Hybridization of *T. brucei* RNA with restriction fragments of kDNA of *T. brucei* or *C. luciliae*. kDNAs of *T. brucei* (427) or *C. luciliae* (0.25 μg and 2.5 μg, respectively) were digested with various restriction endonucleases and the fragments separated by electrophoresis in 0.8% agarose gels. DNA was denatured *in situ* and blotted onto a nitrocellullose filter. The filter-bound DNA was hybridized with ^{32}P-labelled total cell RNA or poly(A)$^{+}$ RNA of *T. brucei* (1 μg). Hybridization conditions were as described. Panel I: Ethidium bromide-stained gel containing kDNA of *T. brucei*, digested with HaeIII, HaeIII+HindIII, HindIII, HindIII+MboI, MboI (lanes 1-5, respectively). Lane 6 contains molecular weight markers (phage T7 DNA, phage Ø29 DNA and phage Ø29 DNA after digestion with EcoRI). Panel II: Autoradiogram of the filter of I, after hybridization with total cell RNA. Panel III: Lanes 7 contain kDNA of *T. brucei* digested with EcoRI and the digest after hybridization with poly(A)$^{+}$ RNA of *T. brucei*, after binding to oligo(dT)-cellulose; lanes 8 show molecular weight markers (phage Ø29 DNA digested with EcoRI). Panel IV: kDNA of *C. luciliae* after digestion with endonuclease HapII (lanes 9) and SalI (lanes 10) and hybridized with ^{32}P-labelled total cell RNA of *T. brucei*. See ref. 10 for further details.

dispensable. In fact, trypanosomes that have apparently irreversibly lost the ability to make functional mitochondria can be maintained in vertebrates if the transmission is carried out in a way that does not require multiplication in an insect vector, which is always dependent on a functional, fully-developed mitochondrion. In the lab this can be done by syringe passage, but also in nature some sub-species of the brucei group are successfully transmitted without insect cycle, e.g. T. equiperdum by horse-horse contact and T. evansi by passive fly transfer. We call such variants insect-minus or I^-.

There is an obvious analogy between the brucei group of trypanosomes and the yeast Saccharomyces. Yeast can also repress mitochondrial biogenesis (under anaeronic conditions). In the presence of glucose, yeast mutants that have lost the ability to make functional mitochondria are fully viable and such mutants may have lost part (rho^- petites) or all (rho^o petites) of their mtDNA. By analogy one may expect some of the I^- variants of T. brucei to contain altered kDNA and the nature of the remaining DNA could provide clues to the function of maxi-circles and mini-circles.

For biochemical studies the I^- criterion is a clumsy one. Most biochemical labs do not keep tse-tse flies, culturing T. brucei is not a routine experiment and both methods are non-quantitative. Fortunately, the yeast analogy was also useful on this point: under complete anaerobiosis, when no functional respiratory chain is made, the three subunits of the ATPase complex specified by mtDNA are still made, resulting in sensitivity of this complex to the inhibitor oligomycin. In petite mutants, oligomycin sensitivity of the ATPase is lost.

In wild-type T. brucei we have likewise found an oligomycin-sensitive ATPase and this has provided us with a simple criterion to distinguish I^+ and I^- trypanosomes (14). A summary of the kDNA status of a series of I^+ and I^- strains is presented in Table II. As predicted by Opperdoes et al. (14), the kDNA in I^- strains is as variable as in yeast petite mutants. At the one extreme end of the spectrum is the kDNA of our strain of T. brucei 31, which is indistinguishable on all criteria from that of I^+ strains. It is, nevertheless, very likely that strain 31 is I^-: the ATPase has been consistently insensitive to olygomycin and attempts by Dr. D.A.Evans (Department of Medical Protozoology, London School of Hygiene and Tropical Medicine, London, U.K.) to bring this strain in culture or passage it through tse-tse flies have failed. We think that this strain has a mutation in a vital maxi-circle gene, but this remains to be proven.

The second category is represented by the T. equiperdum strain (ATCC 30019) studied by Hajduk (15-17). From electron microscopy (18) and restriction enzyme analysis (Hajduk, S.L.,

TABLE II

THE STATE OF kDNA IN TRYPANOZOON SUB-SPECIES THAT HAVE LOST THE ABILITY TO MAKE FUNCTIONAL MITOCHONDRIA (I^- TRYPANOSOMES)

Organism	Strain	Oligo-sens. ATPase	DAPI-stain kDNA	Organized networks present	Maxi-circles (μm)	Mini-circles present	Heterog. circular DNA	Refs
T. brucei[a]	EATRO 427	+	+	+	6	+	-	9
	31	-	+	+	6	+	-	b)
	LUMP 1027	-	-	-	-	-	-	14
T. equiperdum	ATCC 30019	?	+	+	5	+	-	15
	ATCC 30023	?	±	-	5	-	±	15
	Acrifl. dyskin.	?	±	-	5	-	-	15
	Pasteur	?	+	+	?	+	+	18
	Tobie dyskin.	?	±	-	-	-	+	18
	Zwart (+)	-	+	+	-	+	-	9,14[c]
T. evansi	Zwart (+)	-	+	+	-	+	-	1,14
	Zwart (-196)	-	-	-	-	-	-	14[c]
	AnTat-1	?	-	-	-	-	-	c)
	GUP 477	?	?	+	-	+	?	c)

a) I^+ control. b) This paper. c) Unpublished.

Vickerman, K., Hoeijmakers, J.H.J. and Borst, P., unpublished) this strain appears to contain networks with maxi-circles that are only 2/3rd of those in T. brucei I$^+$ strains. Cross-hybridization experiments with cloned T. brucei maxi-circle fragments showed that the second EcoRI fragment (Fig. 4) is entirely missing in the T. equiperdum maxi-circle, whereas at least part of the third fragment is still present. This indicates that maxi-circle deletions are the cause of the I$^-$ state in this T. equiperdum strain. Marked heterogeneity in size of putative maxi-circles has been observed by Riou and Pautrizel (18) in another strain of T. equiperdum.

The next category of I$^-$ strains contains an apparently normal mini-circle network without maxi-circles. Examples are the strain of T. equiperdum (Zwart, +) studied in this lab and a strain of T. evansi (Zwart, +). The networks from these trypanosomes lack edge loops in electron micrographs or maxi-circle fragments in restriction digests.

The fourth category consists of trypanosomes that have lost organized networks, but still contain kDNA components. The presence of DNA of kDNA density in dyskinetoplastic strains (i.e. strains that lack the morphological hall marks of kDNA networks in situ) was first described by Stuart (19) in T. brucei. Hajduk (16) then described small clumps of DNA-like material, stained with DAPI, in two dyskinetoplastic T. equiperdum strains. From both, he isolated 5-μm DNA circles (15). No mini-circles were found in this strain. Riou and Pautrizel (18) have analysed another dyskinetoplastic T. equiperdum strain and found a band of kDNA density in CsCl which contained a very heterogeneous collection of circles, varying in size from 0.9-24 μm. Networks were absent. The simplest interpretation of these results is that the kineto-plastic T. equiperdum strain of Hajduk contains a mini-circle network with deleted 5-μm maxi-circles; in his dyskinetoplastic strains the mini-circle network has been lost and the maxi-circles remain. On the other hand, Riou and Pautrizel's strain contains networks with defective maxi-circles that are heterogeneous in size; when the networks are lost in the dyskinetoplastic strain, heterogeneous maxi-circles remain. This speculative interpretation remains to be verified.

The final category is represented by the four strains studied in this laboratory (unpublished) that do not appear to have any kDNA at all by any of the criteria used. Obviously, the sensitivity of these criteria is limited. Absence of DAPI-staining material could be compatible with the presence of kDNA if the circles are not catenated and spread evenly throughout the mitochondria. The analysis of isolated DNA by restriction enzyme analysis and electron microscopy is more sensitive, but the marked tendency of kDNA to get lost - preferentially during DNA isolation - makes all negative analyses on purified DNA somewhat suspect. kDNA sequences could be present in heterogeneous long, linear molecules that might be interpreted as nuclear DNA in electron micrographs or (less likely) in restriction digests. Nevertheless, the simplest interpretation of these results is that trypanosomes can exist without any kDNA at all, in analogy with the rho° petite mutants in yeast.

We draw three conclusions from this work:

1. Trypanosomes can do without any kDNA.
2. Trypanosomes can retain normal mini-circle networks in the absence of maxi-circle sequences.
3. Trypanosomes can retain maxi-circle sequences in the complete absence of mini-circles.

Conclusions 1 and 3 are speculative, but the evidence is sufficiently suggestive to be taken into account in the discussion of possible functions of mini-circles.

MINI-CIRCLE HETEROGENEITY IS STRONGLY DECREASED IN kDNA NETWORKS FROM TRYPANOSOME STRAINS THAT LACK MAXI-CIRCLES

Fig. 3 shows that mini-circle heterogeneity in the networks from the T. equiperdum and T. evansi strains that lack maxi-circles is very limited. Restriction endonuclease TaqI cuts (nearly) all mini-circles in the network once, other enzymes do not cut (not shown). Nevertheless, the mini-circles in these strains are related to the mini-circles of T. brucei because they still show weak hybridization with T. brucei kDNA (unpublished). In another, recently isolated T. evansi strain (GUP 477) we have found limited mini-circle sequence and size heterogeneity (not shown). On the other hand, Hajduk's T. equiperdum strain (ATCC 30019) shows more extensive heterogeneity (not shown), even though not as pronounced as the T. brucei strains analysed. These results suggest that the presence of maxi-circle sequences - whether functional or not - contributes to sequence heterogeneity in the mini-circles. Maxi-circles are catenated into the mini-circle network and this may require recombination between maxi-circles and mini-circles. If this process is imprecise it could contribute to mini-circle size and sequence heterogeneity. Homology between maxi-circles and mini-circles must be limited, however, because maxi-circles do not show the sequence heterogeneity of mini-circles and because maxi-circle transcripts do not hybridize to mini-circles (Fig. 5; ref. 10).

THE REPETITIVE STRUCTURE OF T. cruzi MINI-CIRCLES

Riou and Yot (20) have observed that digestion of T.cruzi with endonucleases EcoRI and HaeIII gives fragments of 1/4, 1/2 and 3/4 mini-circle length in addition to linearized mini-circles. This suggested a repetitive sequence partially obscured by sequence micro-heterogeneity. In collaboration with Dr. A.C.C.Frasch and Dr. W.Leon we have confirmed some of these results and, in addition, we have found that endonuclease TaqI converts most of the mini-circles into 1/4 molecules, providing direct evidence for a basic repeat equivalent to 1/4 of the contour length.

CONCLUSIONS

Although there is still no direct evidence that the maxi-circle is the trypanosome equivalent of mtDNA, the indirect evidence presented in this paper and elsewhere (2) is rather persuasive and needs no recapitulation. The really vexing problem is the function of mini-circle networks.

Ideas about mini-circle function are of three types (see refs 2, 21 and 22):

1. The mini-circle network has a structural role in cell division, i.e. in the ordered division and segregation of the flagellum-kinetosome (basal body)-mitochondrion complex. This is rendered unlikely by the existence of thriving trypanosomes without mini-circle networks and even without any detectable kDNA at all (Table II).

2. Mini-circles code for an RNA or protein that is required in large amounts. The pronounced sequence heterogeneity, the rapid sequence evolution and the absence of hybridizable transcripts, all argue against this possibility.

3. Mini-circle networks serve to anchor all maxi-circles at a fixed position in the mitochondrion or function in the ordered segregation of maxi-circles after replication. This could require only a small part of the mini-circle sequence, allowing rapid sequence evolution and heterogeneity of the remainder of the sequence. This seems a rather modest function, however, for such a complex structure and it is also unclear why only trypanosomes would require such an elaborate anchor for their mtDNA. It is, therefore, possible that the true function of mini-circles remains to be found.

ACKNOWLEDGEMENTS

We thank Mr. S.L.Hajduk and Dr. K.Vickerman (Department of Zoology, University of Glasgow, Glasgow, U.K.), Dr. W.Leon (Institute of Microbiology, Federal University of Rio de Janeiro, Brasil), Dr. J.James (Histological Laboratory, University of Amsterdam) and Mrs F.Fase-Fowler, Miss A. Snijders, Dr. A.C.C.Frasch and Dr. P.J.Weijers for allowing us to quote unpublished experiments. This work was supported in part by a grant to P.B. from the Foundation for Fundamental Biological Research (BION), which is subsidized by The Netherlands Organization for the Advancement of Pure Research (ZWO).

REFERENCES

1 Williamson, D.H., and Fennell, D.J. (1976). In 'Methods in Cell Biology' (D.M.Prescott, Ed.), pp. 335-351, Vol. 12, Academic Press, New York.

2 Borst, P., and Hoeijmakers, J.H.J. (1979). Plasmid 2, 20-40.

3 Cheng, D., and Simpson, L. (1978). Plasmid 1, 297-315.

4 Kleisen, C.M., Borst, P., and Weijers, P.J. (1976). Europ.J.Biochem. 64, 141-151.

5 Waalwijk, C., and Flavell, R.A. (1978). Nucl.Acids Res. 5, 3231-3236.

6 Steinert. M., Van Assel, S., Borst, P., Mol, J.N.M., Kleisen, C.M., and Newton, B.A. (1973). Exptl.Cell Res. 76, 175-185.

7 Upholt, W.B. (1977). Nucl.Acids Res. 4, 1257-1265.

8 Steinert, M., Van Assel, S., Borst, P., and Newton, B.A. (1976). In 'The Genetic Function of Mitochondrial DNA' (C.Saccone and A.M.Kroon, Eds), pp. 71-81, North-Holland, Amsterdam.

9 Fairlamb, A.H., Weislogel, P.O., Hoeijmakers, J.H.J., and Borst, P. (1978). J.Cell Biol. 76, 293-309.

10 Hoeijmakers, J.H.J., and Borst, P. (1978). Biochim. Biophys.Acta 521, 407-411.

11 Simpson, L., and Simpson, A.M. (1978). Cell 14, 169-178.

12 Nichols, J.M., and Cross, G.A.M. (1977). J.Gen.Microbiol. 99, 291-300.

13 Hanas, J., Linden, G., and Stuart, K. (1975). J.Cell Biol. 65, 103-111.

14 Opperdoes, F.R., Borst, P., and De Rijke, D. (1976). Comp.Biochem.Physiol. 55B, 25-30.

15 Hajduk, S.L., and Cosgrove, W.B. (1979). Biochim.Biophys. Acta 561, 1-9.

16 Hajduk, S.L. (1976). Science 191, 858-859.

17 Hajduk, S.L. (1978). In 'Progress in Molecular and Sub-Cellular Biology' (F.E.Hahn, Ed.), pp. 158-200, Vol. 6, Springer Verlag, Berlin.

18 Riou, G., and Pautrizel, R. (1977). Biochem.Biophys.Res. Commun. 79, 1084-1091.

19 Stuart, K.D. (1971). J.Cell Biol. 49, 189-195.

20 Riou, G., and Yot, P. (1977). Biochemistry 16, 2390-2396.

21 Borst, P., Fairlamb, A.H., Fase-Fowler, F., Hoeijmakers, J.H.J., and Weislogel, P.O. (1976). In 'The Genetic Function of Mitochondrial DNA' (C.Saccone and A.M.Kroon, Eds), pp. 59-69, North-Holland, Amsterdam.

22 Borst, P., and Fairlamb, A.H. (1976). In 'Biochemistry of Parasites and Host-Parasite Relationships' (H.Van den Bossche, Ed.), pp. 169-191, North-Holland, Amsterdam.

REPLICATION AND TRANSCRIPTION OF KINETOPLAST DNA[1]

Larry Simpson, Agda M. Simpson, H. Masuda,[2]
H. Rosenblatt, N. Michael, and G. Kidane

Biology Department and Molecular Biology Institute,
University of California, Los Angeles, Ca. 90024

ABSTRACT The kinetoplast DNA (kDNA) of the hemoflagellate protozoa consists of a single large network of thousands of catenated minicircles and a smaller number of catenated larger maxicircles. In this report we briefly review the known information on the replication of the kDNA and present some recent results on characterization of the maxicircle component of the kDNA. Maxicircle DNA has been isolated by release from the network by cleavage with a restriction enzyme and separation on a buoyant density basis in CsCl in the presence of the dye, Hoechst 33258. In general the maxicircle DNA is higher in %A+T than the minicircle DNA. A restriction map of the maxicircle of *Leishmania tarentolae* has been constructed and the 9 and 12S kinetoplast RNA genes localized on this map. Several maxicircle fragments have been cloned in the bacterial plasmid, pBR322, including a 4.4 x 10 dalton Eco RI/BamHI fragment that contains the 9 and 12S RNA genes. Unit length minicircles have also been cloned in pBR322. The restriction maps of two different semi-homologous minicircle classes have been obtained. Similar studies have been performed on the kDNA of the related species, *Trypanosoma brucei*, which undergoes a cyclical regression and biogenesis of the mitochondrion during its life cycle.

INTRODUCTION

The selection of an appropriate system for study in biological research is of prime importance for answering

[1]This work was supported by research grants from the National Institute of Allergies and Infectious Diseases to L. Simpson.

[2]Present address: Department of Biochemistry, Federal University of Rio de Janeiro, Cidade Universitaria 20 000, Rio de Janeiro, Brasil.

ISBN 0-12-198780-9

particular questions. Occasionally, however, nature provides systems that are so bizarre that they demand to be studied due to their very unusual nature. The kinetoplast DNA (kDNA) of the hemoflagellate protozoa is both an appropriate system for the study of mitochondrial biogenesis and a highly unusual and bizarre system that demands an explanation in molecular terms. This mitochondrial DNA consists of a network composed of thousands of catenated minicircles and a smaller number of catenated larger maxicircles. There is a single extended mitochondrion per cell and a single mitochondrial DNA network per mitochondrion. Some of the presumably more advanced hemoflagellates species undergo a cyclical regression and biogenesis of mitochondrial enzymes during the course of their parasitic life cycle. The maxicircle DNA seems to represent the analogue of the mitochondrial DNA molecules found in other eukaryotes; the function of the minicircle DNA is unknown.

We have approached the study of the kinetoplast DNA in a comparative manner. Several of the more primitive species which do not undergo cyclical changes - *Crithidia fasciculata*, *Phytomonas davidi*, and *Leishmania tarentolae* - have been studied in order to develop appropriate techniques and to provide a baseline of knowledge about this mitochondrial DNA. This knowledge has then been applied in a continuing study of the more advanced species, *Trypanosoma brucei*, which undergoes the cyclical life cycle changes in mitochondrial metobolism mentioned above.

In this paper we will first summarize the current state of knowledge from our laboratory and from other laboratories on the replication of the kDNA. Then we will discuss results from our laboratory on the characterization of the maxicircle component of the kDNA and preliminary results on the use of recombinant DNA technology in an analysis of kDNA.

RESULTS

Replication of kDNA. Table I summarizes various observations on the replication of kDNA and presents the recent unifying hypothesis of Englund (18) which would explain several of the apparently paradoxical observations. This hypothesis invokes a non-specific random removal of closed minicircles from the network, replication in the unattached state, and a specific reattachment of the nicked daughter minicircles to peripheral network sites. This mechanism would allow the replication of all network minicircles as well as account for the reported random distribution of minicircles within the network.

TABLE I
REPLICATION OF KINETOPLAST DNA - SUMMARY

1. Occurs within the nuclear S phase (1-4).

2. The kDNA nucleoid lengthens to twice the G1 length; width remains constant. Division of the nucleoid occurs by constriction of the mitochondrion (5).

3. Networks isolated from cells in different stages of the S phase show a doubling of the surface area in the microscope (6). Replicated networks in G2 contain twice the number of minicircles than networks in G1 prior to replication (7).

4. Apparently segregating kDNA networks from _T. brucei_ and _T. equiperdum_ have a doublet appearance in the electron microscope (8).

5. Density shift experiments show that all minicircles replicate once during one S phase (9, 6). There is also some evidence for recombination between minicircles in _Crithidia_ (9).

6. Digestion of a given minicircle major sequence class causes extensive fragmentation of the network to monomeric minicircles and small catenanes. This implies that minicircles of a given sequence class are not clustered in the network but are distributed in an apparent random manner (18, 21).

7. Labeled network DNA is distributed equally to daughter kinetoplasts for at least five cell divisions in a chase (10).

8. Pulse-labeling of kDNA is limited to two opposite loci on the periphery of the network as shown by autoradiography of 3 minute pulse-labeled DNA (11). Longer pulses yield a complete peripheral labeling pattern (6, 12). Pulse-chase yields an apparent centripetal movement of network label and then a randomization of label throughout the network after one cell generation time (6).

9. Replicated minicircles are nicked and remain as open molecules for several hours until covalent closure of all the network minicircles occurs (11, 7, 13).

10. Non-replicating kDNA networks (form I networks) band in EtdBr-CsCl at the same density as closed monomeric minicircles (13, 7, 14, 15) and consist entirely of covalently closed circular molecules (13). Pulse-labeled replicating networks band in EtdBr-CsCl at an intermediate density between that of closed networks and open circles, and contain a variable proportion of minicircles that are nicked (6, 7). These open minicircles are localized at the periphery of the network (7). Completely replicated networks (form II networks) band at the density of open circles and consist entirely of open minicircles (6, 7, 13). Pulse-chased networks become completely open and then after approximately one cell generation time become completely closed (6, 7).

11. A small percentage of kDNA minicircles do not appear to be attached to the network (16, 17, 18). These free minicircles may be intermediates in the replication of network minicircles, as they show rapid incorporation of ^{3}H thymidine in a pulse and exhibit a turnover of this label (18). Cairns-type free minicircles, which may be replicating intermediates, are seen in Berenil treated kDNA of T. cruzi (19), and a low percentage of Sigma forms are present in the free minicircle kDNA of normal L. tarentolae (16).

12. Englund (18) hypothesizes that there is an enzymatic activity that randomly releases unreplicated closed minicircles from the network and that these free minicircles then proceed to replicate outside the network. A second enzymatic activity is proposed which rejoins the replicated nicked minicircles to the periphery of the network. The pulse-labeled peripheral network sites (6, 11) would then represent specific sites of reattachment of progeny minicircles rather than sites of synthesis. The delayed covalent closure of replicated minicircles may represent a mechanism to insure that each minicircle replicates once and only once per S phase (7).

Nothing is yet known about the replication of the 20-50 copies of the maxicircle component of the kDNA. Rare free maxicircles have been observed in kDNA preparations (20) but there is no evidence that they represent replication intermediates.

The segregation of daughter networks apparently involves the formation of dumbell-shaped "doublet" networks as intermediates in the process at least in *T. brucei* and *T. equiperdum* (8). Nothing is known about the mechanisms involved in the segregation of daughter networks.

Isolation of Maxicircle DNA. A method of general usefulness was developed for the isolation of the maxicircle component of the kDNA based on the fact that in five species examined (*L. tarentolae*, *Leptomonas* sp., *P. davidi*, *C. fasciculata*, and *T. brucei*) the maxicircle DNA contains a higher percentage of A+T than the minicircle DNA (22). The maxicircle DNA is first liberated from the network by cleavage with a single hit restriction endonuclease (Fig. 1) and the linearized molecule is then separated from the minicircles and remaining catenanes by buoyant separation in CsCl in the presence of the AT-binding fluorescent dye, Hoechst 33258 (Fig. 2). The extent of the separation varies with the absolute difference in %A+T between the maxicircle DNA and the minicircle DNA. This isolation method has several advantages over isolation in agarose gels, including large scale quantitative recovery and freedom from gel extractable impurities.

Base Composition of the Maxicircle DNA. The purified Eco RI cleaved maxicircle DNA from *L. tarentolae* (the "RI maxi") showed a monomodal distribution in buoyant CsCl at a density of 1.681 g/cm^3, which implies an A+T content of 79% (22). The buoyant density of purified monomeric minicircles from the same species was 1.705 g/cm^3, and the density of the total network was 1.703 g/cm^3. Purified Pst I-cleaved maxicircle DNA of *T. brucei* kDNA had a density of 1.683 g/cm^3 versus the network density of 1.690 g/cm^3.

An intramolecular base composition heterogeneity in the maxicircle DNA of *L. tarentolae* ranging from 74% A+T to 85% A+T was established by isolation of the three maxicircle Hpa II fragments from digested kDNA and buoyant analysis of these fragments in CsCl (22).

Restriction Mapping of the Maxicircle DNA of *L. tarentolae*. Large quantities of the Eco RI linearized maxicircle DNA of *L. tarentolae* were isolated by the Hoechst dye method for the purpose of restriction mapping and cloning (23). Mapping was performed by comparisons of single and double digests of the RI Maxi and single digests of the intact maxicircle (Fig. 3).

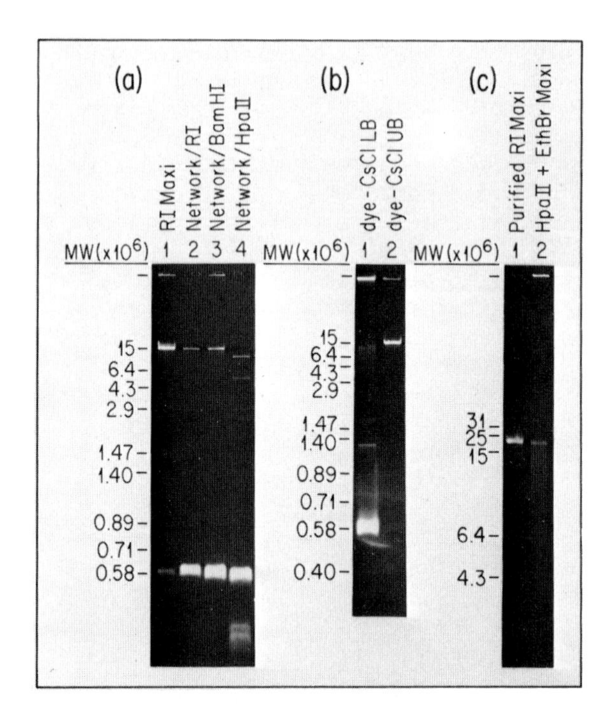

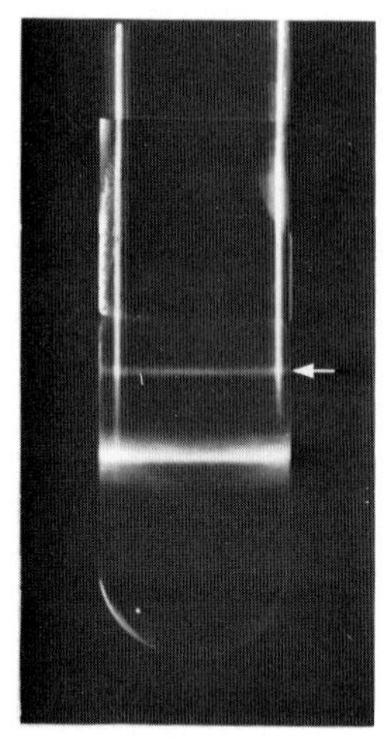

FIGURE 1. Agarose gel electrophoresis. (a) 0.8% agarose gel of RI Maxi DNA and of total network DNA digested with three enzymes. (b) 0.8% agarose gel of the upper and lower bands from the Hoechst dye-CsCl gradient of EcoRI digested kDNA in Fig. 2. (c). 0.5% agarose gel of purified RI Maxi DNA and once cleaved, permuted maxicircle linears produced by digestion of kDNA with Hpa II in the presence of ethidium bromide. Reference DNA fragments were λ/Hind III and ØXRF/Hae III. From (22) with permission.

FIGURE 2. Hoechst 33258-CsCl equilibrium gradient of Eco RI digested L. tarentolae kDNA. Ti60 rotor, 48 hr, 40,000 rpm, 20°C. Long wavelength UV illumination. The arrow points to the maxicircle DNA band. From (22) with permission.

A composite circular map of the restriction sites of 8 enzymes is presented in Fig. 4. The base compositions of the three Hpa II fragments are also indicated.

In order to eliminate the possibility of maxicircle sequence heterogeneity, Bam HI-linearized maxicircle DNA was isolated from network DNA and confirmatory digestions were performed. The predicted fragments were released in each case, making it unlikely that there is a major sequence class of maxicircles lacking an Eco RI site but possessing a Bam HI site (23).

Cloning of Maxicircle Fragments. Two fragments of the maxicircle of *L. tarentolae* have been cloned in the bacterial plasmid, pBR322 (23). The recombinant plasmid,

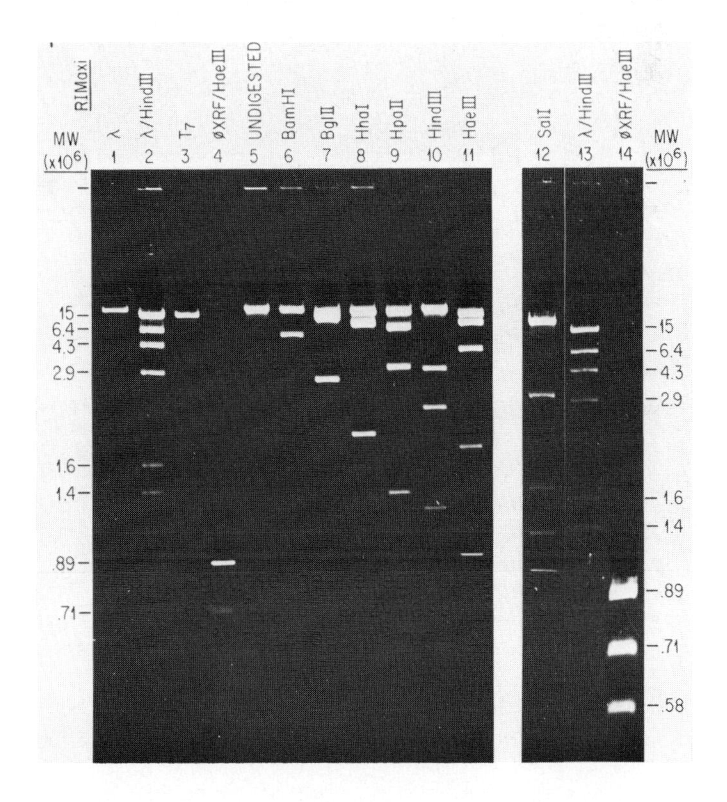

FIGURE 3. Agarose gel electrophoresis of RI Maxi DNA digested with 7 restriction enzymes. Slots 1-11, 1.2% agarose; slots 12-14, 1% agarose. From (23) with permission.

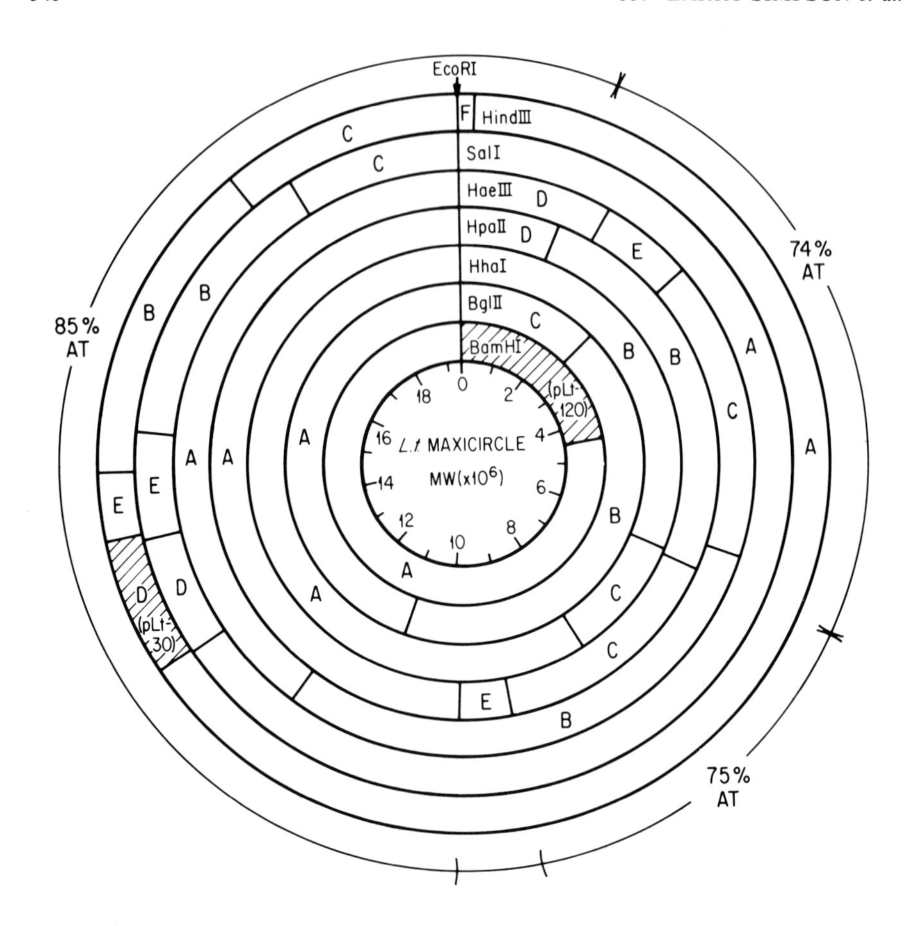

FIGURE 4. Composite circular restriction map of the maxicircle of L. tarentolae kDNA. The base composition of the three Hpa II fragments is indicated and the cloned fragments are indicated by shading. From (23) with permission.

pLt-30, contains the cloned 1.18 x 10^6 dalton maxicircle Hind III fragment D inserted into the Hind III locus in pBR322 (Fig. 5). The recombinant plasmid, pLt-120, contains the 4.4 x 10^6 dalton Eco RI/Bam HI fragment B inserted into the Eco RI/Bam HI locus in pBR322 (Fig. 5). The localizations of the cloned fragments are indicated on the restriction map in Fig. 4. Due to the high content of A+T, the cloned maxicircle inserts could be isolated from the host plasmid by appropriate digestion and buoyant separation in CsCl in the presence of Hoechst 33258 (Fig. 6). The greater relative separation of the pLt-30 insert from the plasmid as compared to the pLt-120 insert implies a higher AT content, which is consistant with the localizations of the fragments in the molecule.

The cloned pLt-120 insert was used to confirm the restriction map which was obtained with uncloned DNA. The fragments released by digestion of this insert with several enzymes were consistant with the map, and an overlap analysis using nick-translated pLt-120 insert DNA as a probe also agreed with the map.

Several fragments of the maxicircle of T. brucei kDNA were also cloned in pBR322. The fragments were inserted at the Hind III locus of the plasmid. The clones were detected by colony hybridization using labeled Bam HI linearized maxicircle DNA as a probe.

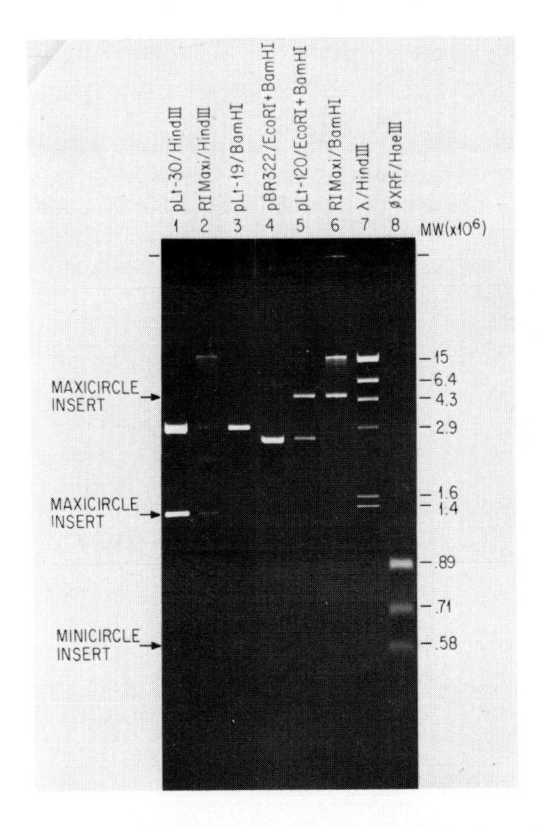

FIGURE 5. Agarose gel electrophoresis of digested recombinant DNA plasmids pLt-30, pLt-19, and pLt-120. RI Maxi DNA digested with Hind III or Bam HI was included as a reference.

Transcription of kDNA. The availability of a method to isolate relatively clean kinetoplast-mitochondrial fractions from hemoflagellate protozoa (24) allowed the isolation and characterization of several kinetoplast RNA species. Two major RNA species, designated the 9 and 12S RNAs, have been isolated from kinetoplast fractions of *L. tarentolae* (25), *P. davidi* (26), and *T. brucei*. The function of these RNAs is uncertain as the isolation and characterization of kinetoplast ribosomes has proved to be a difficult problem. Several additional low molecular weight RNAs have been observed in kinetoplast fractions of *L. tarentolae*, but they probably arise from cytoplasmic RNA contamination. The transcriptional origin of the 9 and 12S kinetoplast RNAs has been shown to be the maxicircle DNA in *L. tarentolae* and *T. brucei*. The *in vivo* transcription of the 9 and 12S RNAs is inhibited selectively by ethidium bromide and rifampin in *L. tarentolae* (25) The 9 and 12S RNAs of *L. tarentolae* possess an unusually high content of A+U (25).

Hoeijmakers and Borst (27) have recently reported similar results with *Crithidia*. *In vitro* labeled total cell RNA was hybridized with Southern blots of agarose gel profiles of total kDNA digested with several restriction enzymes. The most prominant hybridization was obtained with a maxicircle fragment of approximately 2400 base pairs which probably corresponds to the hybridization of the two major kinetoplast RNAs. In addition, however, they observed a lower level of hybridization to other maxicircle fragments representing more than 50% of the molecule, perhaps implying the existance of low frequency maxicircle

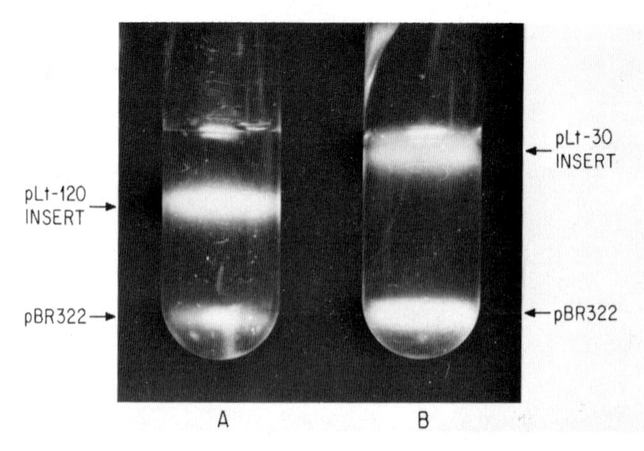

FIGURE 6. Hoechst dye-CsCl gradients of digested recombinant plasmids, pLt-120 and pL5-30. #50 rotor, 48 hr, 40,000 rpm, 20°C.

transcripts in addition to the two major RNA species. No minicircle hybridization was observed.

Mapping of the genes for the kinetoplast 9 and 12S RNAs. A maxicircle localization of the genes for the kinetoplast 9 and 12S RNAs from *L. tarentolae* has previously been demonstrated by hybridization of *in vitro* labeled 9 and 12S RNA to Southern transfers of agarose gel profiles of total kDNA digested with several restriction enzymes (25). A similar maxicircle localization of the genes for the large and small kinetoplast RNAs from *T. brucei* has also been demonstrated by this technique. In addition, a cross-species hybridization of labeled *L. tarentolae* 9 and 12S RNA to a maxicircle fragment of *Phytomonas davidi* kDNA has been demonstrated (26).

In order to map the 9 and 12S RNA genes on the maxicircle of *L. tarentolae*, hybridizations were performed with Southern blots of gel profiles of digested RI Maxi DNA, cloned pLt-120 insert DNA, and total kDNA (23) Figs. 7, 8.

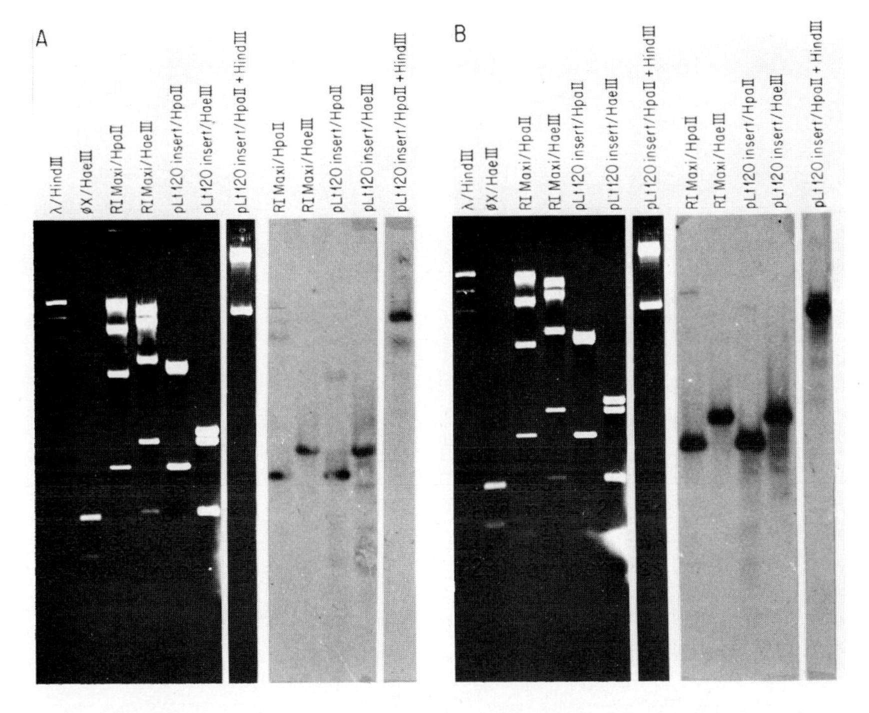

FIGURE 7. Hybridizations of labeled 9 and 12S kinetoplast RNAs from *L. tarentolae* to Southern blots of agarose gel profiles of RI Maxi DNA and pLt-120 insert DNA digested with Hpa II, and Hae III. (a) 9S RNA probe. (b) 12S RNA probe. Reprinted from (23) by permission.

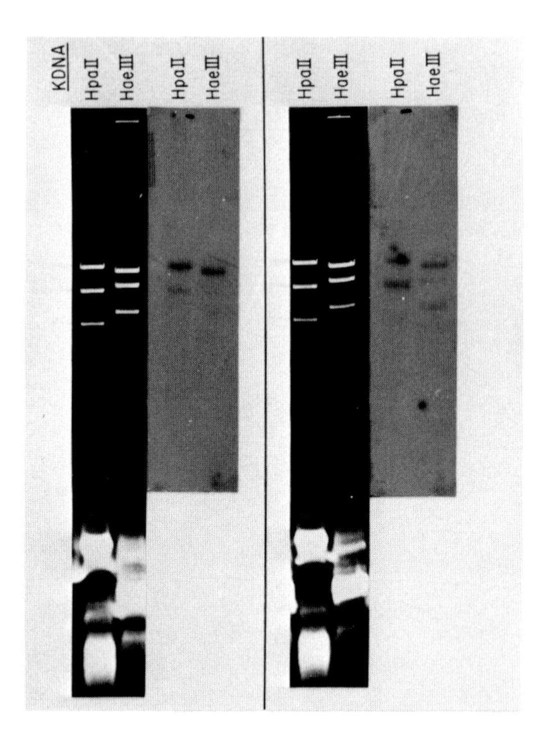

FIGURE 8. Hybridization of labeled 9 and 12S kinetoplast Rans from *L. tarentolae* to Southern blots of total kDNA digested with Hpa II and Hae III. (a) 9S RNA probe. (b) 12S RNA probe. Reprinted from (23) by permission.

In the former two cases, both RNAs hybridized mainly to a subfragment of the Hpa II D fragment. In the case of total kDNA, the 9S RNA hybridized mainly to the Hpa II A fragment (= RI Maxi A+D fragments) and to the Hae III A fragment (= RI Maxi A+D fragments) as expected; the 12S RNA also hybridized to the Hae III A fragment as expected, but approximately equal hybridization was observed to the Hpa II A and B fragments. This apparent discrepancy in the hybridization of the 12S RNA to the RI Maxi and cloned pLt-120 insert as compared to the hybridization to the maxicircle fragments obtained from total kDNA is not understood. It may be an artifact of the Southern transfer method in that large DNA fragments transfer poorly if at all, and therefore the 10.6×10^6 dalton Hpa II A fragment may be underrepresented in the transfer as compared to the Hpa II B fragment. Another explanation could be the existance of maxicircle heterogeneity. In order to test this explanation, the labeled 12S RNA probe was hybridized to a Southern transfer of the purified Bam Maxi DNA digested with several enzymes; the predicted localizations were observed in each case, implying again a lack of maxicircle sequence heterogeneity at this level.

Cloning and Mapping Different Minicircle Sequence Classes. Unit length minicircle clones from both *L. tarentolae* (23) and *T. brucei* were obtained by insertion into the Bam HI and Hind III loci of pBR322. One Bam HI clone from *L. tarentolae*, pLt-19, was selected, and the insert purified by acrylamide gel electrophoresis. This cloned minicircle was labeled by nick translation and used as a probe in colony hybridization to detect non-homologous minicircle clones. Two additional semi-homologous minicircle clones, pLt-26, and pLt-154, both inserted at the Hind III locus, were selected by this method. A labeled "total minicircle" probe was used to identify minicircle clones in general, and a labeled "RI Maxi" probe was used to detect maxicircle fragment clones. The three selected

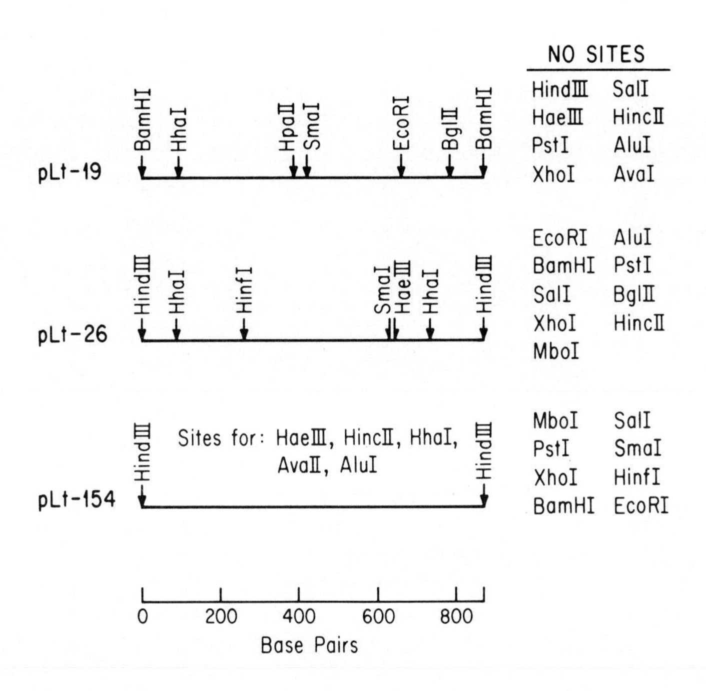

FIGURE 9. Restriction maps of two cloned minicircle classes from the kDNA of *L. tarentolae*.

clones showed different degrees of hybridization with the total minicircle probe, implying that they are present at different frequencies in the network. All minicircle clones however showed a certain amount of sequence homology by colony hybridization.

The three cloned minicircle inserts were isolated and restriction maps were derived by the use of double and triple digestions. The maps of pLt-19 and pLt-26 are shown in Fig. 9. These restriction maps showed no similarities, implying a substantial amount of sequence divergence. Sequencing of these minicircle sequences is now in progress; a comparison of the sequence homologies between these minicircle classes should prove informative in terms of understanding the nature of sequence divergence among the minicircle population.

Hybridization of total labeled minicircles and the cloned minicircles to Southern blots of RI Maxi DNA digested with several enzymes showed no apparent sequence homologies.

No homology was observed between L. tarentolae minicircles and T. brucei minicircles. In the case of T. brucei, unit length minicircle clones were obtained by insertion at both the Hind III locus and the Bam HI locus of pBR322. Selection of different minicircle sequence classes was initiated by use of colony hybridization, using randomly selected cloned minicircle inserts as probes. Restriction mapping of the cloned inserts has proved difficult, as no sites were found for 13 restriction enzymes in the two Bam HI clones selected for study, pTb-5 and pTb-7.

DISCUSSION AND CONCLUSIONS

The application of recombinant DNA technology to the kDNA problem has proved valuable in the study of both the maxicircle and the minicircle components of this system. In addition the study of the maxicircle component of the kDNA Has been aided by the development of a buoyant density technique to isolate large quantities of the linearized molecule. The purified maxicircle of L. tarentolae was mapped with 8 restriction enzymes and transcriptional mapping was initiated with the localization of the genes for the two major kinetoplast RNAs within a 1.1×10^6 dalton Hind III/Hpa II fragment. The availability of a clone which contains this DNA fragment opens up the possibility of sequencing the 9 and 12S kinetoplast RNA genes of L.

tarentolae. The availability of cloned minicircles of different sequence classes should allow the determination of the precise extent and nature of minicircle sequence heterogeneity, and also the unequivocal determination of the existence or absence of any specific minicircle transcript.

The cloning of maxicircle and minicircle fragments from *T. brucei* kDNA is important not only from a comparative point of view, but also in terms of the unique developmental changes that occur in this particular group of hemoflagellates. The transformation from the bloodstream trypomastigotes that contain no mitochondrial cytochromes or other respiratory enzymes into the procyclic culture forms which contain a full complement of mitochondrial enzymes may be an interesting system to study the activation of specific mitochondrial genes as well as the interaction of the mitochondrial and nuclear genomes in the formation of new mitochondrial enzymes. The availability of cloned segments of the maxicircle should aid considerably in this study.

REFERENCES

1. Steinert, M., and Steinert, G. (1967). J. Protozool. 9, 103.
2. Cosgrove, W., and Skeen, M. (1970). J. Protozool. 17, 172.
3. Simpson, L., and Braly, P. (1970). J. Protozool. 17, 511.
4. Steinert, M., and Van Assel, S. (1967). Arch. Int. Phys. Biochim. 75, 370.
5. Simpson, L. (1972). Int. Rev. Cytol. 32, 129.
6. Simpson, L., Simpson, A., and Wesley, R. (1974). Biochim. Biophys. Acta 349, 161.
7. Englund, P. (1978). Cell 14, 157.
8. Fairlamb, A. Weislogel, P., Hoeijmakers, J. and Borst, P. (1978). J. Cell Biol. 76, 293.
9. Manning, J. and Wolstenholme, D. (1976). J. Cell Biol. 70, 406.
10. Simpson, L., and da Silva, A. (1971). J. Mol. Biol. 56, 443.
11. Simpson, A., and Simpson, L. (1976). J. Protozool. 23, 583.
12. Steinert, M., Van Assel, S., and Steinert, G. (1976). In "Biochemistry of Parasites and Host-Parasite Relationships" (H. Van den Bossche, ed.), pp. 193-202. Elsevier Press, Amsterdam.

13. Englund, P., DiMaio, D., and Price, S. (1977). J. Biol. Chem. 252, 6208.
14. Simpson, L., and Berliner, J. (1974). J. Protozool. 21, 382.
15. Simpson, A., and Simpson, L. (1974). J. Protozool. 21, 774.
16. Wesley, R., and Simpson, L. (1973). Biochim. Biophys. Acta 319, 237.
17. Riou, G., and Delain, E. (1969). Proc. Natl. Acad. Sci. U.S. 62, 210.
18. Englund, P. (1979). J. Biol. Chem. In press.
19. Brack, C., Delain, E., and Riou, G. (1972). Proc. Natl. Acad. Sci. U.S.A. 69, 1642.
20. Steinert, M., and Van Assel, S. (1975). Exptl. Cell Res. 96, 406.
21. Weislogel, P., Hoeijmakers, J., Fairlamb, A., Kleisen, C., and Borst, P. (1977). Biochim. Biophys. Acta 478, 167.
22. Simpson, L. (1979). Proc. Natl. Acad. Sci. U.S.A. In press.
23. Masuda, H., Simpson, L., Rosenblatt, H., and Simpson, A. (1979). Gene In press.
24. Braly, P., Simpson, L., and Kretzer, F. (1974). J. Protozool. 21, 782.
25. Simpson, L., and Simpson, A. (1978) Cell 14, 969.
26. Cheng, D., and Simpson, L. (1978). Plasmid 1, 297.
27. Hoeijmakers, J., and Borst, P. (1978). Biochim. Biophys. Acta 521, 407.

MITOCHONDRIAL DNA AND SENESCENCE IN PODOSPORA ANSERINA

Donald J. Cummings[1], Leon Belcour[2] and Claude Grandchamp[2]

[1]Department of Microbiology and Immunology,
University of Colorado Medical Center
Denver, Colorado 80262

[2]Centre de Genetique Moleculaire, CNRS,
Gif sur Yvette, France

ABSTRACT Some twenty years ago, Rizet (1) and Marcou (2) reported that all races of *Podospora anserina* undergo vegetative death or senescence. Furthermore, this senescence is cytoplasmically inherited and appears to involve a transmissible cytoplasmic factor (3,4). More recent work with inhibitors of mitochondrial function and with cytoplasmic mutants implicate mitochondria in the senescent process (5,6). Consequently, we set about to isolate mitochondrial DNA from wild-type, mutants and senescent mycelia and characterize it with respect to density in CsCl, contour length and restriction enzyme analysis. We found that mitochondrial DNA from young wild-type mycelia had a density of 1.694 g/cm^3 and consisted of circular molecules 31 μm in length. Restriction enzyme analysis showed that each of the four races examined of *Podospora anserina* exhibited characteristic EcoRI fragment patterns. No apparent relationship of these patterns were noted with respect to the life span for each of the races. Mitochondrial mutants from race s were studied and their mitochondrial DNA had the same density in CsCl as did wild type. Electron microscopic examination showed that each of the mutants had a characteristic contour length DNA ranging from about 3 μm to 25 μm with no apparent monomer length unit. Restriction enzyme analysis with EcoRI showed that these mutants differed from wild-type in only 0 to 3 bands out of 16 fragments, suggesting that essentially the full complement of DNA was present. Analysis of the mitochondrial DNA from senescent mycelia showed dramatic differences with respect to wild-type and to mutants. First, two density species of DNA were noted, the majority of 1.694 g/cm^3 and a minority at 1.699 g/cm^3. The heavy density population consisted of a multimeric set of circular

ISBN 0-12-198780-9

molecules ranging in size from the monomer 0.9 μm to about 15 μm. EcoRI digestion indicated that the heavy density DNA had no EcoRI sites. Digestion with HaeIII enzyme yielded one fragment of about 2600 base pairs, corresponding to the 0.9 μm contour length monomer circle described above. These results will be discussed with respect to the rho(-)petite mutation in yeast (7).

INTRODUCTION

In addition to the general importance of studying fungal mitochondrial (mt) DNA discussed at length throughout this symposium, there is yet another feature which has not been mentioned: Senescence. It has been known for quite some time that fungi undergo senescence and the best studied example involves the ascomycete, *Podospora anserina*. Rizet (1) and Marcou (2) showed that vegetative cultures of Podospora had a limited life span characterized by a period of active growth, transformation to a senescent state, followed by gradual decline in the ability to divide resulting eventually in death. The life span, transformation rate and median length of growth, etc., are specific for races of different geographic origin. Several studies have shown that senescence was cytoplasmically inherited (2,4) and that a transmissible cytoplasmic factor was involved since senescence could be induced prematurely by anastomosis between young and old mycelia (3). Other workers have shown that inhibitors of mitochondrial function (5) as well as mitochondrial mutants (6,7) also have an effect on the time-course of senescence. From a variety of viewpoints, Belcour and Begel (7) suggested that senescence in Podospora has many similarities with the rho(-) petite mutation in the yeast, *Saccharomyces cerevisiae* (8). Genetic studies as well as physical analysis have shown that large deletions can occur in rho(-) mitochondrial DNA, which can lead to a DNA population consisting of a multimeric set of circular molecules (9-12).

In the present study, we have determined the contour length, density in CsCl and restriction enzyme patterns of mt DNA obtained from young, mutant and senescent mycelia. In many respects, the mt DNA obtained from senescent mycelia resembles the mt DNA observed in yeast rho(-) petite mt DNA (9-12).

MATERIALS AND METHODS

Culture Conditions. *Podospora anserina* races s and A were grown in corn meal extract liquid medium, supplemented with yeast extract (6). Senescent cultures were maintained and transferred on corn-meal extract-agar plates at 30°C, without illumination. After senescence had occurred on plates, plugs were taken 2-3 cm from the edge of growth stoppage and used for inoculating liquid cultures. All liquid cultures were grown by rotation at a 45°C at 27°C for 2-3 days.

Mitochondrial DNA. Mitochondrial DNA was prepared either by sand grinding of whole mycelia at 0°C or by extraction from purified mitochondria. In all cases, at least 100 μM aurintricarboxylic acid (ATA) was present throughout to inhibit nuclease action (14). Mt DNA was separated from nuclear DNA by CsCl density gradient centrifugation in the presence of 0.1 μg/ml diamidino-2-phenyl-indole (DAPI) (15).

Restriction Enzyme Analysis. Mt DNA was dissolved in 0.01 M Tris, 0.001 M EDTA, 0.007 M 2-Mercaptoethanol, pH 7.6 at a concentration of 200 μg/ml. For EcoRI or HaeIII cleavage, 5-10 μl DNA was mixed with 3-6 enzyme units in the appropriate salt conditions and incubated at 37° for 45-90 min (16). DNA fragments were analyzed by electrophoresis on vertical gel slabs in 0.01 M sodium acetate, 0.04 M Tris, 0.002 M EDTA, pH 8.4. For EcoRI, 1.2% agarose gels were used, for HaeIII, 6% acrylamide. After electrophoresis, the gels were soaked in a 1 μg/ml Ethidium bromide solution for 20 min and photographed using Polaroid film. Gels were calibrated using Hind III fragments of bacteriophage λ (17) for EcoRI and HaeIII fragments of ØX 174 RF DNA for the HaeIII; these fragments were gifts from Dr. C. Jacq.

Electronmicroscopy. DNA was examined using a formamide procedure (18,19) with 0.02% cytochrome C and 5 μg/ml DNA. The grids were examined in a Philips model 301 electron microscope. Square grids chosen at random were systematically scanned and all measurable circular molecules were photographed. All length measurements were made with a map measurer (HB. 56 Paris).

RESULTS

Contour Length of Mitochondrial DNA. Mt DNA isolated from young mycelia was examined in the electron microscope. For both races A and s, circular DNA of 29 to 33 μm were observed (Figure 1). About 1% of the mt DNA isolated by sand grinding of whole mycelia consisted of these circules; this amount was increased to about 4% when the mt DNA was isolated from purified mitochondria. These molecules are among the longest mitochondria DNA molecules ever observed. As measured by CsCl density analysis, the density of this mt DNA was 1.694 g/cm^3, compared with 1.713 g/cm^3 for nuclear DNA, and no other density species or circular molecules of different contour length were detected.

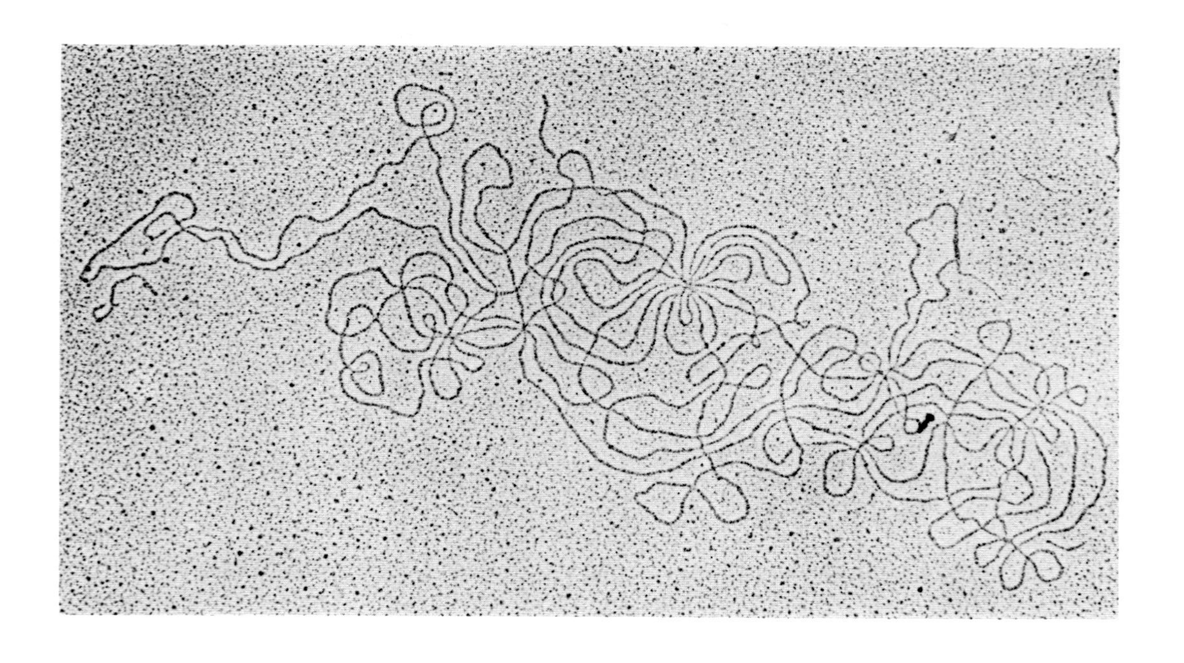

Figure 1. Electron micrograph of a 31 μm circular mt DNA.

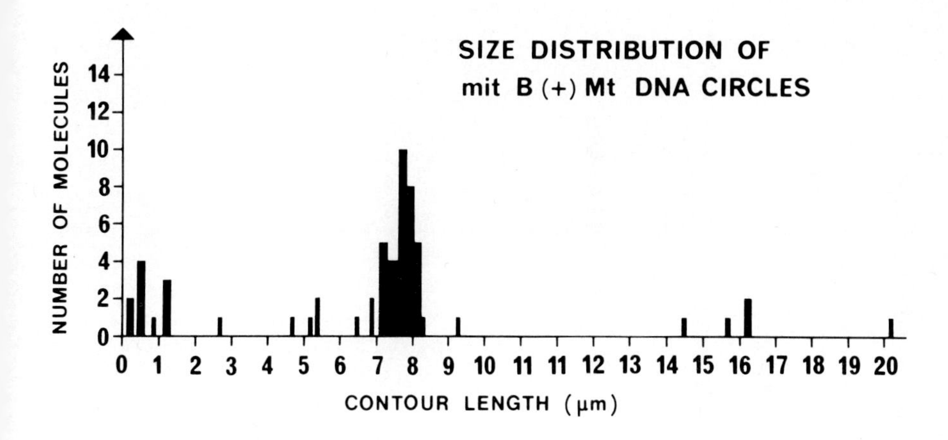

Figure 2. Size distribution of mit B (+) mt DNA circles.

Mt DNA from certain mitochondrial mutants were also examined and here the results were quite different. DNA isolated from mit B (+), a slow growing-senescence delayed mutant, yielded a relatively high number of circular DNA molecules: 3% in a population isolated after sand grinding. None of these molecules were greater than 21 μm, however (Figure 2). Instead, most of the circular molecules were about 7-8 μm, with no apparent monomer length molecule which could account for the size range. Other mutants of interest here (6,7) were certain mutants called spg, for spore germination deficiencies and the progeny emitting from crosses between these made by cytoplasmic mixing, PSW, for pseudo wild-type. Circular DNA molecules were also observed in mt DNA from each of these strains (Figure 3). As with the mit B (+), the contour lengths of these molecules were much different from wild-type. No molecules greater than about 13 μm were found in the PSW strain examined; only one of spg mutants had molecules which approached full length. As can be seen, quite different size classes were observed and although some indication of multiples can be discerned, no monomeric size can be deduced. As with wild-type, the density of the mt DNA from all these mutants was 1.694 g/cm^3, the same as wild type.

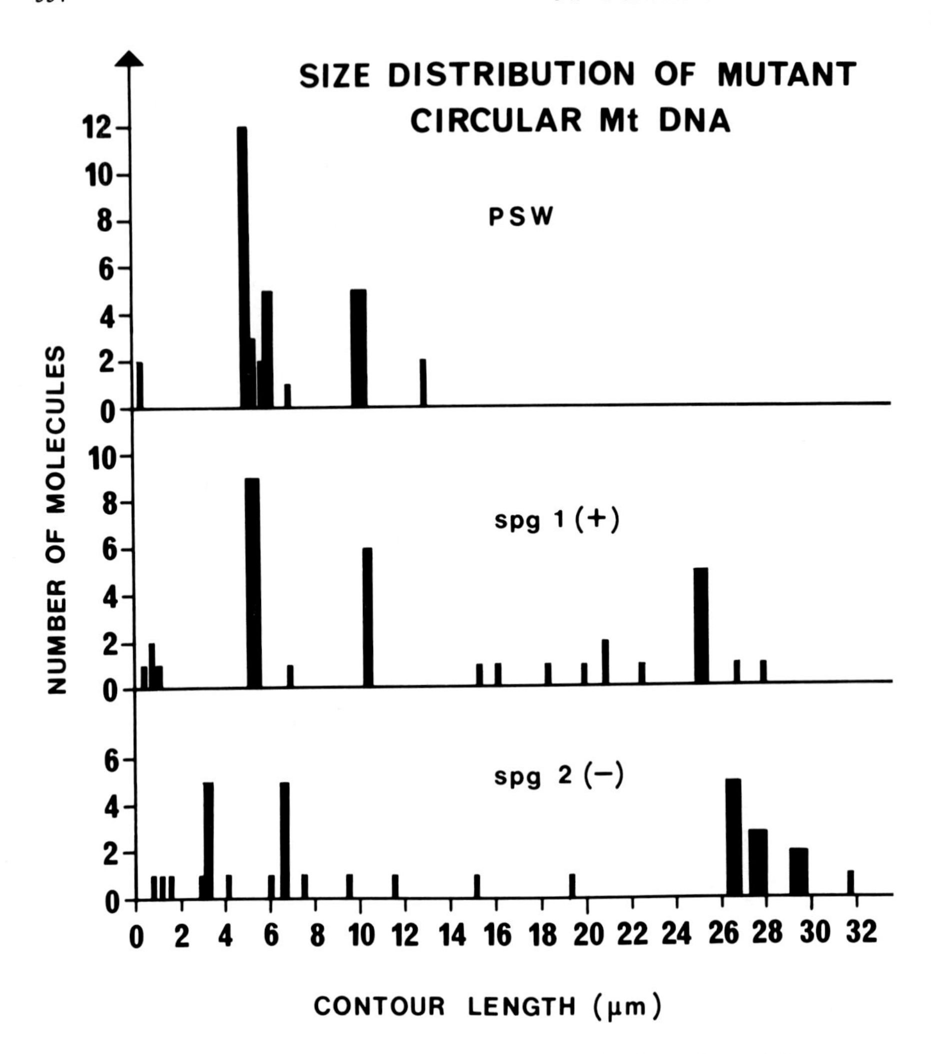

Figure 3. Size distribution of mt DNA circles from PSW, spg 1 and spg 2.

The contour length of mt DNA isolated from senescent cultures of race s (or A) was much different from either wild-type or the mutants. After extraction by sand grinding of whole mycelia, about 10-11% of the mt DNA molecules consisted of a multimeric set of circular molecules ranging in size from about 0.9 μm to 15 μm (Figure 4). Figure 5 is a montage electronmicrograph of some of these molecules. Most of the molecules were in the 1.8 and 2.7 μm classes and only one molecule out of some 200 observed was full-length. CsCl density gradient analysis from DNA isolated by either sand grinding or from purified mitochondria showed two density species. The majority of the DNA (about 80%) had the same density as wild-type, 1.694 g/cm^3. However, the remaining DNA had a density of 1.699 g/cm^3. Analysis of the two DNA density species in the electron microscope showed that the heavy density species contained about 90% of the multimeric set of circular molecules observed in the total mt DNA population.

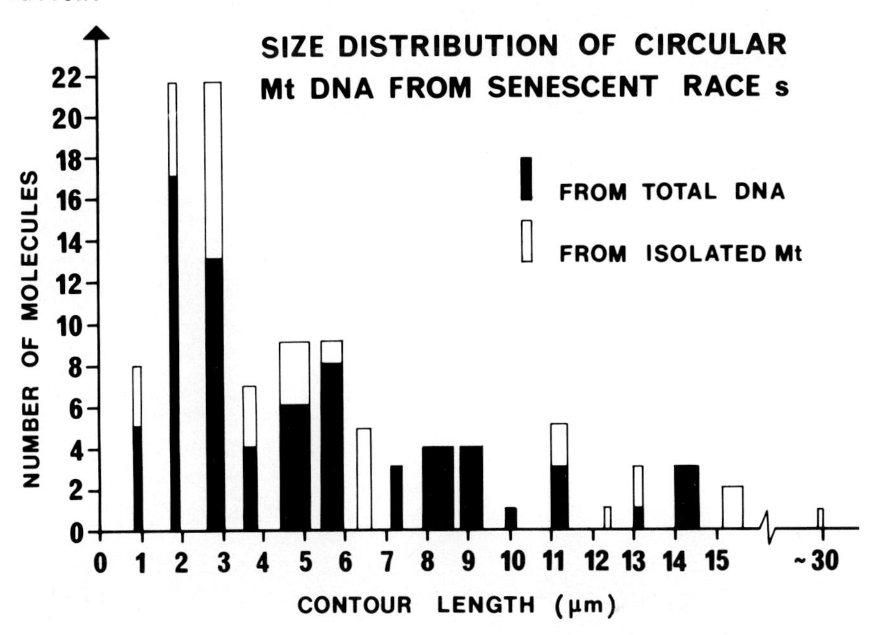

Figure 4. Size distribution of mt DNA from senescent race s.

Restriction Enzyme Analysis. EcoRI digests of mt DNA isolated from young cultures of wild-type mycelia and from mutants are illustrated in Figure 6. It can be noted that wild-type (track a) gave about 16 fragments and except for mit B (+) (track f), the mutants differed by only 1 or 2 fragments, if at all. These results would suggest that the populations of less-than full length circles observed in the mutants are a random arrangement of the genome content rather than a selection of certain portions of the genome. It should be emphasized however, that these restriction enzyme patterns were obtained with the total mt DNA populations and not on isolated circular molecules so definite conclusions about the genomic content of the mutant mt DNA molecules must be delayed.

Restriction enzyme fragments of mt DNA isolated from senescent mycelia were strikingly altered. Analysis of the normal density DNA gave patterns which were basically similar to young wild-type. The only difference was in the intensity

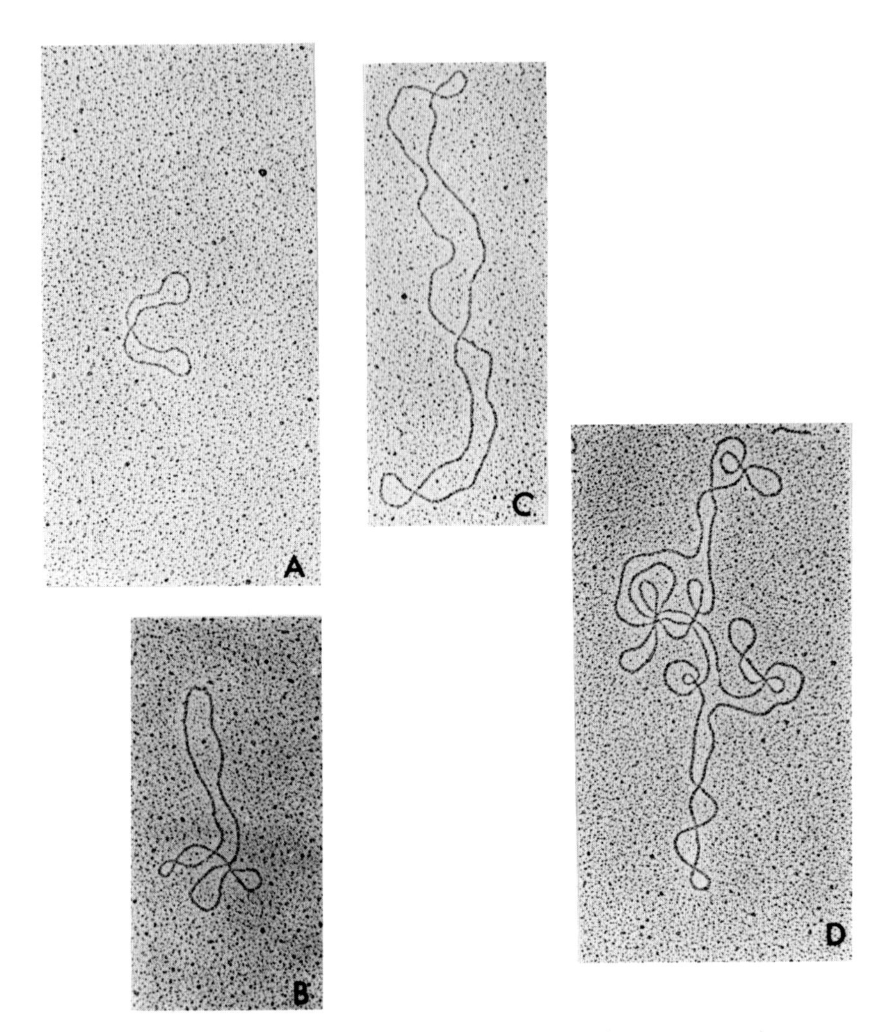

Figure 5. Electronmicrographs of A) 0.9 μm; B) 1.8 μm; C) 2.7 μm; and D) 5.5 μm circular DNA from senescent race s.

of some of the bands. When the so-called heavy DNA was examined, however, we found that it contained no susceptible EcoRI sites. Instead of detecting the several fragments shown in Figure 6, the DNA had the same average molecular weight (~ 20×10^6 daltons) with or without digestion with EcoRI. Analysis with HaeIII, restriction endonuclease, which was expected to yield many more fragments gave the results shown in Figure 7. Only one fragment was observed with the heavy senescent mt DNA (track a). We concluded from this that the heavy mt DNA had an HaeIII site repeated every 2600 bp. This corresponded very well with the 0.9 μm monomer

Figure 6. EcoRI digests of mt DNA from
a) race s (+)
b) spg 2
c) PSW
d) spg 1
e) cap^r (+)
f) mit B (+)

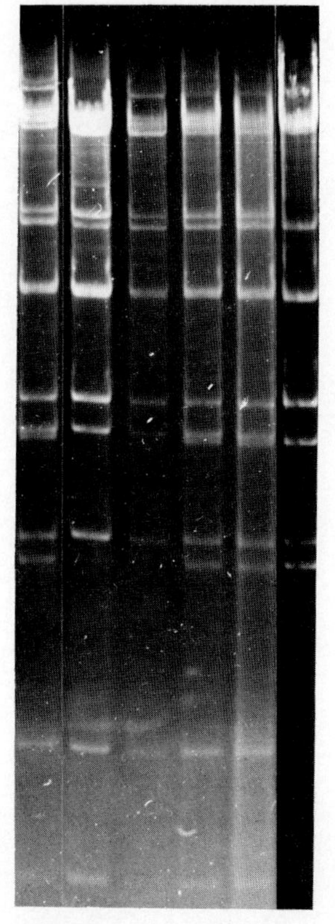

length circle observed by electron microscopy which suggests that the multimeric set of circular molecules observed in senescent mt DNA consists of a tandemly repeated fragment of DNA. As with EcoRI, the normal density senescent mt DNA (track b) had many similarities with young wild-type DNA (track c).

DISCUSSION

Aside from the information presented on the general characteristics of mitochondrial DNA from Podospora anserina, the most important aspect of this report is the effect of mutation and senescence on the structure of mt DNA. The mutants examined were chosen because each had some effect on senescence. In each case, the contour length of the DNA was altered so there may be a correlation between the occurrence of smaller circles and senescence. Without isolating and characterizing the smaller circles separately from the bulk

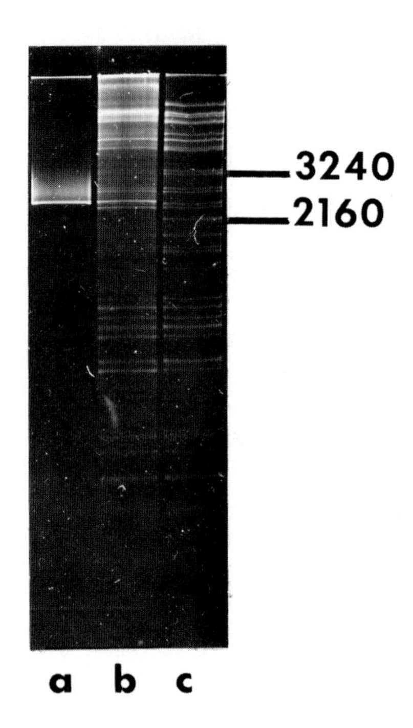

Figure 7. HaeIII digests of race s mt DNA from

a) heavy senescent

b) light senescent

c) young mycelia

Numbers in margin refer to bp in marker DNA.

of the DNA, however it is difficult to be certain. The effect of senescence on the DNA isolated with the mt DNA population was especially dramatic. Earlier we discussed the similarities of senescence with the rho(-) mutation in yeast. In this report we have extended the similarity to including a change in density of the DNA and the occurrence of a multimeric set of circular molecules. It is important to emphasize that our results were obtained with DNA isolated from the total mycelium as well as from purified mitochondria. Moreover, the same set of multimeric circles were observed in 3 independent senescent events for race s and 2 for race A, as well as for duplicate experiments. It would appear that senescence leads to a population of circular molecules of structure and density different from mt DNA contained in young mycelia. At the moment, we are not at all clear on what the similarities between senescence and rho(-) petites signifies. Does it infer that senescence in Podospora, at least is promoted by a deficiency in mitochondrial metabolism or does senescence lead to a deficiency which leads to defective DNA replication? It is also not known which fraction or multiples of the mitochondrial genome is represented in the tandemly repeated units contained in the multimeric set of circles. Establishing the cause and effect relationship of these events may well lead both to an understanding of mt DNA replication in fungi as well as a more basic comprehension of senescence.

REFERENCES

1. Rizet, G. (1953) Comptes Rendus Acad. Sci. 237, 838.
2. Marcou, D. (1961) Ann. Sci. Natur. Botany 11, 653.
3. Marcou, D., and Schecroun, J. (1959) Comptes Rendus Acad. Sci. 248, 280.
4. Smith, J.R., and Rubenstein, I. (1973) J. Gen. Microbiol. 76, 297.
5. Tudzynski, P., and Esser, K. (1977) Molec. gen. Genet. 153, 111.
6. Belcour, L., and Begel, O. (1977) Molec. gen. Genet. 153, 11.
7. Belcour, L., and Begel, O. (1978) Molec. gen. Genet. 163, 113.
8. Slonimski, P.P., and Ephrussi, B. (1949) Ann. Inst. Pasteur. 77, 47
9. Locker, J., Rabinowitz, M., and Getz, G.S. (1974) J. Mol. Biol. 88, 489.
10. Faye, G., Fukuhara, H., Grandchamp, C., Lazowska, J., Michel, F., Casey, J., Getz, G.S., Locker, J., Rabinowitz, M., Bolotin-Fukuhara, M., Coen, D., Deutsch, J., Dujon, B., Netter, P., and Slonimski, P.P. (1973) Biochim., 55, 779.
11. Lazowska, J., Michel, F., Faye, G., Fukuhara, H., and Slonimski, P.P. (1974) J. Mol. Biol. 85, 393.
12. Lazowska, J., and Slonimski, P.P. (1976) Molec. gen. Genet. 146, 61.
13. Luck, D.J.L., and Reich, E. (1964) Proc. Natl. Acad. Sci. U.S. 52, 931.
14. Hallick, R.B., Chelm, B.K., Gray, P.W., and Orozco, J.M., Jr. (1977) Nuc. Acids Res. 4, 3055.
15. Williamson, D.H., and Fennel, D.J. (1975) In Methods in Cell Biology, (ed., D.M. Prescott) Academic Press, New York, pp. 335-351.
16. Roberts, R.J., Breitmeyer, J.B., Taleochnik, N.F., and Myers, P.A. (1975) J. Mol. Biol. 91, 121.
17. Murray, K., Murray, N.E. (1975) J. Mol. Biol. 98, 551.
18. Inman, R.B., and Schnos, M. (1970) J. Mol. Biol. 49, 93.
19. Royer, H.D., and Hollenberg, C. (1977) Molec. gen. Genet. 150, 271.

INDEX

Numbers refer to the chapters in which the entries are discussed